Assigning Structures to Ions in Mass Spectrometry

Assigning Structures to Ions in Mass Spectrometry

John L. Holmes
Christiane Aubry
Paul M. Mayer

CRC Press
Taylor & Francis Group
Boca Raton London New York

CRC Press is an imprint of the
Taylor & Francis Group, an **informa** business

CRC Press
Taylor & Francis Group
6000 Broken Sound Parkway NW, Suite 300
Boca Raton, FL 33487-2742

First issued in paperback 2020

ISBN-13: 978-0-367-57773-5 (pbk)
ISBN-13: 978-0-8493-1950-1 (hbk)

Library of Congress Cataloging-in-Publication Data

Holmes, John L. (John Leonard), 1931-
   Assigning structures to ions in mass spectrometry / John Holmes, Christiane Aubry, and Paul M. Mayer.
        p. cm.
   Includes bibliographical references and index.
   ISBN-13: 978-0-8493-1950-1 (alk. paper)
   ISBN-10: 0-8493-1950-1 (alk. paper)
   1. Mass spectrometry. 2. Ions--Spectra. I. Aubry, Christiane. II. Mayer, Paul M., 1968- III. Title.

QC451.H66 2007
543'.65--dc22   2006021102

Visit the Taylor & Francis Web site at
http://www.taylorandfrancis.com

and the CRC Press Web site at
http://www.crcpress.com

To the late Fred Lossing, *friend, mentor, guide, and fellow enthusiast in the thrill of the scientific chase*

Foreword

In the past 40 or so years, great technical advances have resulted in mass spectrometry becoming the analytical method of choice in many scientific fields. This stems from the very great sensitivity of the method as well as our present ability to convert almost any organic substance into a beam of positively charged ions that can be separated and identified by their mass-to-charge ratio. In addition, the production of smaller, simpler, almost wholly software-controlled instruments has allowed this powerful analytical technique to become accessible to a much wider group of scientists.

The purpose of this book is to be a *vade-mecum* for all users of the technique for whom the structure of the ions present in the mass spectrum of an organic compound is important or of general interest. Indeed it could be argued that all practitioners should have some knowledge of the unique complexities to be found in the gas-phase ion chemistry of even simple organic compounds. Their behavior in turn provides the models for understanding larger polyatomic systems.

Acknowledgments

The authors wish to thank all those who made this task easier and particularly our families for putting up with the apparently never-ending chore.

John L. Holmes thanks all his former students and postdoctorate collaborators who made his laboratory life so stimulating and exciting and especially two colleagues, Hans Terlouw and Peter Burgers, with whom he spent so many happy hours teasing out the intricacies of ion chemistry.

Christiane Aubry wishes to sincerely thank John Holmes, a mentor and dear friend, with whom undertaking a new project has always been a privilege and a great pleasure. She also wishes to thank Orval Mamer from the McGill University Mass Spectrometry Unit for providing access to the electronic resources of the McGill University libraries.

Paul M. Mayer would like to acknowledge four terrific supervisors for their guidance, knowledge, and collegiality over the years: John Westmore, John Holmes, Tom Baer, and Leo Radom.

Additional Credits

Figures 1.1, 1.2, 1.8, 1.9, 1.10, 1.12, 1.14, 1.18, 1.19 and 1.22 are all original drawings.

Figures 1.3, 1.5, 1.6 and 1.7 are from Gross, M.L. and Caprioli, R. (eds.), *Encyclopedia of Mass Spectrometry*, Elsevier, Amsterdam, 2003, Volume 1, Chapter 5; Holmes, J.L. pp. 434–440.

Figure 1.4 is from Burgers, P.C. and Holmes, J.L., *Int. J. Mass Spectrom. Ion Processes*, 1984, *58*, 15.

Figure 1.11 is from Holmes, J.L., *Org. Mass Spectrom.*, 1985, *20*, 169.

Figures 1.13 to 1.17 are from *Encyclopedia of Mass Spectrometry*, (as above) Chapter 3; Holmes, J.L. pp. 91–98.

Figure 1.20 is from *Encyclopedia of Mass Spectrometry*, (as above) Volume 4, Chapter 3, Terlouw, J.K. and Holmes, J.L. pp. 287–297.

Figure 1.21 is from Burgers, P.C., Holmes, J.L., Mommers, A.A. and Szulejko, J.E. *Org. Mass Spectrom.*, 1983, *18*, 596.

Figure 2.1 is from Holmes, J.L., Lossing, F.P., Terlouw, J.K. and Burgers, P.C., *J. Am. Chem. Soc.*, 1982, *104*, 2931.

Figures 2.2 and 2.3 are from Yalcin, T., Khouw, C., Csizmadia, I.G., Peterson, M.P. and Harrison, A.G., *J. Am. Soc. Mass Spectrom.*, 1995, *6*, 1165.

Figures 2.4 to 2.7 are from Holmes, J.L., *Org. Mass Spectrom.*, 1985, *20*, 169.

Authors

John Holmes was born in London, UK, and obtained his PhD from University College, London in 1957 when that powerful chemistry department was in the care of Professors Ingold, Hughes, Nyholm, and Craig. His PhD (gas-phase pyrolysis of alkyl iodides) was under the direction of Professor Allan Maccoll. Professor Holmes spent the postdoctoral years from 1958 to 1960 at the National Research Council of Canada in Ottawa, a laboratory that attracted physical chemists from all around the world. Two years at the University of Edinburgh followed, but the delights of Canada and its great outdoors beckoned Professor Holmes and in 1962, he returned to Canada and the University of Ottawa where he remains to this day, now as Emeritus professor. In 1964, he volunteered to look after the chemistry department's first ever mass spectrometer, a Hitachi RMU 6D, and very quickly became fascinated by the extraordinary and then very poorly understood chemistry that must lie behind the mass spectra themselves.

Professor Holmes has held visiting fellowships and professorships in Australia, Ghana, the Netherlands, and Switzerland. He is a Fellow of the Royal Society of Canada and has received many awards, including the Chemical Institute of Canada Medal. He is the author of some 300 journal publications, editor of the *Encyclopaedia of Analytical Chemistry*, and was North American editor of *Organic Mass Spectrometry* for 17 years, a chore that brought new insights into human nature.

Outside of science he has always sailed as often as possible, cruising across the Atlantic in 2005. For 12 years he was an international judge for yacht racing and has contributed to the training of judges for the sport for over 20 years.

Christiane Aubry obtained a BSc in chemistry from the University of Sherbrooke (Quebec, Canada) in 1989. She pursued her first graduate studies there and received an MSc in organometallic photochemistry in 1991. She joined John Holmes's mass spectrometry group at the University of Ottawa (Ontario, Canada) in 1992 where she undertook research in gas-phase ion chemistry with special interests in the thermochemistry and characterization of small organic ions. She obtained her PhD in 1997. She accepted an industrial position at the Research and Productivity Council in New Brunswick (Canada) where she shared the responsibility of the analytical mass spectrometry facility. In parallel, she conducted research aimed at a better understanding of the de novo production of chlorometabolites by various fungi found in the environment. She returned to the University of Ottawa in 1999 to teach at the undergraduate level and re-joined Professor John Holmes's laboratory to pursue gas-phase ion chemistry. For Dr. Aubry, working on this book has provided much needed light relief to attending to the demands of a growing family.

Paul M. Mayer received a BSc in chemistry at the University of Manitoba in 1990. He spent three summers working with John Westmore in biological mass spectrometry and computational chemistry and is forever indebted to John Westmore for introducing him to the field. Dr. Mayer followed this with a PhD (1990–1994) in gas-phase ion chemistry at Ottawa with John Holmes, a wonderful supervisor and good friend and colleague. He then spent two years (1994–1996) with Tom Baer at UNC Chapel Hill studying the kinetics of unimolecular ion dissociations by TPEPICO spectroscopy. During this work, it became clear that ab initio molecular orbital calculations were essential to understanding ion structures and reactivity, so he then spent a year at the ANU in Canberra, Australia with Leo Radom, learning the fundamentals of modern computational chemistry. He rejoined the chemistry department at the University of Ottawa as assistant professor in 1998 (and in 2003 become associate professor) and has been having a great time exploring the chemistry of all varieties of gas-phase ions by mass spectrometry, spectroscopy, and theory ever since. His interests range from the isomerzation of small electrostatically bound complexes, entropic effects in ion dissociations, and energy transfer in collision processes. He has been the recipient of a Premier's Research Excellence Award in the province of Ontario in 2001 and an Excellence in Education Prize from the University of Ottawa in 2005.

Introduction

The subject of this book, gas-phase ion structures and their chemistry, provides the background material essential for the proper interpretation and understanding of organic mass spectra. It is therefore important for all users of the method, especially those analytical scientists for whom mass spectrometry is a significant technique. Gas-phase ion chemistry is also directly relevant to research in radiation chemistry, upper atmosphere chemistry, and interstellar chemistry.

A good understanding of the characteristics and behavior of organic ions in the gas phase permits one to take full advantage of the wealth of information contained (and more often than not, hidden) in a mass spectrum. This book summarizes our present knowledge of the structures and chemistries of small organic cations (C_1 to C_3) in the gas phase. In a mass spectrometer these ions are wholly isolated, insofar as no solvent or other attached species is present that can affect their stability. The magnitude and location of the charge that ions carry is central to their behavior. The charge may be localized at an atom, or de-localized over many atoms. Throughout the text the symbols $^{1+\bullet}$ and $^{1+}$ are appended to each odd and even electron cation, respectively. This denotes only that the species is an odd or even electron cation and does not show the charge location. Note that determining an ion structure using mass spectrometric methods results in establishing the connectivity of the atoms in the species under investigation and does not include geometric details such as bond lengths or angles.

Traditionally, the interpretation of mass spectra attempted to correlate the most intense ion peaks in a mass spectrum with the structure of the molecule that gave rise to it. This required some ideas as to the structures of the fragment ions themselves and possible mechanisms for their formation. Thus in the early years a mass spectrum was treated as a pattern to be analyzed by noting the most abundant fragment ion peaks, making an educated guess as to their structure, and devising a rational mechanism for their origins. Early attempts to describe the mechanistic chemistry that lay behind mass spectra relied heavily on analogies from conventional solution organic chemistry. By comparing the behavior of homologous or similarly related series of ions, generalized connections between mass spectra and molecular structure were achieved. However, on closer examination, even the simplest of ionized molecules appeared to undergo wholly unpredictable decompositions or rearrangements for which there were no condensed phase analogies. We now know that the structures of many organic cations have no stable neutral counterparts.

A simple illustration of hidden complexities is provided by the mass spectrum of ethylene glycol, $HOCH_2CH_2OH$, where the four most abundant peaks in its electron impact mass spectrum are m/z 29, 31, 32 and 33 (13:100:11:35 are the relative intensities of the corresponding ion peaks). The origin and the structure of m/z 31, the base peak, are straightforward: a simple C–C bond cleavage yielding the $^+CH_2OH$ fragment ion. The presence of the ions m/z 33, assigned to protonated methanol, $CH_3{}^+OH_2$, m/z 32 $[C,H_4,O]^{+\bullet}$ and m/z 29, $[H,CO]^+$ are structure-characteristic for this diol. However, the mechanism by which $CH_3{}^+OH_2$ is produced and the hidden fact that m/z 32 is not ionized methanol but rather its distonic isomer $^\bullet CH_2{}^+OH_2$ are complex problems that cannot be solved by reliance upon analogy with solution chemistry (a distonic ion is a species in which the charge and radical sites are located on different atoms).

With hindsight it is easy to see why the task of uncovering such intricate chemistry was so difficult. First and foremost, when faced with intriguing results, experimentalists had to disregard traditional organic chemistry and harness their imaginations. To find answers to the new questions, experimental techniques had to be developed with the particular aim of assessing ion structure and unraveling the ion chemistry that takes place in mass spectrometers. The advent of computational chemistry and refined theoretical calculations has contributed substantially to our present understanding. Computations have provided thermochemical, structural, and mechanistic information that has challenged, motivated, and complemented the work of experimentalists.

We now take for granted many important items that were not self-evident 40 or so years ago. Some of these may still be unclear to some users, as exemplified by the partial relapse into the publishing of unsound, speculative ion structure assignments (see also Section 2.7). For example, primary carbocations, $RCH_2{}^+$, in which there can be no conjugated stabilization, do not exist as stable species in the gas phase undergoing spontaneous rearrangement to more stable isomers; the keto/enol stability of neutrals is reversed in their gas-phase ionic counterparts (i.e., for an ion, the keto form is generally significantly less thermochemically stable than its enol tautomer); stable ions having unconventional structures, e.g., the distonic ion $^\bullet CH_2{}^+OH_2$, referred to above, and ion-neutral complexes (an ion and a neutral species bound by strong electrostatic forces rather than by covalent bonds) are all commonplace in the gas phase. Although these facts were deduced from studying the behavior of small organic ions, they hold true for the much larger systems of current interest. It is therefore important to emphasize that knowing and understanding such information about gas-phase ion chemistry is essential for making new ion structure assignments.

The topics presented in this book have been chosen to illustrate the tools presently available for solving problems of structure and mechanism in the gas-phase chemistry of organic ions. This is nowadays achieved by combining experimental techniques with computational chemistry to establish the structures of the ions in mass spectra, the mechanism of the reactions that generate them, and also their dissociation pathways. A complete study also provides the potential energy surface for the ion and some or all of its isomers, dissociation limits, and transition state energies. A principal aim of such research is also to identify simple experimental phenomena that will provide the data needed for the unequivocal identification of a given ion. Unfortunately, many readers will not have access to instruments capable of performing some (and possibly most) of the techniques described. However, we strongly believe that the knowledge of their existence and most importantly the significance of the results accumulated from using these techniques are invaluable and provide essential background information for all mass spectrometer users.

Chapter 1 will briefly survey current experimental methods for ion production and separation, followed by a more detailed presentation of the experiments designed to reveal qualitative and quantitative aspects of gas-phase ions. Emphasis will be placed on those methods that are used to probe ion structures, namely, the determination of ionic heats of formation and generalities derived therefrom, controlled experiments on the dissociation characteristics of mass-selected ions, and the reactivity of ions. Because of the increasingly important contribution of computational chemistry to the development of the field, a brief discussion of these methods and how to use them to advantage is also presented. This first section is not intended to be encyclopedic and references to further reading will be given, including inter alia, the new *Encyclopedia of Mass Spectrometry* currently being published by Elsevier.

Chapter 2 will describe five selected case studies. They have been carefully chosen to present the reader with the type and range of difficulties associated with ion structure assignment and thermochemical problems. In each case, sufficient data will be presented and discussed and it will be shown how experiments and when appropriate, molecular orbital theory calculations were used to solve the problem. This will also necessitate some discussion of the mechanisms by which ions dissociate and also will touch on difficulties with reference sources, particularly thermochemical data for ions and neutral species. This chapter concludes with a brief guide as to the best way to assign a structure to a new ion and what pitfalls to avoid.

The last and major section of the book contains the data sufficient for the description and identification of all ions containing C alone and C with H, O, N, S, P, halogen, from C_1 to C_3 and is intended to be a primary source of such information.

Finally, in this book, we have necessarily been very selective in the literature quoted. We have attempted to cover all such ions but, given the magnitude of the task, it is quite possible that we have missed some significant data.

Abbreviations

AE	Appearance Energy
BDE	Bond Dissociation Energy
CI	Chemical Ionization
CID	Collision-Induced Dissociation
CIDI	Collision-Induced Dissociative Ionization
CR	Charge Reversal
CS	Charge Stripping
EI	Electron Impact/Electron Ionization
ESE	Energy Selected Electrons
ESI	Electrospray Ionization
E^*	Ion Internal Energy
E_{rev}	Reverse Energy Barrier
FAB	Fast Atom bombardment
FI(K)	Field Ionization (Kinetics)
FTICR	Fourier Transform Ion Cyclotron Resonance
ICP	Inductively Coupled Plasma
IE_a; IE_v	Adiabatic and Vertical Ionization Energies
IRMPD	Infra-Red Multiphoton Photodissociation Spectroscopy
KER	Kinetic Energy Release
MALDI	Matrix-Assisted Laser Desorption-Ionization
MI	Metastable Ion
MS	Mass Spectrum
NRMS	Neutralization-reionization Mass Spectrometry
PA	Proton Affinity
PES	Potential Energy Surface
PEPICO	Photo-ion Photo-electron Coincidence
PI	Photoionization
RRKM	Rice Ramsperger Kassel Marcus Theory
$T_{0.5}$	Kinetic energy release calculated from the half-height width of a MI peak
TOF	Time of Flight

Table of Contents

Part I

Theory and Methods

1 Tools for Identifying the Structure of Gas-Phase Ions

CONTENTS

1.1   INTRODUCTION

The first section in this chapter (Section 1.2) describes the methods available for the generation of gas-phase organic ions and the way in which they are separated by their mass-to-charge (m/z) ratios. These descriptions are short, but references are provided for further reading whereby details may be sought and found.

The second section (Section 1.3) is devoted to ion (and neutral) thermochemistry. The heat of formation of an ion is an important intrinsic property of the species as well as a useful identifier for its structure. The principles of the experimental methods whereby ion thermochemistries have been evaluated are described in some detail. Included are descriptions of general structure–energy relationships that give insight as to the distribution of charge within an ion and the effects of functional group substitution at or away from the formal charge site.

The third section (Section 1.4) describes experiments for the mass selection of ions and the phenomena associated with their subsequent spontaneous (metastable) dissociations and their fragmentations induced by collision with an inert target gas. Both types of observation can provide structure-defining information. Finally some more esoteric experiments are outlined, in which the mass-selected ions suffer charge permutation (positive-to-negative and vice versa), the identification of the neutral products of an ion's fragmentation by their collision-induced ionization, and also the neutralization and reionization of ions in order to examine the stability of the neutral counterpart of the mass-selected ion.

Two more short sections complete this chapter. The first (Section 1.5) describes how the reactivity of an ion depends on its structure as well as the location of the charge and radical sites therein, a technique especially valuable for distinguishing between isomers. The last section (Section 1.6) outlines how computational chemistry can contribute to ion structure and energy determinations, with particular emphasis on how best to select the level of theory suited to the problem in hand.

1.2   HOW IONS ARE GENERATED AND SEPARATED FOR ANALYSIS

1.2.1   INTRODUCTION

This short section is intended to do not more than outline the principal methods for ion production and separation. Some of the information is described in detail in Section 1.4, particularly the use of tandem sector mass spectrometers that have contributed so greatly to the investigation of ion structures and these will be referred to where appropriate.

1.2.2   IONIZATION METHODS AND ION SOURCES

A very great number of mass spectra have been measured, as exemplified by the huge NIST index that contains over 100,000 mass spectra.[1] The great majority of these spectra have been obtained using electron impact ionization (EI). This ionization method requires that the molecule of interest be volatile (or can be volatilized without decomposition) and this places significant limits on its use. Therefore, this and similar ionization methods (photoionization, chemical ionization (CI), etc.) are restricted to molecules of low molecular weight and consequently exclude the use of mass spectrometry for the analysis of many molecular species of biological interest. Considerable efforts have been made for developing new and more versatile ionization methods suitable to nonvolatile, thermally labile, and/or high molecular weight species. As described below, fast atom bombardment (FAB), electrospray ionization (ESI) and matrix-assisted laser desorption–ionization (MALDI) have contributed greatly to extending the use of mass spectrometry to a much wider range of molecular species.

Electron-Impact Ionization

The idea of using energized electrons for ionizing neutral atoms or molecules dates back to the beginning of the twentieth century when Lenard was one of the first to report the effect of bombarding atoms with low energy electrons.[2] Since then, the interaction of electrons with neutral species has given rise to not only one of the most widely used ionization method in mass spectrometry, but also a wide range of other applications.[3] For a discussion on ionization energies per se, as well as obtaining such data, refer to Section 1.3.4.

Electron-impact ionization,[4] contrary to its name, does not involve an actual collision between an energized electron and a molecule. Rather, as the energized electron comes in close proximity to the neutral species, interaction between the partners takes place, which leads to a transfer of energy to the neutral entity.

Ionization of a molecule takes place in the ion source of the mass spectrometer. Ion sources may vary in size and design but they contain the same core elements (see Figure 1.8): a hollow chamber, usually referred to as the source block, contains the gaseous sample that is introduced by a variable leak valve or other capillary inlet. The electrons are produced outside the source by passing a current through a thin filament of tungsten or other suitable metal, raising its temperature to >2000 K, which is sufficient to produce a good flux of electrons. The filament is maintained at a potential negative with respect to the block (e.g., when 70 eV electrons are used, the filament is held at −70 V with respect to the potential of the block) and therefore the electrons are accelerated toward the source block and enter it through a pinhole. Inside the block, the electrons come in contact with the analyte vapor and ionization takes place. The electrons that travel right across the block are collected at a trap electrode situated just outside the source and the current between it and ground gives a measure of the ionizing electron flux. The ions that are generated are gently pushed toward the source exit by an ion repeller electrode that carries a variable small potential, positive with respect to that of the block. As the ions exit the source they are accelerated toward a grounded exit plate placed very close to the block (e.g., 1 mm or less) and acquire the translational kinetic energy appropriate to that voltage gradient. The ion beam can now be electrostatically focused and subjected to a mass analyzer. The magnitude of the kinetic energy imparted to the ions depends on the type of analyzer used. A typical sector instrument will operate with an ion acceleration potential of 8000 V, whereas a quadrupole mass filter involves much lower values, ranging from 5 to 10 V.

It is noteworthy that the ionization process described above does not deposit a fixed amount of energy into the neutral species, but rather a range of energies, which depends on the kinetic energy of the ionizing electrons (most commonly 70 eV). The ranges of ion internal energies obtained under such conditions can be described by the Wannier threshold law.[5] Note that the ions' overall internal energy will also depend on the internal energy distribution within the neutral molecules before their ionization. Because most ion sources function at very low pressures, the internal energy of the analyte cannot accurately be given a "temperature" described by a particular Boltzmann distribution of internal energies. To circumvent this problem, molecular beam sources, where the analyte is introduced in the source as a molecular beam,[6,7] have been designed. In such a source, the internal degrees of freedom of the sample are first cooled by collisions with a carrier gas (helium or argon) and the resulting gas mixture is then rapidly expanded in the ion source. This results in further cooling of the rotational and vibrational degrees of freedom. In such cases, the "temperature" of the sample will be definable. This together with energy-selected electrons

(ESE) (see Experimental methods for determining appearance energies in Section 1.3.6) can provide accurate thermochemical data.

Photoionization

An alternative method for producing ions is to replace electrons by photons. Photoionization[8] followed by mass analysis of the photoions was first reported in 1929 when potassium vapor was ionized using photons generated from an iron arc.[9] Modern photoionization mass spectrometry using tunable light sources however dates back to the 1950s and 1960s.[10,11] A significant advantage of photoionization over electron impact is that ions of known internal energies can be generated (refer to Measurement of ionization energies in Section 1.3.4).

In order to achieve ionization of most organic compounds, photon energies of more than 8 eV are required, therefore, necessitating the use of vacuum ultraviolet radiation (VUV, 200–10 nm wavelength). One challenge when using such radiation is the need to design completely windowless systems because window materials capable of transmitting wavelengths shorter than 100 nm and sustaining a vacuum are scarce. The light source is, therefore, directly connected to the source (which can have essentially the same design as for EI) and powerful pumping systems are required to keep the mass spectrometer at the appropriate low operating pressure, 10^{-6} Torr or less. Three types of light sources generating VUV radiation can be employed:[12] laboratory discharge lamps, VUV lasers and VUV synchrotron radiation. An alternative to using VUV radiation for ionization is to use a lower-energy light source and achieve photoionization in two steps: the first photon is absorbed and excites the molecules to an intermediate state after which the absorption of a second photon results in ionization. This technique is referred to as resonance-enhanced multiphoton ionization (REMPI).[13] Two advantages of this photoionization technique are that it requires radiation of longer wavelength and also leads to higher ionization efficiency.

Chemical Ionization

Unlike the previously described ionization techniques, chemical ionization[14] relies on chemical reactions taking place within the ion source in order to generate ionic species. This technique was first described by Munson and Field and used for the analysis of hydrocarbons and other petrochemicals.[15]

A reagent gas (e.g., methane, hydrogen, or ammonia) is introduced in large excess along with the sample into a modified electron-impact ion source. The total ion source pressure reaches the 10^{-3} Torr level and so the EI ion sources are modified by making all the apertures much smaller, thus allowing higher source pressures to be achieved. The electron flux is raised to insure that sufficient electrons traverse the source to produce a significant ion flux. Under such conditions the reagent gas undergoes ion–molecule reactions to generate stable reagent ions (e.g., CH_5^+ and $C_2H_5^+$ from methane, H_3O^+

from water), which can react exothermically with the analyte by charge transfer, proton (or other cation), or hydride (or other anion) transfer, to produce the desired ionic species. The subject of chemical ionization mass spectrometry has been well described in the book by Harrison.[14]

Field Ionization

Field ionization (FI) was one of the first "soft" ionization methods (one that produces significant molecular ions and only limited fragmentation) and was originally described by Inghram and Gomer[16] in the mid-1950s and further developed by Beckey and coworkers.[17] The ion source contains a sharp blade or tip carrying a charge of approximately 10^4 V and specially treated to possess an array of tiny dendrites. The high field gradient at these extremities is sufficient for ionizing and activating simple organic molecules.[18] Because very little excess energy is transferred in this process, very little fragmentation of the ionized species is observed. For an application of FI to the kinetics of ion fragmentation refer to Field ionization kinetics in Section 1.4.1.

Fast Atom Bombardment and Secondary Ion Mass Spectrometry

These two methods are applied to samples having little or no vapor pressure or that are otherwise unsuited to introduction into an electron-impact ion source. They can, for example, be used to obtain the mass spectra of sugars, ionic compounds, small proteins, etc. Both methods involve the bombardment of the sample with a beam of small, high energy species. For FAB the particles are usually argon or xenon atoms, whereas for secondary ion mass spectrometry (SIMS), Cs^+ or Ar^+ ions are typically employed. The atoms or ions have translational kinetic energies on the order of kilovolts. The methods have been well described by one of the originators of the technique.[19]

In FAB the sample (analyte) molecules are dissolved in a solvent with a high boiling point, glycerol being the commonest matrix. Many other matrices have been described for particular studies, e.g., 3-nitrobenzyl alcohol can be used for organometallics. Dissolution is important, a very weak solution or a slurry will not produce significant ion yields. The sample solution is placed on a small target plate that is maintained at a high potential so that ions are quickly accelerated away to be focused as a beam and then mass-analyzed.

SIMS uses no solvent, but if a matrix is used, the technique is called liquid SIMS (LSIMS). The two methods have so much in common that it will not be necessary to distinguish between them in what follows.

The positive ions generated by these methods are usually protonated sample molecule (MH^+), metal ion adducts, or in negative ion mode, the deprotonated analyte, $(M–H)^-$. Details of the ionization process are poorly understood but it arises from acid–base or redox reactions within the matrix. Because the ions are produced from a condensed phase their internal energy content is small, much lower than EI-generated ions and so unimolecular ion dissociations are rarely observed. When the ion source is attached to a sector mass spectrometer (see Tandem mass spectrometers and experiments with metastable ion beams in Section 1.4.1) high energy collision-induced dissociation (CID) can be used to investigate the structures of the ions.

Electrospray Ionization

This technique has revolutionized the mass spectrometry of high mass molecules of biochemical significance such as amino acids, proteins and peptides.[20] In this method, the analyte is already protonated by being dissolved in an acidic solution. The solution is forced under pressure through a fine capillary held at several kilovolts above ground potential. The fine droplets that are formed lose solvent by evaporation, leaving singly or multiprotonated ions, which are led into a mass spectrometer for analysis. A related technique, thermospray, uses heat to effect droplet formation and desolvation.[21] Like FAB and SIMS, the two methods are very similar and can be used interchangeably.

The process by which the ions are formed is still not thoroughly understood. The progressive removal of solvent from the ions effectively cools them and so the ions rarely dissociate unimolecularly. Depending on the solution conditions used, the ions can be protonated or cationated (e.g., by Na^+, Ag^+ or Cu^{++}), variations that have extensively been employed in the analysis of biological samples. Negative ions can also be investigated, typically $(M–H)^-$. An especially useful aspect of the electrospray method has been its propensity to produce multiprotonated species from large biomolecules. This gives rise to ions with m/z ratios that can be accommodated by relatively simple mass analyzers, such as quadrupole mass filters. For example, an ion of nominal mass 80,000 Da that carries 40 protons has an m/z ratio of only 2000. Multiply charged ions are often observed as a sequence of analyte ions carrying n, $n-1$, $n-2$, $n-3$, etc. protons. A simple algorithm permits the determination of the molecular weight of the compound. If a pair of closely spaced intense ions $(m/z)_1$ and $(m/z)_2$ are observed, and it is reasonable to assume that they arise from ions carrying n and $(n+1)$ protons, respectively, then solution of the two simultaneous equations

$$(m/z)_1 = [M + nH]/n \quad \text{and} \quad (m/z)_2 = [M + nH + 1]/n + 1$$

will lead to the evaluation of n and M, where M is the molecular mass of the molecule. Confirmation of n is achieved by the selection of other sets of ions.

Matrix-Assisted Laser Desorption–Ionization

In the late 1960s and early 1970s Mumma and Vastola[22] investigated the laser-induced ionization of inorganic solids, followed by others who investigated organic compounds.[23,24] However, it was not until the late 1980s that the technique generated particular interest due to the pioneering work of Karas and Hillenkamp[25] who showed that intense molecular ions of proteins having molecular weights in excess of 10,000 Da could be generated using MALDI.[26,27] Using MALDI, Hillenkamp was able to produce molecular ions of β-D-galactosidase (molecular mass 116,900 Da), which at that time was by far the largest intact molecular ion ever created.[28]

In MALDI, the sample is prepared by making a solution containing the analyte with a light-absorbing matrix material in a molar ratio of approximately 1:5000 (i.e., the matrix is present in very large excess). A drop of the resulting solution is placed on a metal plate and allowed to dry, leaving the sample and matrix together in crystalline form. The plate is then placed in the source housing of the mass spectrometer where it is irradiated (under vacuum) with a laser pulse of a few nanoseconds. Generally, a 337 nm N_2 laser is used for irradiation although some workers have investigated the use of IR light sources.[29] The plate carries a large positive potential and so positively charged species produced are ejected from the ion source, focused, and mass analyzed.

The use of a matrix is essential to the technique and is believed to serve several purposes: First, it ensures strong absorption of the incoming radiation. This absorbed energy is then transferred to the sample via vibrational interactions and is sufficient to induce the ablation of the sample, creating a plume of ejected material. The plume can contain intact matrix, analyte neutrals and ions as well as protonated species. Second, the matrix prevents aggregation of the analyte molecules and is also believed to play an active role in the ionization process itself. Although the ionization step is not fully understood, the sample must acquire its charge from the matrix, most probably by acid–base or redox processes. It is important to emphasize that the overall mechanisms at play in MALDI are still controversial and many key parameters have been investigated in the hopes of producing a comprehensive model of the processes taking place.[30–33]

Inductively Coupled Plasma

The design of inductively coupled plasma (ICP) sources began in the early 1940s and continuous development led to their wide use as excitation sources for analytical atomic emission spectroscopy studies.[34,35] Because ions are present in the plasma (together with electrons and neutral species) their use as sources for mass spectrometry was proposed 40 years later.[36,37]

The ICP source itself typically consists of three concentric quartz tubes that collectively make up the torch of the ICP. On the outside tube is a tightly wound induction coil through which a radio frequency signal is fed to produce an intense oscillating magnetic field. Argon, the support gas, is continuously introduced into the torch and a Tesla unit is used to generate a spark. When the spark passes through the argon gas, ionization takes place and the resulting mixture of ions and electrons are accelerated toward the magnetic field. As the charged species collide with the support gas (neutral argon atoms) a stable, high temperature (6000–10,000 K) plasma is generated. The sample, which consists of a fine aerosol, moves toward the plasma with the carrier gas (Ar) through the inner coil. As it passes through the plasma, the analyte collides with the species present in the plasma and under such extreme conditions, any molecule initially present in the aerosol is decomposed into atomic, ionic and neutral fragments. The charged species are then transmitted to the mass spectrometer for mass separation and detection. Thus ICP is very efficient at producing elemental ions and is often used to determine the concentration of trace metals. It is noteworthy that ICP mass spectra also contain signals corresponding to polyatomic ions but the origin of these is still the source of some debate.[38]

1.2.3   MASS ANALYZERS

That a beam of ions can easily be separated into subbeams, each having a unique mass-to-charge (m/z) ratio, stems directly from the apparatus commonly used to generate an ion beam. In general, using one of the methods described above, ions are produced within a small enclosed space that carries an electrical charge (the ion source). All ions leaving this space are accelerated by falling through the potential gradient between the source and very closely nearby grounded plates, producing an ion beam that can be electrostatically focused easily. Within this beam, all ions that carry the same number of charges, but irrespective of their mass, will have acquired the *same translational kinetic energy*. When such a beam passes through (controllable) homogeneous magnetic or electric fields, the ions will follow different paths depending on either their momentum (magnetic field) or their kinetic energy (electric field). As will be described below, these two intrinsic kinematic properties form the basis of many common mass analyzers.

Magnetic Sector, Electric Sector, and Tandem Mass Spectrometers

These types of instruments are described in some detail in Section 1.4 and so only a few comments are required here. Magnetic sector mass separators (B) have a long association with mass spectrometry[39] and were the mass analysers of choice for many years. The principle of the method is simple: when a beam of ions of various m/z ratios passes through a homogeneous magnetic field they will follow different curved paths whose radii depend on the momentum of the ions; thus mass separation may be achieved.

Mass resolution in such a device is modest, say 3–5000, and is defined with respect to a 10% (or other fraction) valley between adjacent m/z signals. The placing of an electrostatic sector (E) before or after the magnet to achieve greater mass resolution is also described later. E sectors consist of charged, parallel (coaxial) curved metal plates, and ions follow curved paths therein with radii that depend on their translational kinetic energy. Such *BE* or *EB* double-focusing instruments can have mass-resolving power in excess of 150,000. However, the most important results, as far as research in ion structures is concerned, have come from the use of mass spectrometers of *BE* geometry wherein an ion of interest can be separated from all others by the magnet (*B*) and then subjected to a wide range of physical experiments in the field-free region lying between it and the electrostatic sector. Scanning the electrostatic sector permits the analysis of all ionic products from the experiment and in some cases the neutral fragments as well. (See Tandem mass spectrometers and experiments with metastable ion beams in Section 1.4.1 and the following sections.)

Quadrupole Mass Filters and Ion Traps

Ions in quadrupole fields can be induced to adopt stable trajectories and therefore can be mass-selected and controlled. The principle by which such devices operate can briefly be described as follows. An ideal quadrupole mass filter consists of four equally spaced hyperbolic rods, but for practical purposes the rods are usually cylindrical. A beam of ions from a source enters the longitudinal axial space between the rods and mass separation is achieved by a combination of direct current (DC) and radiofrequency (RF) fields of opposite polarities applied to the opposing rods. The RF field gives the ions an oscillatory motion and to achieve mass resolution it is necessary for the ions to experience several RF cycles. For this reason, the ion translational energies in the quadrupole must be quite low, typically of the order of 10 V. The method has been very fully described.[40,41] It should be noted that the instrument can be used as a simple ion guide only, in the RF mode. Ion beams of very low translational energy can be held in a stable path by such means and translational energy dependent cross sections for endothermic ion disso-ciations and ion–molecule reactions can be investigated with precision, lead-ing to valuable thermochemical data.

Experiments with Quadrupole Ion Mass Filters

Quadrupole mass filters are well suited to the study of low energy collision-induced dissociations (see Section 1.4.2) of mass-selected ions. In such experiments a combination of three quadrupole mass filters are placed in series: the first acts as a mass filter, the second, operating in RF mode only (and therefore only an ion transmitter), is the collision zone into which the inert target gas (often argon) is admitted, and the third quadrupole is used to analyze the m/z ratios of the dissociation products. These triple–quadrupole instruments can be used to study ion reactivity, with a reactant molecule replacing the inert collision gas.

An elaboration of the above method is the selected ion flow tube, or SIFT, apparatus.[42] This has been very widely used in the study of bimolecular reactions of ions. The apparatus consists of five regions, an ion source followed by a quadrupole mass filter wherein the reactant ions of interest are mass-selected. The third region, the flow tube, is typically 1 m long and its diameter is approximately 10 cm and it has a number of injection ports through which gaseous reactant molecules may be admitted. Product ions are detected and identified in the fourth region by the quadrupole that precedes the detector.

The Quadrupole Ion Trap

This instrument is the three-dimensional counterpart of the quadrupole ion mass filter. A common geometry consists of two hyperbolic endcap electrodes on either side of a ring electrode. The advantage of this over the quadrupole is that the device can both mass-select and store ions. Details of its operation can be found in the book by March and Todd.[43]

Fourier Transform Ion Cyclotron Resonance Mass Spectrometry

The analysis of ions by Fourier transform ion cyclotron resonance (ICR) mass spectrometry (FT-ICR MS)[44] relies on the discovery made in 1932 by E.O. Lawrence and S. M. Livingston,[45] who observed that the trajectory of charged particles moving perpendicular to a uniform magnetic field is bent into a circular path with a natural angular frequency according to $\omega_0 = qB/m$ (where q is the charge and m the mass of the charged particle and B the static magnetic field). Lawrence received the 1939 Nobel Prize in Physics for the invention and development of the cyclotron and for results obtained with it, especially with regard to the artificial radioactive elements. Although the equation relating the mass-to-charge ratio of a species to its cyclotron frequency (through the strength of the magnetic field) is simple, it took almost 20 years before Hipple, Sommer and Thomas[46,47] first applied the ion cyclotron principle to mass spectrometry. Following its first application, one of the most important technical advances in the early years of ICR MS was the development of the cubic trapped ion cell by McIver in 1970.[48] The new cell could now constrain the ions, preventing them from drifting out of the analyzer region. With trapped ions it became apparent that a greater number of analytical experiments, such as the possibility of tandem mass spectrometry, could be performed using this technique. The other major advance came in 1974 when Comisarow and Marshall[49] applied to ICR the algorithm for Fourier transformation (FT) that Cooley and Tukey[50] had introduced in 1965. By 1978, FT-ICR MS was established as a powerful, ultrahigh-resolution method capable of investigating complex ion chemistry.[51,52]

The cubic trapping cell consists of six plates arranged in the form of a cube (although other arrangements are also used).[44] One of the plates is punctured, allowing ions generated outside the cell to enter it and also permitting photon and/or electron beams to enter the cell. The cell is positioned such that the applied magnetic field, usually of 6 T or more, is parallel to the aperture and along the z axial direction. As stated above, under the influence of a static magnetic field, the ions within the cell will acquire a cyclotron motion and describe orbits (in the xy plane) characterized by a cyclotron frequency. Typically, for thermalized ions, cyclotron radii range from 0.1 to 2 mm, the smaller radii being obtained with higher magnetic fields.[44] In order to prevent the ions from escaping along the z axial direction, a small DC voltage (ca 1 V) is applied to the two endcap electrodes. Under such conditions, in the absence of any RF excitation pulse, the ions can be trapped for extended periods of time (hours). To prevent unwanted collisions that would affect the cyclotron motion and in turn the trapping efficiency of the cell, the instrument is held at a high vacuum, typically of the order of 10^{-8} Torr. Note that the presence of both the magnetic and the electric fields induces on the ions a magnetron motion. The magnetron frequency serves no analytical purpose as it is a consequence of the curvature of the trapping field due to the finite length of the cell electrodes.[44]

In order to accelerate ions coherently toward larger and therefore detectable orbital radii or to increase their kinetic energy to observe dissociation and/or ion–molecule reactions or to eject unwanted ions from the trap, an oscillating electric field is applied between the excitation plates that lie in the yz plane of the cell. When the frequency of the RF component of the electric field matches the cyclotron frequency of ions of a given m/z ratio, these ions are excited, coherently producing a packet of ions and are pushed outward to occupy larger orbits inside the cell. These ions therefore pass in close proximity to the detector electrodes (that are placed in the xz plane) and induce an alternating current in the external circuit that joins them. This alternating current is referred to as the image current and it produces a sinusoidal image signal that is acquired over a given period of time as the ion packet naturally decays back to incoherent motion. The image current, which is a time-dependent signal, can be Fourier-transformed to produce a frequency spectrum. When a singular excitation frequency is used, the resulting frequency spectrum contains one component corresponding to the cyclotron frequency of the excited ions. Using the FT capability, ions of many m/z ratios can be detected simultaneously by applying a broadband excitation pulse. The resulting frequency spectrum is converted to a mass spectrum by applying a calibration.

The main uses of the FT-ICR method, besides its capacity for ultrahigh mass resolution (as high as 10^6), have been the study of ion–molecule reaction kinetics and mechanisms and especially the analysis of large biomolecules. The reader is particularly directed to the journal *Mass Spectrometry Reviews*, where many articles about the specialized uses of FT-ICR MS can be found.

Time-of-Flight Mass Spectrometers

This method for separating ions is perhaps the simplest of all and was among the earliest of mass spectrometers. A good review is to be found in Ref. [53]. Singly charged ions that exit a typical mass spectrometer ion source have all been accelerated through the same potential gradient and so irrespective of their mass, will have the same kinetic energy. This is represented by the equation $eV = mv^2/2$, where e is the electron charge, V is the ion acceleration potential, m is the ion mass and v its velocity. It follows that the time taken for an ion to travel a given distance depends on its mass and the time t_d required for an ion to travel a distance d is given by

$$t_d = \frac{d}{(m/2eV)^{1/2}}$$

If d is the distance between the source and the detector, typically about 1 m, then an 8 kV ion of $m/z = 100$ will take about 8.1 μs to traverse the apparatus and it will be separated from an ion of $m/z = 101$ by approximately 0.04 μs. Clearly, very fast electronics will be needed to detect these sequential events and provide unit mass resolution. There are other factors that weigh against mass resolution, such as the velocity distribution among the ions within the source; this will be superimposed on the exiting ions' velocities, thus blurring the otherwise sharp velocity distribution, i.e., it broadens the time-of-flight (TOF) distribution for each ion mass. These and related matters have been described in detail by Wiley and McLaren.[54]

The resolution of a TOF mass spectrometer can be improved by adding energy focusing to help obviate the translational energy spread. This is achieved in a reflectron instrument, where the ions' trajectory is reversed by an ion mirror (a curved charged plate) placed at the end of the flight tube. Ions with a higher kinetic energy penetrate the field of the mirror farther than those of lower energy and the net result is an improved focusing (in time) of the ion beam. Moreover, the flight path is roughly doubled by the reflectron, again increasing mass resolution.

1.3 THERMOCHEMISTRY; ITS ROLE IN ASSIGNING ION STRUCTURES

1.3.1 INTRODUCTION

The measurement of the heat or enthalpy of formation, $\Delta_f H$, is unlikely to be used as a key structure-identifying feature of a neutral molecule, but for gas-phase ions, with their surprisingly large numbers of stable isomers, such data have provided significant support for the identification or confirmation of a given ion's structure. On occasion, new thermochemical data have sparked the discovery of an unsuspected ion structure and in some instances the heat of formation of the ion has been the most reliable identifier of a given structure. Moreover, a wide knowledge of neutral and ionic heats of formation permits the relative stabilities of isomeric species to be established as well as defining the energy requirements of ion dissociation processes.

This section describes how such data for ions and neutrals are obtained and used, and also illustrates the strikingly simple empirical relationships that exist between ion enthalpies of formation and their structural features.

The existence of such correlations between structure and $\Delta_f H$ values greatly enlarges the scope of such information, permitting the reliable estimation of unknown ionic heats of formation and also giving insight into the location of greatest charge density in series of cations. It is these simple structure–energy relationships that can sometimes remove the need to perform high level computational chemistry.

1.3.2   ION STABILITY

Before proceeding further, it is important to define the meaning of the expression "ion stability" as used in this book. An ion is considered to be *stable* only if it lies in a potential energy well. An ion A^+ is declared to be more stable than ion B^+ if $\Delta_f H(A^+)$ is less than $\Delta_f H(B^+)$. Stability is thus not to be confused with reactivity.

1.3.3   HEATS OF FORMATION OF NEUTRAL MOLECULES AND FREE RADICALS

For neutral molecules containing the elements C, H, O, N, S and halogens, there is a useful amount of reliable data for molecular heats of formation[55] and they are also readily available from the NIST WebBook.[56] However, in toto there are only some 3500 such values for organic compounds and data for organic free radicals are in even shorter supply. See, for example, the recent book by Luo,[57] which includes about 400 carefully selected values for free radical heats of formation and also group additivity values (GAVs, see below), which permit the estimation of enthalpies of formation for unknown radicals. The dearth of more experimental data is not a major problem because both organic free radical and molecular heats of formation are additive properties, as described hereinafter.

The usually quoted standard heat or enthalpy of formation of a species M is $\Delta_f H_{298}^0(M)$, which is defined as the heat absorbed or released when 1 mole of the substance M is formed from its elements in their standard state at 298 K and 1 bar pressure (101,325 Pa). This can conveniently be represented by an equation, as given in the following:

$$2C(s) + 2H_2(g) \rightarrow C_2H_4(g)$$
$$\Delta H_{\text{reaction}} = \Delta_f H_{298}^0(C_2H_4) = 52.5\,\text{kJ/mol}$$

(1.1)

The terms in parentheses represent the phase in which the reactants and product are stable at 298 K and 1 bar pressure, namely the standard state of the species. Except when it is necessary to carefully define an enthalpy term, the symbol $\Delta_f H$ will be used in this book for the standard heat of formation.

The lack of experimental data for $\Delta_f H$ for organic compounds provided an excellent incentive for the discovery of a simple method for their estimation. It was realized long ago that the heats of formation of homologous series of organic compounds, e.g., the alkanes, were additive, in that the addition of each successive –CH_2– unit caused a constant change in $\Delta_f H$. Tables of these incremental changes have been developed for specific groups of atoms, most notably by Benson.[58] These tables contain the average GAVs for a very wide range of molecular components. For the above –CH_2– group, when attached to saturated C atoms, it is represented by the term symbol C–$(C)_2(H)_2$, and each such group contributes -20.9 kJ/mol to the total $\Delta_f H$. A simple example of the use of GAV terms follows:

For ethanol, CH_3CH_2OH, the terms are C–$(C)(H)_3$, C–$(H)_2(O)(C)$, and O–$(H)(C)$, and their Benson GAVs are -42.7, -33.5 and -158.6 kJ/mol, respectively. Their sum gives $\Delta_f H(C_2H_6O) = -234.8$ kJ/mol, which is in excellent agreement with the reference value -235.2 kJ/mol.[55] By such means many heats of formation can be estimated using simple arithmetic. Correction terms have also been developed to allow for such phenomena as geometric isomerism, ring strain, and a wide variety of steric effects.[58,59]

Where there are no GAV terms available for the molecule in question, there are three courses of action: One is to perform an experiment, i.e., undertake an accurate calorimetric study of a reaction involving the molecule, such as determining its heat of combustion. (This is no light undertaking; moreover, it is unlikely that a research grant would be awarded for such a mundane, albeit important study.) The second is to calculate a heat of formation value using a high level of computational chemistry theory, such as the composite G3 method or employing B3-LYP-based density functional theory (DFT) with a suitably large basis set (see Section 1.6). The third is to use chemical intuition, or an argument based on a sound analogy, to create a new, provisional GAV term.

Thus it is seldom the case that one cannot produce a good to fair value for the $\Delta_f H$ of a new molecule. Organic free radicals can be similarly treated and the latest series of GAV tables for them can be found in the book by Luo[57] together with a complete survey of homolytic bond strengths.

1.3.4 HEATS OF FORMATION OF IONIZED MOLECULES

The heat of formation of an ionized molecule $AB^{+\bullet}$ (an odd-electron ion) is given by the equation:

$$\Delta_f H(AB^{+\bullet}) = IE_a(AB) + \Delta_f H(AB) \qquad (1.2)$$

where IE_a is the *adiabatic ionization energy* of the molecule AB and is defined as the energy required to remove an electron from the *ground state* of the molecule to produce the *ground state* of the ion. It is important to note that the geometry of the ground state of the ion and that of the neutral are not necessarily identical and so one usually finds two ionization energies quoted: IE_a as defined above and IE_v the *vertical ionization energy*. The latter is the energy difference between the ground state of the neutral and the position on the potential energy surface of the ion that has the same geometry as the neutral's ground state. These properties are illustrated in Figure 1.1.

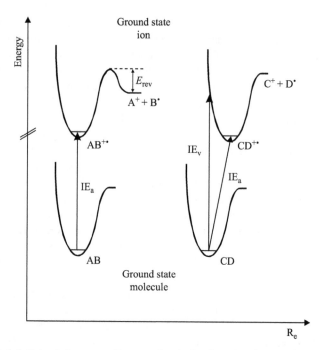

FIGURE 1.1 Potential energy diagram for ionization showing vertical (IE_v) and adiabatic (IE_a) ionization energies. Also shown are dissociations involving a reverse energy barrier (E_{rev}) and none for the lowest energy dissociation of the ions.

Note that IE_v represents a "Franck–Condon" excitation, i.e., no nuclear displacement takes place in the excitation time. In general, for many molecules, $IE_a \approx IE_v$ is not an unreasonable first assumption.

The heats of formation of many organic ions are to be found in Chapter 3, which describes individual ion structures. See also the NIST WebBook[56] and the collection of Lias et al.[60]

Measurement of Ionization Energies

The measurement of an ionization energy consists of precisely determining the energy threshold at which the ionized species of interest is generated from its neutral counterpart. Ionization can be induced by either photons or electrons and the most commonly used methods for determining ionization energies are briefly described below.

Photoionization

Ionization by photons is a strictly vertical process and so yields IE_v values. Photoionization and photoelectron spectra[61] (see below) can often give IE_a values, but where there is a large geometry difference between the molecule/radical and the ion, obtaining an IE_a value can sometimes be very challenging and so may even require ingenious experimental approaches. For example, this is the case for NO_2 for which the ground state is bent whereas for the ion it is linear. However, the problem can be circumvented by measuring the IE_a in a two-photon experiment, where the first excitation raises the molecule to an excited state having a geometry the same as or very close to that of the ground state of the ion and the second photon achieves ionization thereto.

Electron-Impact (EI) Ionization

EI is a compromise method, in that the approach of an ionizing electron can partially distort the neutral before its electron loss, thereby reducing the $IE_v - IE_a$ difference. In other words, EI is not a strictly vertical process. Therefore the values obtained by EI are deemed to be satisfactory provided that the difference in the two ionization energies is, say, less than 0.3 eV. In general, the cross-section for electron ionization falls quite rapidly away from the Franck–Condon (vertical) region.[62] Note that in extreme cases, vertical ionization can lead to a point on the ion's potential energy surface above its dissociation limit; this will be responsible for the absence of a molecular ion in the normal EI mass spectrum of the molecule. A good example of a simple molecule that displays no molecular ion is the alcohol tertiary butanol, $(CH_3)_3COH$. Its molecular ion is greatly distorted, and can be visualized as a methyl radical bound electrostatically via a long bond to protonated acetone.[63]

Photoelectron Spectroscopy[64,65]

Photoelectron spectroscopy (PES) is a valuable method. If a molecule is excited by a high energy photon, such as those having wavelengths in the vacuum ultraviolet (VUV) region of the spectrum and thus possessing sufficient energy to ionize the molecule, the excited species will eject an electron. PES is the analysis of the kinetic energies of these electrons. For a given photon energy, the distribution of kinetic energies among the ejected electrons reflects the distribution of accessible energy levels in the ion. In terms of an equation then

$$E_{ion} = h\nu - E_{electron} = h\nu - 1/2m_e v_e^2 \qquad (1.3)$$

where E_x is the energy of species x, h is Planck's constant, ν is the photon frequency, m_e is the electron mass and v_e its velocity.

If the electron energy distributions are highly resolved, then the orbitals from which the electrons were lost can be identified as well as vibrational progressions for the excited ionic states. The most commonly used photons are from the He(I) line $(1s^12p^1 \longrightarrow 1s^2)$ from a helium discharge lamp. Their wavelength, 58.43 nm, corresponds to an energy of 21.22 eV, well above the ionization energies of organic compounds, that typically lie in the 8–12 eV range. The value of IE_a is obtained from the onset in the plot of E_{ion} vs. electron yield. A small section of the (smoothed) PES for 1-propanol is shown in Figure 1.2, where both IE_v and IE_a are indicated.

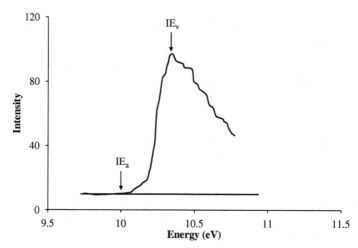

FIGURE 1.2 Portion of the photoelectron spectrum for 1-propanol showing the adiabatic (IE_a) and vertical (IE_v) ionization energies.

The difference between the measured IE_v and IE_a values results from the geometry of ionized 1-propanol being not the same as for the neutral molecule; the ion has an elongated α-C—C bond. See the discussion of this ion in Part 2.

Recent implementations of PES usually focus on the analysis of threshold electrons (TPES) and employ tunable light sources such as those found at synchrotron sources and they allow IE values to be measured with greater precision. Zero kinetic energy pulsed field ionization (ZEKE-PFI)[66] experiments provide perhaps the most accurate IE data but they tend to be limited to molecules having long-lived Rydberg states with high principal quantum numbers, (n). Occasionally, the difference between IE_a and IE_v is so large that none of the previously described methods can yield accurate IE_a values.

Charge Transfer Equilibria and Bracketing Methods
A variety of methods based on charge transfer experiments have been described.[60] If a high-pressure ion source or ICR mass spectrometer is used, the equilibrium constant (k) for the process

$$A^+ + M \leftrightarrow A + M^+$$

can be measured as a function of temperature and application of the Van't Hoff equation $\{-RT\log(k) = \Delta G = \Delta H - T\Delta S\}$ then permits an accurate evaluation of the difference in ionization energy of A and M.

Alternatively, if it is not possible to achieve equilibrium in the system, the method of bracketing is used. The ion of interest is allowed to undergo charge exchange with a series of molecules, chosen such that their IE lies close above and below that of the target molecule. Note that if the ion–molecule reaction were to involve proton transfer, the proton affinity (PA), qv can be determined (see Measurement of proton affinities; equilibria and bracketing studies in Section 1.3.8).

1.3.5 ESTIMATION OF IONIZATION ENERGIES OF MOLECULES

As outlined above, because of the possible experimental difficulties in determining them, the number of accurate IE_a values is alarmingly small and so any versatile estimation scheme for extending the good data to a wider variety of molecules will find a use.

For cations, a simple additivity scheme cannot work because the IE_a of molecules and free radicals, even in homologous series, are not directly proportional to, for example, the number of $-CH_2-$ groups therein. Therefore Equation 1.2 cannot be transformed into a simple additive function. However, with the sufficient good data available, it has been possible to devise some useful empirical relationships.

The first such scheme for estimating IE_a values was that of Bachiri et al.[67] They designed the equation given below:

$$\log \{IP[R_1XR_2] - IP_\infty\}/\{IP_0 - IP_\infty\} = 0.106\{I(R_1) + I(R_2)\} \qquad (1.4)$$

where X is a functional group, e.g., alkene, alkyne, aldehyde, ketone, mercaptan or thioether and R_1 and R_2 are alkyl groups, IP_0 is the IE_a of the reference compound, where $R_1 = R_2 = H$. $I(R_n)$ is a constant, characteristic of each alkyl group and IP_∞ is the value obtained by extrapolating IE_a to infinite ion size. The latter value is not difficult to determine because IE_a values reach an asymptote when plotted against increasing molecular size.

Equation 1.4 works well, but it is cumbersome to use and does not lend itself to any form of graphical presentation. In addition, it does not provide the user with any clear physicochemical insights about the ionized molecules.

A simpler equation was developed by Holmes et al.[68] This followed from the observation that for homologous series of compounds, IE_a is a linear function of the reciprocal of their size. For homologs, size is adequately represented by the number of atoms in the molecule, $\underline{n}$. These results are closely similar to those obtained for molecular electron polarization, where the refractive index (η) functions $(\eta^2 - 1)/(\eta^2 + 2)$ for homologous series are also a linear function of $1/\underline{n}$. It is worth noting that when appropriately scaled, these two types of $1/\underline{n}$ plot, for example, alkanes, alkanols and bromides, have the same relative slopes.

The combination of this relationship between IE_a and $1/\underline{n}$ for the ion, with the additivity of $\Delta_f H$ for the neutral species M, gives rise to the general equation:

$$\Delta_f H(M^{+\bullet}) = A - B\underline{n} + C/\underline{n} \qquad (1.5)$$

where the constants A and B apply to the neutral species and C is for the ion.[68] Their values have been determined for homologous alkanes, alkenes, alkynes, alkanols, aliphatic ethers, aldehydes and ketones, alkanoic acids, alkyl chlorides, bromides and iodides. Correction terms were also developed for chain branching, double bond position, and asymmetry effects.[68] As explained above, it is generally possible to estimate the heat of formation of a neutral and so the linearity of the IE_a vs $1/n$ plots is particularly useful[69] to estimate and verify IE values. By combining these values, $\Delta_f H$ for the ion is obtained. Examples of IE_a vs $1/n$ plots are shown in Figure 1.3.

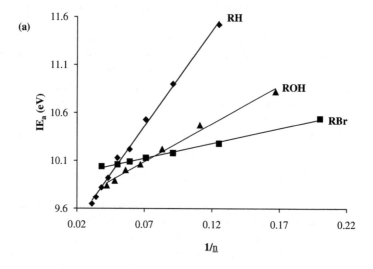

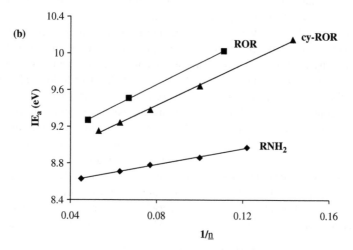

FIGURE 1.3 Plots of adiabatic ionization energy (IE_a) vs the reciprocal of the number of atoms in the species ($1/n$) for some homologous series: (a) alkanes (RH), primary alkanols (ROH) and bromides (RBr); (b) di-n-alkyl ethers (ROR), cyclic ethers (cy-ROR) and alkylamines (RNH_2).

The lines have different slopes and these are in keeping with the relative abilities of the ionized functional group to localize the charge; the gentler the slope, the stronger the charge localization. Thus, for example, localization of charge is greatest in bromides, less in alkanols, and least in alkanes,[70] (see Figure 1.3a). This, of course, is in keeping with chemical intuition; the ionization of bromocompounds takes place <u>at</u> the halogen atom by loss of a lone pair electron, and their IE_a is low, relative to hydrocarbons. Such plots also led to the conclusion that ionization at a double bond results in a localization of charge thereat, similar to that for RBr^{71} and so on. The only reported exception to this general behavior is that of the cycloalkanes, whose IE_a values are approximately constant, and not a function of increasing ring size. However, the substitution of alkyl groups thereon, remarkably, reintroduces the typical $1/\underline{n}$ ion size effect.[71]

Figure 1.3b includes data for n-alkylamines and also shows that charge localization is essentially the same for homologous di-n-alkyl ethers and cyclic ethers for which the lines are parallel, showing that charge stabilization (delocalization) is the same irrespective of whether an extra $—CH_2—$ group is added to the linear chain or to increase the size of the ring.[70] Note too that the results for the cyclic ethers show that the ring strain energy, which is appreciable for the first three members (113, 107 and 24 kJ/mol, respectively, for oxirane (C_2H_4O), oxetane and tetrahydrofuran),[58] has no apparent effect on the ionization energy. General correlations of this kind can provide a quick and easy method for producing new IE_a data, correcting erroneous values, as well as giving some useful insight into the degree of charge delocalization in molecular ions.

1.3.6 HEATS OF FORMATION OF EVEN ELECTRON IONS

Even electron ions (ionized free radicals) comprise the majority of peaks observed in the normal EI mass spectrum of most organic compounds and so to aid in the identification of the structure of such ions, their heats of formation need to be known. Although they are ionized free radicals, the direct measurement of their IE_a is usually not possible owing to the difficulty of finding a good experimental gas-phase source of such highly reactive neutral species. Pioneering work was however done by Lossing and co-workers on the direct measurement of IE_a for many free radicals produced by an appropriate pyrolysis.[72] In his method nitrite esters of formula RCH_2ONO thermally decompose at moderate temperatures by the consecutive losses of NO and CH_2O, to produce the radical $R^\bullet$. Direct IE_a measurements using energy selected electrons (ESE) have provided many reliable heats of formation.[73]

The use of the relationship given by Equation 1.2 to obtain the heats of formation for even electron cations is therefore limited by the general lack of $IE_a(R^\bullet)$ values. Heats of formation of even electron ions ($\Delta_f H(A^+)$) are mostly determined experimentally by measuring the minimum energy required for their observation in the mass spectrum of a suitable precursor molecule, AB. This measurement is referred to as an appearance energy (AE) measurement. Preferably A^+ is generated *directly* from ionized AB by a simple bond cleavage:

$$AB \rightarrow (AB^{+\bullet}) \rightarrow A^+ + B^\bullet \tag{1.6}$$

The AE of A^+ is related to $\Delta_f H(A^+)$ by the following equation:

$$AE(A^+) = \Delta_f H(A^+) + \Delta_f H(B^\bullet) - \Delta_f H(AB) \tag{1.7}$$

Thus in principle, $\Delta_f H(A^+)$ can be obtained from an AE measurement, provided that the ancillary thermochemical data are known or can be estimated.

Experimental Methods for Determining Appearance Energies

AE values are the energy threshold at which a given fragment ion is produced from a particular precursor molecule. The dissociative ionization that leads to the generation of the desired fragment ion can be induced either by electron or by photon ionization (EI or PI).

With appropriately constructed apparatus to produce monochromatic VUV radiation or ESE beams, AE values can be determined with fair to good precision, say to within ± 5 kJ/mol. The former method is well described in the publications by Traeger[74] and the latter apparatus was elegantly developed by Maeda and Lossing.[75] Note that the timescales for photoionization (VUV) and EI (ESE) measurements are similar because similarly low drawout voltages from the ion source to the mass analyzer were employed. Thus the limiting observable rate constant for the two methods is typically 10^2–10^3 s^{-1} and so *kinetic shift effects* (see description below in AE measurements and the kinetic shift in Section 1.3.6) will be minimized.

Metastable ions (see Section 1.4.1) also dissociate in a time frame well defined by the mass spectrometer that is used and so metastable ion peak appearance energies also can be easily related to the limiting observable (minimum) rate constant. In general, MI AE values agree well with those from the other two methods. Details of the technique used to measure MI AE values have been described.[76]

An important but relatively less employed method for AE measurements is photoelectron–photoion coincidence spectroscopy (PEPICO). The topic has been reviewed by Dannacher[77] and by Baer.[78] This method was first described in 1972 by Eland.[79] The principle of the method is as follows.

The sample molecule is photoionized using either a fixed wavelength in the VUV or monochromated radiation from the hydrogen discharge spectrum or a synchrotron. For the former the He(I) line (21.22 eV) is employed. The photon energy used exceeds that to ionize and to dissociate the molecular ions. The energy content of the ions can be determined by analyzing the energies of the electrons emitted from the excited molecules. If a given molecular ion is mass-selected in coincidence with an electron of measured translational energy, then the internal energy of the ion is the difference between the electron's energy and that of the exciting photon (Equation 1.3). In addition, the dissociation rate constants (k) of the ion of interest can be observed on a timescale from about 1–100 μs by monitoring the appearance of the appropriate product ion. The rate data are analyzed using RRKM unimolecular reaction rate theory using calculated vibrational frequencies for the ion and a carefully chosen model for the transition state(s). The frequencies are chosen such that the calculated rates resulting from the model fit the experimental $\ln(k)$ vs E^* (the internal energy content of the ion) plots for each dissociation channel observed.

AE values can also be obtained from the measurement of fragment ion thresholds as a function of collision energy with an inert target gas. These reactions are studied in quadrupole mass spectrometers. The method suffers from the same disadvantages, described in the ensuing sections, as do the other threshold studies mentioned above. Key articles are to be found in Ref. [80].

In principle it would appear that the direct measurement of an AE value should be no more difficult than obtaining an IE. However, this is not so and the following problems must be addressed whenever $\Delta_f H$ for a fragment ion is determined from an AE experiment.

AE Measurements and Fragment Ion Structures

The first issue that must be addressed is the structure of the fragment ion. Any AE measurement, as accurate as can be, is useful for determining the heat of formation of a fragment ion *only* if the structure of the fragment ion and that of the cogenerated neutral are both known. For many ion dissociations, the identity of the fragment ion, at least provisionally, can be assumed on the basis of chemical common sense. However, because the rearrangement of a precursor molecular ion that leads to the transition state of lowest AE can often be profoundly complex, the structures of the dissociation products may turn out to be very different from those predicted and based on intuition alone and therefore cannot be taken for granted. If the reaction that produces the desired ion is simple enough, then the structure of the neutral product will not be in doubt because small molecules and radicals have no isomers. However, in rare cases the neutral product is not as predicted. For example, the $[C,H_3,O]^\bullet$ radical directly produced from the dissociation of ionized methyl acetate ($CH_3C(O)OCH_3^{1+\bullet}$) was shown to be not methoxy as expected, but instead the lower energy $^\bullet CH_2OH$ isomer.[81] In contrast, other ionized methyl alkanoates lose $CH_3O^\bullet$, as might reasonably have been proposed. A good example of unexpected product ion generation is provided by the isomeric C_6H_{10} hydrocarbons. All of these ions, irrespective of their initial neutral structure, have been found to yield at the threshold for the loss of $CH_3^{|\bullet}$ (a major fragmentation pathway) the most stable $C_5H_7^{|+}$ fragment ion, namely, the allylic, cyclopentenium cation.[82] Therefore, to be certain, it is worthwhile to identify the product ion by a structure-elucidating experiment, such as isotopic labeling or collision-induced dissociation mass spectrometry. These methods and their application are described in Section 1.4.

AE Measurements and Reverse Energy Barriers

If the reaction in Equation 1.6 has a reverse energy barrier (E_{rev}) (i.e., the reverse ion-plus-radical reaction has a nonzero activation energy, see Figure 1.1), then the AE for A^+ will be too high by that amount. The possible presence of a reverse energy barrier for the reaction can be signaled by the shape and width of the metastable ion peak for the dissociation of the $AB^{+\bullet}$ ions to yield A^+. (For full details see the section on metastable ions, Section 1.4.1.) In the absence of any appreciable E_{rev}, the translational kinetic energy released in the dissociation of a metastable ion is small, typically 0–20 meV. A large kinetic energy release signals the likely presence of a reverse energy barrier. In general however, there is no sound method for a *quantitative* correction for the $E_{rev} \neq 0$ problem using MI peak data. Note that the kinetic energy released in the dissociation, whose magnitude and distribution can readily be determined from the metastable ion peak shape (qv), is an unknown fraction of E_{rev} and therefore no simple correction procedure follows.

To account for the contribution of the reverse energy barrier in the calculation of the heat of formation of the fragment ion, Equation 1.7 now becomes

$$AE(A^+) = \Delta_f H(A^+) + \Delta_f H(B^\bullet) - \Delta_f H(AB) + E_{rev} \qquad (1.7a)$$

AE Measurements and the Kinetic Shift

Consideration must also be given to the timescale on which the AE measurement is made. The residence time of molecular and fragment ions in the ion source of a typical sector mass spectrometer is generally of the order of microseconds. $AB^{+\bullet}$ ions of longer lifetime will therefore not dissociate in the ion source but may do so en route to the detector, or not at all. The limiting smallest first-order rate constant for the observation of A^+ in an AE experiment with ion lifetimes of microseconds will be approximately 10^3 s^{-1}. The rate constant of a unimolecular dissociation is a function of the internal energy of the ion, i.e., higher rate constants are accessed by ions with higher internal energy. Therefore AE values will in turn depend on the timescale of the experiment (the range of rate constants sampled). The kinetic shift (KS) is the excess energy above the thermochemical threshold that is required for the dissociation to take place with a rate constant large enough for the products to be observed. When the curve describing the relationship between the rate constant and the internal energy is steep, the difference between the thermo-chemical threshold and the observed threshold is small and therefore the kinetic shift too is small. However, a slowly rising curve will lead to a significant kinetic shift. Figure 1.4 shows how some typical first-order ion dissociation rate constants vary with internal energy. As can clearly be seen, the larger the observable rate constant, the larger the AE will be. The steepest curve shown, that for iodobenzene, thus has the smallest kinetic shift.

In many reactions that involve simple bond cleavages and for which the activation energy (E_a) is small (say, less than 150 kJ/mol), the kinetic shift will be insignificant, with the ln k vs internal energy curves being very steep. However, for reactions involving a large E_a (and hence a large density of states for $AB^{+\bullet}$ at the transition state thus reducing the dissociation rate), the kinetic shift can be as much as 0.5 eV (48 kJ/mol). The data shown in Figure 1.4 illustrate this well. The loss of the halogen atom from the three ionized halobenzenes (C_6H_5X, X = Cl, Br, I) and the kinetic shifts associated with these reactions have been measured by PEPICO mass spec-trometry.[83–86] The magnitude of the KS was found to decrease in the order

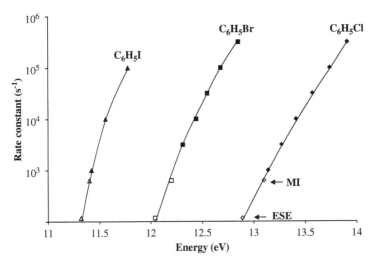

FIGURE 1.4 Illustration of the kinetic shift (KS). Dissociation rate constants (k) for the loss of the halogen atom from ionized phenyl iodide, bromide and chloride vs ion internal energy (E^*). Dark symbols are from PEPICO whereas open symbols are from metastable ion (MI) and energy-selected electron (ESE) measurements, respectively.

$C_6H_5Cl>C_6H_5Br>C_6H_5I$, consistent with the decreases in the activation energy for the C–X bond cleavage, which are approximately 320, 270 and 230 kJ/mol for the chloride, bromide and iodide, respectively. The relationship between these results and those from ESE-impact AE values and metastable ion phenomena has also been described[87] and the corresponding data are shown in Figure 1.4. The metastable ion AE values lie close to the PEPICO data at the lowest rate constants, while the ESE data reflect the longer source residence time for such apparatus of approximately 35 μs. Further details of the phenyl halides' dissociation can be found in a report by Klippenstein.[31]

To account for the kinetic shift then, Equation 1.7a now requires another term:

$$AE(A^+) = \Delta_f H(A^+) + \Delta_f H(B^\bullet) - \Delta_f H(AB) + E_{rev} + KS \qquad (1.7b)$$

AE Measurements and the Energy Content of the Products

The final problem with these measurements concerns the internal and translational energy contents of the product ion and neutral, E_{prods}. Equation 1.7 implies that the products have the same "temperature," i.e., internal and translational energy content, as the reactant molecules that are often at the ambient laboratory temperature of approximately 298 K. In general this assumption may not be correct. Traeger[88] emphasized this problem in 1982 and in all his photon-induced AE measurements assumes that the products are at 0 K and is careful in the identification of the true onset for the ion signal. This assumption necessitates the addition of heat capacity terms to Equation 1.7b to include the energy required to raise the products from 0 K to 298 K. In the authors' view, this is likely an overcorrection because 0 K products would only result from a situation of low probability where, in the transition state for the observed limiting rate constant, all of the internal (rotational and vibrational) energy of $AB^{+\bullet}$ is located in the reaction coordinate. Thus corrections for the products' energy cannot at present be made with absolute certainty because their temperature remains essentially unknown.

When the 0 K assumption is included, there remains another minor correction[88] of $-(5/2)R\Delta T$ ($\Delta T = 298$ K) to account for the energy involved in the translational degrees of freedom of the products. The final equation is then Equation 1.7c:

$$\begin{aligned}AE(A^+) = \; & \Delta_f H(A^+) + \Delta_f H(B^\bullet) - \Delta_f H(AB) + E_{rev} \\ & + KS - Cp(A^+, B^\bullet)\Delta T - (5/2)R\Delta T \end{aligned} \qquad (1.7c)$$

To a very useful first approximation and for simple ions, the last three terms tend to cancel each other out and the only major problem remaining is that associated with E_{rev}.

The Competitive Shift

This topic deserves brief mention. In general it is *not* satisfactory to extend AE measurements much beyond the lowest energy fragmentations of the ionized molecule or radical being studied. Such AE data will be too high as a result of the internal energy necessary for the chosen reaction to compete kinetically with the much faster lower energy processes. The competitive shift is therefore the excess energy above the threshold, for a given high energy reaction, required for the products to be observed on the mass spectrometer timescale.

1.3.7 CORRELATION BETWEEN ION HEAT OF FORMATION AND ION SIZE; SUBSTITUTION EFFECTS AT THE CHARGE-BEARING SITE

It was found that the effect on $\Delta_f H$(ion) resulting from the substitution of a functional group, e.g., $-CH_3$, $-OCH_3$, $-OH$, $-NH_2$, at the formal charge-bearing sites in both odd and even electron ions, was for each group a simple function of ion size.[69] For example, with the series of carbocations CH_3^{1+}, $C_2H_5^{1+}$, $(CH_3)_2CH^{1+}$ and $(CH_3)_3C^{1+}$ (which represents sequential methyl substitution in the methyl cation) the plot of $\Delta_f H$(ion) versus ln $\underline{n}$ is an excellent straight line. $\underline{n}$, as before, see Section 1.3.5, is the number of atoms in the ion. Figure 1.5 shows that this empirical correlation works well for the above series

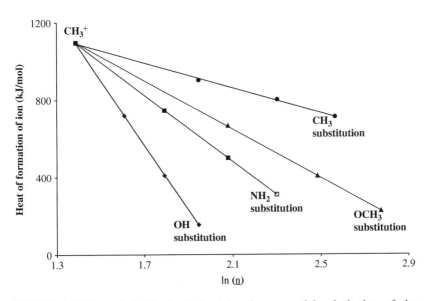

FIGURE 1.5 Plots of $\Delta_f H$(ion) obtained by the sequential substitution of the functional groups, $-CH_3$, $-OCH_3$, $-NH_2$ and $-OH$ in the methyl cation as a function of ln($\underline{n}$). Note that only two values are known for $-NH_2$ substitution, the open square (for three $-NH_2$ groups) is estimated.

and also for the successive substitutions of $-OH$, $-NH_2$ and $-OCH_3$ in the methyl cation. Note that there are no experimental values for $\Delta_f H(^+C(NH_2)_3)$ but by extrapolating the line defined by the first three members of the series, we may obtain an estimate for its $\Delta_f H$ value (see open square in Figure 1.5). Such estimated quantities are likely to be correct to within a few kilojoules per mole.

This empirical approach is strongly supported by the series of similar plots shown in Figure 1.6 for $-CH_3$ substitution at a variety of formal charge-bearing sites, the resonance-stabilized allyl cation and at C in $^+CH_2NH_2$ and $^+CH_2OH$. (The effect of methyl substitution in the methyl cation itself is shown in Figure 1.5). Note that all the lines are almost parallel, showing that the effect of methyl substitution is essentially non-site-specific. The value for $(CH_3)_2 CNH_2{}^{1+}$ is a recent datum for the substituted amino-methyl cations.[89]

In addition to providing a method by which missing thermochemical data may be estimated, these plots provide physicochemical information. Thus, for example, from these observations the location of greatest charge density in an ion may be assigned. In many textbooks, the ions $^+CH_2OH$, $CH_3{}^+CHOH$ and $^+CH_2OCH_3$ are displayed as oxonium ions, with the formal charge placed at oxygen, e.g., $CH_2 = O^+H$. However, the $\Delta_f H$ values for the above three ions are 711, 592 and 667 kJ/mol, respectively (see $[C,H_3,O]^+$ and $[C_2,H_5,O]^+$ sections in Part 2), showing that the stabilizing effect of methyl substitution at *carbon* in $^+CH_2OH$ is much greater than that at *oxygen*. It follows therefore that the charge density is greater at C than at O. Indeed, as can be seen in

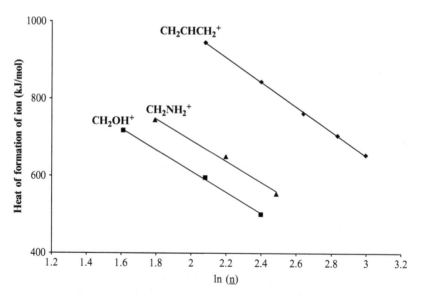

FIGURE 1.6 The effect of methyl substitution at a variety of localized and delocalized charge centers. Note the parallel lines.

Figure 1.6, the third ion above is perhaps better perceived as arising from methoxy substitution in a methyl cation, i.e., as $^+CH_2OCH_3$, rather than methyl substitution at oxygen in $CH_2=O^+H$.

Finally, there is a caveat; the above plots cannot be directly applied to mixed substituents, for example, to estimate the $\Delta_fH(^+C(OH)_2NH_2)$ by drawing in the line for one $-NH_2$ substitution from $H^+C(OH)_2$ and that for two $-OH$ substitutions from $^+CH_2NH_2$. These two lines do not meet at a single point. The reason for this almost certainly lies in that the assumption that the number of atoms can be used to represent "size" is only valid within a single family of substituents and ions such as those depicted in the plots. Corrections can be made for each substituent type by adjusting the value of $\underline{n}$ using an appropriate modifier but these can only be relative to a selected standard. Such exercises do not lead directly to any new physicochemical insights and will not be discussed further.

1.3.8 PROTON AFFINITIES AND THE HEATS OF FORMATION OF PROTONATED MOLECULES

Before we conclude the discussion on ion enthalpies, it is necessary to consider the heats of formation of protonated molecules. The PA of a species is defined as the negative of the enthalpy change accompanying the reaction (Δ_rH):

$$M + H^+ \rightarrow MH^+ \tag{1.8}$$

$$PA(M) = -\Delta_rH_8 = -[\Delta_fH(MH^+) - (\Delta_fH(M) + \Delta_fH(H^+))] \tag{1.8a}$$

Reference PA values can be found in the NIST WebBook.[56] As might be expected, PA values also vary linearly with the reciprocal of molecular size $(1/\underline{n})$ for homologous series. The slopes of such plots are *negative* because the PA increases with increase in molecular size. Some plots of recent PA data vs. $1/\underline{n}$ are shown in Figure 1.7.

Such straight-line plots can be used to assess the reliability of old and new data, as well as to provide PA values for unmeasured species. These graphs for protonation are comparable with those for ionization energies, with the slopes of the lines indicating the degree of charge delocalization in the protonated and molecular ions, respectively. Figure 1.7b shows the plots analogous to Figure 1.3b for di-*n*-alkyl ethers and cyclic ethers. The open square is the reference value[56] for the $C_5H_{10}O$ cyclic ether tetrahydropyran (PA = 823 kJ/mol). This is virtually the same as that given for its lower homologue, tetrahydrofuran (PA = 822 kJ/mol) and so the former value appears to be too low. A measurement in this laboratory using the kinetic method (see Measurement of proton affinities; the kinetic method in Section 1.3.8) raises the value for the $C_5H_{10}O$ species from 822 to 828 kJ/mol,

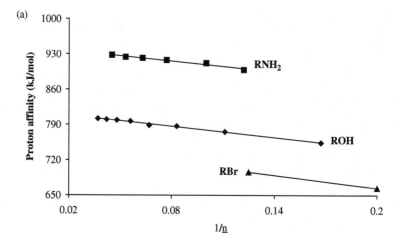

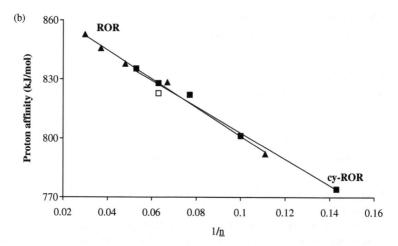

FIGURE 1.7 Plots of proton affinity (PA) vs $1/\underline{n}$ for a variety of homologous molecules: (a) primary alkylamines (RNH$_2$), alkanols (ROH) and bromides (RBr); (b) di-*n*-alkyl ethers (ROR) and cyclic ethers (cy-ROR). See text for discussion of the open square in (b).

mol, placing the data point (dark square) on the (now slightly steeper) line. This provides a good example of the utility of such graphs.

With reliable PA values and provided that the ancillary thermochemical data are available, the heats of formation of protonated molecules are given by rearranging Equation 1.8a:

$$\Delta_f H(MH^+) = \Delta_f H(M) + \Delta_f H(H^+) - PA(M) \qquad (1.8b)$$

Note that in 1972 Aue et al.[90] showed, not surprisingly, that PA values for alkylamines correlated closely with their solution basicity as well as with their ionization energy.

Measurement of Proton Affinities; Equilibria and Bracketing Studies

Proton affinities are best measured by studying the temperature dependence of the equilibrium constant (K_{eq}) for the following reaction:

$$AH^+ + B \leftrightarrow A + BH^+ \tag{1.9}$$

From the Van't Hoff equation,

$$\Delta G = -RT \log(K_{eq}) = \Delta H - T\Delta S \tag{1.10}$$

both enthalpy and entropy of protonation may be obtained provided that either PA(A) or PA(B) is known. This is achieved by using a high-pressure ion source with a sector instrument[91] or by ICR mass spectrometry.[92] The results of these equilibrium experiments however give PA (and entropy) *differences* and so some absolute PA values are required in order that a scale of values can be constructed. These have come, for example, from precise $\Delta_f H$ data for selected ions such as t-butyl$^+$ (PA of methylpropene), sec-C$_3$H$_7^+$ (PA of propene) and HCO$^+$ (PA of CO), obtained from accurate PA measurements[91] and from high level computational chemistry.[93] From these anchor points the PA scale can be constructed.

The bracketing method uses the same apparatus as above and is similar to the method employed for the measurement of ionization energies. Again, it is for systems in which equilibrium is not achieved.[60] The molecule of interest (B) is allowed to exchange protons with molecules (A) having known PA values that bracket that of the unknown B. The result is to give an upper and lower limit for PA(B); careful selection of molecules A will narrow the bracket to better than, say, 15 kJ/mol.

Measurement of Proton Affinities; the Kinetic Method

This approach[94] measures the product ion peak ratios for the competing dissociations of proton-bound pairs:

$$AH^+B \rightarrow AH^+ + B \tag{1.11}$$

$$AH^+B \rightarrow A + BH^+ \tag{1.12}$$

and then equates $[AH^+]/[BH^+]$ to k_{11}/k_{12}. Usually the PA for molecule B is unknown and a series of molecules A are standards of known PA.

The measurements are made for mass-selected AH^+B ions, their metastable and collision-induced dissociations, in both sector and quadrupole mass spectrometers. In its simplest form, the logarithm of the above ratio is plotted against the known PA(A) values. Linear graphs are obtained and so PA(B) can be read off from the graph:

$$\ln[AH^+]/[BH^+] = \ln(k_{11}/k_{12}) = [PA(A) - PA(B)]/RT_{eff} \qquad (1.13)$$

The basic difficulty with the method is that the experiment takes one just to the transition states for the competing reactions and so the unknown effects of entropy or reverse energy barriers can seriously compromise the observations. Also, the value of T_{eff}, the ion temperatures (internal energies) for both MI and CID studies, are unknown. Moreover, the effect of the metastable ion "time window" (see Section 1.4.1) distorts the distribution of observable rate constants and exaggerates the product ion ratios at the highest internal energies accessed by the window.

However, in spite of these inherent difficulties, the method appears to work quite well, particularly for simple homologous series of molecules where entropy effects tend to cancel out.[95] When choosing partner molecules for such a PA determination for a new species, the mixing of molecular types having PA values similar to that expected for the unknown is singularly unreliable because of the widely different partner-dependent entropy effects that may result. In recent years, a variety of correction methods for the above entropy effect have been proposed and used in selected examples.[96–99]

1.3.9 BOND STRENGTHS IN PROTON-BOUND MOLECULAR PAIRS

Closely related to PA values are the strengths of bonds in symmetric and unsymmetric proton-bound molecular pairs, AH^+A and AH^+B, respectively. These bond strengths are a measure of the electrostatic forces that hold the species together. The appropriate competing dissociations of AH^+B are thus to AH^+ and BH^+. The pairs of molecules may have a common heteroatom, e.g., oxygen or nitrogen, or even mixed pairs containing these atoms.

For $O-H^+-O$-bound species, the bond dissociation energy relationship is as follows:[100]

$$D(AH^+ - B) = 0.46[PA(B) - PA(A)] + 129 \pm 8 \text{ kJ/mol} \qquad (1.14)$$

and for $O-H^+-N$ systems the equation is as given below:[101]

$$D(AH^+ - B) = 0.26[PA(B) - PA(A)] + 126 \pm 6 \text{ kJ/mol} \qquad (1.15)$$

Note that these equations also permit estimates to be made for ions in which the proton binds a molecule and a free radical, provided that the required free radical PA value is available. The multiplicands for the PA differences in the above equations were selected to reflect the best current PA values; they may well be modified as better data accrue. For example, for Equation 1.14, a preferred coefficient is 0.34.[102]

1.3.10 BOND STRENGTHS IN IONS

It is worthwhile to ask whether the bond strength in an ion can easily be related to that of the same bond in the neutral counterpart. The key determining feature for this quantity is the proximity of the bond of interest to the formal charge site in the ion. Two examples will suffice to illustrate this.

The distonic isomer of ionized methanol, $^\bullet CH_2{}^+OH_2$ (an α-distonic ion, qv), can be viewed as the product of two reactions: first the protonation of methanol followed then by the homolytic cleavage of a C–H bond (Equation 1.16 and Equation 1.17):

$$CH_3OH + H^+ \rightarrow CH_3{}^+OH_2 \tag{1.16}$$

$$-PA(CH_3OH) = \Delta_f H(CH_3{}^+OH_2) - \{\Delta_f H(CH_3OH) + \Delta_f H(H^+)\} \tag{1.16a}$$

$$CH_3{}^+OH_2 \rightarrow H^\bullet + {}^\bullet CH_2{}^+OH_2 \tag{1.17}$$

$$D(C - H) = \Delta_f H(H^\bullet) + \Delta_f H(^\bullet CH_2{}^+OH_2) - \Delta_f H(CH_3{}^+OH_2) \tag{1.17a}$$

Combining Equation 1.16a and Equation 1.17a yields Equation 1.18:

$$\begin{aligned} \Delta_f H(^\bullet CH_2{}^+OH_2) = {} & D(C - H) - \Delta_f H(H^\bullet) - PA(CH_3OH) \\ & + \Delta_f H(CH_3OH) + \Delta_f H(H^+) \end{aligned} \tag{1.18}$$

The available thermochemical data give the following results.[56] Assuming that $D(C–H)$ in $CH_3{}^+OH_2$ is the same as in neutral methanol ($D(C–H)=402$ kJ/mol[57]) and using the well-established reference values for $\Delta_f H(H^\bullet)$, $PA(CH_3OH)$, $\Delta_f H(CH_3OH)$ and $\Delta_f H(H^+)$, namely, 218, 754, -201 and 1530 kJ/mol, respectively, the derived $\Delta_f H(^\bullet CH_2{}^+OH_2) = 759$ kJ/mol. Comparison with the well-known experimental (and calculated) $\Delta_f H(^\bullet CH_2{}^+OH_2) = 816$ kJ/mol clearly shows that the above assumption is incorrect and that the C–H bond in the protonated ion is *stronger* by some 60 kJ/mol than that in the neutral analog. Similar situations arise whenever the bond to be broken is α- to a *charge-bearing site*.

However, if the bond of interest is one or more atoms further off, then the bond strength for the neutral analog now is appropriate for the calculation of $\Delta_f H$ for a distonic ion. For example, consider the γ-distonic isomer of ionized propanol, $^\bullet CH_2CH_2CH_2{}^+OH_2$. The required data are $D(C–H)$ at methyl in the propanol molecule, 406 kJ/mol,[57] $\Delta_f H(H^\bullet)$, $PA(C_3H_8O)$, $\Delta_f H(C_3H_8O)$ and $\Delta_f H(H^+)$, which are 218, 787, -255 and 1530 kJ/mol, respectively.[56] Using an equation similar to Equation 1.18 results in $\Delta_f H(^\bullet CH_2CH_2CH_2{}^+OH_2) =$ 676 kJ/mol, satisfactorily close to the most recent computed value, 688 kJ/mol (see Part 2). (Note that the adiabatic IE for propanol is 10.0 eV, significantly lower than the IE reported in the NIST database.[56] This ion has a very long $(C_\alpha – C_\beta)$ bond, hence the difference $IE_a – IE_v = 0.22$ eV.)

The above arguments concerning bond strengths apply well to related systems; consider for example the reaction of protonated propanal:

$$CH_3CH_2{}^+CHOH \rightarrow CH_3{}^{]\bullet} + CH_2CHOH^{]+\bullet} \qquad (1.19)$$

Here the charge site in the reactant ion is at the α-carbon. The energy to be assessed is the C–C bond strength in the ion and is given by

$$D(C-C) = \Delta_f H(CH_3{}^{]\bullet})$$
$$+ \Delta_f H(CH_2CHOH^{]+\bullet}) - \Delta_f H(CH_3CH_2{}^+CHOH) \qquad (1.20)$$

Substituting the reference values[56,57] for the respective heats of formation gives $D(C–C) = 147 + 771 – 557 = 361$ kJ/mol, in excellent agreement with 365 kJ/mol,[57] the listed C–C bond strength in the analogous neutral propanol molecule.

A similar approach is to be seen in the estimation of $\Delta_f H$ values for the McLafferty rearrangement (see Section 2.3), for example, in methyl propyl ketone, where the distonic intermediate ion has a lower $\Delta_f H$ than its keto counterpart.

1.3.11 THERMOCHEMICAL DATA IN THIS BOOK

As will be seen in the data sections of Part 2, selected $\Delta_f H$ data from experiment and from computational chemistry are given for every ion that has been characterized by an experimental method. Where high level calculations have been performed for other unobserved but stable isomers, their $\Delta_f H$ values or those relative to the most stable isomer, are also included. Wherever possible, the validity of the data has been checked by means of an appropriate thermochemical cycle. When ambiguities remain, the reader is given some guide as to the most reliable value.

1.4 EXPERIMENTS WITH MASS-SELECTED IONS

1.4.1 THE STUDY OF METASTABLE IONS

Metastable ions were a topic of particular interest in the period from about 1960 to 1985, a time during which most mass spectrometry laboratories possessed at least one simple magnetic sector instrument. The magnetic sector was often combined with one (or more) electric sectors, making a variety of tandem mass spectrometers. As is described below, the metastable ion signals (or peaks) observed in such instruments have characteristic shapes and an analysis of these can provide remarkable insight into many physicochemical aspects of ion fragmentation chemistry. The subject has produced a significant book[103] and some lengthy review articles.[104–106]

Historical Background

In normal electron-impact-induced mass spectra of simple organic compounds obtained with a single-focusing magnetic sector instrument (Figure 1.8) weak, diffuse signals can be observed. Their intensity is of the order of less than 0.1–1% of the base peak (the most intense peak in the mass spectrum). They are broader than the other fragment ion peaks, often of approximately Gaussian profile, and they generally appear at a nonintegral mass.

In 1946, Hipple et al.,[107] provided the first explanation for the origin of these weak signals. Their hypothesis, which proved to be correct, was that the observed signals were due to the fragmentation products of ions with a lifetime sufficiently long to exit the ion source and be accelerated but not long enough to reach the detector intact.

"Metastable" is the adjective given to those ions that dissociate during their flight through the mass spectrometer, i.e., between the grounded plate outside the ion source (exit slit in Figure 1.8) and the detector. Note that the term metastable applies strictly only to those ions that undergo unimolecular dissociations by virtue of the internal energy they have acquired in the ion source itself and not as a result of any energy received postsource, e.g., from a collision- or radiation-induced event.

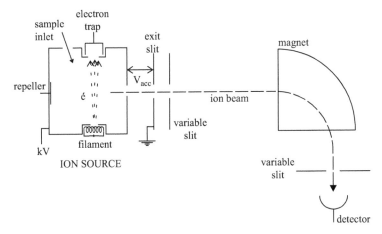

FIGURE 1.8 The layout (not to scale) of a typical, single-focusing magnetic sector mass spectrometer.

Metastable Ion Phenomena

In a single-focusing magnetic sector instrument (Figure 1.8), singly charged ($z = 1$) molecular or fragment ions are uniformly accelerated when exiting the ion source. The ion acceleration zone is small, between the source block, at a high positive potential (usually kilovolts in magnitude), and a closely proximate grounded exit slit. Therefore, after passing the exit slit, all ions irrespective of their mass, will have the same translational kinetic energy. This can be defined by the following equation for singly charged ions:

$$eV_{acc} = \frac{1}{2}m_1v_1{}^2 = \frac{1}{2}m_2v_2{}^2 = \frac{1}{2}m_nv_n{}^2 \qquad (1.21)$$

where m_n is the ion mass, v_n is its velocity, e is the electronic charge and V_{acc} is the ion source acceleration voltage.

A variable magnetic field separates the ions by their momentum according to the equation

$$m/z = eB^2r^2/2V_{acc} \qquad (1.22)$$

where B is the magnetic field strength and r is the radius of curvature of the flight tube in the magnetic field.

Thus, as the magnetic analyzer field strength is raised from zero, ions of increasing mass-to-charge (m/z) ratio will be transmitted to the detector giving rise to well-defined, sharp signals at m/z values corresponding to each of the fragment ions and ending, in most cases, with the molecular ion at the highest mass. Note that following ionization, not all molecules will produce a molecular ion. For example, the ionized molecule could be thermodynamically unstable and dissociate without activation or it could be formed with excess energy because of geometry constraints and be kinetically unstable and dissociate (see vertical and adiabatic ionization energy in Heats of formation of ionized molecules in Section 1.3.4). In both of these cases no molecular ion would be observed.

When a given ion $M_1{}^+$ is metastable and, therefore, fragments during flight in the region after the ion source and before the magnetic analyzer (Figure 1.8), its translational kinetic energy is partitioned between the product ion $M_2{}^+$ and the neutral fragment M_3. The fragment ion $M_2{}^+$ will possess the same velocity as $M_1{}^+$ (its precursor) but only a fraction of the kinetic energy that it would have possessed had it been generated before acceleration from the ion source ($\frac{1}{2}m_2v_1{}^2 < \frac{1}{2}m_2v_2{}^2$). Therefore, this ion will not be seen at the m/z value for $M_2{}^+$ given by Equation 1.22, but rather at an *apparent mass*, m^*. Equation 1.21 can be rearranged to show that $v_1 = (m_2/m_1)^{1/2} v_2$ and the energy of the fragment ion produced from metastable $M_1{}^+$ therefore, becomes $\frac{1}{2}m_2(m_2/m_1)v_2{}^2$. From this, the apparent mass (m^*) is deduced to be $m_2{}^2/m_1$, a quotient that is generally not an integer.

The great majority of organic ions display only a few metastable decompositions, rarely exceeding three or four, and for the most part they involve the loss of small molecules or radicals, e.g., H_2, H_2O, CO, CO_2, CH_4, C_2H_4, NH_3, $H^\bullet$, $CH_3^{|\bullet}$, $C_2H_5^{|\bullet}$, halogen atom.

Ions that dissociate elsewhere in flight (e.g., in the magnetic analyzer) are not focused and so are not observed as defined signals. They rather provide a weak background continuum that can be evident from a close scrutiny of the base line of the mass spectra.[104]

Possibly the most useful primary information derived from MI signals is that their apparent mass provides *proof* that a given fragmentation $M_1^+ \longrightarrow M_2^+ + M_3$ has indeed taken place. As a simple example Figure 1.9 shows two typical metastable ion peaks that appear in the normal mass spectrum of diethyl ether, $(C_2H_5)_2O$. These signals arise from the dissociation of the major fragment ion, $CH_3CH_2O^+CH_2$ m/z 59.

The first peak at $m/z = 28.5$ results from the loss of a water molecule (18 Da) from the m/z 59 ion, giving the apparent mass $41^2/59 = 28.49$. This cannot be a simple bond cleavage and so the presence of the peak alerts the observer to the presence of a rearrangement process. The peak at $m/z = 16.3$ comes from the loss of C_2H_4 (28 Da) from the m/z 59 ion, e.g., $31^2/59 = 16.3$.

Possible ambiguities in the molecular formula assigned to the fragment ion can arise; for example, the loss of CO from an ion is isobaric with the loss of C_2H_4. The processes can however easily be distinguished by a simple experiment with a tandem mass spectrometer, as will be described below (Collision-induced dissociative ionization mass spectra in Section 1.4.2), or indeed by isotopic labeling, also discussed later (Isotopic labeling and metastable ions in Section 1.4.1).

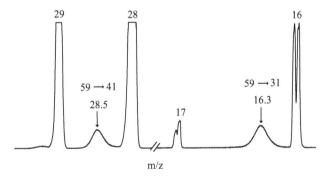

FIGURE 1.9 Sections of the electron-impact mass spectrum of diethylether ($(CH_3CH_2)_2O$) obtained from a single magnetic sector mass spectrometer showing metastable ion (MI) peaks for the fragmentations of $CH_3CH_2O^+CH_2$ ions (m/z 59) to $C_3H_5^{|+}$ (m/z 41) at $m^* = 28.5$ and $^+CH_2OH$ (m/z 31) at $m^* = 16.3$.

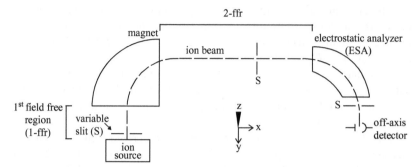

FIGURE 1.10 The layout (not to scale) of a tandem mass spectrometer of BE geometry. Note the axial directions.

Tandem Mass Spectrometers and Experiments with Metastable Ion Beams

Detailed metastable ion studies are best performed with multisector mass spectrometers such as a tandem mass spectrometer with "*BE*" geometry (magnetic analyzer (*B*) before an electrostatic sector (*E*)). This configuration, conventionally known as "reversed" or "inverse Nier–Johnson" geometry, is shown in Figure 1.10. The original purpose in making mass spectrometers to this design was to use their double-focusing properties to give greatly enhanced mass resolution. However, they also provide a unique method for studying in detail the dissociation behavior of metastable ions of a particular m/z ratio by using the magnet for their mass separation and then performing experiments on these mass-selected ions within the second field-free region (2-ffr) of the apparatus.

Before discussing the details of MI mass spectra, Figure 1.11 should first be introduced. It summarizes the processes that take place in a *BE* mass spectrometer.

The figure is self-explanatory, showing the timescale of the events that take place in the instrument from the generation of an ion to its arrival at the detector. Note the field-free regions, the MI and collision-induced dissociation ions (refer to Section 1.4.2 for the description of collision-induced dissociation experiments) and the different timescales on the two x axes.

Measuring Metastable Ion Peaks with a Tandem Mass Spectrometer

Suppose the magnetic analyzer is set at the appropriate field strength to transmit M_1^+ ions and that these undergo metastable dissociation in the second field-free region (2ffr) of the instrument, i.e., between the magnet

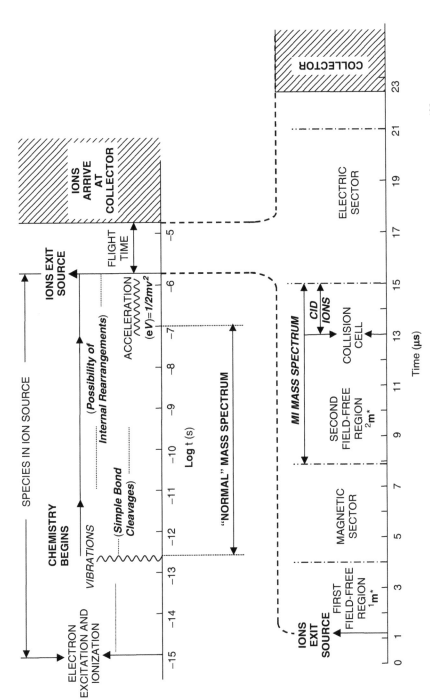

FIGURE 1.11 The timescales of physicochemical events in a BE tandem mass spectrometer, from ionization to detection.[135] Note the change of timescale.

and the electric sector. The fragments produced from metastable ions will, as described above, possess a fraction of the kinetic energy of their precursor ion.

The purpose of the electric sector is to separate ions according to their kinetic energy, as described by the following equation:

$$\frac{1}{2}mv^2 = V_{acc} = E_sR_e/2 \tag{1.23}$$

which defines the condition under which ions having kinetic energy V_{acc} are transmitted through the electric sector. The field between the plates is E_s and their radius of curvature, and hence the flight path, is R_e.

Therefore, if E_1 is the electrostatic sector potential that transmits mass-selected ions M_1^+ onward to the detector (approximately 99% of the total ion flux), then lower potentials, namely, $m_2/m_1 \times E_1$, $m_3/m_1 \times E_1$, will be required to transmit and focus the lighter fragment ions (M_2^+, M_3^+, etc.) generated from the dissociation of *metastable* M_1^+ in the 2-ffr. These fragment ions have a lower translational kinetic energy because of the mass loss. Thus, sweeping the sector voltage from E_1 to zero will transmit all fragment ions produced by *metastable* M_1^+ ions. The resulting spectrum is known as a mass-analyzed ion kinetic energy (MIKE) mass spectrum. As noted above, the number of such fragment ion peaks will be small. Figure 1.12 displays the MIKE mass spectrum of $C_2H_5O^+CH_2$ ions described above.

The selected M_1^+ ions should normally have a very narrow range of translational kinetic energies when they exit the ion source, preferably less than about 0.025% of E_1 (say 2 V in 8000 V), which can be easily obtained with a well-designed instrument. Energy resolution is achieved by narrowing the

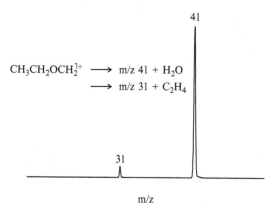

$$CH_3CH_2OCH_2^{\cdot+} \longrightarrow m/z\ 41 + H_2O$$
$$\longrightarrow m/z\ 31 + C_2H_4$$

FIGURE 1.12 The MIKE mass spectrum of $CH_3CH_2O^+CH_2$ ions showing the same two dissociations as in Figure 1.9.

variable ion source and collector slits (Figure 1.10) until the observed profile of the mass-selected ion beam no longer changes. Under such conditions the shape observed for the fragment ion peak in a MIKE mass spectrum can provide information about the dissociation process (this is addressed later).

Energy-resolved metastable ion peaks appear in two basic shapes: Gaussian and flat-topped or dish-topped, Figure 1.13a and Figure 1.13b, respectively. For comparison, the profile of the energy-resolved main ion beam is also shown in Figure 1.13c.

Note that some double-focusing mass spectrometers have the opposite geometry: electrostatic sector before the magnet (*EB*), so-called normal geometry. Separation and analysis of metastable ion peaks can be performed with this type of instrument but the methods are less versatile.[103]

The particular problem, referred to above (Metastable ion phenomena in Section 1.4.1), of isobaric neutral fragment losses (e.g., CO vs C_2H_4 or C_3H_6 vs CH_2CO) can be easily solved by transmitting the ion one mass unit higher than M_1^+ into the field-free region. This ion will (in general) contain *one* randomly positioned ^{13}C atom and so the statistical consequence for metastable ion peak abundances for the losses of CO and ^{13}CO, relative to the losses of $^{13}CCH_4$ and C_2H_4, will result in different observations. For example, metastable ionized cyclopentanone, $C_5H_8O^{]+\bullet}$ ions (m/z 84) lose 28 Da, which could be either CO or C_2H_4, to give a fragment ion at $m/z = 56$. By mass-selecting m/z 85, $^{13}CC_4H_8O^{]+\bullet}$ ions, the metastable dissociation can now proceed by retention or loss of the ^{13}C label. The calculated ratio (see Isotopic labeling and metastable ions in Section 1.4.1) for the statistical loss of ^{13}CO (producing m/z 56) to that of ^{12}CO (producing m/z 57) is 1:4, whereas that for the loss of $^{13}CCH_4:C_2H_4$ is 2:3 and the metastable ion

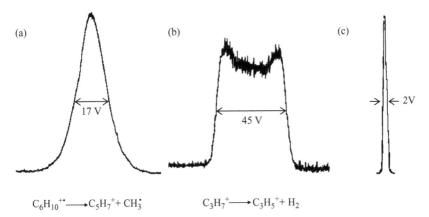

(a)   (b)   (c)

17 V

45 V

→ ←2V

$C_6H_{10}^{+\bullet} \longrightarrow C_5H_7^+ + CH_3^\bullet$ $C_3H_7^+ \longrightarrow C_3H_5^+ + H_2$

FIGURE 1.13 (a) Typical Gaussian and (b) dished metastable ion (MI) peaks. The signal (c) is for a beam of mass-selected, nonfragmenting ions and shows their much smaller translational kinetic energy spread. The peaks were all measured at an ion acceleration potential of 8000 V.

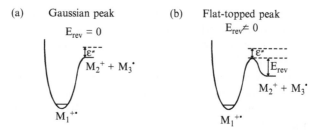

FIGURE 1.14 Potential energy curves for (a) an ion dissociation having no reverse energy barrier ($E_{rev} = 0$) and thus giving rise to a Gaussian peak and (b) that giving rise to a flat-topped (or dished) metastable ion (MI) peak ($E_{rev} \neq 0$). The kinetic energy release derives from $\varepsilon^{\neq}$, termed the "nonfixed energy," and from E_{rev}, named the "fixed energy."

peak abundances must reflect either of these ratios. The experimental result gives a metastable ion peak ratio of 2:3 and, therefore, the neutral lost is C_2H_4, not CO.

Note that when measuring the relative abundances of energy resolved MI peaks, *peak areas* should be recorded. Without energy resolution, *peak heights* are appropriate.

Measuring Kinetic Energy Release

For a given metastable dissociation to be observed, the fragmenting ion must possess sufficient internal energy. Both the excess energy of the activated complex ($\varepsilon^{\neq}$, the nonfixed energy) and the reverse energy of activation (E_{rev}, the fixed energy) will contribute to the energy available for dissociation (Figure 1.14).

The observed peak broadening referred to above is due to the conversion of a fraction of the excess energy of the fragmenting ion into the translational degrees of freedom of the charged and neutral products. The release of translational kinetic energy is isotropic and consequently the fragment ion peaks generally show a wider kinetic energy distribution than the peak for the precursor ion. However, the amount of excess energy that is converted in the dissociation process is generally not known in any detail because it will depend on the nature of the transition state and the number of internal degrees of freedom that participate therein.

It is the convention that the kinetic energy released in a fragmentation is characterized by $T_{0.5}$, derived from the peak width at half-height ($W_{0.5}$) measured in electric sector or ion acceleration volts. The relationship[103] for calculating the kinetic energy release is as follows:

$$T_h = (m_1^2)(W_h^2 - W_m^2)/16m_2m_3V_{acc} \qquad (1.24)$$

where h is the height, expressed as a fraction of the total height (h_0) at which the peak width (W_h) is measured, W_m is the corresponding width of the main ion beam, and m_1, m_2 and m_3 are the masses of the precursor ion and reaction products, respectively. T_h is usually reported in milli-electronvolt (meV) units. Note that the voltage scale most commonly used in meta-stable ion peak measurements is usually a V_{acc} value. (Recall that $V_{acc} = \text{constant} \times E_1$ (Equation 1.23), the constant depending only on the radius of the electric sector).

Because most metastable ion peaks have an exactly Gaussian, or close to Gaussian profile, they therefore can be reproduced by a Gaussian-type expression of the form

$$h = \exp\left(-\alpha W^n\right) \tag{1.25}$$

where h is again the height at which the peak width (W) is measured. In the Gaussian equation, $n = 2$ and $\alpha = \ln 2$. Most observed peaks of Gaussian form can be fitted to the above expression with n as the only variable and having a small range of values, from about 1.4 to 2.1.[106] These n values thus, serve as an additional descriptor of the peak: as n decreases from approximately 2, the peak develops a broadening skirt and the effect has been illustrated in an early review article.[2] The peak shape can also be mathematically analyzed and the distribution of released kinetic energies obtained.[108] Note that the distribution of released energies derived from the exact Gaussian function is of Maxwell–Boltzmann form.[109] For a peak of Gaussian profile ($n = 2$), the average kinetic energy release (T_{ave}) is given by $T_{ave} = 2.16 \times T_{0.5}$ and this quantity[109] is sometimes reported instead of $T_{0.5}$. In rare instances, the kinetic energy release (KER) corresponding to the maximum of the KER distribution is reported as T (most probable).

Physicochemical Aspects of Metastable Ion Dissociations

A Gaussian-type signal generally arises from reactions having little or no reverse energy of activation (E_{rev}). In this case, the excess energy to be partitioned among the translational degrees of freedom of the products will be quite small (Figure 1.14). For a simple bond cleavage, $T_{0.5}$ values are typically in the range of 10–25 meV. A representative example is shown in Figure 1.13a for the metastable ion dissociation $C_6H_{10}^{1+\bullet} \longrightarrow C_5H_7^{1+} + CH_3^{1\bullet}$ for which the kinetic energy release is calculated to be as follows:

$$T_{0.5} = 82^2(17^2 - 2^2)/16.67.15.8000 = 14.9 \text{ meV}$$

In some instances, the kinetic energy release is so small that it is difficult to measure it accurately, i.e., the metastable peak is barely wider than that of the precursor ion beam. In such cases, the fragment ion and neutral product have been shown (by computational chemistry) to be only electrostatically bound in the metastable ion near its dissociation limit, with the neutral free to roam around the incipient fragment ion. Consequently, there is no defined reaction coordinate for the departure of the neutral species, in contrast with the situation where a specific covalent bond is elongated in the dissociation. Vibronic coupling between the two electrostatically bound species is very weak and so they separate without any appreciable release of kinetic energy. A simple example is the $[C_2,H_6,O]^{+\bullet}$ isomer $[CH_2CH_2]^{+\bullet} \cdots OH_2]$, which loses water as its sole metastable fragmentation and with a kinetic energy release of only 0.2 meV.[110]

A "dished" (or flat-topped) peak, such as that illustrated in Figure 1.13b, indicates that a large amount of translational kinetic energy has been released. Such peaks are generated from fragmentations that have an appreciable reverse energy of activation (Figure 1.14). This generally arises from, and is indicative of, a molecular elimination reaction (e.g., the loss of H_2, HCl) or an ion rearrangement having a large energy requirement resulting in the transition state lying well above the product energies. In these cases, the $T_{0.5}$ values are much larger, from about 100 meV up to one or more electronvolts.

It is worth remarking that there may be a few anomalies to this behavior. For example, see under ionized CH_3CN (see $[C_2,H_2,N]^+$ ions, in Part 2) and CH_3CDO (see $[C_2,H_4,O]^{+\bullet}$ Part 2). These metastable ions lose a H(D) atom, producing a broad and a flat-topped signal, respectively. The most recent calculations indicate that there is no reverse energy barrier for these fragmentations and so another reason for the large kinetic energy release must be found for such examples. A possibility is that more than one vibrational degree of freedom may participate in H(D) atom loss processes.

Note that the dish in the center of broad metastable ion peaks is an experimental artifact. It arises from the energy-resolving slit having a limited height in the z axial direction (see Figure 1.10). Some of the fragment ions produced with a large kinetic energy release will have correspondingly great components in the z axial direction and so such ions will fail to pass this slit because of its finite height. These lost ions consequently have only small x (and y) axial components and so they will be observed as missing from the *center* of the MI peak. Note that the MIKE experiment involves the sweep of the sector voltage in the xy plane (Figure 1.10). In the absence of z axial discrimination these peaks would have flat summits. Some instruments have a variable z axial slit. For kinetic energy release of the order of volts, when the z slit is very greatly narrowed, the resulting peak can be reduced to a pair of sharp signals, equidistant from the prescribed center. In addition to the above types of signal, composite metastable ion peaks are also observed and two examples are shown in Figure 1.15.

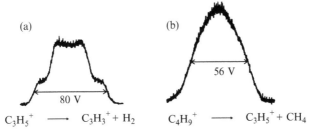

$$C_3H_5^+ \quad \longrightarrow \quad C_3H_3^+ + H_2 \qquad\qquad C_4H_9^+ \quad \longrightarrow \quad C_3H_5^+ + CH_4$$

FIGURE 1.15 Composite metastable ion (MI) peaks. Peak (a) is a double "dished" peak for the reaction shown and peak (b) is a mixed Gaussian and dished peak.

These can arise from the following circumstances:

(1) A single metastable ion structure dissociates via two competing reactions to produce isomeric fragment ions. In Figure 1.15a, the $C_3H_5^{1+}$ ion dissociates by loss of H_2 to give isomeric $C_3H_3^{1+}$ ions, namely, cyclopropenium (the broad component) and the propargyl cation $HCCCH_2^{1+}$ (the narrow component), both at energies far above the minimum for each process.[111,112] It is worth noting that the loss of C_2H_2 from metastable $C_5H_5^{1+}$ ions produces a composite signal resulting from the generation of the same two $C_3H_3^{1+}$ isomers.[112]

(2) The fragmenting ion beam consists of a mixture of two stable isomeric forms, both of which are metastable, and so competitively fragment via different transition states to give product ions of either the same or different structures. Figure 1.15b shows the metastable ion peak for the loss of CH_4 from $C_4H_9^{1+}$ ions. The shape of this complex (two component) signal is essentially independent of the structure of the butyl group in the precursor molecule whose dissociative ionization was used to generate the $C_4H_9^{1+}$ ions. Thus $C_4H_9^{1+}$ ions, irrespective of their origin, dissociate via different transition states to produce the 2-propenyl cation $(CH_3CCH_2^{1+})$ plus methane. The broad component is for the loss of CH_4 from the tertiary ion $(CH_3)_3 C^{1+}$ and the narrow peak arises from the secondary carbocation, $CH_3CH^+CH_2CH_3$.[113] Note that the thermodynamically more stable allyl cation, $CH_2CHCH_2^{1+}$, is not generated in these dissociations.

The Relationship Between the Internal Energy of a Metastable Ion and Its Dissociation Rate Constant; log *k* vs *E** Curves

As described above, for a given MI dissociation to be observable (AE measurements and the kinetic shift in Section 1.3.6), the fragmenting ion must possess sufficient excess internal energy to dissociate in the appropriate time frame. The relationship between this excess energy, the rate constants for the dissociation and the magnitude of the KER and how their observation depends on the geometry and setting of the instrument all deserve comment.

For sector mass spectrometers, V_{acc} is usually in the range 4–10 kV. Typical residence times for ions in the ion source are 1–2 μs. Ion flight-tube dimensions, from the source to the magnet, are of the order of 30 cm and second field-free regions are up to 1 m in length. Metastable ions of $m/z \approx 100$ will therefore have a range of lifetimes of approximately 5–50 μs (based on their dissociations obeying first-order reaction kinetics) depending on V_{acc} and the dimensions of the mass spectrometer. These lifetime intervals, therefore, define the metastable ion observation "window" and consequently establish the range of ion internal energies and dissociation rate constants that can be sampled in metastable ion observations.

This is illustrated in Figure 1.16, where the log k vs E^* curves (k being the unimolecular rate constant for the dissociation and E^* the internal energy content of the ions) are shown for two dissociation processes.

The range of k values that can be accessed is calculated by means of the first-order rate equation, inserting the times for the mass-selected ions' entry to and exit from the appropriate field-free region. Rate constant ranges are shown for two typical commercial sector mass spectrometers having different geometries and hence slightly different time windows for metastable ion dissociations. The curves show the fraction of the ions, $\delta n/n_0$, that fragment in the designated time interval. For the first window, the time that the ions spend in the field-free region is 30–40 μs; for the second window, it is 2–3 μs. The ion internal energy ranges, ε_1^* and ε_2^*, that correspond to each ffr are also shown. These are quite different, with the larger internal energies corresponding to the shorter time window. This results in the kinetic energy release values for metastable peaks being inversely dependent on the magnitude and the length of the time window.

It must be emphasized that the time taken for the ions to reach the field-free regions is long relative to vibrational processes that have frequencies of approximately 10^{12} s^{-1} and so there is ample time for metastable ions to structurally rearrange before they dissociate. Indeed, metastable ions often convert into isomeric forms before they fragment and moreover, the resulting product ion is often found to be the isomer of greatest stability, i.e., that having the lowest heat of formation. A major challenge in gas-phase ion chemistry is to unravel such rearrangements, which may not be readily evident.

Rearrangements of Metastable Ions; Their Reacting Configuration and Ion Structure Assignments

In 1960s, the first semiquantitative relationship between ion structure and metastable ion phenomena was proposed.[114] It was known as the "metastable ion peak abundance ratio rule." It was based on the following simple premise.

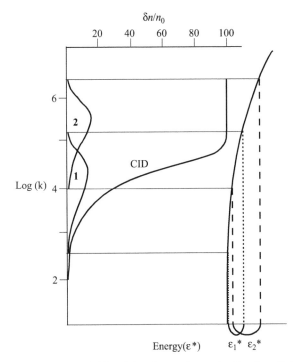

FIGURE 1.16 The left hand side of the figure shows the fraction (δn) of the total number of ions (n_0) that fragment in two different metastable ion (MI) observation time windows, as a function of the first-order rate constant (k) for their dissociation. Window 1 is from 30 to 40 μs and window 2 is from 2 to 3 μs. The plot also shows the dissociation of ions energized by collisions (collision-induced dissociation, CID) in the collision cell contained in the second field-free region (Figure 1.18). Note that these ions fragment *completely* when k exceeds 10^5 s^{-1}. The right hand side shows the log k vs internal energy (E^*) curve for the ion. The symbols ε_1^* and ε_2^* denote the ranges of ion internal energies that are sampled by the two time windows. Note that the minimum threshold energy for the reaction lies close to the low energy end of ε_1^*. It is also worth noting that the minimum energy at which the fragment ion first will be recorded depends on the observational timescale of the experiment (see kinetic shift, in AE measurements and the kinetic shift in Section 1.3.6).

Suppose that a metastable ion (M_1^+) of known or assumed structure, produced from a particular molecule or precursor ion, undergoes two or more competing fragmentations having a measurable abundance ratio. If another metastable ion of the same molecular formula, but generated from another precursor, displays the same relative MI peak abundances, then it must have the same structure as M_1^+.

The difficulty with this simple criterion was how to decide when it failed. What departure from equal ratios was significant? Remember that in principle it is also possible that the fragmenting species is actually a mixture of ions and that the measured peak abundance ratios could well depend on the composition of the ion beam. It was soon realized that the ratio can also be susceptible to differences in internal energy distributions among the fragmenting ions. However, in spite of difficulties, it was found that many small even electron cations derived from a wide range of precursor organic molecules had closely similar metastable ion peak abundance ratios.

In general, the abundance ratio test can be made more stringent if energy-resolved peaks are examined, leading to the use of relative abundances or peak shapes ($T_{0.5}$ values) as the key structure-characteristic feature. The use of the $T_{0.5}$ value as a structure identifier was proposed by Jones et al. in 1973[115] and this clearly was an advantage when the metastable ion had only one significant dissociation channel. Again however, it was not clear as to how large differences in $T_{0.5}$ must be for a peak to be declared as arising from two (or more) isomers.

The use of the complete metastable ion peak *shape* as the principal ion structure-identifying feature initially appeared to hold considerable promise,[116] as illustrated by the peaks shown in Figure 1.17.

This shows the metastable ion peaks for the loss of a hydrogen atom from three $[C_2,H_4,O]^{+\bullet}$ isomers: (a) ionized oxirane, (b) ionized acetaldehyde and (c) ionized vinyl alcohol. Note that the latter is easily and unequivocally produced by the loss of C_2H_4 from ionized butanal or cyclobutanol.[116] The three peaks are significantly different and so may serve to identify each isomer. Indeed, the great differences for $CH_3CHO^{+\bullet}$ and $CH_2CHOH^{+\bullet}$ allowed one to show that the loss of CH_4 from ionized 2-propanol $CH_3CH(OH)CH_3^{+\bullet}$ produced the isomeric $[C_2,H_4,O]^{+\bullet}$ fragment ions, $CH_3CHO^{+\bullet}$ and $CH_2CHOH^{+\bullet}$, from two competing 1,3-methane eliminations[116] involving the hydroxyl hydrogen and methyl hydrogen, respectively. Because their differences are so striking, the foregoing describes a single example of the use of peak shapes. However,

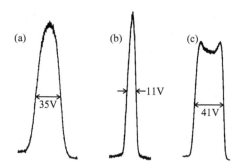

FIGURE 1.17 Metastable ion (MI) peaks for the loss of a H atom from three $[C_2,H_4,O]^{+\bullet}$ ions: (a) ionized oxirane (cy–C(H)$_2$C(H)$_2$O–$^{+\bullet}$), (b) ionized acetaldehyde (CH$_3$CHO$^{+\bullet}$) and (c) ionized vinyl alcohol (CH$_2$CHOH$^{+\bullet}$).

the majority of energy-resolved metastable ion peaks are of simple Gaussian shape and do not encompass a very wide range of $T_{0.5}$ values.

By the late 1970s, it began to be realized that the assumed structure for an ion (e.g., usually the structure of the species, a molecular or fragment ion, from which the metastable ion was derived) could well be incorrect, with unobservable and often complex ion isomerization taking place before metastable dissociation. A striking example of ease and universality of rearrangements that precede fragmentation was found with the ionized C_6H_{10} isomers, 35 of which, irrespective of their initial structure, were shown to produce the cyclopentenium cation $(C_5H_7^{1+})$ when they lose a $CH_3^{1\bullet}$ radical.[117–119] That cyclopentenium was the common fragment ion coming from the observation that in all dissociations the Δ_fH values of the product ion,[118] determined from AE measurements, were the same and corresponded to that of the cyclopentenium ion. The shapes of the Gaussian metastable ion peaks for the methyl loss from some of the $[C_6,H_{10}]^{+\bullet}$ precursors were isomer-independent (i.e., cyclohexene, 1- and 3-methyl cyclopentenes, methylenecyclopentane, bicyclo[3.1.0.]-cyclohexane and 2-methylpenta-1,4-diene), indicating that the different rearrangement pathways likely led to a common structure (reacting configuration (RC), see definition below) before the $CH_3^{1\bullet}$ loss. For these precursors, the cyclopentenium ion is produced at the thermochemical threshold. However, the range of $T_{0.5}$ values from all the 35 isomers was from 17 to 46 meV, the highest being associated with C_6H_{10} alkynes. The higher $T_{0.5}$ values were all associated with dissociations that showed a reverse energy barrier. It is believed that the 3-methylcyclopentene ion is the unique structure that leads directly to the common product ion.[118] Indeed, this example shows that the shape of the MI peak may not only depend on the initial structure of the fragmenting ion, but also reflects the reaction pathway leading to the dissociation.

It is now clear that in order to completely understand a metastable ion's dissociation it is necessary to identify the *reacting configuration* of the fragmenting ion. The reacting configuration is the ion structure that leads directly to the transition state for the dissociation and may well not be the "initial" structure of the ion. For example, it is known that although the $CH_2CHOH^{1+\bullet}$ ion loses only the hydroxyl H atom to give the signal shown in Figure 1.17, it is the acetyl fragment ion that is produced, CH_3CO^{1+}. To achieve this, the vinyl alcohol ion rearranges to the reacting configuration, the methyl hydroxy carbene ion $(CH_3COH^{1+\bullet})$ before the H atom loss.[120] Thus, the dashed peak in Figure 1.17 not only identifies the vinyl alcohol ion, but also characterizes the carbene ion because of the "hidden" 1,2-H atom migration in $CH_2CHOH^{1+\bullet}$ before its dissociation.

In view of the above possibilities, we wish to reemphasize that the behavior of a metastable ion reflects that of the reacting configuration for each dissociation channel. A reacting configuration does not necessarily have the same connectivity as the ground state of the mass-selected, stable precursor ion, because the latter may rearrange before its MI fragmentations. Therefore, the use of MI mass spectra for structure identification usually requires ancillary experimental results (e.g., thermochemical data, labeling results, collision-induced dissociation (CID) mass spectra) before structure assignments can confidently be made.

Isotopic Labeling and Metastable Ions

In the preceding example, deuterium labeling was used to show which H atom was lost from metastable $CH_3CHO^{]+\bullet}$ and $CH_2CHOH^{]+\bullet}$ ions. The combination of isotopic labeling with metastable ion studies has proved to be a powerful tool for unraveling the complexities of gas-phase ion chemistry. Two further examples are given below:

The first is a salutary reminder that gas-phase ions can undergo very great rearrangement, the presence of which can only be described by isotopic labeling. It was shown that the loss of $H^{13}CN$ from labeled metastable benzonitrile ions, $C_6H_5{}^{13}CN^{]+\bullet}$, indeed took place, but only to the extent of 5–7% of the total, i.e., mostly HCN was lost. It was found inter alia that linear isomers of the benzonitrile ion were involved.[121]

The normal electron-impact mass spectrum of benzoic acid, $C_6H_5C(O)OH$, is very simple with $C_6H_5C(O)OH^{]+\bullet}$, $C_6H_5CO^{]+}$ and $C_6H_5{}^{]+}$ as the three dominant peaks. Thus, the two major fragment ions apparently arise only from simple bond cleavages, the losses of $\bullet OH$ and $\bullet C(O)OH$, respectively. Metastable $C_6H_5C(O)OH^{]+\bullet}$ ions show an intense peak for $\bullet OH$ loss. However, $C_6H_5C(O)OD^{]+\bullet}$ unpredictably yields metastable ion peaks for the loss of $\bullet OH$ and also $\bullet OD$, in a ratio of 2:1, respectively,[122] whereas benzoic acid ortho-d_2 also displays the same losses but now in the ratio of 1:2. Thus it was shown that an unexpected, hidden hydrogen exchange had taken place between the carboxyl group and the *ortho*-hydrogens. When D atoms were placed at the meta and para positions no loss of label (i.e., no $\bullet OD$ loss) was observed.[123]

Before concluding the discussion on isotopic labeling it is important to note that in many ions, especially hydrocarbon ions, all the carbon and hydrogen atoms can completely lose their positional identity in the rearrangements that precede fragmentation. This is known as "atom scrambling." Thus for example, the loss of carbon-13-labeled methane from $C_4H_9{}^{]+}$ ions produced by the loss of $X^\bullet$ from ionized simple *t*-butyl derivatives, $(CH_3)_3{}^{13}CX$, was statistical, $CH_4{:}^{13}CH_4 = 3{:}1$.[124] A similar result was obtained for methane loss from ionized $CD_3(CH_3)_2CX$.[22] For the C_6H_{10} isomers referred to above,[117] complete loss of H atom position precedes the loss of $CH_3{}^{]\bullet}$ to give the cyclopentenium ion. Observations such as these give no insight into reaction mechanisms other than to indicate that a very wide range of stable isomeric forms of the metastable ions are accessible below the dissociation energy.

The relative unlabeled-to-labeled peak abundance ratios for "atom scrambling" can be calculated by simple statistics.[125] The number of permutations obtained when r objects are sampled, without replacement and irrespective of the order, from a set of n objects is given by $n!/r!(n-r)!$. Applied to the above example of $CD_3(CH_3)_2C^{1+}$ ions ($m/z = 60$) the ratios for the statistical loss of labeled and unlabeled methanes, CH_4, CH_3D, CH_2D_2 and CHD_3, leading to $C_3(H,D)_5^{1+}$ ions is, therefore, calculated to be $6!/4!2! = \mathbf{15}$; $[6!/3!3!] \times [3!/2!] = \mathbf{60}$; $[6!/4!2!] \times [3!/2!] = \mathbf{45}$ and $6!/5! = \mathbf{6}$, i.e., $\mathbf{5:20:15:2}$ for the fragment ions $C_3D_3H_2^{1+}$, $C_3D_2H_3^{1+}$, $C_3DH_4^{1+}$ and $C_3H_5^{1+}$, m/z 44, 43, 42 and 41, respectively. Experimental observations usually closely match these ratios, although isotope effects may bias the numbers in favor of the loss of H-rich species.

Field Ionization Kinetics

This section of the book would not be complete without a very brief discussion of this important topic, which alas, has almost disappeared from the literature probably because it is suited to older but versatile mass spectrometers that were not (wholly) software-controlled and that could readily be modified in the laboratory to cover a wider range of experimental techniques. Field ionization kinetics (FIK) was thoroughly reviewed in 1984 by Nibbering.[179]

The advantage of this technique is that it enables an ion's dissociation to be studied from the picosecond to the microsecond (MI) timescale, seven orders of magnitude of rate constant. The mass spectrometer for such studies usually had EB (or "normal") geometry. The ion source contains a sharp blade or tip carrying a charge of approximately 10^4 V and specially treated to possess an array of tiny dendrites at which the great field gradient will suffice to ionize and activate simple organic molecules.[18] The ions are repelled from the tip and cross a gap of variable dimensions to a slotted, earthed cathode, an acceleration region that encompasses an ion-flight timescale from approximately 10^{-11} to 10^{-9} s. Ions generated at the tip will have a kinetic energy equal to that of the source voltage and so will pass an electric sector set to transmit them. Subsequent mass analysis will produce the FI mass spectrum of these very short-lived ions that fragment close to the tip. Fragment ions generated in fast dissociations between the tip and the cathode can be made to pass the electric sector (kept at the same potential) by raising the anode voltage. Metastable ions that fragment in the field-free region between the cathode and the electric sector (MI.1) will have lifetimes of approximately 10^{-7} s and dissociations in the region between the ESA and the magnet (MI.2) will have longer, conventional MI lives. Note that the kinetic energy releases for the MI.1 ions is significantly greater than for MI.2 ions. The analysis of all the results is described in both of the above references.[18,179]

The greatest strength of the FIK method lies in its ability to observe the fate of an isotope label in the lowest energy dissociations of an ionized molecule as a function of time, from pico- to microseconds. An excellent example is provided by the investigation of deuterium-labeled cyclohexenes.[180] In cyclohexene-3,3,6,6-d_4 only the losses of C_2H_4, C_2H_3D, $C_2H_2D_2$ and CHD_3 were observed. At the shortest times, approximately 10^{-11} s, half of the ethene lost was C_2H_4 but by approximately 10^{-9} s near-statistical loss of labeled ethenes was found. The results were explained by invoking a series of 1,3-allylic H/D shifts preceding the ion's fragmentation by a retro-Diels–Alder reaction. Three such shifts are required for the loss of C_2H_3D. FIK studies also played a significant role in the elucidation of the reaction mechanism of the McLafferty rearrangement[181] (see also Section 2.3).

1.4.2 Collision Experiments; the Key Ion Structure Tool

The experiments that have led to the greatest body of information concerning ion structures are the high energy collision-induced dissociation mass spectra of mass-selected ions, measured with a double-focusing mass spectrometer. As will be described below, CID mass spectra relate more readily to ion structures than do MI mass spectra not least because the ions that are subjected to a collision experiment are those with insufficient energy to decompose on the timescale of the instrument, typically microseconds. Thus, an ion subjected to CID will have retained its structural identity provided that there is no rearrangement pathway whose energy requirement lies below the ion's lowest dissociation limit. Also, the CID mass spectrum will generally contain as many peaks as are found in a normal electron-impact mass spectrum and many of these will be structure-significant. The structural feature most securely identified by these experiments is the atom *connectivity* of the ion. The collision-induced dissociation of ions at high and at low translational kinetic energies has given rise to a book[126] and useful review articles.[127–129]

Historical Background

It is interesting to note that the phenomena associated with the collision-induced dissociation of ion beams were observed essentially by accident. Two reports in 1968[130,131] described how the leakage of gas into a field-free region of a tandem mass spectrometer gave rise to extensive additional fragmentation of the ion beams that traversed them, thus interfering with the observation of metastable ion peaks. In 1973, a key paper by McLafferty[132] showed how these hitherto "unwanted" effects could be put to invaluable use, containing as they do, information concerning ion structure.

This book has focused on collision events observed at high ion kinetic energies, typically 5–10 kV, such as obtained in tandem sector mass spectrometers. There is some mention of low energy collision experiments (0–40 V) conducted in multiple quadrupole mass spectrometers. The latter experiments have analytical value, particularly in the identification of isomeric molecules, but have not contributed greatly to the establishing of ion structures. They will, however, feature in ion reactivity experiments (see Section 1.5). (The low collision energy CID mass spectra of isomeric molecular ions can of course be used to distinguish between them, as indeed can the normal EI mass spectra. However, the CID mass spectra of ions having high translational energy are much more structure-sensitive and provide the principal method of choice.)

Collision-Induced Dissociation Phenomena

With an ion source acceleration potential of 8000 V (which is typically used in sector instruments), the ion velocities are very great. For example, an ion of $m/z = 100$ travels at approximately 4.5×10^4 m s^{-1} and thus, approximately 10^{-14} s are required for the ion to traverse a distance of molecular dimensions. Therefore, in the collision experiment (which will be described in greater detail in the following section), the mass-selected beam of ions traveling at a high velocity will encounter a neutral collision (target) gas. The collision event takes place in a small and restricted zone usually in the second field-free region of the instrument (Figure 1.18). Collision-activated mass spectra are enormously useful as a structure-identifying tool. However, the physicochemical nature of the collision event between the incoming ions and the stationary target gas is not understood in detail.

The collision encounters that lead to mass spectra are those that involve no momentum transfer between the ion and the target molecule. Ions thus

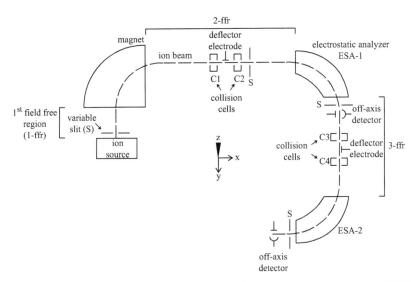

FIGURE 1.18 The layout (not to scale) of a typical three sector instrument of BEE geometry showing collision cells and deflector electrodes.

excited are therefore not scattered from their trajectory by the collision. The energy deposited in the ion under typical conditions comprises a broad range, peaking at a few electronvolts and with a very long tail extending to some tens of electronvolts. The most widely accepted concept describing the collision event is that of curve crossing. This mechanism of collisional activation of kilovolt ions has been described in some detail by Lorquet et al.[133] In this picture, the ground state of the collision complex (the ion and the target gas) is expected to interact (by curve crossing) with higher excited states. As the ion pursues its trajectory, the collision complex separates, leaving the reactant (the ion) in excited states. It is noteworthy that during the collision event some undefined part of the translational energy of the ion is converted to internal energy in the collision complex. When the ion is small enough, the excited states are most likely to be unstable and isolated and, therefore, dissociate without randomization of energy. However, when the ion is large enough and by consequence possesses a large number of degrees of freedom, radiationless transitions to the ground electronic state of the ion results in randomization of energy before dissociation. Thus, as with EI, the energized ions behave statistically, undergoing unimolecular dissociation predominantly from excited vibrational states of the ground electronic state. The quasi-equilibrium theory[134] thus adequately describes their behavior.

It is noteworthy that Lorquet et al. do emphasize that a simple Franck–Condon excitation description, a strictly vertical process where no nuclear vibrations take place, is "superficial and misleading."[133] Because of the ion's charge, the interaction distance between the relatively stationary neutral target gas and the ion projectile is, however, much larger than molecular dimensions, allowing some time for the nuclei to vibrate and so the encounter is only quasi-vertical (compare with section on Electron-impact (EI) ionization).

To put in perspective the timescales of MI and CID events in an instrument such as that in Figure 1.18, please refer to Figure 1.11. As for the relationship between rate constant, lifetime and energy of these processes, Figure 1.16 clearly illustrates the differences between MI and CID mass spectra recorded in the same field-free region of a tandem mass spectrometer. For ions activated by collision, the timescale extends from zero (the moment of collision) to the microsecond time frame and so a much wider range of dissociation rates are sampled. Figure 1.16 also shows that for ion internal energies corresponding to rate constants greater than 10^5 s^{-1}, all ions activated by collision will fragment.

Measuring Collision-Induced Dissociation Mass Spectra

The collision cell in this apparatus (Figure 1.19) is a strong differentially pumped cell or region at or near the focus of the magnetic field and is usually located over the throat of a diffusion or turbomolecular pump. The cell is typically only 2–3 cm long and is placed above the pump, so that gas escaping

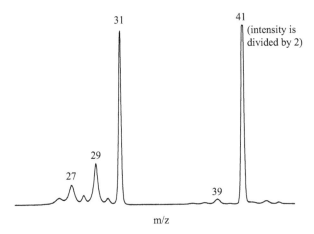

FIGURE 1.19 The collision-induced dissociation (CID) mass spectrum of ion source–generated $CH_3CH_2O^+CH_2$ ions. Note that the m/z 41 peak was *unaffected* by the collision gas; see text for discussion.

from it is swiftly pumped away with very little diffusing into the main field-free region. Thus the collision zone is restricted in size.

To obtain CID mass spectra, the target gas (usually helium) is admitted to the cell at a pressure sufficient only to reduce the main ion beam (mass selected ion) by approximately 10%. The gas density within the cell is controlled by a needle valve at the collision gas inlet and is maintained at a pressure of approximately 10^{-3} mbar, well above the ambient pressure in the rest of the field-free region, typically 10^{-8} mbar or better. Under so-called single-collision conditions[135] i.e., when the collision gas density reduces the mass-selected ion beam intensity by approximately 10%, on average the ions predominantly only suffer single encounters with the target gas. By scanning the electrostatic analyzer (ESA) downward from E_1 (which when the experiment takes place in the second field-free region of the instrument corresponds to the accelerating voltage), the fragment ions will appear at a sector voltage value very close to that appropriate to their masses, i.e., at $E_2 = m_2/m_1 \times E_1$, etc.

The CID experiment usually produces an intense, complex mass spectrum containing as many signals as in a normal EI mass spectrum and usually swamping the MI signals. Figure 1.19 shows the CID mass spectrum for the source generated $CH_3CH_2O^+CH_2$ ions discussed earlier.

Note that the m/z 41 peak is collision insensitive (see discussion on the effect of collision gas on metastable ion peak shapes and abundances in Section 1.4.2), confirming that (not surprisingly in this case) considerable rearrangement must precede the loss of H_2O, while the m/z 31 signal has increased about fivefold under single-collision conditions. Little or no rearrangement of the $CH_3CH_2O^+CH_2$ ions is required before the loss of C_2H_4.

The presence of fragment ions in the region m/z 25–30 shows that some $C_2H_n^{1+}$ fragments are also generated by collision.

Under single-collision conditions, the CID mass spectra recorded are reproducible (a very important requirement in any experiment) and collectively they contain the most information about mass-selected ions. Multiple collision conditions, beam reductions of 20% or greater, only increase the fragment ion yield and bias the peak intensities toward lower mass ions and provide no new structure-diagnostic features.

It is noteworthy that since some of the ions' translational kinetic energy will have been converted by the encounter into internal energy (as mentioned above), the product ion peaks will be centered slightly below the calculated E_2 value. The voltage difference is indeed small, being typically less than 5 V for singly charged ion products when $V_{acc} = 8000$ V. In principle, the above energy defect potentially provides a method for measuring the second ionization energy (IE_2) of a molecule or radical, when the exact position (the measured ESA voltage (E^{++}) at which the ion M^{++} is transmitted) of the doubly charged mass-selected ion is recorded, thus $IE_2 = (E_1/2 - E^{++})$. The phenomenon is known as charge stripping and was first reported by Cooks et al. in 1972.[136] The method is inexact because there are problems analogous to those pertaining to adiabatic and vertical ionization energies. The geometry of the doubly charged ion usually is significantly different from that of its singly charged counterpart and so the resultant second ionization energy can be difficult to define. This particular method has not found universal favor but it has been well reviewed together with other charge permutation processes.[137,138]

Collision (Target) Gases

A number of collision gases are commonly used, with He being the most popular. It can be obtained in a high state of purity and is fast to pump away. In general, high translational energy CID mass spectra are almost wholly independent of the target gas except for the case of oxygen.[139] This gas has the unique ability to enhance specific ion fragmentations that lie within a relatively narrow band of activation energies above the ground state of the mass-selected ion, namely, in the range 450–600 kJ/mol. It has been proposed[139] that the O_2 molecules are excited by the collision from the ground $^3\Sigma_g^-$ to the excited $^1\Sigma_u^-$ state (that gives rise to the Schumann–Runge bands in the UV spectrum of molecular oxygen) and that this state can transfer energy quanta back to the target ion allowing its quasi energy-selected excitation. This argument is in keeping with the picture of CID presented by Lorquet.[133] These events must take place in a very short time frame, $<10^{-12}$ s. Oxygen is also particularly adept at producing an enhanced yield of doubly charged ions and this is attributed to the molecule having a significant electron affinity, namely 0.45 eV.[56]

The Effect of Collision Gas on Metastable Ion Peak Shapes and Abundances

As mentioned earlier, poor background pressure in the field-free regions resulting from inefficient pumping can, therefore, lead to unwanted and potentially misleading CID contributions to MI mass spectra. However, a very simple experimental test for the "purity" of MI mass spectra has been described.[140] The MI peak for loss of $CH_3^{1\bullet}$ from the ion CH_3CO^{1+} (produced from ionized acetone) is very narrow ($T_{0.5} = 1.5$ meV) and the signal is extremely collision gas sensitive.[141] In contrast, the corresponding CID peak is much broader, $T_{0.5} = 20$ meV.[140] Therefore, the lack of a very good vacuum is thus easily observed by the presence of the broad CID peak below the narrow MI signal.

The effect of collision gas on every metastable ion peak is an observation that deserves particular care because it can provide useful information on the reacting configuration of the ion. When a very small pressure of collision gas is first admitted to the collision cell, its effect on all MI peaks should be carefully noted. If a peak is very sensitive to collision gas then it is likely that the fragmentation reaction involves a simple bond cleavage in the probably nonrearranged ion, i.e., the reacting configuration for the dissociation is the same as the original ground state structure of the ion. The dramatic increase in signal intensity results from a significantly larger number of ions having now acquired, via the collision with the target gas, sufficient internal energy to dissociate. In this case, the fragment ion peak will also be greatly broadened by the collision gas[140,141] because the range of internal energies sampled is now so much greater (see Figure 1.16).

However, if the MI peak is insensitive to collision gas or only weakly affected thereby, the MI process *must* proceed from a stable isomer of the ion. This applies when the energy barrier for the rearrangement to the new isomer lies close to but below the energy threshold for the metastable ion dissociation. Moreover, the isomeric ion that is produced should lie in a potential well of similar or greater depth than that of the original ion structure. The last criterion assures that at the dissociation limit the density of states is large or larger for the rearranged species; in turn, the rate of dissociation from this ion will be small (dissociation rate is inversely proportional to the density of states).

This observation (whether an MI peak is collision sensitive or not) therefore serves as a useful test for the integrity of an ion's structure before its metastable decay. One example has already been described; see Figure 1.19 and the associated text. Another good example is provided by the loss of water from ionized propanol, $CH_3CH_2CH_2OH^{1+\bullet}$ (m/z 60), for which a partial potential energy diagram is shown in Figure 1.20.

This fragmentation produces an intense Gaussian metastable peak (m/z 42) that is unaffected by collision gas.[142] The CID mass spectrum of $CH_3CH_2CH_2OH^{1+\bullet}$ (m/z 60) shows m/z 31, $^+CH_2OH$, as the most intense

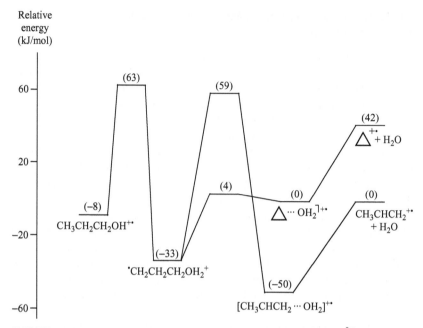

FIGURE 1.20 The partial potential energy surface for isomeric $C_3H_8O^{1+\bullet}$ ions showing the rearrangement of ionized propanol over a high energy barrier.

CID signal. It results from the simple C1–C2 bond cleavage in the unrearranged propanol ion, the species lying in the left-hand-side potential well. To account for the observed collision gas insensitivity of the metastable ion peak corresponding to loss of water (m/z 42), the isomeric distonic ion $^\bullet CH_2CH_2CH_2{}^+OH_2$, which is thermodynamically more stable than the conventional ion and which is reached over an energy barrier close to the dissociation limit (Figure 1.20) is proposed to be the reacting configuration for this dissociation. The structure of this distonic ion close to the dissociation limit is best represented as a cyclopropane ion associated with a water molecule, an ion–dipole complex. This remarkably complicated ion chemistry has recently been reviewed and updated.[143] (See also the discussion for $[C_3,H_8,O]^{+\bullet}$ ions in Part 2.)

Separating Metastable Ion Peaks from Collision-Induced Dissociation Mass Spectra

It is also useful to be able to separate the few MI peaks from the many in a CID mass spectrum. This is easily done provided that the collision cell is electrically isolated and thus, can have a significant potential safely applied to it. For example, when a potential of $+500$ V is applied to the cell, the incoming positive ion beam will be decelerated (by 500 V) on entering the cell and accelerated (by 500 V) on exiting the cell. Therefore, the kinetic energy of all fragment ions M_n^+ of mass m_n, generated from metastable M_1^+ ions before or after the cell, will have unchanged E values, namely $m_n V_{acc}/m_1$. However, all product ions generated *within* the cell by collision or by MI dissociation, will however have an enhanced translational kinetic energy (E). Ions M_n^+, derived from the CID of M_1^+ inside the cell, will have an energy on leaving the cell given by $[(m_n/m_1)(V_{acc}-500)+500]$ V. If V_{acc} is 8000 V, m_1 is 100 Da and m_n is 72 Da, the MI peak appears at 8000 $(72/100)=5760$ V, whereas the CID peak (fragmentation inside the cell) appears at $[(72/100)7500+500]$ V $=5900$ V. All CID peaks will therefore be displaced to higher energies whereas the MI peaks are unmoved. Recall that the length of the collision cell itself is only a small fraction of that of the 2-ffr, typically about 2%, and so corrections to CID mass spectra are rarely necessary for the few metastable ion dissociations that take place within the cell.

Ion Structure Assignments from Collision-Induced Dissociation Mass Spectra

The first assessment of a new CID mass spectrum requires one to search for obvious atom connectivities, which will be signaled by fragmentations likely resulting from direct bond cleavages. In general, CID favors simple bond cleavages (fast reactions) over rearrangement processes, thus aiding isomer differentiation. An excellent example is provided by the CID mass spectra of the $[C,H_4,O]^{+\bullet}$ isomers,[144–146] ionized methanol, and the α-distonic ion (or ylidion), $^\bullet CH_2^+OH_2$ (see the case study Section 2.2).

A less striking but more typical example is provided by the CID MS of the $[C_3,H_5]^+$ isomers $(m/z = 41)$, the resonance-stabilized allyl cation, $CH_2CHCH_2^{+}$, and the 2-propenyl species, $CH_3CCH_2^{+}$, shown in Figure 1.21.[147]

Here, isomer differentiation depends on fragmention peak ratios, specifically m/z 27:26, corresponding to the losses of CH_2 and $CH_3^\bullet$, respectively. As might be surmised, the ratio is greater than unity for the allyl cation (a) and less than one for its isomer (c). To minimize the internal energy of the ions subjected to collisional activation, those generated by the metastable dissociation of the iodo-analogs $(CH_2CHCH_2I$ and $CH_3C(I)CH_2$, $m/z = 168)$ were used for the CID experiments, items (a) and (c), respectively, in Figure 1.21. This was achieved by transmitting the species with an apparent mass of $41^2/168 = 10.0$ from the first field-free region through the magnet; these ions have a low translational energy, $41 \times 8000/168 = 1952$ V.

In the above example, the critical m/z 27:26 ratio depends on the translational kinetic energy of the ions as well as their origin and internal energy.[147] The ions in Figure 1.21b are source-generated $C_3H_5^{+}$ ions from CH_2CHCH_2I. They undoubtedly have the allyl structure but the peak ratios

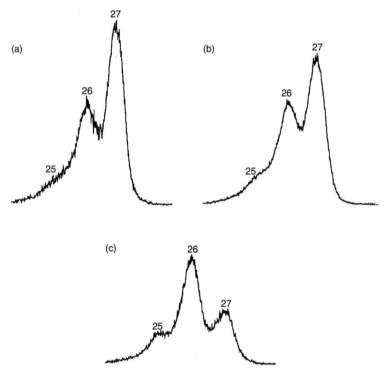

FIGURE 1.21 Partial collision-induced dissociation (CID) mass spectra of the allyl ($CH_2CHCH_2^{1+}$) and 2-propenyl ($CH_3CCH_2^{1+}$) cations showing their structure characteristic differences: (a) $CH_2CHCH_2^{1+}$ cations produced from metastable $CH_2CHCH_2I^{1+\bullet}$ (b) ions generated in the ion source from $CH_2CHCH_2I^{1+\bullet}$ and (c) $CH_3CCH_2^{1+}$ cations generated from metastable $CH_3C(I)CH_2^{1+}$.

have changed because of their greater translational kinetic energy (8000 V vs $41 \times 8000/168 = 1952$ V for the ions from the metastable dissociation).

Note also that although these isomeric ions occupy independent potential energy wells, they can interconvert freely at an energy that lies below their first dissociation limits, namely, the competing losses of H_2 to yield the aromatic cyclopropenium cation and its propargyl isomer. Both of these dissociations involve a reverse energy barrier; see the discussion of composite metastable ion peak shapes above (Physicochemical aspects of metastable ion dissociations in Section 1.4.1). It is worth noting that the $C_3H_5^{1+}$ ions ($m/z = 41$) referred to in Figure 1.9 and Figure 1.12 were shown to have the allyl structure by CID of the metastably generated ($m/z = 41$) ions.[148]

A third example is the $[C_2,H_3,O]^+$ isomers,[141] the acetyl cation CH_3CO^{1+}, the 1-hydroxyvinyl cation CH_2COH^{1+} and the oxiranyl cation cy–$C(H)_2C(H)O–^{1+}$. For this set of isomers the peak abundance ratios m/z 29:28

and those for the doubly charged ions (produced by charge stripping collisions) m/z 21.5:21 (43^{++}:42^{++}) serve as the key structure-identifying features (See Part 2). The m/z 29:28 ratios for ions generated from metastable precursors are 0.45, 2.5 and 5.0 for the CH_3CO^{1+}, CH_2COH^{1+} and cy–$C(H)_2C(H)O–^{1+}$, ion structures, respectively. For the doubly charged ions the corresponding m/z 21.5:21 ratios are 0.24, 25 and 0.04.[141] The acetyl ion is fairly easy to identify by the above means but as can be seen from the original report,[141] generation of the isomers in a pure state was far from simple and moreover, the results are complicated by translational and internal energy effects.

In general, it can be confidently stated that a careful selection of precursor molecules has led to the acquisition of reproducible, reliable CID MS for a very wide range of simple, isomeric organic cations. Critically evaluated results are to be found in Part 2, which forms the bulk of this book.

Other Collision Experiments

Collision-Induced Dissociative Ionization Mass Spectra

When metastable ions fragment, a neutral radical or molecule is lost. In general, the identity of this species is not in doubt because, being chemically simple, it cannot have isomeric forms. The problem of distinguishing between isobaric species, e.g., CO/C_2H_4 or $HCO^{1\bullet}/C_2H_5^{1\bullet}$ was discussed previously (Metastable ion phenomena and Measuring kinetic energy release in Section 1.4.1) but an alternative and more powerful technique is introduced here.

In Figure 1.18, in front of the collision cell in the 2-ffr, a beam deflector electrode is indicated. If it is positively charged, say, to $+500$ V, then all ions will be deflected away from the beam path and only neutral species, generated from the metastable ions in the zone before the electrode, can enter the collision cell. Herein, the neutrals, which like their ionic precursors have kilovolt translational kinetic energies, will be ionized by collisions with the target gas. A scan of the ESA will therefore produce the collision-induced dissociative ionization (CIDI) mass spectrum of the neutrals. Such mass spectra are almost indistinguishable from the normal electron-impact mass spectra of the molecules or radicals.[149] Clearly, the CIDI mass spectral differences between the above isobaric neutrals will definitively identify them.

In some cases however, the identity of the neutral is unexpected or controversial. Examples are the loss of HCN from ionized pyridine, the loss of HNC, rather than the thermochemically more stable HCN, from metastable ionized aniline[150,151] (their $\Delta_f H$ values are 135^{56} and 208 ± 12^{152} kJ/mol, respectively (also see $[H,C,N]^{+\bullet}$ in Part 2)), and the wholly unexpected loss[153] of $^\bullet CH_2OH$ ($\Delta_f H = -19 \pm 1$ kJ/mol[154]) rather than the "obvious" but less stable $CH_3O^\bullet$ radical ($\Delta_f H = +17$ kJ/mol[56]) from metastable ionized methyl acetate, $CH_3C(O)OCH_3^{1+\bullet}$. In contrast, ionized methyl propanoate ($CH_3CH_2C(O)OCH_3^{1+\bullet}$) lost only a $CH_3O^\bullet$ radical. A complete solution to the mechanism for the former reaction still awaits discovery (see also Section 2.5). In all these examples the CIDI mass spectra of the isomeric neutrals differ greatly.

Collision-Induced Charge Reversal Mass Spectra

A useful method for obtaining the CID mass spectrum of an otherwise elusive or intrinsically unstable cation (e.g., $HCCN^{\cdot+}$ or CH_3O^+) is to use the collision-induced charge reversal (CR) of the corresponding anion, a method particularly explored by Bursey et al.[155] These are produced by appropriate chemistry in the ion source of the mass spectrometer with all polarities reversed. The anions are mass-separated by the electromagnet, also with a reversed magnetic field, allowed to undergo (single) collisions with a target gas in the second field-free region and the positive ion products are mass analyzed in the usual way by the electric sector. The excitation, $M^- \longrightarrow M^+ + 2e$, which is a quasivertical collision process, is often performed with oxygen as the target gas. The modest electron affinity of O_2 (0.45 eV[56]) is a good argument in support of its efficacy.

In the detailed discussions of isomeric ions it will be apparent that CR mass spectrometry of the appropriate anion can unequivocally produce the cation having the desired structure.

Neutralization-Reionization Mass Spectrometry

Although this experimental technique has had relatively few direct applications to ion structure determinations, its development required double- and on occasion triple-focusing mass spectrometers and so will be briefly described here. Neutralization-reionization mass spectrometry (NRMS) has been reviewed in detail several times[156–159] and so only an outline of the method will be given. Figure 1.18 shows a three-sector mass spectrometer with two collision cells in the second and third field-free regions.

In order to perform an NRMS experiment the mass-selected ion encounters an electron transfer target gas in the first cell. Such a target can be selected so that its ionization energy either exceeds, is roughly equal to, or is significantly less than the neutralization energy of the mass-selected ion, i.e., the negative of the ionization energy of the corresponding molecule (odd electron ion) or free radical (even electron ion). In the above cases the process will be endothermic, approximately thermally neutral and exothermic, respectively. Given the great translational kinetic energy of the mass-selected ions, any energy deficit of the neutralization reaction is easily overcome and the experiment even works (although with very low efficiency) with He (IE He = 24.6 eV[56]) as the neutralization target. For organic ions the most useful (albeit expensive) target is xenon (IE Xe = 12.1 eV[56]), a gaseous reagent available in high purity and speedy to pump away. Mercury vapor is closer in ionization energy to many organic species (IE Hg = 10.4 eV[56]) but is not so easy to control in and remove from the mass spectrometer. A wide variety of other target atoms and apparatus for their management has been described.[160–162] Sodium vapor, for example (IE Na $= 5.1$ eV[56]), has been a popular exothermic neutralization target.

The electron transfer process is only quasi-vertical and so similar geometry constraints as were discussed for ionization processes are obtained here. A more detailed description of the physicochemical events accompanying neutralization can be found in a paper by Lorquet et al.[163] Figure 1.22 summarizes the situation.

After the neutralization step, the mixture of ionic and neutral species leaving the first collision cell are immediately separated by the beam deflector electrode and only the neutrals continue onward to be reionized by collision in the second cell (see Figure 1.18). Oxygen is frequently selected as the reionization target gas. When the experiment is successful, the resulting NR mass spectrum contains a recovery peak, arising from ions having the same m/z ratio as the original mass-selected ion, their dissociation products, and also very minor contributions from the products of the metastable dissociations of the mass-selected ion. As will be illustrated below, great care must be taken in interpreting a NR mass spectrum but at

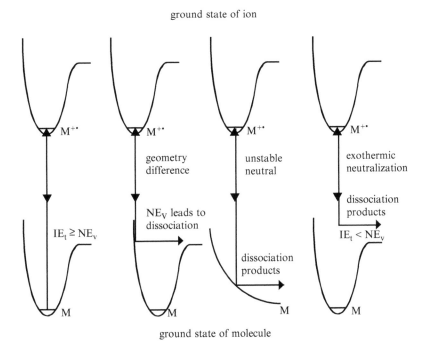

FIGURE 1.22 The neutralization of ions with different target gases showing the effects of verticality and endothermic, thermoneutral and exothermic processes. IE_t and NE_v are the vertical ionization energy of the target and the vertical neutralization energy of the mass-selected ion, respectively.

least the presence of a recovery ion peak indicates that the neutral counterpart of the mass-selected ion very likely exists on the timescale of the experiment, typically microseconds. It is often necessary to confirm the identity of the reionized species itself and this can be achieved by transmitting it to the third field-free region of a multisector instrument (Figure 1.18) where it can be subjected to collision-induced dissociation. The CID mass spectrum then recorded should be the same as that of the original mass-selected ion, thus providing unequivocal confirmation of the neutral's structure.

One system that exemplifies the interest in and difficulties pertaining to such studies was the investigation of the possible existence of the neutral counterparts of small distonic ions such as $^{\bullet}CH_2{}^+XH$, where X = halogen, OH, NH_2, etc. Although initially they appeared to be isolable species in the gas phase,[164] the very high electron multiplier gains necessarily used in these experiments resulted in the great amplification of artifacts that give rise to a misinterpretation of the observations. For example, although the distonic ion $^{\bullet}CH_2{}^+ClH$ is efficiently generated by the dissociative ionization of $ClCH_2C(O)OH^{]+\bullet}$ by the loss of CO_2, a very small yield of the conventional ion $CH_3Cl^{]+\bullet}$ is cogenerated. After careful analysis, it was concluded that the recovery signal arose from the latter conventional ion and not from the neutral ylide.[165] Moreover, computations by Radom et al.[166] showed that the ground state neutral occupied so small a well on the potential energy surface that it would not be produced by vertical neutralization of the geometrically different cation. Nevertheless, a concurrent publication[167] in which Hg vapor was the target gas produced a mixture of the two $[C,H_3,Cl]$ isomers.

Other problems that have to be considered in addition to the above, where the mass-selected ion had the correct molecular formula but was impure in structure, are the presence of isobaric ions of different molecular formulae or natural isotopic components from ions of lower mass, e.g., the ion $CH_5{}^{]+}$, generated from the protonation of methane, can always be expected to be contaminated by the ion $^{13}CH_4{}^{]+\bullet}$, a problem that arose when the possible stability of the hypervalent $CH_5{}^{]\bullet}$ radical was investigated. The difficulty was overcome using $CD_5{}^{]+}$ ions, but the clear recovery signal from the neutral hypervalent radical is now ascribed to Rydberg states of the species.[168,169] Here the transferred electron resides in an orbital, in a heavy atom in the ion, having a high principal quantum number, i.e., an orbital near the ionization threshold. The core of such a species is the stable mass-selected ion, the received electron occupying no more than a "spectator" position. The whole is electrically neutral and so is transmitted to the reionization cell and produces a recovery ion and a complete mass spectrum. Another striking example is provided by the hypervalent radical $(CH_3)_2OH^{]\bullet}$ generated by the neutralization of protonated $(CH_3)_2O$.[170] Finally, it is worth remarking that the NRMS method, although fascinating in its ability to generate unexpected species, has not played a very significant role in ion structure determinations.

Uses for Multiple Sector Mass Spectrometers

The addition of another (third) field-free region equipped with collision cells and followed by an electrostatic sector and detector, as displayed in Figure 1.18, gives rise to the possible recording of the CID mass spectra of ions generated by MI, CID, or as described above, by NR in the second ffr and so this affords a method for examining and/or confirming their structure.

Emission of Radiation

This topic deserves a few words. It might be supposed that ions subjected to high energy collisions will emit radiation. Experiments in the UV and visible regions of the spectrum to test this hypothesis were described in 1991 and the initial results appeared to hold considerable promise for ion structure determinations.[171,172] However, more detailed investigations showed that the only radiation observed in the above wavelength regions came from the atomic and bi- or triatomic species produced by the kilovolt collisions[173] and not from any polyatomic intermediates.

1.4.3   INFRARED MULTIPHOTON PHOTODISSOCIATION SPECTROSCOPY

This powerful technique has been in place for many years as an alternative to collision-induced dissociation (CID) experiments, particularly with the use of ion-trapping instruments. An infrared photon is absorbed by the ion, exciting it to a higher vibrational energy level. If photons are absorbed at a rate faster than that of internal energy relaxation, the ion eventually will be excited to above its dissociation threshold. The resulting mass spectra are usually similar to low energy CID mass spectra, but tend to contain more structurally informative fragmentation channels. This is because the primary fragment ions can themselves absorb photons and dissociate; in contrast, collisional excitation in a trapping instrument is usually limited to the ion whose trajectory has been resonantly excited.[174,175] This feature has made infrared multiphoton photodissociation spectroscopy (IRMPD) a popular complement to CID mass spectra in peptide ion sequencing.

A more recent development in IRMPD that has greatly expanded its value for ion structure determination is the use of variable wavelength light sources. Due to the low ion densities in mass spectrometers, intense light sources are required, such as those from a free-electron laser facility. As the wavelength of the light is changed, it will come into resonance with the vibrational energy levels of the various normal vibrational modes of the ion. When in resonance, excitation and then dissociation will be facilitated and a maximum will occur in the plot of total fragment ion abundance versus wavelength (i.e., the IRMPD spectrum). In this manner an IR spectrum of the ion is obtained, which can then be compared with the calculated IR spectra of isomeric ion structures.[176–178] Ions with closely similar structures may have very distinct IR spectra. However, as with collision-based experiments, the presence of a mixture of ion structures makes the IR spectra difficult to interpret because there is no straightforward way to combine computed IR spectra to simulate the experimental spectrum. This arises from the (great) number of ill-defined phenomena that make up the IRMPD process (for example, energy relaxation rates and the dissociation kinetics).

1.5 REACTIVITY OF IONS; ANOTHER TOOL FOR ION STRUCTURE DETERMINATION

1.5.1 INTRODUCTION

The study of gas-phase ion–molecule reactions is a relatively new and active research area that began, however, almost a century ago when discovered by Thompson.[39] The confirmation that ion–molecule reactions indeed took place in the ion source of mass spectrometers came with Dempster's proposal[182] that the peak at m/z 3 in the mass spectrum of hydrogen resulted from the reaction between $H_2^{+\bullet}$ and H_2.

The development of chemical ionization[15] has without doubt significantly contributed both to our interest and to our understanding of gas-phase ion–molecule reactions. It is however the development of techniques/instruments (discussed briefly in Section 1.2) such as SIFT,[183] the quadrupole ion-trap,[43,184–186] ICR mass spectroscopy and FT-ICR MS[44,187,188] that has allowed this particular field of gas-phase ion chemistry to evolve so dramatically. The reason why these techniques/instruments are so useful for studying ion–molecule reactions lies in their capacity to trap ions for long periods of time during which the ions can be thermalized and then reacted with a selected neutral reagent at controlled concentrations and temperatures. Thus reaction kinetics, as well as simple reactivity studies, can be achieved.

A prime interest in studying reactions taking place in the gas-phase is that they indicate the intrinsic reactivities of the reaction partners without any effects from ion pairing or solvation contributing to the observed chemistry. The study of gas-phase ion–molecule reactions has provided a wealth of information in areas such as the investigation of their mechanistic pathways, the determination of thermochemical quantities such as proton affinities, gas-phase basicities and acidities, the modeling of upper atmospheric and interstellar chemical reactions, the identification of protonation or binding sites in amino acids and peptides and the investigation of host–guest chemistry (molecular recognition).[189–191] The latter two examples indicate the many applications of gas-phase ion chemistry to the study of biological systems.

It is not within the scope of this book to present details of the above methods but rather to focus on the use of gas-phase ion–molecule reactions as a tool for distinguishing between isomeric ions. In these cases, differentiation can be made based either on the rate at which the reaction takes place (isomer-dependent rate constants) or by the products generated by a given reaction (isomer-dependent products or branching ratios). Two examples are described below.

1.5.2 DISTINGUISHING BETWEEN HCO^{1+} AND HOC^{1+} ISOMERS

The isomeric pair of ions, HCO^{1+} and HOC^{1+}, has received considerable attention over the decades as these ions have been identified as interstellar species[192,193] and shown to be ubiquitous in flames and plasma.[194] The dissociative ionization of several aldehydes generates essentially pure formyl cations, HCO^{1+}, whereas dissociative ionization of CD$_3$OH (by the consecutive losses of D$_2$ and D$^\bullet$)[195] mainly produces the isoformyl cation, HOC^{1+}, with HCO^{1+} being cogenerated in small amounts.[195,196]

Their mass spectrometric identification is somewhat difficult because of reliance on the abundance ratio m/z 12:13 of the fragment peaks in the collision-induced dissociation mass spectra of the ions.[195,196] This ratio has been shown to depend on their translational kinetic energy,[197] likely because the isomers can interconvert below their first dissociation limit (refer to Part 2 for a complete discussion of these ions). Therefore, any other method allowing their differentiation is to be welcomed.

In the presence of neutral reagents, both HCO^{1+} and HOC^{1+} undergo proton transfer. However, the rates at which the reactions take place are isomer dependent. For example, when mixtures of the two isomeric ions are allowed to react with N$_2$O, the plot of the m/z 29 ion current (sum of HCO^{1+} and HOC^{1+}) as a function of concentration of N$_2$O shows a marked curvature. An initial steep decrease of the m/z 29 ion current is observed followed by a slow decay as the concentration of N$_2$O is further increased.[198] The two different slopes indicate the characteristic reactivity of the two isomers toward proton transfer to N$_2$O (note that the PA of N$_2$O at N and O are, respectively, 548.9 and 575.2 kJ/mol[56]). Given the PA of N$_2$O, and because the PA of CO at oxygen is considerably lower than at carbon (456.3 and 594 kJ/mol, respectively),[56] HOC^{1+} cations will more readily transfer a proton to N$_2$O than will the HCO^{1+} ions. HOC^{1+} reacts almost three orders of magnitude faster than its isomer.[198]

It is important to note however, that under certain conditions, e.g., particular reagent molecules, some isomers may show very similar reactivities that a priori could suggest that only a single ion structure is present. This is the case for the isomeric pair CH$_3$CHO$^{1+\bullet}$ (ionized acetaldehyde) and CH$_3$COH$^{1+\bullet}$ (the methylhydroxycarbene radical cation, generated by the loss of CO$_2$ from metastable ionized pyruvic acid (CH$_3$C(O)C(O)OH)[199]).

Indeed, their reaction with water proceeds at almost the same rate and yields essentially the same products.[200] Further investigation has shown that in the presence of water the carbene ion is converted to the more stable conventional ionized aldehyde, the water acting as a catalyst for the 1,2-hydrogen shift.[200] Therefore, the reactivity results are not characteristic of the carbene ion. This process, where the high energy isomer is converted to the lower energy species in the presence of a small neutral molecule, is now known as "proton-transport catalysis," and it and related processes are not uncommon.[201]

1.5.3 Distinguishing Between Distonic Ions and Conventional Radical Cations

As will be seen in Part 2, distonic ions (odd electron ion structures in which the charge and radical sites are located on different atoms, such as the ions $^\bullet CH_2{}^+OH_2$ and $CH_3{}^+C(OH)CH_2{}^\bullet CH_2$) are very commonly produced in the dissociative ionization of molecules. As shown in the case studies presented in Section 2.2 and Section 2.3, these ions are quite widespread and also have been identified as stable intermediates in the well-known McLafferty rearrangement. In most cases, distonic ions can be distinguished from their conventional isomer by their different heats of formation and their mass spectral characteristics. However, the different reactivities of distonic and conventional ions toward several carefully chosen neutral reagents such as dimethyl disulfide, $(CH_3SSCH_3)^{202}$ and dimethyl diselenide, $(CH_3SeSeCH_3)^{203}$ and others[204,205] also permit their identification. In general, distonic ions will predominantly show free radical type reactivity and so will act as radical traps. Therefore, the addition of $CH_3S^\bullet$ or $CH_3Se^\bullet$ to distonic ions is generally a major reaction product. In contrast, the conventional isomer will usually react by electron transfer from the reagant molecule.

Most of these experiments have been performed using FT-ICR mass spectrometers (Section 1.2), which allow ions to be trapped for various reaction times and temperatures. FT-ICR mass spectrometry also has the advantage, over other trapping instruments, of being able to perform exact mass measurements, which are very helpful when precise identification of the reaction products is necessary. Several reviews of the subject have been published.[206-208] Examples are given below.

The dissociative ionization of 1,4-dioxane (cy–OC(H)$_2$C(H)$_2$OC(H)$_2$C(H)$_2$–) produces, by loss of CH_2O, a species to which the connectivity of the distonic ion $^\bullet CH_2CH_2O^+CH_2$ has been assigned.[209] It corresponds to the C–C opened form of the trimethylene oxide radical cation (cy–C(H)$_2$C(H)$_2$C(H)$_2$O–$^{1+\bullet}$). Calculations at the RHF/4-31G//RHF/STO-3G level of theory predict that the distonic ion is stable and lies 10 kJ/mol lower in energy than ionized trimethylene oxide.[210] When the two isomers are reacted (reaction time = 300 ms) with the CH_3SSCH_3 molecule the following products and relative ratios are obtained:[202]

$^\bullet CH_2CH_2O^+CH_2 + CH_3SSCH_3$: $\rightarrow$ m/z 105 $[M + CH_3S^\bullet]$ 31

$\rightarrow$ m/z 94 $CH_3SSCH_3^{]+\bullet}$ 44

$\rightarrow$ m/z 75 $[M + CH_3S^\bullet - CH_2O]$ 22

$\rightarrow$ m/z 58 $^\bullet CH_2CH_2O^+CH_2$ 100

$cy\text{-}C(H)_2C(H)_2C(H)_2O\text{-}^{]+\bullet} + CH_3SSCH_3$ $\rightarrow$ m/z 94 $CH_3SSCH_3^{]+\bullet}$ 100

$\rightarrow$ m/z 58 $cy\text{-}C(H)_2C(H)_2C(H)_2O\text{-}^{]+\bullet}$ 26

The reaction products from the two isomers are clearly different, with the distonic ion showing a marked preference for the addition of a CH_3S- group (m/z 105 and m/z 75) whereas the conventional isomer exclusively undergoes charge exchange with the neutral reagent (m/z 94). It is noteworthy that a priori the distonic ion would not have been expected to react to such an extent by charge transfer (m/z 94). However, a recent reinvestigation of the nature of the m/z 58 ions generated from the dissociative ionization of 1,4-dioxane has clearly shown that a second isomeric ion, the methyl vinyl ether radical cation ($CH_3OCHCH_2^{]+\bullet}$), is cogenerated and accounts for approximately 7% of the m/z 58 ion beam.[211] Therefore this latter ion could be responsible, at least in part, for the observed charge transfer product generated from the distonic ion.

It is noteworthy that differentiation of distonic/conventional isomers based on their reaction with CH_3SSCH_3 is not always straightforward. For example, the reactions of the two α-distonic ions (charge and radical sites on adjacent atoms) $^\bullet CH_2^+PH_3$ and $^\bullet CH_2^+SH_2$ with CH_3SSCH_3 have also been investigated. Both distonic ions are generated by dissociative ionization (from ionized $HOCH_2CH_2SH$, by loss of $CH_2O^{212,213}$ and from ionized n-hexylphosphine, by loss of C_5H_{10}, respectively[214,215]) and ionization of the corresponding neutral molecules CH_3PH_2 and CH_3SH produces the conventional radical cations (for a complete discussion of these isomers refer to Part 2). For $^\bullet CH_2^+PH_3$, both the addition product $CH_3SCH_2^+PH_3$ and the charge transfer ion ($CH_3SSCH_3^{]+\bullet}$) are observed in a ratio of 50:50, whereas the conventional isomer $CH_3PH_2^{]+\bullet}$ yields the charge transfer ion exclusively. However, when $CH_3SeSeCH_3$, which is more reactive than CH_3SSCH_3, is used, the distonic ion reacts typically by the exclusive addition of CH_3Se- and the conventional ion only by charge transfer.[216] In contrast, when the nitrogen analogs are examined, the distonic ion reacts exclusively by charge transfer as does its conventional isomer, albeit at different rates.[217] Because the measured reaction rates for the charge transfer processes are different for the two isomers, isomerization before reaction can be ruled out. In order to account for the observations, the authors proposed that isomerization does not take place before reaction but does occur, intermolecularly, within the collision complex itself.[217]

1.6   USE OF COMPUTATIONAL CHEMISTRY IN ION STRUCTURE DETERMINATION

1.6.1   INTRODUCTION

The use of computational methods to aid in assigning ion structures has exploded in the past 10 years, as will be evident on examining Part 2 of this book. It is now an integral part of the exploration of new ion structures. One of the case studies presented herein (Section 2.2) describes the discovery of distonic ions that were first "observed" by calculation. This incident highlights one of the main strengths of theoretical methods: they are limited only by one's imagination and not by what one has available in the laboratory. Note that this can also be theory's greatest weakness. In general, for ion structure determination, theoretical methods provide two very important pieces of information: ion thermochemistry and transition state structures for the interconversion of isomeric ions. Together with the isomeric structures themselves, calculations can provide what no experiment can, a complete potential energy surface for an ion's decomposition and isomerization pathways. To do this it is necessary to engage in three basic computational exercises: geometry optimization of equilibrium and transition state structures, vibrational frequency analysis, and the determination of reliable thermochemical values for all species. Each of these will be discussed below.

1.6.2   COMPUTATIONAL METHODS

It is not our intention to give a complete description of the various computational methods, rather it is to outline the basic steps in choosing a method for ion structure determination. There are two main components to any computational method: a method for treating electron correlation and a basis set for describing molecular orbitals (MOs). Together they comprise a "level of theory." An example is MP2/6-31+G(d). The MP2 stands for Møller–Plesset 2,[218] a method for incorporating electron correlation using second-order perturbation theory and differs from the Hartree Fock (HF) level calculations, which ignore electron correlation. We have reproduced what has come to be known as the Pople diagram in the figure below.[219]

Pople Diagram

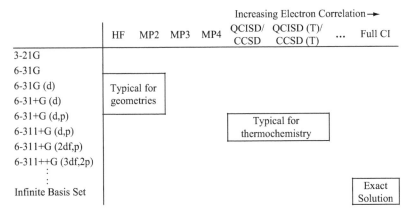

It ranks the various *ab initio* methods according to the complexity with which they incorporate electron correlation, with full configuration interaction (CI) as the ultimate, but usually unattainable, goal.

The 6–31+G(d) is the notation for a Pople basis set.

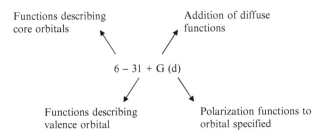

It is a split valence basis set in that the valence orbitals are divided into two parts described by different functions denoted by (31). Polarization (in the form of d-orbitals) and diffuse functions (+) have been added to non-H atoms to increase flexibility. Core orbitals are described by six functions. Each atom has a unique set of functions in this basis set and together they are combined to form MOs. Other basis sets have been developed, including those of Dunnings[220,221] and Stevens and coworkers.[222] A full description of basis sets available to the user can be found in the manual for the corresponding software package. The Pople diagram illustrates the increasing complexity of basis sets for those with Pople notation (it is straightforward for the reader to insert Dunning basis sets into this list), with an infinite basis set being the ultimate but yet again unattainable goal.

Another group of methods employs density functional theory (DFT).[223] Rather than use MOs to describe electrons and wavefunctions, DFT calculates the electron density. It was only relatively recently that practical implementations of DFT were introduced by the development of methods for treating electron correlation and electron exchange. Thus, most DFT methods used in ion structure determination are described by their correlation and exchange functionals: B-LYP uses Becke's exchange functional[224] and a correlation functional developed by Lee, Yang and Parr.[225] Probably the most widely used DFT methods are B3-LYP and B3-PW91, both of which use an exchange functional developed by Becke that is a hybrid between HF and DFT.[226] The great advantage of DFT over ab initio methods is speed. B3-LYP is intermediate in speed and computational resources between HF and MP2, yet has just as good a treatment for electron correlation as MP2. For this reason, DFT calculations have almost become the norm in ion structure determination but care must be taken, as will be discussed later under assessments. One drawback to DFT is that there is no well-defined relationship between the different methods as we have seen in the Pople diagram for ab initio methods. Each must be taken on its own and tested for reliability.

1.6.3   GEOMETRY OPTIMIZATION

For an ion structure to be significant for mass spectrometry or any other enterprise, it must represent a stationary point on the potential energy surface. In other words, it must represent either an equilibrium state (a minimum lying in a potential energy well) or a transition state (a first-order saddle point on the potential energy surface). The process of identifying such structures involves the optimization of bond lengths, angles and dihedral angles to minimize the energy of the species in question.

For transition states, all degrees of freedom are optimized save one, the degree responsible for motion over the barrier that connects one equilibrium structure to one other. The two minima that are connected by a transition state structure are confirmed by performing what is usually referred to as an "intrinsic reaction coordinate" (IRC) calculation that follows the negative eigenvalue (that characterizes a transition state) in both directions to find the corresponding minima. Sometimes it is possible to verify a TS just by examining the motion involved in the imaginary vibrational frequency that goes with the negative eigenvalue, but great care must be taken when doing this. Processes other than the simplest of reactions (such as H shift reactions) can have unpredictable final structures and IRC calculations need to be performed.

1.6.4   VIBRATIONAL FREQUENCIES

The calculation of vibrational frequencies serves two purposes: to verify that a species is either an equilibrium structure or a transition state and to calculate the zero-point vibrational energy (ZPE) for the optimized structure. The latter is needed in any determination of relative energies. These values are normally scaled to improve agreement with experiment.[227]

1.6.5   ION THERMOCHEMISTRY

The most fundamental information obtained for equilibrium and transition state species are their relative energies. All calculations will yield an absolute energy for a species but this absolute energy is so rife with errors from the assumptions made in the theoretical model that it is often valueless. However, the difference between the calculated absolute energies of two isomeric ions will be satisfactory due to the cancellation of these errors. For a given potential energy surface, every level of theory employed will produce a unique set of relative energies. It is up to the user to decide which are adequate for the task at hand (see Section 1.6.6). In general, reliable relative energies (and subsequently, heats of formation) require higher levels of theory than do geometry optimizations or vibrational frequency analyses. Indeed, the levels have to be so high that they are often not practical. To get around this problem, composite methods have been developed. A composite method uses several low level calculations to additively produce a result that approximates to a high level of theory.

An example of such a composite approach is the popular G3 method[228] developed by Curtiss and coworkers. G3 attempts to approximate the QCISD(T)/6–311+G(3df,2p) level of theory by making a series of additive corrections to a base MP4/6–31G(d) energy. After inclusion of a ZPE and a higher level correction (to account for residual basis set errors), the final G3 energy is generally thought to be accurate to within 8 kJ/mol. It is often better than this but, since the method was not optimized for transition states (for which no highly accurate experimental energies exist), it could be less accurate for these species. Other comparable composite methods include the CBS family of methods by Petersson and coworkers[229] and Martin's model chemistries.[230]

1.6.6   WHICH LEVEL OF THEORY TO USE AND RELIABILITY ASSESSMENT STEPS

This is perhaps the most asked question when it comes to ion structure determination. How can the best level of theory be selected? The answer is not straightforward, at least until it has been determined how the structures, vibrational frequencies and thermochemistry for a system change as the level of theory is altered. Therefore, one cannot arbitrarily select one level of theory without first confirming that it will give reliable results. The assessment of the reliability of computational approaches has several well-defined steps and can be time-consuming but will pay off in the end. After all, for example, when a new apparatus for measuring ionization energy (IE) values of molecules is constructed, it must first be tested on compounds with reliable IE data before venturing to previously unstudied compounds.

The reliability assessment steps are as follows:

1. Selection of a prototype chemical system. This is the key first step because it may be undesirable to do the calculations described later for the ion system of interest due to their large size. Find a suitably smaller ion or family of ions that are chemically equivalent to the target ion family. Thought should be given to potential isomeric structures that should be included in the prototype system.

2. Geometry assessment. In this step, the sensitivity of the various geometric parameters of the ionic system to level of theory is tested. The ions are optimized at HF, MP2 and B3-LYP, each with a wide range of basis sets, for example, 6-31G(d), 6-31+G(d), 6-31G(d,p), 6-311G(d), 6-311G(2d,p), 6-311+G(3df,2p). Each basis set differs from the previous one by only *one* factor, so in this way it can be determined if, for example, diffuse functions have a large influence on bond length or if going from double-split to triple-split valence orbitals is important. HF is chosen because of its speed, and MP2 and B3-LYP because they are very common for geometry optimizations. The next step is to select a modest basis set, say 6-31G(d), and optimize the ion(s) at HF, MP2, QCISD or CCSD, QCISD(T) or CCSD(T), B3-LYP and B3-PW91. This will show how important the electron correlation treatment is in defining good ion structures. One question remains: what is implied by the word "good"? Obviously, for gas-phase ions, there are very few experimentally determined structures in the literature, so it is necessary to rely on an examination of the "convergence" of theory. Assuming that QCISD(T) or CCSD(T) are the "best" treatments of electron correlation and that large basis sets such as 6-311+G(3df,2p) provide the "best" descriptions of MOs, those will be used for comparison. It can be helpful to make a plot of, for example, bond length vs basis set. Hopefully, the value of the bond length will come to a plateau, converging to the 6-311+G(3df,2p) value. What is sought is the lowest level of theory that comes close to this plateau. More than one can be selected and the final decision to be made is as described in Section 1.6.6 (5) later. A similar approach will be used in examining electron correlation treatments. Often, quite low levels of theory such as MP2/6-31+G(d) or B3-LYP/6-31G(d,p) yield excellent geometries, thus saving considerable time. As an example, consider the proton-bound HCN dimer, $(HCN)_2H^+$. The N–H bond in this pair was calculated at the following levels of theory.[231]

MP2/	N—H Bond Length (Å)
6-31G(d)	1.142
6-31+G(d)	1.132
6-311G(d)	1.267
6-311G(d,p)	1.260
6-311+G(d,p)	1.261
6-311G(df,p)	1.261
6-311G(2df,p)	1.262
6-311+G(2df,p)	1.262

Clearly, the bond converges to a value of 1.262 Å with large basis sets. In addition, diffuse functions decrease the bond length when using the 6-31G(d) basis set but have little effect with triple-split basis sets. Having identified an adequate basis set one can test the effect of the electron correlation method on the N–H bond length.

6-311+G(d,p)	N—H Bond Length (Å)
HF	1.083
MP2	1.261
B3-LYP	1.266
QCISD	1.183
CCSD	1.183
QCISD(T)	1.263

This limiting value of 1.262 Å (MP2/6-311+G(2df,p)) is close to the QCISD(T)/6-311+G(d,p) limit of 1.263 Å. The HF result is obviously poor, but the MP2 and B3-LYP treatments appear to give better agreement with the QCISD(T)/6-311+G(d,p) level of theory than QCISD and CCSD. This type of agreement may be fortuitous.

3. ZPE assessment. Vibrational frequencies would have been calculated for each of the geometries probed in Section 1.6.6 (2). Although there may be problems with the QCISD and CCSD frequencies, these levels are seldom needed to get good ZPE values. Note that these values have recommended scaling factors.[227] It is necessary to make sure that the ZPE values for the geometry (or geometries) that were selected in Section 1.6.6 (2) are consistent with the best values.

4. Relative energies. All of the above calculations also provide energies (and thus relative energies) for all the species in the prototype system. A comparison of all the relative energies for each level of theory (do not forget to include ZPE) will quickly show that some treatments are horrible, while others tend to agree.

5. Absolute thermochemistry. In this phase, the best relative energies and absolute heats of formation for the prototype system will be determined. Using one or several composite methods, the energies for each

of the suitable geometries that were found in Section 1.6.6 (2) are calculated. This will show how sensitive the final absolute $\Delta_f H^\circ$ is to the level of theory used in the geometry optimization. By comparing several composite methods for the same geometry, one can get a feel, for example, if G3 is chronically underestimating the heats of formation. Any comparisons with experimental data are valuable at this point. It may be useful to compare these relative energies to those from Section 1.6.6 (4). If heats of formation are not required it may be suffice to use a lower level of theory.

Again $(HCN)_2H^+$ provides an example. The $\Delta_f H^\circ$ for this proton-bound pair was calculated at the G2 level of theory using the optimized geometries listed below.

Geometry	G2 $\Delta_f H^\circ$
MP2/6-31G(d)	966.5
/6-31+G(d)	966.9
/6-311G(d,p)	964.9
/6-311+G(d,p)	965.0
/6-311+G(2df,p)	964.0
HF/6-311+G(d,p)	977.0
B3-LYP/6-311+G(d,p)	964.7
QCISD/6-311+G(d,p)	963.6
QCISD(T)/6-311+G(d,p)	964.2

Convergence is seen in the $\Delta_f H^\circ$ to ~964 kJ/mol and many of the geometries are suitable. So if $\Delta_f H^\circ$ values (and not accurate geometries) are needed, geometry optimization for this and related proton-bound pairs could potentially be done at the inexpensive MP2/6–31G(d) or B3-LYP/6–31G(d) levels of theory. The previously observed poor performance of HF theory also manifests itself in a poor G2 $\Delta_f H^\circ$ value.

Once the choice is made the target system can be attacked. This process will also indicate the accuracy consequences of lowering the level of theory further and be useful information for the future study of even larger, analogous systems. If the chemical make-up of the system is changed significantly a new assessment is required. All of these steps will lead to a level of theory in which one can have confidence, at least insofar as it will be close to the best theoretical value one can get.

Cross-Referenced Index
to Introduction and Chapter 1

2 What and What Not to Expect from Gas-Phase Ions

CONTENTS

2.1 INTRODUCTION

One of the challenges always faced by scientists is to "think outside the box." Often, what we learn in our formative years as chemists can limit our imagination and how we believe that our branch of natural science works. Fortunately, the barriers of tradition and of authority are not wholly unyielding and are always susceptible to the revelations of new and unexpected experimental results. This was certainly true of mass spectrometry, when for the better part of two decades, from 1965 to 1985, new experiments began to break the traditional links between it and solution chemistry.

The five examples described later have been chosen to illustrate this and to show how past problems in gas-phase ion structure determinations have been solved by experiment, by computational chemistry and by the appropriate reevaluation of existing data. They were also selected to show that even the simplest of systems are worthy of further experimentation and that the subject is still very likely to hold discoveries and surprises for future workers.

2.2 IDENTIFICATION OF A NEW CLASS OF IONS: DISTONIC IONS

Up until the early 1980s, mass spectra were for the most part interpreted using ion connectivities (the way in which the atoms are connected) that corresponded to neutral structures known to be stable. For example, when a fragmentation product ion in an organic mass spectrum appeared with m/z 32, it was quite reasonably assumed to have the connectivity of the methanol molecule. However, a number of workers, such as Morton and Beauchamp, began to consider unconventional ion structures that corresponded to protonated free radicals, with the proton adjacent to or farther removed from the radical site, as possible intermediates in ion–molecule reactions.[232] Bursey even used such ions to design ion–molecule reactions specific for nitriles.[233] However, it was not until Radom and coworkers used computational chemistry to explore the thermochemistry and interconversion of these protonated free radicals that the stability and likely ubiquity of these structures became apparent.[234]

They used the ideas of the above-mentioned authors to postulate the existence of the O-protonated $^\bullet CH_2OH$ radical, $^\bullet CH_2{}^+OH_2$, an isomer of $CH_3OH^{]+\bullet}$. Their calculations showed that the $^\bullet CH_2{}^+OH_2$ ion was thermodynamically stable and quite surprisingly, even more so than the conventional isomer, ionized methanol. The most recent calculations[235,236] predict that $^\bullet CH_2{}^+OH_2$ is 29 kJ/mol lower in energy than $CH_3OH^{]+\bullet}$. The calculated isomerization barrier between the two species (108 kJ/mol above $CH_3OH^{]+\bullet}$)[235,236] is sufficiently high to prevent them from freely interconverting in the ion source of a mass spectrometer. This barrier lies above the first dissociation limit of the conventional ion (loss of H$^\bullet$) and therefore both ions should likely be generated by experiment. Bouma, MacLeod and Radom indeed generated the $^\bullet CH_2{}^+OH_2$ ion by the dissociative ionization of ethylene glycol $HOCH_2CH_2OH^{]+\bullet}$ that loses CH_2O) and compared its collision-induced dissociation (CID) mass spectrum with that of $CH_3OH^{]+\bullet}$ (see Figure 2.1 and refer to Section 1.4.2 for discussion of collision-induced dissociation).[237]

The main distinguishing feature was a much greater m/z 14:15 ratio for $^\bullet CH_2{}^+OH_2$ than for $CH_3OH^{]+\bullet}$, consistent with their respective connectivities. The name distonic ion, from the Greek *diestos* and the Latin *distans* meaning separate, was given to those species in which the radical and ionic sites are located on different atoms. Note, however, that the $CH_3OH^{]+\bullet}$ ion produces peaks at m/z 18 and 19 in its collision-induced dissociation mass spectrum, indicating that some complex chemistry also accompanies the high energy fragmentations of this simple ion.

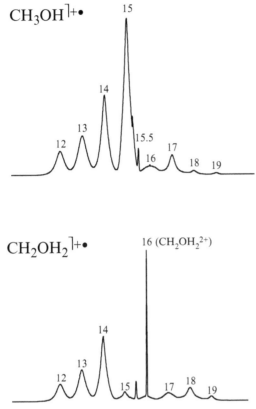

FIGURE 2.1 The collision-induced dissociation mass spectra of the conventional ion $CH_3OH^{+\bullet}$ and its distonic isomer $^{\bullet}CH_2{}^+OH_2$.

At the same time, the group of Holmes, Lossing, Terlouw and Burgers characterized a whole variety of distonic ions of general formula $^{\bullet}CH_2{}^+RH$ (R = NH$_2$, OH, SH, Cl and Br), particularly by their charge-stripping mass spectra (Figure 2.1; note the narrow peaks at m/z 15.5 and 16).[238] Generally, distonic ions[239] have a larger cross section for electron detachment than the corresponding conventional ions and so usually exhibit a pronounced doubly charged ion peak in their collision-induced dissociation mass spectra. This is certainly the case for the $CH_2OH_2{}^{2+}$ ion, which is isoelectronic with ethene and hence is a particularly stable species. Using energy-selected electrons from an electron monochromator and the appearance energy (AE) method (section on Experimental methods for determining appearance energies, in Chapter 1), they determined the heat of formation of the $^{\bullet}CH_2{}^+OH_2$ ion to be 816 ± 8 kJ/mol.[238] This places $^{\bullet}CH_2{}^+OH_2$ 29 kJ/mol lower in energy than $CH_3OH^{+\bullet}$ ($\Delta_f H = 845$ kJ/mol deduced from $\Delta_f H(CH_3OH) = -201$ kJ/mol and its ionization energy, $IE = 10.84 \pm 0.01$ eV),[56] which is in excellent agreement with the calculated relative energies of the isomers.[235,236]

Distonic ions have now been shown to be significant isomers of many ionized molecules and the majority of them are more stable than their conventional counterparts. The earlier work was comprehensively reviewed in 1988 by Hammerum.[239]

2.3 THE McLAFFERTY REARRANGEMENT: A CONCERTED OR STEPWISE PROCESS?

Probably the best-known ion decomposition pathway in the whole of mass spectrometry is the so-called McLafferty rearrangement.[240,241] This reaction involves a γ-hydrogen transfer, usually from an alkyl carbon to a keto-oxygen, followed by a β-C–C bond cleavage leading to the elimination of an olefin, as shown in the following example:

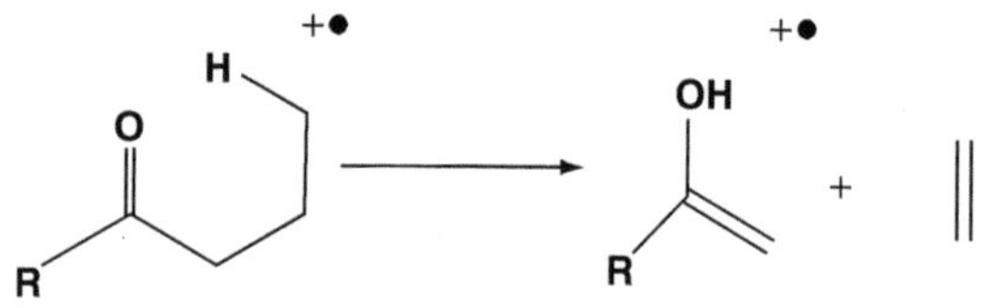

SCHEME 2.1

This type of reaction has a very long history, beginning with the development of the photochemistry of simple molecules containing a carbonyl group such as aldehydes, ketones and acids for example, and having at least one hydrogen atom γ- to the carbonyl group. In such photochemical studies, the reaction was known as the "Norrish type II" rearrangement.[242] Since its discovery in mass spectrometry in the 1950s and until quite recently, there was an ongoing debate about the nature of the mechanism responsible for the olefin loss: Was it a *concerted* process or did it take place in a *stepwise* fashion involving a more-than-transient or even stable intermediate? These rival processes are illustrated in Scheme 2.2.

This debate resulted in many investigators devising elegant mass spectrometric experiments to try to distinguish between these two mechanisms and if appropriate, to determine the nature of the intermediate in the stepwise reaction. This section is not a complete review of the topic and some significant work in the area may have therefore been omitted. Our aim is to highlight the key experiments that contributed to the definition of the ion structures involved in the McLafferty rearrangement.

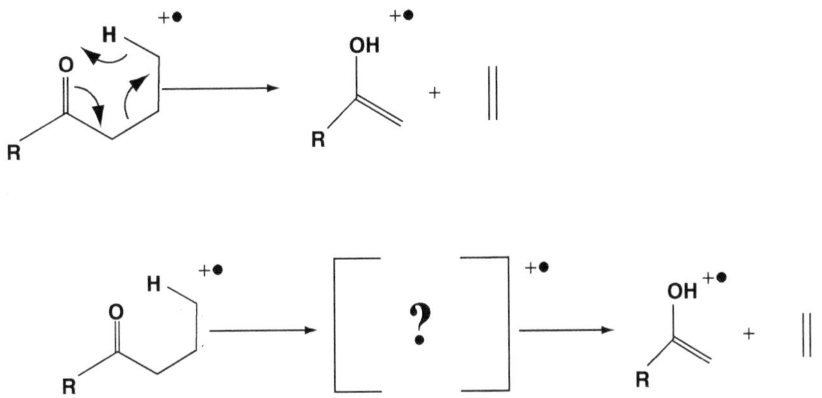

SCHEME 2.2

The stepwise nature of the McLafferty rearrangement was proven most convincingly by deuterium labeling studies. In the concerted process shown in Scheme 2.2, when the γ-hydrogen is replaced by deuterium, the resulting ion must contain the label atom, $RC(OD)=CH_2^{1+\bullet}$. Indeed, for the fast reactions taking place in the ion source of a mass spectrometer (those having rate constants $>10^{11}$ s^{-1}) or species observed on the picosecond timescale (field ionization kinetics) this is usually what is observed.[243] However, when the reaction is allowed to proceed on a longer timescale, which is the case for the metastable ions (MI) studied in a double-focusing (tandem) mass spectrometer (refer to Section 1.4.1 for discussion of metastable ions), both product ions, $RC(OD)=CH_2^{1+\bullet}$ and $RC(OH)=CH_2^{1+\bullet}$ are observed.[243–245] For example, in ionized butanoic acid ($R=OH$), the label exchange took place between the hydroxyl and the β- and γ-hydrogens.[246–248] For this exchange of hydrogen and deuterium to occur there must be at least one intermediate ion structure in the mechanism to allow reversibility in the initial hydrogen transfer step. The nature of this intermediate was originally postulated by Djerrasi in a study of isotopically labeled aliphatic ketones.[249] Referred to as a "protonated rearrangement species"[249] or "radical intermediate,"[243] it was an excellent example of what is now called a distonic ion.

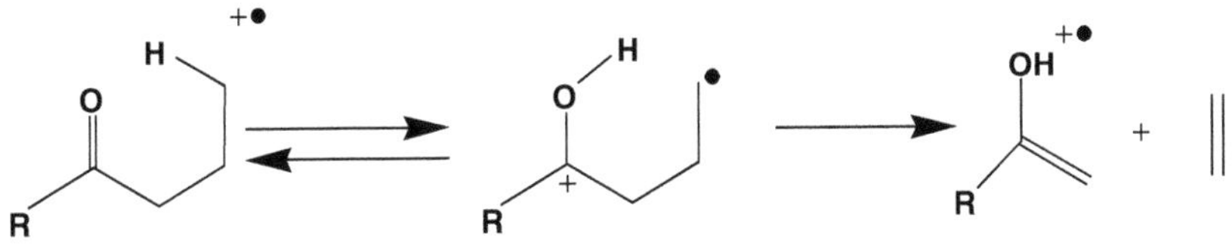

SCHEME 2.3

A definitive isotopic labeling study was performed in 1992 by the groups of Bowie and Derrick[250] using both ^{13}C and D labeling in a wide variety of aromatic and aliphatic ketones, all of which showed a McLafferty rearrangement involving the loss of ethene. Primary and secondary deuterium isotope effects at the γ- and β-positions, respectively, and large ^{13}C effects at the β-positions and for the γ-sites in 3-ethylpentan-2-one and heptan-4-one as well as confirmatory double isotope effects, all showed that the reaction must involve a stepwise mechanism and therefore involve an intermediate distonic ion.

In general, distonic ions such as the McLafferty intermediates can be produced independently from an appropriate ionized cyclic alkanol that undergoes a reversible α-cleavage upon ionization, yielding the distonic ion.[251]

It is worth noting that the intermediate distonic ions generally have lower heats of formation than their keto precursors. Several such distonic ions and therefore McLafferty rearrangement intermediates, have been generated and characterized and their heats of formation are known from either experiment or theory. However, in cases where the heat of formation of a distonic ion is unknown, the general scheme described below allows this quantity to be satisfactorily estimated. For example, the McLafferty rearrangement of methyl propyl ketone involves the $CH_3{}^+C(OH)CH_2CH_2CH_2{}^\bullet$ distonic ion intermediate. In order to estimate its heat of formation, this distonic ion can be considered as resulting from the following two reactions:

$$CH_3C(O)CH_2CH_2CH_3 + H^+ \rightarrow CH_3{}^+C(OH)CH_2CH_2CH_3 \quad (2.1)$$

$$CH_3{}^+C(OH)CH_2CH_2CH_3 \rightarrow CH_3{}^+C(OH)CH_2CH_2CH_2{}^\bullet + H^\bullet \quad (2.2)$$

Thus the heat of formation of the distonic ion can be estimated from the sum of the corresponding heats of reactions, where $\Delta H_{reaction} = \Delta H_r = \Sigma \Delta_f H$ (products) $- \Sigma \Delta_f H$(reactants):

$$\Delta H_r(1) = -PA = \Delta_f H(CH_3{}^+C(OH)(CH_2)_2CH_3) - [\Delta_f H(CH_3C(O)(CH_2)_2CH_3) + \Delta_f H(H^+)]$$

$$\Delta H_r(2) = BDE = \Delta_f H(CH_3{}^+C(OH)(CH_2)_2CH_2{}^\bullet) + \Delta_f H(H^\bullet) - \Delta_f H(CH_3{}^+C(OH)(CH_2)_2CH_3)$$

$$\Delta_f H(CH_3{}^+C(OH)(CH_2)_2CH_2{}^\bullet) = BDE - PA - \Delta_f H(H^\bullet) + \Delta_f H(H^+) + \Delta_f H(CH_3C(O)(CH_2)_2CH_3),$$

where PA refers to the proton affinity of methyl propyl ketone (832.7 kJ/mol[252]) and BDE to the terminal C–H homolytic bond dissociation energy. The latter value is not affected by the protonation, the charge site being remote from the C–H bond to be broken, and therefore is that for a typical primary C–H bond in molecular species (421 kJ/mol[253]) (see also Section 1.3.10, where bond dissociation energies in ions are discussed). The ancillary thermochemical data are $\Delta_f H(H^\bullet) = 218$ kJ/mol, $\Delta_f H(H^+) = 1530$ kJ/mol, and $\Delta_f H(CH_3C(O)CH_2CH_2CH_3) = -259$ kJ/mol.[56] From this, $\Delta_f H(CH_3{}^+C(OH)CH_2CH_2CH_2{}^\bullet) = 641$ kJ/mol and as expected for distonic ions, is slightly but significantly lower in energy than that of the conventional isomer, $\Delta_f H(CH_3C(O)CH_2CH_2CH_3]^{+\bullet}) = 646$ kJ/mol, $(\Delta_f H(M^{+\bullet})) =$ ionization energy $+ \Delta_f H(M)$; $IE(CH_3C(O)CH_2CH_2CH_3) = 9.38 \pm 0.06$ eV and $\Delta_f H(CH_3C(O)CH_2CH_2CH_3) = -259.1 \pm 1.1$ kJ/mol).[56]

2.4 PEPTIDE ION FRAGMENTATION: THE STRUCTURE OF b-TYPE IONS

One of the most significant recent developments in mass spectrometry has been the advent of electrospray ionization[254] (ESI) and matrix-assisted laser desorption–ionization[25,255] (MALDI). These two techniques provide an efficient means to ionize (by protonation) large nonvolatile molecules and therefore allow mass spectrometry to be applied to biological samples such as peptide fragments from protein digests. In particular, the collision-induced dissociation (refer to Section 1.4.2 for a discussion on collision-induced dissociation) of mass-selected peptide ions usually results in a series of fragment ion peaks that correspond to the sequence of amino acids in the molecule. Therefore, mass spectrometry has become a powerful and widely used technique in proteomics, the sequencing of peptides and proteins.

As can be seen in Scheme 2.4, fragment ions generated by the dissociation of ionized peptides can contain either the N-terminal or the C-terminal amino acid. To identify the fragment ions, Roepstorff and Fohlman[256] and Johnson et al.[257,258] devised a notation, which is shown in Scheme 2.4, classifying these ions based on where the cleavage takes place in the ionized peptide's backbone. Note that the subscript n indicates the number of amino acids contained in the fragment ion.

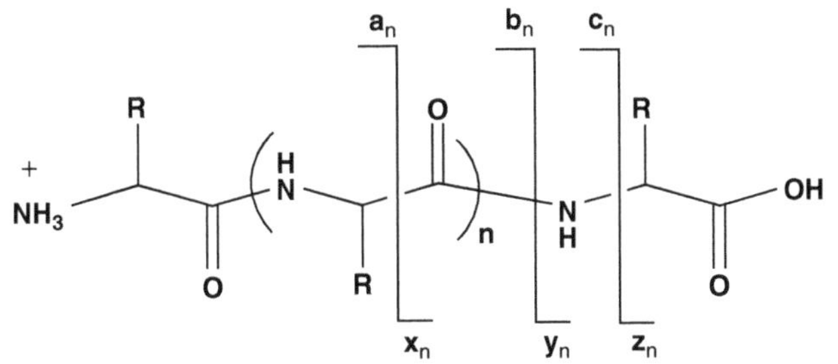

SCHEME 2.4

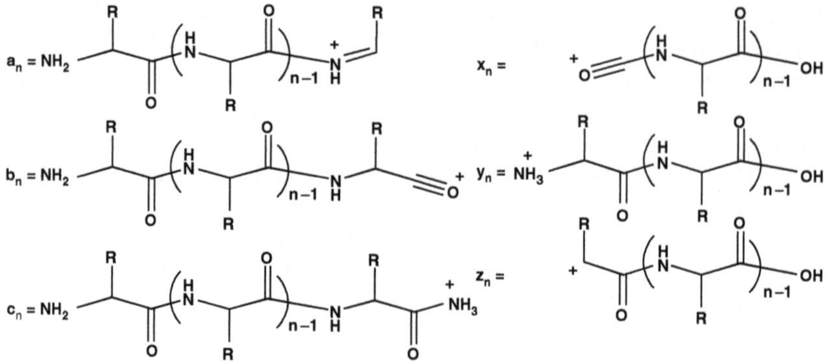

$a_n = NH_2$

$x_n =$

$b_n = NH_2$

$+ \ y_n = NH_3$

$c_n = NH_2$

$+ \ z_n =$

$+ \ NH_3$

SCHEME 2.5

On the basis of the atom connectivity in the precursor peptide, each fragment ion was initially assumed to have the structures shown in Scheme 2.5.

However, conflicting experimental evidence has led to a necessary revision of some of these structures, as will be described in the following example.

As shown earlier, the b-type ions were originally believed to have the acylium structure.[256,259,260] Acylium ions are ubiquitous in the mass spectra of protonated carboxylic acids and their methyl esters, being produced by the loss of water or a methanol molecule respectively,[261–266] and so to assume that these ions were generated in the dissociation of peptides was quite reasonable. It is common to choose smaller compounds as models when studying peptides and other large systems. For example, amino acids have been investigated initially and/or in parallel in the hope that they provide information that can be extended to larger systems. Therefore, the dissociations that yield b-type acylium ions from protonated amino acids, as well as those from larger di- and tripeptides, were investigated.

It was observed that when metastable protonated amino acids $(NH_2-CH(R)-CO_2H)H^+$ dissociate (refer to Section 1.4.1 for a discussion on the characteristics of metastable ions), the fragmentation products observed are $RCH=NH_2^{1+}+CO+H_2O$ and never the simple b-type acylium ion $NH_2CH(R)^+CO$ that would have resulted from the loss of H_2O alone.[262,267] The above dissociation is accompanied by a large kinetic energy release, as shown by $T_{0.5}$ values of typically 360–390 meV.[267] This observation strongly indicates that dissociation by loss of CO and H_2O takes place over a high reverse energy barrier, resulting in the partitioning of the large excess energy into the translational degrees of freedom of the departing fragments (refer to Physicochemical aspects of metastable ion dissociations in Section 1.4.1 for a discussion on kinetic energy release). When the system is extended to di- and tripeptides the situation changes and stable b-type fragment ions are now observed.[256,268–271] In addition, Harrison particularly noted that the dissociation of metastable di- and tripeptides to form b-type ions was accompanied by only a small kinetic energy release ($T_{0.5}$ values of 20–30 meV) typical of simple bond cleavages.[267] However, when the b-type ions formed as fragmentation products of peptides, nominally acylium ions, are mass selected and allowed to undergo metastable dissociation, loss of CO is observed and the kinetic energy release associated with it is large ($T_{0.5}$ values of 340–570 meV).[267]

It was clear from these results that the dissociation characteristics of metastable b-type ions derived from the di- and tripeptides were similar to those of the protonated amino acids. Consequently, these ions could not have the acylium ion structure because, as seen in the investigation of the protonated amino acids, these were found to be kinetically unobservable. This showed that at best the acylium ion structure is a high energy intermediate on the potential energy surface (the reacting configuration leading to the exothermic loss of CO) and that stable b-type ions must have an alternative structure.[267]

An alternative structure might be formed, if ring closure takes place through the interaction of the carbonyl and the terminal amino group. Harrison et al.[267] tested this hypothesis by comparing the mass spectrum of the b-type ion derived from protonated Gly-Leu-Gly and of the analog amide Gly-Leu-NH$_2$ with that of the protonated diketopiperazine (cy-[Leu-Gly]H$^+$), which would result from such a proposed cyclization in Gly-Leu-Gly and Gly-Leu-NH$_2$.

The mass spectra are completely different (Figure 2.2), thus confirming that b-type ions resulting from the dissociation of the protonated peptides do not have the above ion structure. This conclusion was also reached by Wesdemiotis and coworkers from a related experiment.[272]

Another possible alternative ion structure was a five-membered ring isomer resulting from the interaction of the acylium carbon with the carbonyl group, one amino acid residue away, to form an oxazolone (Scheme 2.6).[272,273]

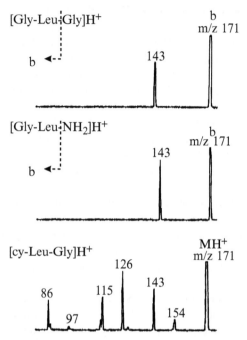

FIGURE 2.2 The collision-induced dissociation mass spectra of the b-type ions generated from protonated Gly-Leu-Gly and the analog amide Gly-Leu-NH$_2$ with that of protonated diketopiperazine ([cy-Leu-Gly]H$^+$).

To verify this hypothesis, the collision-induced dissociation mass spectrum of protonated 2-phenyl-5-oxazolone (a) was compared with that of the b-type ion from the dissociation of C$_6$H$_5$C(O)GlyGly-OH (b) (Scheme 2.7):[267]

The resulting mass spectra are virtually identical (Figure 2.3). Moreover, the two metastable ions exhibited broad, dished-topped peaks resulting from CO loss and having similar kinetic energy release values of 340 and 310 meV for the protonated oxazolone and the b-type ion, respectively.

Harrison has extended these results and showed that the oxazolone structure for the b-type ions is indeed general.[274] Calculations[267,274] at the HF level of theory showed the acylium ion structure to be a transition state on the potential energy surface, lying high above the dissociation products, fully consistent with the large kinetic energy release values observed for the dissociation.

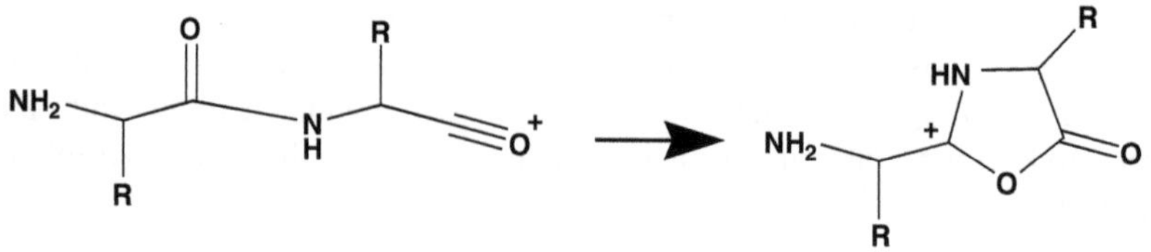

SCHEME 2.6

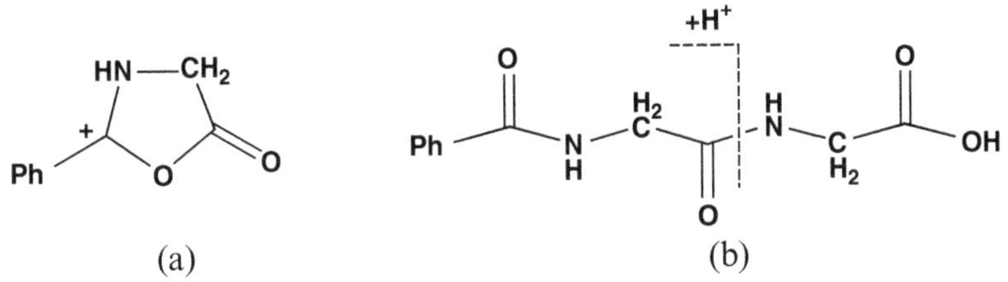

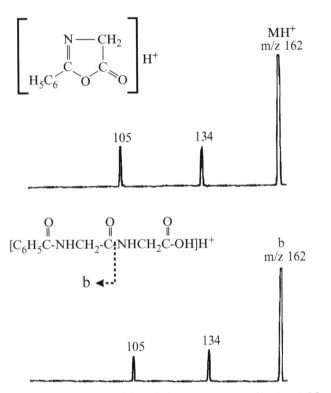

SCHEME 2.7

FIGURE 2.3 The collision-induced dissociation mass spectra of protonated 2-phenyl-5-oxazolone (top) and the b-type ion obtained from the dissociation of protonated $C_6H_5C(O)GlyGlyOH$ (bottom).

2.5 THE NEED FOR COMPUTATIONAL CHEMISTRY: THE METHYLACETATE ION STORY

Keto–enol isomerization is a topic that has garnered considerable interest in both solution and gas-phase studies. Again however, there are some general aspects of keto–enol tautomers that are unique to gas-phase ions. For example, neutral acetone is considerably more thermodynamically stable than its enol isomer, propen-2-ol,[275] the two being separated by a high energy barrier for the 1,3-hydrogen shift.[276] In contrast, the relative stability of the corresponding radical cations is reversed, the enol being markedly lower in energy,[277] and a high energy barrier for the 1,3-hydrogen shift prevents ionized acetone from tautomerizing to $CH_2 = C(OH)CH_3^{]+\bullet}$ at energies below the ion's dissociation by losses of $CH_3^{]\bullet}$ and $CH_4.^{276}$ The general topic of keto–enol tautomers has been reviewed by Bouchoux.[277]

The tautomerization of ionized methyl acetate to its enol $CH_2 = C(OH)OCH_3^{]+\bullet}$ has generated considerable interest. To investigate mechanism and structure, mass spectrometry relies heavily on observations of ion dissociations and so this tautomerization was initially probed by examining the fragmentation behavior of the two isomeric ions, namely, $CH_3C(O)OCH_3^{]+\bullet}$ and $CH_2 = C(OH)OCH_3^{]+\bullet}$.

The main dissociation channel for both metastable ions (metastable ions possess narrow range of internal energies close to that of the dissociation limit, refer to Section 1.4.1 for a discussion of MI characteristics of metastable ions) is the formation of the acetyl cation, $CH_3CO^{]+}$, and a $[C,H_3,O]^\bullet$ radical. The kinetic energy release associated with this dissociation (unlike acetone above) is identical for both isomers,[278] thus proving that the two isomers can freely interconvert before their dissociation on the microsecond timescale. However, the collision-induced dissociation mass spectra (refer to Section 1.4.2 for a discussion of collision-induced dissociation) of the two ions are distinct, showing that low energy ions retain their structure and hence the two isomers represent discrete minima on the potential energy surface. Both these observations are consistent with the tautomerization of the two structures proceeding via a 1,3-hydrogen shift (Scheme 2.8) but close to the dissociation limit.

If a specific 1,3-hydrogen shift is indeed the mechanism by which tauto-merization takes place, substitution of the hydrogen atoms on the methoxy group with deuterium in either ion ($CH_3C(O)OCD_3^{]+\bullet}$ and $CH_2 = C(OH)OCD_3^{]+\bullet}$) should result in the loss of *only* $[C,D_3,O]^\bullet$ radicals. However, significant H/D scrambling was observed in the dissociation of these labeled ions, negating the simple 1,3-H shift mechanism. The interconversion of the two ions was then proposed instead to take place via two sequential 1,4-hydrogen shift reactions, from ester methyl to keto-oxygen and then from acid methyl to methylene (see later), rather than the more direct 1,3-hydrogen shift route.[278] The intermediate was assigned to be the distonic ion A shown in Scheme 2.9.

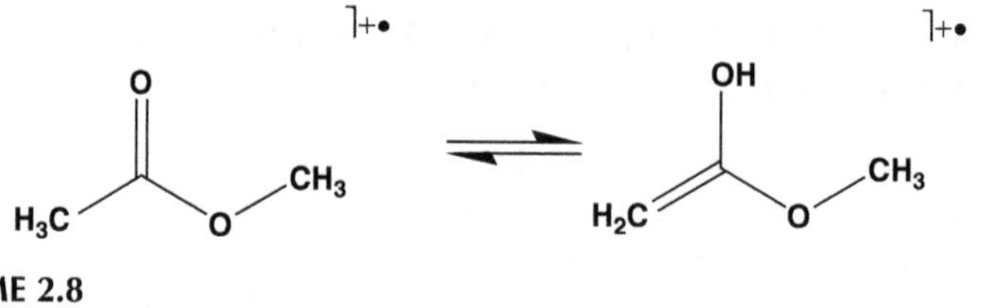

SCHEME 2.8

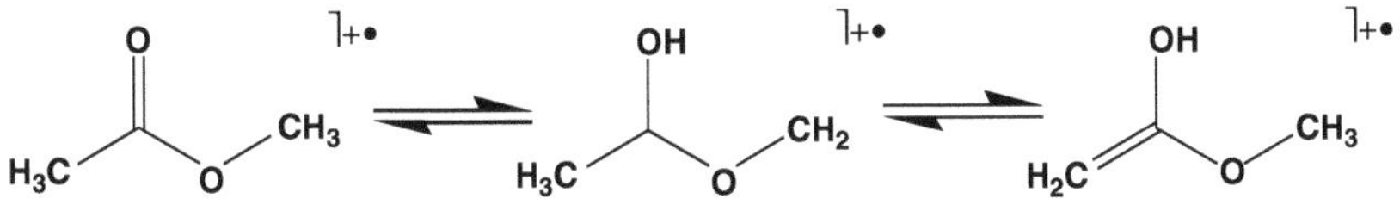

A

SCHEME 2.9

This new mechanism nicely accounted for the mixing of H/D atoms in the two methyl groups. Wesdemiotis et al.[279] were able to independently prepare the distonic intermediate A and show that it was a distinct species on the potential energy surface.

It is tempting at this point to believe that the story is complete. However, further experiments investigating the neutral $[C,H_3,O]^{\bullet}$ radicals co-generated from these ions together with $CH_3CO^{\uparrow+}$ provided an alarming complication. Rather than generating exclusively the methoxy radical, $CH_3O^{\bullet}$, collision-induced dissociative ionization (CIDI) mass spectrometry, an experiment which allows characterization of the neutrals cogenerated in metastable dissociations (refer to Section 1.4.2 for a discussion on collision-induced dissociative ionization), showed that ionized methylacetate produced a mixture of $CH_3O^{\bullet}$ and $^{\bullet}CH_2OH$ radicals, with different research groups finding different proportions of the two isomers.[280–283] To form the $^{\bullet}CH_2OH$ radical, further isomerization is necessary. A second intermediate distonic ion, intermediate B, was next proposed to be the reacting configuration for the dissociation that produced the hydroxy-methyl radical.[284]

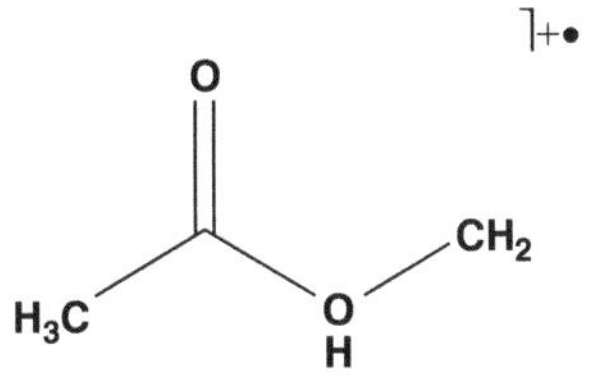

B

Breaking the central bond would yield the $^{\bullet}CH_2OH$ radical. The next problem was that this ion proved to be difficult to make independently and its participation in the ion chemistry therefore remained speculative. The issue was only resolved by computational chemistry.[285] Calculations at the HF/3–21G* level of theory showed that ionized methylacetate, the distonic ion A and the enol ion $CH_2=C(OH)OCH_3^{\uparrow+\bullet}$ can indeed interconvert on the microsecond timescale. However, the $^{\bullet}CH_2OH$ loss was shown to occur only from A by first forming the ion–molecule complex C and not the closely related ion B.

$$[H_3CCO \cdots H \cdots OCH_2]^{+\bullet}$$
$$C$$

Ion C then dissociates via an avoided crossing with an excited state having the correct electron distribution.[285] A recent threshold photoelectron–photoion coincidence (TPEPICO) study of the dissociation kinetics of ionized methylacetate has confirmed the overall mechanism and has also established that the distonic isomer B does not participate in the loss of $^\bullet CH_2OH$.[283] The present potential energy surface[283] is shown in Scheme 2.10. Relative energies are in kJ/mol.

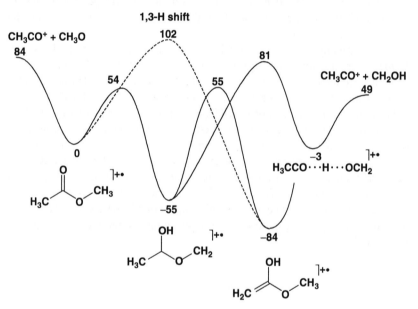

SCHEME 2.10

The chemistry of ionized methyl acetate is thus much richer than originally believed and this example highlights the surprising complexity of hidden ion dissociation mechanisms. It provides an excellent illustration when experiment alone was unable to provide a complete picture of the ion rearrangements that precede dissociation. Rather, it was only by including computations that a full understanding was obtained.

2.6   A PROBLEM WITH NEUTRAL THERMOCHEMISTRY: $\Delta_f H(CH_3CH_2OOH)$ REVISITED

This example arises from our review of the thermochemistry of neutral molecules and ions containing the peroxy function (ROOR′) and it was in the discussion of the ion^+CH_2OOH that the difficulty first arose (refer to $[C,H_3,O_2]^+$ ions in Part 2). In the evaluation of $\Delta_f H(^+CH_2OOH)$ it was necessary to have a reliable $\Delta_f H$ value for its precursor molecule, CH_3CH_2OOH, and so the available data had to be critically assessed. This molecule is a good example of how care must be taken before uncritically accepting $\Delta_f H$ values from major reference works and emphasizes that thermochemical data such as heats of formation of ions, neutral molecules, and radicals should never be considered as independent values; they are all interrelated. This interdependency allows one to verify whether or not a given thermochemical datum is reasonable.

By combining the measured AE value for the generation of the $^+CH_2OOH$ ion from CH_3CH_2OOH with the NIST WebBook[56] values for $\Delta_f H(CH_3CH_2OOH) = -210$ kJ/mol and $\Delta_f H(CH_3^{1\bullet}) = 146 \pm 1$ kJ/mol, we get $\Delta_f H(^+CH_2OOH) = 744$ kJ/mol. However, this value is considerably lower than that predicted by theory[286,287] (782 kJ/mol) and obtained from reactivity experiments[288] (774 ± 10 kJ/mol).

The reliability of the heat of formation of an ion necessarily depends on the accuracy of both the AE measurement and the ancillary thermochemical data for the neutral species (i.e., in the present example, $\Delta_f H(CH_3^{1\bullet}) = 146 \pm 1$ kJ/mol[56] and that chosen for $\Delta_f H(CH_3CH_2OOH)$). The original data for the AE measurement were reviewed by the present authors and deemed to be satisfactory. There is no doubt as to the accuracy of $\Delta_f H(CH_3^{1\bullet})$. Therefore, the NIST WebBook value for $\Delta_f H(CH_3CH_2OOH) = -210$ kJ/mol deserves to be reconsidered.

The calorimetry experiment that yielded this result dates back to 1940[289] and in the original paper the value given was -199 kJ/mol, which is also the value given in the Pedley collection.[290] No explanation for the alteration of this value to -210 kJ/mol is given in the NIST compilation. Moreover, a markedly lower value of -174 ± 6 kJ/mol was obtained for $\Delta_f H(CH_3CH_2OOH)$ from computational chemistry.[291]

Using existing reliable thermochemical data it is possible to estimate a good value for $\Delta_f H(CH_3CH_2OOH)$. The first step is to identify molecules of known heat of formation in which a functional group can be replaced to give the species of interest. For example, replacing one $-OCH_2CH_3$ group in the molecule $CH_3CH_2OOCH_2CH_3$ by $-OH$ yields CH_3CH_2OOH. Similarly, the desired species is also obtained by consecutively replacing one $-OCH_3$ by $-OH$ and the other by $-OCH_2CH_3$ in the molecule CH_3OOCH_3 or by replacing $tert$-C_4H_9O- by CH_3CH_2O- in $tert$-C_4H_9OOH. Because the heat of formation of an organic molecule is an additive property,[58] each functional group therein contributes a characteristic increment to the overall heat of formation. The second step is to determine the changes in molecular $\Delta_f H$ values resulting from the above substitutions. This is achieved by comparing molecules of similar structure having a reliably known heat of formation in which the same substitution is made. The principle of additivity then allows the simple arithmetic to be done with confidence. Thus for example, replacing CH_3CH_2O- by $-OH$ gives the following:

$\Delta_f H(CH_3OCH_2CH_3) = -216\,kJ/mol$
$\Delta_f H(CH_3OH) = -201\;kJ/mol$ $\Delta\Delta_f H = +15\,kJ/mol$

$\Delta_f H(CH_3CH_2OCH_2CH_3) = -253\,kJ/mol$
$\Delta_f H(CH_3CH_2OH) = -235\;kJ/mol$ $\Delta\Delta_f H = +18\,kJ/mol$

$\Delta_f H(CH_3CH_2CH_2OCH_2CH_3) = -272\,kJ/mol$
$\Delta_f H(CH_3CH_2CH_2OH) = -256\,kJ/mol$ $\Delta\Delta_f H = +16\,kJ/mol$
         $\Delta\Delta_f H_{ave} = +16\,kJ/mol$

$\therefore$	$CH_3CH_2OOCH_2CH_3$	$\rightarrow$	CH_3CH_2OOH
$\Delta_f H =$	-193	$+16$	$-177\,kJ/mol$

By analogy, the changes in $\Delta_f H$ resulting from replacing CH_3O- by $-OH$, CH_3O- by CH_3CH_2O- and $tert\text{-}C_4H_9O-$ by CH_3CH_2O- are estimated to be -17, -33 and $+76$ kJ/mol, respectively, and therefore:

$\therefore$	CH_3OOCH_3	$\rightarrow$	CH_3OOH	$\rightarrow$	CH_3CH_2OOH
$\Delta_f H =$	-126	-17	-143	-33	$-177\,kJ/mol$
	$tert\text{-}C_4HOOH$		$\rightarrow$		CH_3CH_2OOH
$\Delta_f H =$	-246		$+76$		$-177\,kJ/mol$

Combining the above three estimates gives $\Delta_f H(CH_3CH_2OOH) = -175 \pm 4$ kJ/mol. This value is now in good agreement with the calculations[291] and is justified by the above thermochemical arguments, indicating that the NIST value must be revised and that the original experimental results were flawed.

Note that using the latter heat of formation and the following ancillary thermochemical data[253] we can also calculate the homolytic O–O bond dissociation energy, another valuable thermochemical check:

$D(CH_3CH_2O - OH) = \Delta_f H(CH_3CH_2O^\bullet) + \Delta_f H(^\bullet OH) - \Delta_f H(CH_3CH_2OOH)$
$= \quad\quad -16 \quad\quad\quad\quad 39 \quad\quad\quad\quad -175$

This bond strength, 198 kJ/mol, is in good agreement with the current BDE values for alkyl hydroperoxides, e.g., $D(CH_3O–OH) = 195 \pm 6$ kJ/mol,[253] and provides additional support for the above estimate and again illustrates that the NIST $\Delta_f H$ value (-210 kJ/mol) is unsatisfactory. Finally, using this estimated datum together with the AE result gives $\Delta_f H(^+CH_2OOH) = 779$ kJ/mol, which is in satisfactory agreement with the computed[286,287] and reactivity[288] results.

2.7 SOME ADVICE WHEN ASSIGNING STRUCTURES TO GAS-PHASE IONS

This section of the book contains advice for realizing the full potential of the methods described in the foregoing chapters. The experiments and the significance of the observations derived therefrom are summarized by means of the four "strategy diagrams" that begin this section.

2.7.1 STRATEGY DIAGRAMS FOR EXPERIMENTAL WORK

These four diagrams provide a shorthand description of the experiments used to establish and identify ion structures and the way in which the experimental results are interrelated. It is assumed that the experimenter has access to a three-sector mass spectrometer of the kind shown in Figure 1.18.

The first diagram (Figure 2.4) is for thermochemical data and the use of metastable ion phenomena (or their absence) to aid the interpretation of ionization and appearance energy (AE) measurements for establishing values for $\Delta_f H$(ion).

The second diagram (Figure 2.5) explains how metastable ion phenomena can be used as a comparative technique for the identification of ion structures and also indicates the particular limitations of the method. The use of isotopic labeling can be essential, in that it immediately shows the effect of "atom scrambling," i.e., the complete loss of positional identity of the label atoms before the ion's fragmentations of lowest energy requirement. The link between metastable ion data and ion thermochemistry is indicated.

The third diagram (Figure 2.6) shows how the presence of a composite metastable ion peak and isotopic labeling experiments may assist in the unraveling of ion fragmentation mechanisms. The ancillary use of collision-induced dissociation mass spectra is indicated.

The fourth diagram (Figure 2.7) shows how collision-induced dissociation mass spectra contribute to the establishing of ion structures. The connections with ion thermochemistry and with metastable ion and related data from the other diagrams are shown. Note that the collision-induced dissociation MS of metastably produced ions may be performed in either the second field-free region (transmit m* using the magnet) or the third field-free region (sector voltage selection).

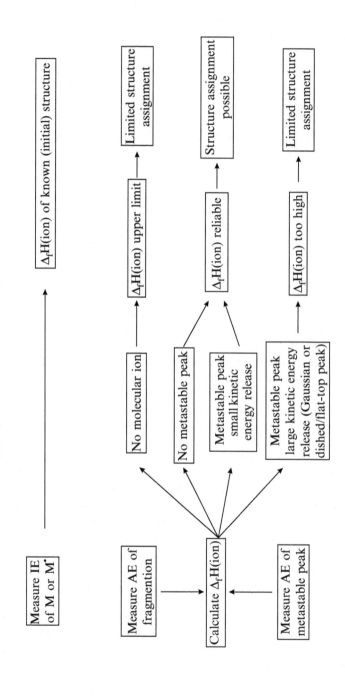

EXPERIMENTS   ANCILLIARY OBSERVATIONS   CONCLUSIONS

Measure IE of M or M' → $\Delta_f H$(ion) of known (initial) structure

Measure AE of fragmention → Calculate $\Delta_f H$(ion) ← Measure AE of metastable peak

No molecular ion → $\Delta_f H$(ion) upper limit → Limited structure assignment

No metastable peak

Metastable peak small kinetic energy release → $\Delta_f H$(ion) reliable → Structure assignment possible

Metastable peak large kinetic energy release (Gaussian or dished/flat-top peak) → $\Delta_f H$(ion) too high → Limited structure assignment

Ionization energy = IE= $\Delta_f H$(ion) + $\Delta_f H$(neutral precursor)
Appearance energy = AE = $\Delta_f H$(ion) + $\Delta_f H$(neutral fragment) − $\Delta_f H$(neutral precursor)

FIGURE 2.4 Strategy diagram for thermochemical data.[135]

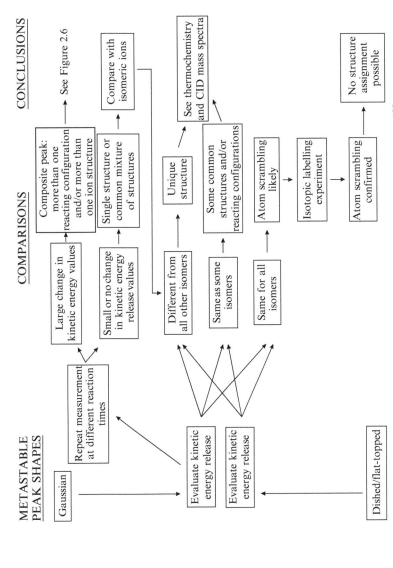

FIGURE 2.5 Strategy diagram for metastable ion phenomena: Gaussian and dished/flat-topped peaks.[135]

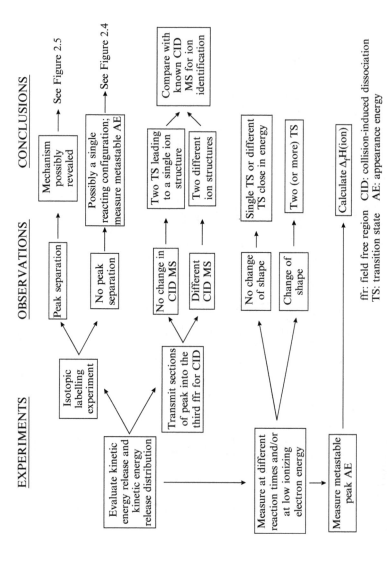

FIGURE 2.6 Strategy diagram for metastable ion phenomena: composite peaks.[135]

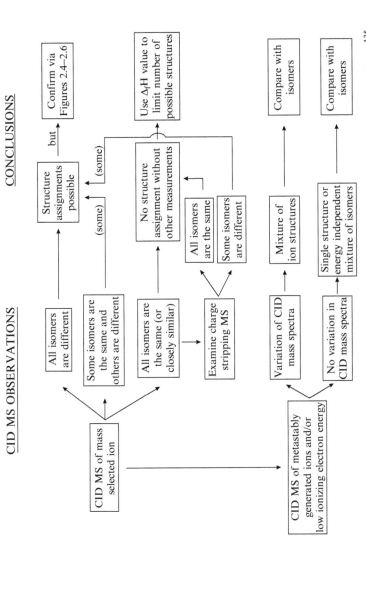

FIGURE 2.7 Strategy diagram for collision-induced dissociation mass spectra (CID MS) of isomeric mass-selected ions.[135]

2.7.2 SOME GENERAL CAVEATS ABOUT SCIENTIFIC PITFALLS OF WHICH THE READER MUST BE AWARE

(1) *Thermochemistry*: A particular number given in reference works may come from a good experimental result, but might be flawed by the use of outdated or inaccurate ancillary data. Such corrections are often not performed when, years later, the data are assembled for publication in a book. Scientists should be alert to this possibility, especially if the assigned value appears to be difficult to reconcile with data for chemically related species. In this book we have tried to live up to this standard, but we cannot guarantee all the data.

(2) *Metastable ion mass spectra*: The reader should recognize that published metastable ion MS may contain peaks arising not from metastable ions but from collision-induced dissociation processes that have a large cross section. This occurs when the pressure in the field-free region is too high because of inadequate pumping. An additional clue is if the results include an ion showing more than three competing metastable ion processes; this can happen but it is rare. For reactions to compete effectively in the metastable ion time frame, their activation energies must cover only a small span, in general not exceeding 35–40 kJ/mol.

2.7.3 USING THE DATA IN THIS BOOK

In its present edition, this book only covers ions having up to three carbon atoms. Therefore, if you know the molecular formula of your ion of unknown structure, it is a simple matter to find in Part 2 a description of the (many) stable isomeric structures that it may assume. Fortunately, it is a priori likely that your ion is the isomer of lowest heat of formation (i.e., the most stable structure). That should be your starting point. If however you only know the m/z ratio of the unknown ion then the problem is more complex, requiring some assumptions or another experiment to help assign a molecular formula. However in contrast, if you want to make some educated guess as to the structure of a new fragment ion or a rearranged molecular ion, there are some simple and general principles that should be applied. A short list is given below.

(1) *Primary alkyl carbocations* do not exist as stable species unless they can be stabilized by conjugative delocalization of the charge, e.g., as in the allyl or benzyl cations.[56] Otherwise they undergo very facile 1,2-hydrogen shifts to produce (stable) secondary or tertiary cations, as given in the following example,

$$CH_3CH_2{}^+CH_2 \rightarrow (CH_3)_2{}^+CH$$

(2) *Alkoxy ions* similarly do not exist; again a 1,2-hydrogen shift to oxygen produces a much more thermochemically stable species, an OH-substituted carbocation,[56] as given in the following example,

$$CH_3CH_2O^+ \rightarrow CH_3{}^+CHOH$$

(3) *Ring strain* in an ionized molecule containing a three- or four-membered ring is usually large enough for it to be immediately relieved by a rearrangement to produce an acyclic isomer or by a facile fragmentation in which the ring is opened. In general, strained-ring ions are not generated by an ion's dissociation. A possible exception is when an aromatic species may result, e.g., the aromatic cy-$C_3H_3{}^{1+}$ ion is more stable by over 100 kJ/mol than its isomer, the resonance-stabilized propargyl cation, $HCCCH_2{}^{1+}$ ($\Delta_f H$(cy-$C_3H_3{}^{1+}$)) = 1075 kJ/mol vs $\Delta_f H$($HCCCH_2{}^{1+}$) = 1179 kJ/mol (refer to [C_3,H_3]$^+$ section in Part 2). Note that although methyl substitution in an ion at the charge-bearing site is significantly stabilizing, for the given example only the single methyl derivative remains the most stable isomer, i.e., methyl cyclopropenium is only slightly more stable than its closest rival, the again resonance-stabilized buta-2-dienyl cation ($\Delta_f H$(cy–C(H)C(H)C(CH_3)–$^{1+}$) = 992 kJ/mol vs $\Delta_f H$(CH_2 $CCHCH_2{}^{1+}$) = 1029 = 1029 kJ/mol).[287] The more highly methyl-substituted cyclopropenium ions are less stable than some of their nonaromatic isomers, e.g., the allylic cyclopentenyl cation, cy-$C_5H_7{}^{1+}$, is the global minimum (most stable isomer) on the [C_5,H_7]$^+$ potential energy surface, significantly more stable than its aromatic isomer dimethyl-cyclopropenium cation, cy–C(H)C(CH_3)C(CH_3)–$^{1+}$(ΔH_f(cy-$C_5H_7{}^{1+}$) = 830 kJ/mol vs $\Delta_f H$(cy–C(H)C(CH_3)C(CH_3)_2–$^{1+}$) = 908 kJ/mol); these and related matters have been addressed in some detail elsewhere.[292] Thus as a good general rule, the introduction of small (especially unsaturated) strained rings into proposed fragment ion structures is very likely wrong and so is strongly to be discouraged.

(4) *Hydrogen atoms* migrate to positions of greatest electron density, i.e., to radical sites in preference to positive charge sites. A good example is the McLafferty rearrangement (refer to Section 2.3) for, say, methyl propyl ketone. The charge in the molecular ion resides largely at the keto-carbon, leaving a formal radical site at oxygen. Note that the γ-hydrogen shift from propyl group to oxygen significantly enhances the stability of the resulting ion by converting the erstwhile carbonyl carbon into a triply substituted carbocation center, a change sufficient to overcome the destabilizing effect that results from the generation of the new odd electron site at the end of the propyl group (for further details refer also to the thermochemistry of distonic ions in Section 1.3.10).

REFERENCES TO PART 1

1. *NIST Mass Spectral Library*, National Institute of Standards and Technology: Gaithersburg MD, **2005**.
2. Lenard, P. *Ann. Phys.* **1902**, *8*, 149.
3. Mark, T., Illenberger, E. *Int. J. Mass Spectrom.* **2001**, *205*, 1.
4. Mark, T.D. *Int. J. Mass Spectrom. Ion Phys.* **1982**, *45*, 125.
5. Wannier, G.H. *Phys. Rev.* **1953**, *90*, 817.
6. Miller, D.R. In *Atomic and Molecular Beam Methods*, Scoles, G., Ed., Oxford University Press: New York, **1988**.
7. DePaul, S., Pullman, D., Friedrich, B. *J. Phys. Chem.* **1993**, *97*, 2167.
8. Ng, C.-Y. *Int. J. Mass Spectrom.* **2000**, *200*, 357.
9. Ditchburn, R.W., Arnot, F.L. *Proc. R. Soc.* **1929**, *A123*, 516.
10. Hurzeler, H., Inghram, M.G., Morrison, J.D. *J. Chem. Phys.* **1958**, *28*, 76.
11. Villarejo, D., Herm, R., Inghram, M.G. *J. Chem. Phys.* **1967**, *46*, 4995.
12. Ng, C.-Y. *Annu. Rev. Phys. Chem.* **2002**, *53*, 101.
13. Ashfold, M.N.R. *Nonlinear Spectroscopy for Molecular Structure Determination*, Field, F.H., Ed., Blackwell: Oxford, **1998**, p. 127.
14. Harrison, A.G. *Chemical Ionization Mass Spectrometry*, CRC Press: Boca Raton, **1992**.
15. Munson, M.S.B., Field, F.H. *J. Am. Chem. Soc.* **1966**, *88*, 2621.
16. Inghram, M.G., Gomer, R.J. *J. Chem. Phys.* **1954**, *22*, 1279.
17. Beckey, H.D. *Principles of Field Ionization and Field Desorption Mass Spectrometry*, Pergamon Press: Oxford, **1977**.
18. Derrick, P.J., Robertson, A.J.B. *Proc. Roy. Soc. London Ser. A* **1971**, *324*, 491.
19. Barber, M., Bordoli, R.S., Elliot, G.J., Sedgwick, R.D., Tyler, A.N. *Anal. Chem.* **1989**, *54*, 645A.
20. Fenn, J.B., Mann, M., Mang, C.K., Wong, S.F., Whitehouse, C.M. *Science* **1989**, *246*, 64.
21. Arpino, P.J. *Mass Spectrom. Rev.* **1990**, *9*, 631.
22. Mumma, R.O., Vastola, F. *J. Org. Mass Spectrom.* **1972**, *6*, 1373.
23. Hardin, E.D., Vestal, M.L. *Anal. Chem.* **1981**, *53*, 1492.
24. Hillenkamp, F. *Second International Workshop on Ion Formation from Organic Solids*, Springer-Verlag: New York, **1983**.
25. Karas, M., Hillenkamp, F. *Anal. Chem.* **1988**, *60*, 2299.
26. Stump, M.J., Fleming, R.C., Gong, W.-H., Jaber, A.J., Jones, J.J., Surber, C.W., Wilkins, C.L. *Appl. Spectrosc. Rev.* **2002**, *37*, 275.
27. Murgasova, R., Hercules, D.M. *Int. J. Mass Spectrom.* **2003**, *226*, 151.
28. Hillenkamp, F. *Adv. Mass Spectrom.* **1989**, *11A*, 354.
29. Dreisewerd, K., Berkenkamp, S., Leisner, A., Rohlfing, A., Menzel, C. *Int. J. Mass Spectrom.* **2003**, *226*, 189.
30. Vertes, A., Gijbels, R., Levine, R.D. *Rapid Commun. Mass Spectrom.* **1990**, *4*, 228.
31. Dreisewerd, K., Schurenberg, M., Karas, M., Hillenkamp, F. *Int. J. Mass Spectrom. Ion Processes* **1995**, *141*, 127.
32. Johnson, R.E. *Large Ions: Their Vaporization, Detection and Structural Analysis*, Powis, I., Ed., Wiley: New York, **1996**.

33. Beavis, R.C., Chen, X., Carroll, J.A. *J. Am. Mass Spectrom. Soc.* **1998**, *9*, 885.
34. Beauchemin, D. *Anal. Chem.* **2004**, *76*, 3395.
35. Fassel, V.A. *Science* **1978**, *202*, 183.
36. Gray, A.L. *Analyst* **1975**, *100*, 289.
37. Houk, R.S., Fassel, V.A., Flesch, G.D., Svec, H.J., Gray, A.L., Taylor, C.E. *Anal. Chem.* **1980**, *52*, 2283.
38. Taylor, V.F., March, R.E., Longerich, H.P., Stadey, C.J. *Int. J. Mass Spectrom.* **2005**, *243*, 71.
39. Thompson, J.J. *Rays of Positive Electricity and Their Applications to Chemical Analysis*, Longmans Green and Co.: London, **1913**.
40. Dawson, P.H. *Quadrupole Mass Spectrometry and Its Applications*, Elsevier: Amsterdam, **1976**.
41. March, R.E., Hughes, R.J. *Quadrupole Ion Storage Mass Spectrometry*, Wiley Interscience: New York, **1989**.
42. Smith, D., Adams, N.G. *Advances in Atomic and Molecular Physics*, Academic Press: New York, **1988**.
43. March, R.E., Todd, J.F.J. *Practical Aspects of Ion Trap Mass Spectrometry*, CRC Press: Boca Raton, **1995**.
44. Marshall, A.G., Hendrickson, C.L., Jackson, G.S. *Mass Spectrom. Rev.* **1998**, *17*, 1.
45. Lawrence, E.O., Livingston, M.S. *Phys. Rev.* **1932**, *40*, 19.
46. Hipple, J.A., Sommer, H., Thomas, H.A. *Phys. Rev.* **1949**, *76*, 1877.
47. Sommer, H., Thomas, H.A., Hipple, J.A. *Phys. Rev.* **1951**, *82*, 697.
48. McIver, R.T. *Rev. Sci. Instrum.* **1970**, *41*, 555.
49. Comisarow, M.B., Marshall, A.G. *Chem. Phys. Lett.* **1974**, *25*, 282.
50. Cooley, J.W., Tukey, J.W. *Math. Comp.* **1965**, *19*, 9.
51. Comisarow, M.B., Marshall, A.G. *J. Mass Spectrom.* **1996**, *31*, 581.
52. Amster, I.J. *J. Mass Spectrom.* **1996**, *31*, 1325.
53. Guilhaus, M. *J. Mass Spectrom.* **1995**, *30*, 1519.
54. Wiley, W.C., McLaren, I.H. *Rev. Sci. Instr.* **1955**, *26*, 1150.
55. Pedley, J.B., Naylor, R.D., Kirby, S.P. *Thermochemical Data of Organic Compounds*, 2nd ed., Chapman and Hall: London, **1986**.
56. *NIST Chemistry WebBook, NIST Standard Reference Database Number 69*, National Institute of Standards and Technology: Gaithersburg MD, **2005**.
57. Luo, Y.-R. *Handbook of Bond Dissociation Energies in Organic Compounds*, CRC Press: Boca Raton, **2002**.
58. Benson, S.W. *Thermochemical Kinetics*, 2nd ed., Wiley: New York, **1976**.
59. Cohen, N., Benson, S.W. *Chem. Rev.* **1993**, *93*, 2419.
60. Lias, S.G., Bartmess, J.E., Liebman, J.F., Holmes, J.L., Levin, R.D., Mallard, W.G. *J. Phys. Chem. Ref. Data* **1988**, *17 Suppl. 1*.
61. Turner, D.W., Baker, A.D., Baker, C., Brundle, C.R. *Molecular Photoelectron Spectroscopy*, Wiley Interscience: London, **1970**.
62. Morrison, J.D. In *Mass Spectrometry*, Maccoll, A., Ed., Butterworths: London, **1972**, Vol. 5.
63. Beveridge, W., Hunter, J.A., Johnson, C.A.F., Parker, J.E. *Org. Mass Spectrom.* **1992**, *27*, 543.
64. Kimura, K., Katsumata, S., Achiba, Y., Yamazaki, T., Iwata, S. *Handbook of He(I) Photoelectron Spectra of Fundamental Organic Molecules*, Halstead Press: New York, **1981**.

65. Eland, J.H.D. *Photoelectron Spectroscopy, an Introduction to UV Photoelectron Spectroscopy in the Gas Phase*, 2nd ed., Butterworths: London, **1983**.

66. Muller-Dethlefs, K., Sclhag, E.W. *Annu. Rev. Phys. Chem.* **1991**, *42*, 109.

67. Bachiri, M., Mouvier, G., Carlier, P., Dubois, J.E. *J. Chim. Phys. Phys. Chim. Bio.* **1980**, *77*, 899.

68. Holmes, J.L., Fingas, M., Lossing, F.P. *Can. J. Chem.* **1981**, *59*, 80.

69. Holmes, J.L., Lossing, F.P. *Can. J. Chem.* **1982**, *60*, 2365.

70. Aubry, C., Holmes, J.L. *Int. J. Mass Spectrom.* **2000**, *200*, 277.

71. Holmes, J.L., Lossing, F.P. *Org. Mass Spectrom.* **1991**, *26*, 537.

72. Lossing, F.P., de Sousa, J.B. *J. Am. Chem. Soc.* **1959**, *81*, 281.

73. Schultz, J.C., Houle, F.A., Beauchamp, J.L. *J. Am. Chem. Soc.* **1984**, *106*, 3917.

74. Traeger, J.C. *Int. J. Mass Spectrom. Ion Processes* **1984**, *58*, 259.

75. Maeda, K., Semeluk, G.P., Lossing, F.P. *Int. J. Mass Spectrom. Ion Phys.* **1968**, *1*, 395.

76. Holmes, J.L., Burgers, P.C. *Org. Mass Spectrom.* **1982**, *17*, 123.

77. Dannacher, *J. Org. Mass Spectrom.* **1984**, *19*, 253.

78. Baer, T. *Adv. Chem. Phys.* **1986**, *64*, 111.

79. Eland, J.H.D. *Int. J. Mass Spectrom. Ion Phys.* **1972**, *8*, 143.

80. Armentrout, P.B. *Encyclopedia of Mass Spectrometry*, Gross, M.L., Caprioli, R.M., Eds., Elsevier: Amsterdam, **2003**, Vol. 1.

81. Holmes, J.L., Hop, C.E.C.A., Terlouw, J.K. *Org. Mass Spectrom.* **1986**, *21*, 776.

82. Holmes, J.L., Wolkoff, P., Lossing, F.P. *Can. J. Chem.* **1980**, *58*, 251.

83. Rosenstock, H.M., Stockbauer, R.L., Parr, A.C. *J. Chem. Phys.* **1979**, *71*, 3708.

84. Rosenstock, H.M., Stockbauer, R.L., Parr, A.C. *J. Chem. Phys.* **1980**, *73*, 773.

85. Pratt, S.J., Chupka, W.A. *Chem. Phys.* **1981**, *62*, 153.

86. Dannacher, J., Rosenstock, H.M., Buff, R., Parr, A.C., Stockbauer, R.L., Bombach, R., Stadelmann, J.P. *Chem. Phys.* **1983**, *75*, 23.

87. Burgers, P.C., Holmes, J.L. *Int. J. Mass Spectrom. Ion Processes* **1984**, *58*, 15.

88. Traeger, J.C., McLoughlin, R.G. *J. Am. Chem. Soc.* **1981**, *103*, 3647.

89. Lossing, F.P. Unpublished data, **1999**.

90. Aue, D.H., Webb, H.M., Bowers, M.T. *J. Am. Chem. Soc.* **1972**, *94*, 4726.

91. Szulejko, J.E., McMahon, T.B. *J. Am. Chem. Soc.* **1993**, *115*, 7839.

92. Meot-Ner, M., Sieck, L.W. *J. Am. Chem. Soc.* **1991**, *113*, 4448.

93. Smith, B.J., Radom, L. *J. Am. Chem. Soc.* **1993**, *115*, 4885.

94. Harrison, A.G. *Mass Spectrom. Rev.* **1997**, *16*, 201.

95. Cao, J., Aubry, C., Holmes, J.L. *J. Phys. Chem. A* **2000**, *104*, 10045.

96. Bouchoux, G., Sablier, M., Berruyer-Penaud, F. *J. Mass Spectrom.* **2004**, *39*, 986.

97. Wesdemiotis, C. *J. Mass Spectrom.* **2004**, *39*, 998.

98. Ervin, K.M., Armentrout, P.B. *J. Mass Spectrom.* **2004**, *39*, 1004.

99. Drahos, L., Peltz, C., Vekey, K. *J. Mass Spectrom.* **2004**, *39*, 1016.

100. Larson, J.W., McMahon, T.B. *J. Am. Chem. Soc.* **1982**, *104*, 6255.

101. Meot-Ner, M. *J. Am. Chem. Soc.* **1984**, *106*, 1257.

102. Cao, J. In *Chemistry*, Ph.D. Thesis; Proton Affinities of Organic Molecules by the Kinetic Method and a Mass Spectrometric Study of Immonium Ions by Tandem Mass Spectrometry; University of Ottawa: Ottawa, **2001**, pp. 260.

103. Cooks, R.G., Beynon, J.H., Caprioli, R.M., Lester, G.R. *Metastable Ions*, Elsevier: Amsterdam, **1973**.

104. Holmes, J.L., Benoit, F. In *MTP International Review of Science*, Maccoll, A., Ed., Butterworths: London, **1972**, Vol. 5, p. 259.

105. Derrick, P.J., Donchi, K.F. In *Comprehensive Chemical Kinetics*, Bamford, C.H., Tipper, C.F.H., Eds., Elsevier: Amsterdam, **1983**, Vol. 24, p. 53.
106. Holmes, J.L., Terlouw, J.K. *Org. Mass Spectrom.* **1980**, *15*, 383.
107. Hipple, J.A., Fox, R.E., Condon, E.U. *Phys. Rev.* **1946**, *69*, 347.
108. Laskin, J., Lifshitz, C. *J. Mass Spectrom.* **2001**, *36*, 459.
109. Holmes, J.L., Osborne, A.D. *Int. J. Mass Spectrom. Ion Phys.* **1977**, *23*, 189.
110. Ruttink, P.J.A. *J. Phys. Chem. A* **1987**, *91*, 703.
111. Holmes, J.L., Osborne, A.D., Weese, G.M. *Org. Mass Spectrom.* **1975**, *10*, 867.
112. Burgers, P.C., Holmes, J.L., Mommers, A.A., Szulejko, J.E. *J. Am. Chem. Soc.* **1984**, *106*, 521.
113. Aubry, C., Holmes, J.L. *J. Phys. Chem. A* **1998**, *102*, 6441.
114. Shannon, T.S., McLafferty, F.W. *J. Am. Chem. Soc.* **1966**, *88*, 5021.
115. Jones, E.G., Baumann, L.E., Beynon, J.H., Cooks, R.G. *Org. Mass Spectrom.* **1973**, *7*, 185.
116. Holmes, J.L., Terlouw, J.K. *Can. J. Chem.* **1975**, *53*, 1076.
117. Wolkoff, P., Holmes, J.L. *Can. J. Chem.* **1979**, *57*, 348.
118. Wolkoff, P., Holmes, J.L., Lossing, F.P. *Can. J. Chem.* **1980**, *58*, 251.
119. Bouchoux, G., Salpin, J.-Y., Yanez, M. *J. Phys. Chem. A* **2004**, *108*, 9853.
120. Terlouw, J.K., Wezenberg, J., Burgers, P.C., Holmes, J.L. *J. Chem. Soc. Chem. Commun.* **1983**, 1121.
121. Molenaar-Langeveld, T.A., Fokkens, R.H., Nibbering, N.M.M. *Org. Mass Spectrom.* **1986**, *21*, 15.
122. Beynon, J.H., Job, B.E., Williams, A.E. *Z. Naturforsch.* **1965**, *20a*, 883.
123. Meyerson, S., Corbin, J.L. *J. Am. Chem. Soc.* **1965**, *87*, 3045.
124. Davis, B., Williams, D.H., Yeo, A.N.H. *J. Chem. Soc. B* **1970**, 81.
125. Hogg, R.V., Tanis, E.A. *Probability and Statistical Inference*, 2nd ed., Macmillan: New York, **1983**.
126. Cooks, R.G. *Collision Spectroscopy*, Plenum Press: New York, **1978**.
127. Lorquet, J.C., Levsen, K., Schwarz, H. *Mass Spectrom. Rev.* **1983**, *2*, 77.
128. Todd, P.J., McLafferty, F.W. *Tandem Mass Spectrometry*, Wiley Interscience: New York, **1983**.
129. McLuckey, S. *J. Am. Soc. Mass Spectrom.* **1999**, *3*, 599.
130. Jennings, K.R. *Int. J. Mass Spectrom. Ion Phys.* **1968**, *1*, 227.
131. Haddon, W.K., McLafferty, F.W. *J. Am. Chem. Soc.* **1968**, *90*, 4745.
132. McLafferty, F.W., Bente III, P.F., Kornfeld, F.C., Tsai, S.-C., Howe, I. *J. Am. Chem. Soc.* **1973**, *95*, 2120.
133. de Froidement, Y., Lorquet, A.J., Lorquet, J.C. *J. Phys. Chem.* **1991**, *95*, 4220.
134. Rosenstock, H.M., Wallenstein, M.B., Wahrhaftig, A.L., Eyring, H. *Proc. Natl. Acad. Sci. USA* **1952**, *38*, 667.
135. Holmes, J.L. *Org. Mass Spectrom.* **1985**, *20*, 169.
136. Cooks, R.G., Beynon, J.H., Ast, T. *J. Am. Chem. Soc.* **1972**, *94*, 1004.
137. Harris, F.M. *Int. J. Mass Spectrom. Ion Processes* **1992**, *120*, 1.
138. Appelle, J. *Collision Spectroscopy*, Plenum Press: New York, **1978**.
139. Aubry, C., Holmes, J.L. *J. Am. Soc. Mass Spectrom.* **2001**, *12*, 23.
140. Cao, J., Aubry, C., Holmes, J.L. *J. Phys. Chem. A* **2000**, *104*, 10045.
141. Burgers, P.C., Holmes, J.L., Szulejko, J.E., Mommers, A.A., Terlouw, J.K. *Org. Mass Spectrom.* **1983**, *18*, 254.

142. Holmes, J.L. In *Encyclopedia of Mass Spectrometry*, Elsevier: Amsterdam, **2003**, Vol. 1, p. 91.

143. Holmes, J.L., Terlouw, J.K. In *Encyclopedia of Mass Spectrometry*, Elsevier: Amsterdam, **2005**, Vol. 4, p. 287.

144. Bouma, W.J., MacLeod, J.K., Radom, L. *J. Am. Chem. Soc.* **1982**, *104*, 2930.

145. Bouma, W.J., Nobes, R.H., Radom, L. *J. Am. Chem. Soc.* **1982**, *104*, 2929.

146. Holmes, J.L., Lossing, F.P., Burgers, P.C., Terlouw, J.K. *J. Am. Chem. Soc.* **1982**, *104*, 2931.

147. Burgers, P.C., Holmes, J.L., Mommers, A.A., Szulejko, J.E. *Org. Mass Spectrom.* **1983**, *18*, 596.

148. Holmes, J.L., Wang, X. Unpublished data, **2005**.

149. Burgers, P.C., Holmes, J.L., Mommers, A.A., Terlouw, J.K. *Org. Mass Spectrom.* **1984**, *19*, 442.

150. Lifshitz, C., Gotchiguian, P., Roller, R. *Chem. Phys. Lett.* **1983**, *95*, 106.

151. Burgers, P.C., Holmes, J.L., Mommers, A.A., Terlouw, J.K. *Chem. Phys. Lett.* **1983**, *102*, 1.

152. Wenthold, P.G. *J. Phys. Chem. A* **2000**, *104*, 5612.

153. Holmes, J.L., Hop, C.E.C.A., Terlouw, J.K. *Org. Mass Spectrom.* **1986**, *21*, 776.

154. Holmes, J.L., Traeger, J.C. *J. Phys. Chem.* **1993**, *97*, 3453.

155. Bursey, M.M., Hass, J.R., Harvan, D.J., Parker, C.E. *J. Am. Chem. Soc.* **1979**, *101*, 5485.

156. Holmes, J.L. *Mass Spec. Rev.* **1989**, *8*, 513.

157. McLafferty, F.W. *Science* **1990**, *247*, 925.

158. Schalley, C.A., Hornung, G., Schroder, D., Schwarz, H. *Chem. Soc. Rev.* **1998**, *27*, 91.

159. Turecek, F. *Org. Mass Spectrom.* **1992**, *27*, 1087.

160. Gellene, G.I., Porter, R.F. *Int. J. Mass Spectrom. Ion Processes* **1985**, *64*, 55.

161. Danis, P.O., Feng, R., McLafferty, F.W. *Anal. Chem.* **1986**, *58*, 348.

162. Blanchette, M.C., Bordas-Nagy, J., Holmes, J.L., Hop, C.E.C.A., Mommers, A.A., Terlouw, J.K. *Org. Mass Spectrom.* **1988**, *23*, 804.

163. Lorquet, J.C., Leyh-Nihant, B., McLafferty, F.W. *Int. J. Mass Spectrom. Ion Processes* **1990**, *91*, 465.

164. Wesdemiotis, C., Feng, R., Danis, P.O., Williams, E.R., McLafferty, F.W. *J. Am. Chem. Soc.* **1986**, *108*, 5847.

165. Terlouw, J.K., Kieskamp, W.M., Holmes, J.L., Mommers, A.A., Burgers, P.C. *Int. J. Mass Spectrom. Ion Processes* **1985**, *64*, 245.

166. Yates, B.F., Bouma, W.J., Radom, L. *J. Am. Chem. Soc.* **1987**, *109*, 2250.

167. Wesdemiotis, C., Feng, R., Baldwin, M.A., McLafferty, F.W. *Org. Mass Spectrom.* **1988**, *23*, 166.

168. Selgren, S.F., Gellene, G.I. *J. Chem. Phys.* **1987**, *87*, 5804.

169. Holmes, J.L. *Mass Spectrom. Rev.* **1989**, *8*, 513.

170. Sadilek, M., Turecek, F. *J. Phys. Chem.* **1996**, *100*, 9610.

171. Holmes, J.L., Lossing, F.P., Mayer, P.M. *J. Am. Chem. Soc* **1991**, *113*, 9405.

172. Holmes, J.L., Mayer, P.M., Mommers, A.A. *Org. Mass Spectrom.* **1992**, *27*, 537.

173. Holmes, J.L., Mayer, P.M. *Eur. Mass Spectrom.* **1995**, *1*, 23.

174. Little, D.P., Speir, J.P., Senko, M.W., O'Connor, P.B., McLafferty, F.W. *Anal. Chem.* **1994**, *66*, 2809.

175. Colorado, A., Shen, J.X., Vartanian, V.H., Brodbelt, J. *Anal. Chem.* **1996**, *68*, 4033 and references therein.
176. Chiavarino, B., Crestoni, M.E., Fornarini, S., Dopfer, O., Lemaire, J., Maitre, P. *J. Phys. Chem. A* **2006**, *110*, 9352.
177. Fridgen, T.D., McMahon, T.B., Maitre, P., Lemaire, J. *Phys. Chem. Chem. Phys.* **2006**, *8*, 2483.
178. Dunbar, R.C., Moore, D.T., Oomens, J. *J. Phys. Chem. A* **2006**, *110*, 8316.
179. Nibbering, N.M.M. *Mass Spectrom. Rev.* **1984**, *3*, 445.
180. Derrick, P.J., Falick, A.M., Burlingame, A.L. *J. Am. Chem. Soc.* **1972**, *94*, 6794.
181. Derrick, P.J., Falick, A.M., Burlingame, A.L., Djerassi, C.J. *J. Am. Chem. Soc.* **1974**, *96*, 1054.
182. Dempster, A.J. *Phil. Mag.* **1916**, *31*, 438.
183. Squires, R.R. *Int. J. Mass Spectrom. Ion Processes* **1992**, *118/119*, 503.
184. March, R.E. *J. Mass Spectrom.* **1997**, *32*, 351.
185. March, R.E. *Int. J. Mass Spectrom.* **2000**, *200*, 285.
186. Gronert, S. *Mass Spectrom. Rev.* **2005**, *24*, 100.
187. Buchanan, M.V. *Fourier-Transform Mass Spectrometry, Evolution, Innovation and Applications*, American Chemical Society: Washington DC, **1987**.
188. Nibbering, N.M.M. *Acc. Chem. Res.* **1990**, *23*, 279.
189. Nibbering, N.M.M. *Int. J. Mass Spectrom.* **2000**, *200*, 27.
190. Wenthold, P.G., Munsch, T.E. *Annu. Rep. Prog. Chem. Sect. B* **2003**, *99*, 420.
191. Wenthold, P.G., Munsch, T.E. *Annu. Rep. Prog. Chem. Sect. B* **2004**, *100*, 377.
192. Woods, R.C., Dixon, T.A., Saykally, R.J., Szanto, P.G. *Phys. Rev. Lett.* **1975**, *35*, 1269.
193. Guedeman, C.S., Woods, R.C. *Phys. Rev. Lett.* **1982**, *48*, 1344.
194. Bohme, D.K., Goodings, J.M., Ng, C.-W. *Int. J. Mass Spectrom. Ion Phys.* **1977**, *24*, 25.
195. Burgers, P.C., Holmes, J.L. *Chem. Phys. Lett.* **1983**, *97*, 236.
196. Illies, A.J., Jarrold, M.F., Bowers, M.T. *J. Am. Chem. Soc.* **1983**, *105*, 2562.
197. Burgers, P.C., Holmes, J.L., Mommers, A.A. *J. Am. Chem. Soc.* **1985**, *107*, 1099.
198. Freeman, C.G., Knight, J.S., Love, J.G., McEwan, M.J. *Int. J. Mass Spectrom. Ion Processes* **1987**, *80*, 255.
199. Terlouw, J.K., Wezenberg, J., Burgers, P.C., Holmes, J.L. *J. Chem. Soc. Chem. Commun.* **1983**, 1121.
200. Nedev, H., van der Rest, G., Mourgues, P., Audier, H.E. *Eur. J. Mass Spectrom.* **2003**, *9*, 319.
201. Bohme, D.K. *Int. J. Mass Spectrom. Ion Processes* **1992**, *115*, 95.
202. Stirk, K.M., Orlowski, J.C., Leeck, D.T., Kenttamaa, H.I. *J. Am. Chem. Soc.* **1992**, *114*, 8604.
203. Thoen, K.K., Beasley, B.J., Smith, R.L., Kenttamaa, H.I. *J. Am. Soc. Mass Spectrom.* **1996**, *7*, 1245.
204. Leeck, D.T., Stirk, K.M., Zeller, L.C., Kiminkinen, L.K.M., Castro, L.M., Vainiotalo, P., Kenttamaa, H.I. *J. Am. Chem. Soc.* **1994**, *116*, 3028.
205. Nelson, E.D., Li, R., Kenttamaa, H.I. *Int. J. Mass Spectrom.* **1999**, *185–187*, 91.
206. Stirk, K.M., Kiminkinen, L.K.M., Kenttamaa, H.I. *Chem. Rev.* **1992**, *92*, 1649.
207. Kenttamaa, H.I. *Org. Mass Spectrom.* **1994**, *29*, 1.
208. Gronert, S. *Chem. Rev.* **2001**, *101*, 329.

209. Wittneben, D., Grutzmacher, H.-F. *Int. J. Mass Spectrom. Ion Processes* **1990**, *100*, 545.
210. Bouma, W.J., MacLeod, J.K., Radom, L. *J. Am. Chem. Soc.* **1980**, *102*, 2246.
211. Thissen, R., Audier, H.E., Chamot-Rooke, J., Mourgues, P. *Eur. J. Mass Spectrom.* **1999**, *5*, 147.
212. Holmes, J.L., Lossing, F.P., Terlouw, J.K., Burgers, P.C. *Can. J. Chem.* **1983**, *61*, 2305.
213. Terlouw, J.K., Heerma, W., Dijkstra, G., Holmes, J.L., Burgers, P.C. *Int. J. Mass Spectrom. Ion Phys.* **1983**, *47*, 147.
214. Weger, E., Levsen, K., Ruppert, I., Burgers, P.C., Terlouw, J.K. *Org. Mass Spectrom.* **1983**, *18*, 327.
215. Keck, H., Kuchen, W., Tommes, P., Terlouw, J.K., Wong, T. *Angew. Chem. Int. Ed. Engl.* **1992**, *31*, 86.
216. Schweighofer, A., Chou, P.K., Thoen, K.K., Nanayakkara, V.K., Keck, H., Kuchen, W., Kenttamaa, H.I. *J. Am. Chem. Soc.* **1996**, *118*, 11893.
217. Chou, P.K., Smith, R.L., Chyall, L.J., Kenttamaa, H.I. *J. Am. Chem. Soc.* **1995**, *117*, 4374.
218. Møller, C., Plesset, M.S. *Phys. Rev.* **1934**, *46*, 618.
219. Hehre, W.J., Radom, L., Schleyer, P.V.R., Pople, J.A. *Ab Initio Molecular Orbital Theory*, Wiley: New York, **1986**.
220. Dunning, T.H., Haley, P.J. In *Modern Theoretical Chemistry*, Schaeffer III, H.F., Ed., Plenum: New York, **1976**, pp. 1–28.
221. Dunning, T.H. *J. Chem. Phys* **1989**, *90*, 1007.
222. Stevens, W., Basch, H., Krauss, J. *J. Chem. Phys* **1984**, *81*, 6026.
223. Koch, W., Holthausen, M.C. *A Chemist's Guide to Density Functional Theory*, Wiley-VCH: New York, **2000**.
224. Becke, A.D. *Phys. Rev. A* **1988**, *38*, 3098.
225. Lee, C., Yang, W., Parr, R.G. *Phys. Rev. B* **1988**, *37*, 785.
226. Becke, A.D. *J. Chem. Phys.* **1993**, *98*, 5648.
227. Scott, A.P., Radom, L. *J. Phys. Chem.* **1996**, *100*, 16502.
228. Curtiss, L.A., Raghavachari, K., Refern, P.C., Rassolov, V., Pople, J.A. *J. Chem. Phys.* **1998**, *109*, 7764.
229. Ochterski, J.W., Petersson, G.A., Montgomery, J.A. *J. Chem. Phys.* **1996**, *104*, 2598.
230. Martin, J.M.L., Oliveira, G.D. *J. Chem. Phys.* **1999**, *111*, 1843.
231. Mayer, P.M. *J. Chem. Phys.* **1999**, *110*, 7779.
232. Morton, T.H., Beauchamp, J.L. *J. Am. Chem. Soc.* **1975**, *97*, 2355.
233. Busch, K.L., Nixon, W.B., Bursey, M.M. *J. Am. Chem. Soc.* **1978**, *100*, 1621.
234. Bouma, W.J., Nobes, R.H., Radom, L. *J. Am. Chem. Soc.* **1982**, *104*, 2929.
235. Gauld, J.W., Audier, H.E., Fossey, L., Radom, L. *J. Am. Chem. Soc.* **1996**, *118*, 6299.
236. Gauld, J.W., Radom, L. *J. Am. Chem. Soc.* **1997**, *119*, 9831.
237. Bouma, W.J., MacLeod, J.K., Radom, L. *J. Am. Chem. Soc.* **1982**, *104*, 2930.
238. Holmes, J.L., Lossing, F.P., Terlouw, J.K., Burgers, P.C. *J. Am. Chem. Soc.* **1982**, *104*, 2931.
239. Hammerum, S. *Mass Spectrom. Rev.* **1988**, *7*, 123.
240. McLafferty, F.W. *Anal. Chem.* **1956**, *28*, 306.
241. McLafferty, F.W. *Anal. Chem.* **1959**, *31*, 82.

242. Pitts, J.N. *J. Chem. Educ.* **1957**, *34*, 112.
243. Derrick, P.J., Falick, A.M., Burlingame, A.L., Djerassi, C. *J. Am. Chem. Soc.* **1974**, *96*, 1054.
244. Yeo, A.N.H., Cooks, R.G., Williams, D.H. *J. Chem. Soc. Chem. Commun.* **1968**, 1269.
245. Yeo, A.N.H., Williams, D.H. *J. Am. Chem. Soc.* **1969**, *91*, 3582.
246. McAdoo, D.J., Witiak, F.W., McLafferty, F.W., Dill, J.D. *J. Am. Chem. Soc.* **1978**, *100*, 6639.
247. McAdoo, D.J., Hudson, C.E. *J. Am. Chem. Soc.* **1981**, *103*, 7710.
248. Weber, R., Levsen, K., Wesdemiotis, C., Weiske, T., Schwarz, H. *Int. J. Mass Spectrom. Ion Phys.* **1982**, *43*, 131.
249. Carpenter, W., Duffield, A.M., Djerassi, C. *J. Am. Chem. Soc.* **1968**, *90*, 160.
250. Stringer, M.B., Underwood, D.J., Bowie, J.H., Allison, C.E., Donchi, K.F., Derrick, P.J. *Org. Mass Spectrom.* **1992**, *27*, 270.
251. McAdoo, D.J., Hudson, C.E., Witiak, D.N. *Org. Mass Spectrom.* **1979**, *14*, 350.
252. Hunter, E.P., Lias, S.G. *J. Phys. Chem. Ref. Data* **1998**, *27*, 413.
253. Luo, Y.-R. *Bond Dissociation in Organic Compounds*, CRC Press: Boca Raton, **2003**.
254. Gaskell, S.J. *J. Mass Spectrom.* **1997**, *32*, 677.
255. Busch, K.L. *J. Mass Spectrom.* **1995**, *30*, 233.
256. Roepstorff, P., Fohlman, J. *Biomed. Mass Spectrom.* **1984**, *11*, 601.
257. Johnson, R.S., Martin, S.A., Biemann, K., Stults, J.T., Watson, J.T. *Anal. Chem.* **1987**, *59*, 2621.
258. Johnson, R.S., Martin, S.A., Biemann, K. *Int. J. Mass Spectrom. Ion Processes* **1988**, *86*, 137.
259. Biemann, K. *Methods Enzymol.* **1990**, *193*, 455.
260. Papayannopoulos, A. *Mass Spectrom. Rev.* **1995**, *14*, 49.
261. Weinkam, R.J. *J. Am. Chem. Soc.* **1974**, *96*, 1032.
262. Tsang, C.W., Harrison, A.G. *J. Chem. Soc. Perkin Trans. 2* **1975**, 1718.
263. Weinkam, R.J., Gal, J. *Org. Mass Spectrom.* **1976**, *11*, 188.
264. Ichikawa, H., Harrison, A.G. *Org. Mass Spectrom.* **1978**, *13*, 389.
265. Middlemiss, N.E., Harrison, A.G. *Can. J. Chem.* **1979**, *57*, 2827.
266. Harrison, A.G., Kallury, R.K.M.R. *Org. Mass Spectrom.* **1980**, *15*, 277.
267. Yalcin, T., Khouw, C., Csizmadia, I.G., Peterson, M.P., Harrison, A.G. *J. Am. Soc. Mass Spectrom.* **1995**, *6*, 1165.
268. Hunt, D.F., Yates III, J.R., Shabanowitz, J., Winston, S., Hauer, C.R. *Proc. Natl. Acad. Sci. USA* **1986**, *83*, 6233.
269. Biemann, K., Martin, S. *Mass Spectrom. Rev.* **1987**, *6*, 75.
270. Biemann, K. *Biomed. Environ. Mass Spectrom.* **1988**, *16*, 99.
271. Biemann, K. *Mass Spectrometry Methods in Enzymology*, McCloskey, J.A., Ed., Academic Press: San Diego CA, **1990**, Vol. 193, Chapters 18, 25.
272. Cordero, M.M., Houser, J.J., Wesdemiotis, C. *Anal. Chem.* **1993**, *65*, 1594.
273. Arnott, D., Kottmeir, D., Yates, N., Shabanowitz, J., Hunt, D.F. In *Proceedings of the 42nd ASMS Conference on Mass Spectrometry*: Chicago, **1994**, p. 470.
274. Harrison, A.G., Csizmadia, I.G., Tang, T.-H. *J. Am. Soc. Mass Spectrom.* **2000**, *11*, 427.
275. Turecek, F., Hanus, V. *Org. Mass Spectrom.* **1984**, *19*, 631.
276. McAdoo, D. *J. Mass Spectrom. Rev.* **2000**, *19*, 38.

277. Bouchoux, G. *Mass Spectrom. Rev.* **1998**, *7*, 1 and 203.
278. Vajda, J.H., Harrison, A.G., Hirota, A., McLafferty, F.W. *J. Am. Chem. Soc.* **1981**, *103*, 36.
279. Wesdemiotis, C., Csencsits, R., McLafferty, F.W. *Org. Mass Spectrom.* **1985**, *20*, 98.
280. Burgers, P.C., Holmes, J.L., Mommers, A.A., Szulejko, J.E., Terlouw, J.K. *Org. Mass Spectrom.* **1984**, *19*, 492.
281. Wesdemiotis, C., Feng, R., Williams, E.R., McLafferty, F.W. *Org. Mass Spectrom.* **1986**, *21*, 689.
282. Holmes, J.L., Hop, C.E.C.A., Terlouw, J.K. *Org. Mass Spectrom.* **1986**, *21*, 776.
283. Mazyar, O.A., Mayer, P.M., Baer, T. *Int. J. Mass Spectrom. Ion Processes* **1997**, *167/168*, 389.
284. Burgers, P.C., Holmes, J.L., Hop, C.E.C.A., Terlouw, J.K. *Org. Mass Spectrom.* **1986**, *21*, 549.
285. Heinrich, N., Schmidt, J., Schwarz, H., Apeloig, Y. *J. Am. Chem. Soc.* **1987**, *109*, 1317.
286. Cheung, Y.-S., Li, W.-K. *Chem. Phys. Lett.* **1994**, *223*, 383.
287. Cheung, Y.-S., Li, W.-K. *Chem. Phys. Lett.* **1995**, *333*, 135.
288. Van Doren, J.M., Barlow, S.E., Depuy, C.H., Bierbaum, V.M., Dotan, I., Fergueson, E.E. *J. Phys. Chem.* **1986**, *90*, 2772.
289. Stathis, E.C., Egerton, A.C. *Trans. Faraday Soc.* **1940**, *36*, 606.
290. Pedley, J.B., Naylor, R.D., Kirby, S.P. *Thermochemical Data of Organic Compounds*, 2nd ed., Chapman and Hall: London, **1986**.
291. Lay, T.H., Krasnoperov, L.N., Venanzi, C.A., Bozzelli, N.V. *J. Phys. Chem.* **1996**, *100*, 8240.
292. Lossing, F.P., Holmes, J.L. *J. Am. Chem. Soc.* **1984**, *106*, 6917.

Part II

Ions Containing C and Polyatomic Ions Containing One to Three C Atoms

3 Ion Structures

CONTENTS

3.1 INTRODUCTION

This, the largest section of the book, contains selected mass spectral and thermochemical data for ionized carbon clusters and ions containing one to three carbon atoms and any of the elements H, N, O, S, P, and halogen (F, Cl, Br, and I). Not all of the huge array of data is presented but those given are the authors' choice of data that are believed to be the most reliable. This selection therefore is inevitably subjective in part. The ions are initially classified only by their empirical formulae but the descriptions that follow give the salient characteristics of the isomeric species.

We have focused on three objectives: First, to describe how a given ion structure (namely, an isomer) may be generated in a mass spectrometer's ion source, or by an associated experiment involving mass selected ion fragmentations or ion–molecule reactions. Second, to provide experimental criteria by means of which the different isomeric species may be identified and confirmed. Third, to give the best available thermochemical data obtained from experiment, by estimate or by computation (see comment below).

It is important to emphasize at this point that because of the very great volume of literature pertaining to small isomeric ionic species as well as the challenges imposed by searching among such a wealth of experimental and computational data, it is to be expected that some data will have escaped our net.

Note that when a complex ion is drawn, e.g., $[CH_3\cdots H_3O]^+$, the atoms specifically involved in the electrostatic bond may not be indicated. **The reader should note that sets of computed structures show only the atom connectivities and not the bond multiplicities or the charge/radical sites.**

3.2 THERMOCHEMICAL DATA

Something must also be said concerning the thermochemical data supplied in this book. At present there are a number of sources available on web sites and in compendia, e.g., for neutral molecules and radicals, the texts by Luo (*Handbook of Bond Dissociation Energies in Organic Compounds*, CRC Press, Boca Raton, 2003) and Pedley, Naylor and Kirby (*Thermochemical Data of Organic Compounds*, 2nd ed., Chapman and Hall, London, 1986) and for ions (and neutrals), gas-phase ion and neutral thermochemistry, *J. Phys. Chem. Ref. Data* 1988, 17, Supplement 1, and the current NIST WebBook.

The data in these and in the scientific journals are, alas, not always compatible. Wherever possible we have struggled to steer a sensible path between rival data and to give a considered opinion as to the likely best, or at least the most consistent value. Where such situations arise it will be evident from the text that difficulties have been encountered and solutions proposed.

Note in particular that the results of computations often give the energies of the isomeric ions relative to some dissociation products whose $\Delta_f H$ values may be well established. In such cases, the reader can therefore estimate an absolute $\Delta_f H$(ion) using reliable data, obtaining them from this book or from sources such as those indicated above. In some instances, we have already performed this operation! This practice should be limited however to relative energies obtained only from high levels of theory.

3.3 IONS CONTAINING ONLY CARBON ATOMS

Carbon chains, rings, and fullerenes are generated by a variety of techniques, especially laser ablation of graphite, pyrolysis of graphite, and the electron-impact degradation of halocarbons.[1] They are intermediates in combustion and are reactive interstellar species.[2,3]

Mass spectra of carbon clusters (C_n^{1+}) generated by laser desorption of graphite typically have a bimodal distribution with progressions of ions between C_3^{1+} and C_{32}^{1+} and between C_{32}^{1+} and C_{80}^{+}.[4] The general structures for C_n^{1+} cations can be classified as chains ($n = 2$–9), rings ($n = 10$–30) and fullerenes ($n \geq 32$). On the whole, there have been relatively few isomeric structures found for individual C_n^{1+} ions. Some of these are noted below. In general, the mass spectrometry of the odd-numbered carbon chains is dominated by C_3 loss,[5] whereas the even-numbered fullerenes lose C_2 due to the stability of the even-numbered fullerene product ions.[6,7]

3.4 ISOMERS/IDENTIFICATION

3.4.1 $C_3^{]+}$ – m/z 36

The two isomers of $C_3^{]+}$ are linear and cyclic. High-level ab initio calculations by Scuseria[8] at the CCSD(T)/[5s44p3d2f1g] level of theory show that the bent, or cyclic, isomer ($<CCC = 67.8°$) is more stable than the collinear ion by 28.5 kJ/mol. It is unlikely that mass spectrometry could be used to distinguish these two isomers.

3.4.2 $C_4^{]+}$ – m/z 48

Charge-reversal experiments with ^{13}C-labeled linear C_4 anions produced $C_4^{]+}$ species that showed some positional mixing of ^{12}C and ^{13}C atoms before dissociating to $C_3^{]+}$ and $C_2^{]+}$ product ions.[9] The linear anions were easily produced in the mass spectrometer ion source from bis(trimethylsilyl)butadiene. Calculations at the B3LYP/6-31G(d) level of theory for geometries and using CCSD(T)/aug-cc-pVDZ for single-point energies show that the linear form is stable but was not the global minimum.[6] The latter is a rhombic species, some 15 kJ/mol lower in energy, but separated from the linear ion by a barrier of 84 kJ/mol. It was concluded that the above two isomers were involved in the isotope mixing. A third structure, a triangle with an attached C atom, was calculated to lie 170 kJ/mol above the rhombic form. References in this study show that previous calculations had identified only one stable $C_4^{]+}$ conformer.

3.4.3 $C_7^{]+}$, $C_8^{]+}$ AND $C_9^{]+}$ – m/z 84, 96 AND 108

The $C_n^{]+}$ ions with $n = 7$ to 9 were assumed to be primarily linear chains terminated by carbene carbons. These linear chains exhibit a variety of ion–molecule reactions that distinguish them from the higher order cyclic structures. McElvany and coworkers examined the ion–molecule reactions of ions from $C_3^{]+}$ to $C_{20}^{]+}$ with D_2,[10] O_2,[10] HCN[11] and the hydrocarbons (CH$_4$, C_2H_2, and C_2H_4)[12] in a Fourier transform ion cyclotron resonance (FT-ICR) cell. Three types of reactions were observed: hydrogen abstraction, addition of a carbon atom(s), and elimination of one or three carbon atoms.[12] For reactions with methane, the cyclic structures ($n \geq 10$) did not react. The measured ion–molecule reaction rate constant for $C_7^{]+}$ was unusually low compared with the other linear chains, leading to the conclusion that two isomers of this ion were present, a linear chain and a ring structure. Further evidence was found in the reactions with ethyne. Cyclic structures with $n \geq 10$ exhibited only addition reactions, whereas $C_3^{]+}$–$C_6^{]+}$ ions only eliminated carbon. The presence of both channels for $C_7^{]+}$–$C_9^{]+}$ ions indicated that all three had both linear and cyclic isomers. McElvany was able to isolate the cyclic C_7^{1+} structure by a series of tandem ion–molecule reactions.[12] C_7^{1+} (linear + cyclic) was first isolated in the trapping cell of an FT-ICR mass spectrometer. Reactions with D_2 were used to generate two products, C_7D^{1+} (a reaction that only linear chains can do) and unreacted C_7^{1+}, which must then have the cyclic structure. The now "pure" cyclic ion was reacted with C_2H_2 to form only one product, $C_9H_2^{1+}$.

3.4.4 C_{10}^{1+} – m/z 120

The C_{10}^{1+} ion is a ring structure having special stability. Mass-analyzed ion kinetic energy (MIKE) mass spectra of small carbon chains and rings from C_5^{1+} to C_{16}^{1+} are dominated by the loss of C_3. C_{11}^{1+}, however, loses C to form C_{10}^{1+}, an indication of this product's special stability.[5,7]

3.4.5 C_{58}^{1+}, C_{60}^{1+} AND C_{62}^{1+} – m/z 696, 720 AND 744

C_{60} is the most stable of the fullerene structures as evidenced by its intensity in the mass spectra of fullerenes generated by virtually any method.[4] It is also an ion of special stability. Radi et al.[13] performed MIKE experiments on C_{60}^{1+}, C_{62}^{1+} and C_{58}^{1+}. One observation was that metastable C_{62}^{1+} ions lose C_2 (as do most even-numbered fullerene ions) to produce C_{60}^{1+}. What was unusual about this mass spectrum was that the fragment ion was actually more intense than the precursor ion, showing the kinetic instability of C_{62}^{1+} vs C_{60}^{1+}. Of the three ions, C_{60}^{1+} had the smallest unimolecular decay rate constant, again indicative of its stability.

Thermochemistry

The thermochemistry of carbon cluster ions has not been well established due to the difficulty of their preparation. Most widely studied have been their ionization energies and the data are summarized in the following table:

Experimental and Calculated Ionization Energies for C_n Species

Species	Experiment		Theory	
	IE_a (eV)	IE_v (eV)	IE_a (eV)	IE_v (eV)
C_2	11.1 ± 0.1 (E)[14]	9.98–11.61 (P)[15]	12.1[a]	
C_3	12.97 ± 0.1 (B)[16]	9.98–11.61 (P)[15]	11.4[a]	12.95 MRD-CI[19]
	12.6 ± 0.6 (E)[17]			12.7 Eq-of-mo/4-31G[20]
	12.1 ± 0.3 (E)[18]			12.21 MRCI/6-31G(sp)[21]
	11.1 ± 0.1 (E)[14]			

(Continued)

Species	Experiment		Theory	
	IE$_a$ (eV)	IE$_v$ (eV)	IE$_a$ (eV)	IE$_v$ (eV)
C$_4$	12.54 ± 0.35 (B)[16]	9.98–11.61 (P)[15]	10.5[a]	
	12.6 (E)[17]			
C$_5$	12.26 ± 0.1 (B)[16]	9.98–11.61 (P)[15]	10.7[a]	
	12.5 ± 0.1 (E)[17]			
	12.7 ± 0.5			
	9.98–11.61			
C$_6$	9.70 ± 0.2 (B)[16]	9.98–12.84 (P)[15]	9.8 (lin)[b]	
	9.6 ± 0.3 (B)[22]			
	12.5 ± 0.3 (E)[23]			
C$_7$	8.09 ± 0.1 (B)[22]	9.98–12.84 (P)[15]		10.0 (lin)[c]
C$_8$	8.76 ± 0.1 (B)[22]	9.98–12.84 (P)[15]	9.2 (lin)[b]	
C$_9$	8.76 ± 0.1 (B)[22]	9.98–12.84 (P)[15]		9.4[c]
C$_{10}$	9.08 ± 0.1 (B)[22]	9.98–12.84 (P)[15]		
C$_{11}$	7.45 ± 0.1 (B)[22]	9.98–12.84 (P)[15]		
C$_{12}$	8.50 ± 0.1 (B)[22]	6.42–12.84 (P)[15]		
C$_{13}$	8.09 ± 0.1 (B)[22]	6.42–12.84 (P)[15]		
C$_{14}$	8.50 ± 0.1 (B)[22]	6.42–12.84 (P)[15]		
C$_{15}$	7.2 ± 0.1 (B)[22]	6.42–12.84 (P)[15]		
C$_{16}$	8.09 ± 0.1 (B)[22]	6.42–12.84 (P)[15]		
C$_{17}$	8.09 ± 0.1 (B)[22]	6.42–12.84 (P)[15]		
C$_{18}$	8.09 ± 0.1 (B)[22]	6.42–12.84 (P)[15]		
C$_{19}$	7.45 ± 0.1 (B)[22]	6.42–12.84 (P)[15]		
C$_{20}$	8.17 ± 0.2 (B)[22]	6.42–12.84 (P)[15]		
C$_{21}$	8.17 ± 0.2 (B)[22]	6.42–12.84 (P)[15]		
C$_{22}$	8.17 ± 0.2 (B)[22]	6.42–12.84 (P)[15]		
C$_{23}$	7.2 ± 0.3 (B)[22]	6.42–12.84 (P)[15]		
C$_{24}$	7.9 ± 0.2 (B)[22]	6.42–12.84 (P)[15]		
C$_{48}$	7.17 ± 0.1 (B)[24]	≤4.99[15]		
C$_{50}$	7.61 ± 0.11 (B)[24]	≤4.99[15]		
C$_{52}$	7.07 ± 0.1 (B)[24]	≤4.99[15]		
C$_{58}$	7.07 ± 0.1 (B)[24]	≤4.99[15]		
C$_{60}$	7.61 ± 0.11 (B)[24]	≤4.99[15]		
	7.61 ± 0.02 (PE)[25]			
	7.75 ± 0.01 (P)[26]			
	7.6 (E)[27]			
	7.8 (E)[28]			
	7.6 ± 0.5 (E)[29]			
	7.54 ± 0.04 (P)[30]			
	7.58 ± 0.03 (P)[31]			

(continued)

(Continued)

Species	Experiment		Theory	
	IE_a (eV)	IE_v (eV)	IE_a (eV)	IE_v (eV)
C_{62}	7.07 ± 0.1 (B)[24]	≤ 4.99[15]		
C_{66}	7.27 ± 0.14 (B)[24]	≤ 4.99[15]		
C_{68}	7.27 ± 0.14 (B)[24]	≤ 4.99[15]		
C_{70}	7.61 ± 0.11 (B)[24]	≤ 4.99[15]		
C_{72}	6.97 ± 0.1 (B)[24]	≤ 4.99[15]		
C_{74}	6.84 ± 0.1 (B)[24]	≤ 4.99[15]		
C_{80}	6.84 ± 0.1 (B)[24]	≤ 4.99[15]		
$C_{100-200}$	$6.20–6.45$ (B)[24]			

IE_a = adiabatic IE; IE_v = vertical IE; (B) = bracketing charge transfer reactions; (E) = electron impact ionization threshold; (P) = photoionization threshold; (PE) = photoelectron spectroscopy; (lin) = linear form.

[a]CCD(ST)/6-31G*//HF/6-31G(2d) level of theory.[32]
[b]CCD(ST)/6-31G*//HF/6-31G* level of theory.[32]
[c]CCD(ST)/6-31G*//HF/6-31G* level of theory,[32] IE determined by Koopman's theorem.

The trends observed by McElvany, Eyler and coworkers[33,34] in the ionization energies of C_n clusters demonstrate the stability of the C_{50}, C_{60} and C_{70} neutral fullerenes, these molecules having significantly larger IEs than the other fullerenes.

The only experimental heats of formation that have been reported in the literature were based on EI thresholds using non-energy resolved electrons and thus are quite unreliable.[23] Generally, the neutral $\Delta_f H$ values for C_n species below C_8 are known and can be combined with the IE values to give ion $\Delta_f H$ values.[35]

REFERENCES

1. McIlvany, S.W. *Int. J. Mass Spectrom. Ion Processes* **1990**, *102*, 81.
2. Kroto, H.W., McKay, K. *Nature* **1988**, *331*, 328.
3. Bettens, R.P.A., Herbst, J. *Astrophys. J.* **1997**, *478*, 585.
4. McElvany, S.W., Ross, M.M., Callahan, J.H. *Acc. Chem. Res.* **1992**, *25*, 162.
5. Lifshitz, C., Peres, T., Agranat, I. *Int. J. Mass Spectrom. Ion Processes* **1989**, *93*, 149.

6. Raghavachari, K., Binkley, S. *J. Chem. Phys.* **1987**, *87*, 2191.
7. Radi, P.P., Bunn, T.L., Kemper, P.R., Molchan, M.E., Bowers, M.T. *J. Chem. Phys.* **1988**, *88*, 2809.
8. Scuseria, G.E. *Chem. Phys. Lett.* **1991**, *176*, 27.
9. Blanksby, S.J., Schroder, D., Dua, S., Bowie, J.H., Schwarz, H. *J. Am. Chem. Soc.* **2000**, *122*, 7105.
10. McElvany, S.W., Dunlap, B.I., O'Keefe, A. *J. Chem. Phys.* **1987**, *86*, 715.
11. Parent, D.C., McElvany, S.W. *J. Am. Chem. Soc.* **1989**, *111*, 2393.
12. McElvany, S.W. *J. Chem. Phys.* **1988**, *89*, 2063.
13. Radi, P.P., Hsu, M.-T., Rincon, M.E., Kemper, P.R., Bowers, M.T. *Chem. Phys. Lett.* **1990**, *174*, 223.
14. Gupta, S.K., Gingerich, K.A. *J. Chem. Phys.* **1979**, *71*, 3072.
15. Rohlfing, E.A., Cox, D.M., Kaldon, A. *J. Chem. Phys.* **1984**, *81*, 3322.
16. Ramanathan, R., Zimmerman, J.A., Eyler, J.R. *J. Chem. Phys.* **1993**, *98*, 7838.
17. Drowart, J., Burns, R.P., Demaria, G., Inghram, M.G. *J. Chem. Phys.* **1959**, *31*, 1131.
18. Kohl, F.J., Stearns, C.A. *J. Chem. Phys.* **1970**, *52*, 6310.
19. Romelt, J., Peyerimhoff, S.D., Buenker, R. *J. Chem. Phys. Lett.* **1978**, *58*, 1.
20. Williams, G.R.J. *Chem. Phys. Lett.* **1975**, *33*, 582.
21. Sunil, K.K., Orendt, A., Jordan, K.D. *Chem. Phys.* **1984**, *89*, 245.
22. Bach, S.B.H., Eyler, J.R. *J. Chem. Phys.* **1990**, *92*, 358.
23. Dibeler, V.H., Reese, R.M., Franklin, J.F. *J. Am. Chem. Soc.* **1961**, *83*, 1813.
24. Zimmerman, J.A., Eyler, J.R. *J. Chem. Phys.* **1991**, *94*, 3556.
25. Lichtenberger, D.L., Jatcko, M.E., Nebesny, K.W., Ray, C.D., Huffmann, D.R., Lamb, L.D. *Chem. Phys. Lett.* **1991**, *176*, 203.
26. Yoo, R.K., Ruscic, B., Berkowitz, J. *J. Chem. Phys.* **1992**, *92*, 911.
27. Muigg, D., Scheier, P., Becker, K., Mark, T.D. *J. Phys. B* **1996**, *29*, 5193.
28. SaiBaba, M., Lakshmi Narasimhan, T.S., Balasubramanian, R., Mathews, C.K. *J. Phys. Chem.* **1995**, *99*, 3020.
29. Scheier, P., Dunser, B., Worgotter, R., Lezius, M., Robl, R., Mark, T.D. *Int. J. Mass Spectrom. Ion Processes* **1994**, *138*, 77.
30. Hertel, I.V., Steger, H., DeVries, J., Weisser, B., Menzel, C., Kamke, B., Kamke, W. *Phys. Rev. Lett.* **1992**, *68*, 784.
31. DeVries, J., Steger, H., Kamke, B., Menzel, C., Weisser, B., Kamke, W., Hertel, I.V. *Chem. Phys. Lett.* **1992**, *188*, 159.
32. Raghavachari, K., Binkley, J.S. *J. Chem. Phys.* **1987**, *87*, 2191.
33. McIlvany, S.W. *J. Chem. Phys.* **1991**, *94*, 3556.
34. Bruce, J.E., Eyler, J.R. *J. Chem. Phys.* **1993**, *98*, 7838.
35. NIST Chemistry WebBook, NIST Standard Reference Data Base Number 69, Mallard, W.G., Linstrom, P.J., eds, National Institute of Standards and Technology: Gaithersburg, MD, **2005**.

3.5 IONS CONTAINING ONE CARBON ATOM

3.5.1 $[C,H_n]^{+/+\bullet}$ – m/z 12, 13, 14, 15, 16 AND 17

Ions

The ground-state geometries for $CH_2^{+\bullet}$ and CH_3^{+} are bent[1] and planar D_{3h},[2] respectively. These small cations are involved in the first stages of molecular synthesis in interstellar clouds.[3,4]

Upon ionization, methane is subject to Jahn–Teller distortions and so the ion is not tetrahedral but rather C_{2v} (with two different C−H distances) and fluxional in nature.[5–7] The properties of $CH_4^{+\bullet}$ are of particular interest because this ion has been identified in the upper atmosphere ($CH_4^{+\bullet}$ is produced in the radiolysis of methane, which is believed to have been present in the early planetary atmosphere) and also has been proposed to be involved in the chemical evolution that must have preceded the origin of life.[8]

The CH_5^{+} ion was first observed as a stable species in the high-pressure mass spectrometry of methane[9] and is also produced in the reaction of CH_3^{+} with H_2.[10] It possesses a C_s structure with a 2-electron, 3-center bond and is considered to be the parent of nonclassical carbocations.[11]

Thermochemistry

Heat of Formation Values for $[C,H_n]^{+/+\bullet}$ Ions Where $n \leq 5$

Ions	Δ_fH(neutral) (kJ/mol)	IE (eV)	Δ_fH(ion)^a (kJ/mol)
C	716.68 ± 0.45	11.26030	1803
CH	594.13	10.64 ± 0.01	1621
	(587.73)		
CH_2	386.39	10.396 ± 0.003	1389
	(385.22)		
CH_3	145.69	9.84 ± 0.01	1095
	(143.97)		
CH_4	-74.87	12.618 ± 0.004^6	1142
CH_5	—	—	912^b

All values are taken from the NIST WebBook,[12] or otherwise stated; parentheses indicate a computed value (G3MP2B3).[13]

$^a\Delta_fH$(ion) $=$ IE $+ \Delta_fH$(neutral).
$^b\Delta_fH$(ion) $= \Delta_fH$(neutral) $+ \Delta_fH(H^+)$ − PA, where PA(CH_4) $= 543.5$ kJ/mol, $\Delta_fH(CH_4) = -74.87$ kJ/mol and $\Delta_fH(H^+) = 1530$ kJ/mol.[12]

3.5.2 $[C,N]^{+}$ – m/z 26

Combining the heat of formation and the ionization energy of the cyano radical (435.14 kJ/mol, in agreement with the computed value of 436.53 kJ/mol (G3MP2B3)[13]) and 13.598 eV, respectively[12] gives $\Delta_fH(CN^{1+}) = 1747$ kJ/mol.

3.5.3 [C,H,N]$^{+\bullet}$ – m/z 27

Isomers

The two possible isomers are HCN$^{1+\bullet}$, the hydrogen cyanide radical cation, and its tautomer HNC$^{1+\bullet}$, hydrogen isocyanide. The neutral counterparts of these ions are both of astrophysical interest.[14]

Identification

Dissociative ionization of several nitrogen-containing molecules such as CH$_3$NH$_2$, CH$_3$CN, HCONH$_2$, aniline, and benzonitrile generate HNC$^{1+\bullet}$ whereas ionization of hydrogen cyanide and dissociative ionization of pyridine both produce mostly HCN$^{1+\bullet}$ ions.[15–18] The m/z 12 to m/z 15 regions of the CID mass spectra are structure-characteristic and consistent with the respective connectivities of the ions. HNC$^{1+\bullet}$ ions show a more intense m/z 15 fragment than its isomer (m/z 15:12 = ~0.4 and 0.15 for HNC$^{1+\bullet}$ and HCN$^{1+\bullet}$, respectively) whereas HCN$^{1+\bullet}$ ions produce more m/z 13 than its isomer (m/z 13:12 = ~0.4 and 1 for HNC$^{1+\bullet}$ and HCN$^{1+\bullet}$, respectively).[17] The two ions have also been differentiated by their reactivities toward CF$_4$, the product distribution with SF$_6$,[18] and their charge exchange reaction with methanol.[16] Note that the neutral isomers have equally distinctive collision-induced dissociative ionization mass spectra.[15,19]

Thermochemistry

Using 135.14 kJ/mol and 13.60±0.01 eV for the heat of formation and ionization energy of HCN[12] leads to $\Delta_f H$(HCN$^{1+\bullet}$) = 1447 kJ/mol. The heat of formation of HNC$^{1+\bullet}$ and the ionization energy of the corresponding neutral have been measured to be 1349±4 kJ/mol and 12.04±0.01 eV,[20] respectively, yielding a $\Delta_f H$(HNC) = 187 kJ/mol. A calculated barrier for isomerization of only 79 kJ/mol separates HCN$^{1+\bullet}$ from its isomer.[21] Note that this lies far below the threshold for the lowest energy dissociation to H$^\bullet$+ CN^{1+} ($\Sigma \Delta_f H_{products}$ = 1965 kJ/mol),[12] therefore allowing isomerization to take place before this dissociation.

3.5.4 [C,H$_2$,N]$^+$ – m/z 28

Isomers

Three isomers have been identified by experiment, namely, the iminomethylene cation (HCNH^{1+}), the aminomethylidene cation (CNH$_2$$^{1+}$) and the methylenenitrenium cation (H$_2$CN^{1+}). These cations are of particular interest since they have been identified as the precursors for interstellar HCN and HNC.[22]

Identification

Protonation of HCN yields the isomer HCNH[1+],[17,23] whereas the CNH$_2$[1+] cation is generated in the dissociative ionization of 1-amino-2-methoxy-cyclobutenedione according to the following reaction scheme:[24]

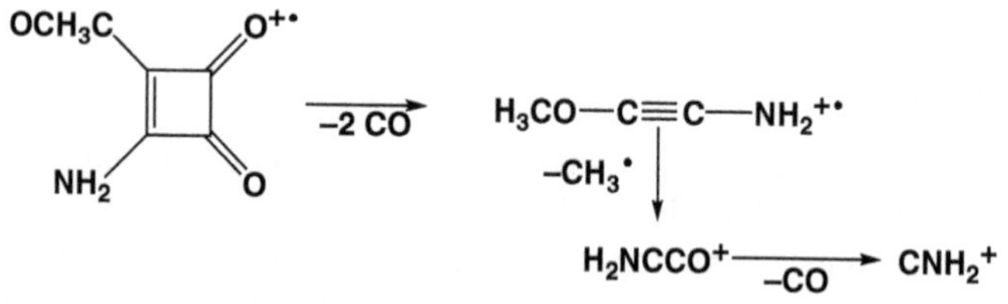

The dissociative ionization of nitrogen-containing molecules such as CH$_3$NH$_2$ and CH$_3$CN yields a mixture of isomeric species.[23]

The more elusive isomer H$_2$CN[1+] can be generated by the charge reversal of the corresponding anion.[23] The observation of a peak at m/z 28 in the charge reversal mass spectrum suggests that ions possessing the H$_2$CN connectivity occupy a potential energy well. Differentiation of the isomers is based on the relative fragment peak intensities in the m/z 12 to m/z 16 regions of their respective CID mass spectra.

Partial CID Mass Spectra of the [C,H$_2$,N]$^+$ Isomers[23,24]

	m/z				
Ions	12	13	14	15	16
HCNH	68	100	28	78	13
CNH$_2$	100	57	22	47	58
H$_2$CN	15	23	100	8	—

Note that for HCNH[1+] ions, a very intense narrow peak is also observed at m/z 14, corresponding to HCNH[2+].[17,23,24]

Thermochemistry

From the proton affinity of HCN, $\Delta_f H(\text{HCNH}^{1+}) = 952$ kJ/mol (using PA(HCN) = 712.9 kJ/mol, $\Delta_f H(\text{H}^+) = 1530$ kJ/mol and $\Delta_f H(\text{HCN}) = 135.14$ kJ/mol).[12] This ion is the global minimum.

Calculations (MP4 SDTQ/6-311G(d,p)//MP26-31G(d)) show that on the singlet potential energy surface, the CNH$_2$[1+] ion is the only other species lying in a potential energy well, 218 kJ/mol above HCNH[1+], therefore leading to $\Delta_f H(CNH_2^{1+}) = 1170$ kJ/mol.[25] This is significantly higher than the measured $\Delta_f H(CNH_2^{1+}) = 1109 \pm 38$ kJ/mol.[23] The calculated (MP4 SDTQ/6-311G(d,p)//MP26-31G(d)) barrier for the isomerization of this ion to HCNH^{1+} is 84 kJ/mol.[25]

H_2CN^{1+} is a saddle point on the singlet potential energy surface and is estimated to lie 310 kJ/mol above the linear isomer.[25] However, this ion is a minimum on the *triplet* potential energy surface (502 kJ/mol above HCNH^{1+}) and most likely is the ion formed in the charge reversal experiment.[25] For an analogous system, refer to the methoxy cation described below.

3.5.5 [C,H$_3$,N]$^{+\bullet}$ – m/z 29

Isomers

There are two stable isomers, the aminocarbene radical cation (HCNH$_2^{1+\bullet}$) and ionized formaldimine (CH$_2$NH$^{1+\bullet}$). A third isomer, the methylnitrenium ion (CH$_3$N$^{1+\bullet}$), was shown by calculations to collapse without activation to CH$_2$NH$^{1+\bullet}$.[26]

Identification

Dissociative ionization of cyclopropylamine and azetidine (cy–C(H)$_2$C(H)$_2$ C(H)$_2$N(H)–) generate (loss of C$_2$H$_4$ in both cases) essentially pure HCNH$_2^{1+\bullet}$ and CH$_2$NH$^{1+\bullet}$ fragment ions, respectively.[23,27] Other amines such as methyl[23] and isopropylamine[28] generate the aminocarbene radical cation, HCNH$_2^{+\bullet}$. The m/z 12 to m/z 16 regions in the ions' CID mass spectra are structure-indicative and therefore allow their differentiation.[23,27]

Only the carbene ion shows a doubly charged ion at m/z 14.5 corresponding to HCNH$_2^{1 2+}$.

Partial CID Mass Spectra of the [C,H$_3$,N]$^{+\bullet}$ Isomers[27]

			m/z			
Ions	12	13	14	15	16	17
HCNH$_2$	43	100	45	50	93	10
CH$_2$NH	31	56	100	67	14	—

Thermochemistry

The aminocarbene radical cation HCNH$_2^{1+\bullet}$ is the lower energy isomer with a calculated $\Delta_f H(HCNH_2^{1+\bullet}) = 1030 \pm 10$ kJ/mol (UQCISD(T)/6-311++ G(3df,2p), RCCSD(T)/cc-pVTZ and G2MP2).[28,29] Theory also predicts $\Delta_f H(CH_2NH^{1+\bullet}) = 1046 \pm 10$ kJ/mol (UQCISD(T)/6-311++G(3df,2p),

RCCSD(T)/cc-pVTZ, G2(MP2), and CBS-Q).[29,30] This value agrees well with that obtained from an appearance energy measurement of 1046 ± 8 kJ/mol.[31] The two ions are separated by a large barrier to their interconversion, 211 kJ/mol above $HCNH_2^{]+\bullet}$, effectively preventing their isomerization before dissociation by $H^\bullet$ loss (UQCISD(T)/6-311++G(3df,2p)//UMP2/6-31G(d,p) + ZPE).[29]

3.5.6 $[C,H_4,N]^+$ – m/z 30

Isomers

Only one isomer has been investigated by experiment, namely, the methaniminium (or methylenimine) cation, $CH_2NH_2^{]+}$.

Identification

The metastable ion dissociates by loss of H_2 with a large kinetic energy release, in a manner similar to its oxygen analog, $^+CH_2OH$. Deuterium labeling experiments show that scrambling of the hydrogen atoms does not occur.[32]

Thermochemistry

The heat of formation of the ion has recently been calculated to be 751 ± 2 kJ/mol (Weizmann-2 thermochemical approach),[33] in agreement with earlier calculations[34,35] and with an experimental appearance energy value, 745 ± 8 kJ/mol.[36] This is also consistent with 760 kJ/mol, obtained by combining the ionization energy and heat of formation of the neutral $CH_2NH_2^{]\bullet}$ (IE $= 6.29 \pm 0.03$ eV[12] and a computed $\Delta_f H(CH_2NH_2^{]\bullet}) = 153.49$ kJ/mol (G3MP2B3)[37]).

3.5.7 $[C,H_5,N]^{+\bullet}$ – m/z 31

Isomers

Both the conventional and distonic ions, $CH_3NH_2^{]+\bullet}$ (methylamine radical cation) and $CH_2NH_3^{]+\bullet}$ (methylenammonium radical cation), respectively, have been investigated by experiment.

Identification

Ionization of methylamine generates the $CH_3NH_2^{]+\bullet}$ radical cation whereas the distonic isomer, $CH_2NH_3^{]+\bullet}$, is readily produced by the dissociative ionization of ethanolamine (loss of H_2CO).[38] The m/z 12 to m/z 18 regions of the CID mass spectra of the ions are characteristic of their respective connectivities.[38,39]

Partial CID Mass Spectra of the [C,H$_5$,N]$^{+\bullet}$ Isomers[38]

Ions	m/z						
	12	13	14	15	16	17	18
CH$_3$NH$_2$	9.5	17	38	100	48	11	1.9
CH$_2$NH$_3$	28	41	100	31	41	86	17

Note that for the distonic ion, the base peak in the m/z 12 to m/z 18 region is the *doubly charged ion* CH$_2$NH$_3$$^{]2+}$ at m/z 15.5.[38] This is similar to the behavior of its oxy-analog. This ion also exhibits a doubly charged ion at m/z 15 (CH$_2$NH$_2$$^{2+(\bullet)}$) whereas for the conventional isomer, doubly charged ions are seen at m/z 14.5 and 15.5.

Thermochemistry

Combining the heat of formation and ionization energy of methylamine (-23.0 ± 0.4 kJ/mol and 8.97 eV, respectively)[40] gives $\Delta_f H$(CH$_3$NH$_2$$^{]+\bullet}$) = 842 kJ/mol. The heat of formation of the distonic isomer has been calculated to be 851 kJ/mol (G2),[41] in good agreement with earlier calculations[34,42] and an experimental value of <841 kJ/mol.[40] A substantial barrier to isomerization, 168 kJ/mol, is calculated to separate CH$_3$NH$_2$$^{]+\bullet}$ ions from CH$_2$NH$_3$$^{]+\bullet}$ (MP3/6-31G**//HF/6-31G* with ZPVE correction).[42]

3.5.8 [C,H$_6$,N]$^+$ – m/z 32

The protonation of methylamine yields the CH$_3$NH$_3$$^{]+}$ cation. Its heat of formation, 608 kJ/mol, is obtained from the proton affinity and the heat of formation of methylamine (PA(CH$_3$NH$_2$) = 899 kJ/mol,[12] $\Delta_f H$(CH$_3$NH$_2$) = -23.0 kJ/mol,[40] and $\Delta_f H$(H$^+$) = 1530 kJ/mol[12]).

3.5.9 [C,N$_2$]$^{+\bullet}$ – m/z 40

Isomers

There are two isomers: cyanonitrene (NCN$^{]+\bullet}$) and diazocarbene (CNN$^{]+\bullet}$).

Identification

These ions have been investigated mostly by theory, although the NCN$^{]+\bullet}$ radical cation has been generated by the charge reversal of the corresponding anion. The cation so produced dissociates to yield m/z 28 (N$_2$$^{+\bullet}$), 26 (CN$^+$), 14 (N$^+$) and 12 (C$^{+\bullet}$).[43]

Thermochemistry

By combining the current recommended values for the neutral heats of formation of NCN and CNN (465 kJ/mol[44] and 584.51 kJ/mol,[12,45] respectively) with their calculated ionization energies (12.52 ± 0.07 eV (RCCSD(T)/aug-cc-pvtz//B3LYP/6-31G + G(d)[43] and G2[46]) and 11.01 ± 0.07 eV (G2[46]), respectively): $\Delta_f H(NCN^{1+\bullet}) = 1673$ kJ/mol and $\Delta_f H(CNN^{1+\bullet}) = 1647$ kJ/mol are obtained. Note that these values should be considered as estimates because data available are limited and the agreement between the reported values, especially for the heat of formation of the neutrals, is not very good.[12,44,47,48]

3.5.10   [C,H,N$_2$]$^+$ – m/z 41

All recent ab initio calculations agree that there are two isomeric [CH,N$_2$]$^\bullet$ radicals: the diazomethyl radical ([HCNN]$^\bullet$) and the cyclic analog (the diazirinyl radical) with calculated heats of formation of 460 ± 8 and 485 ± 8 kJ/mol, respectively (CBS-4, CBS-Q and CBS-QCI/APNO).[49]

The corresponding cations have been investigated solely by theory. Recently, calculations at the CBS-4, CBS-Q and CBS-APNO levels of theory have shown that ionization of the [HCNN]$^\bullet$ (diazomethyl) radical could generate the singlet (planar) and triplet (linear) states of the cation, the latter being lower in energy.[47] The corresponding ionization energies have been computed to be 10.33 ± 0.04 eV and 9.63 ± 0.04 eV, respectively.[47] Combining the heat of formation and ionization energies leads to $\Delta_f H(^1HCNN^{1+}) = 1457$ kJ/mol and $\Delta_f H(^3HCNN^{1+}) = 1414$ kJ/mol. Early computations, using a modest level of theory (6-31G*//STO-3G), have suggested that the lowest energy form of the cation is the linear $^3HNCN^{1+}$ with the isomers $^3HCNN^{1+}$ and cy–$^1C(H)NN–^{1+}$ predicted to lie 51 and 154 kJ/mol, respectively, above $^3HNCN^{1+}$.[50]

Since no mass spectral characteristics have been reported, it is unclear as to which ion is produced in the dissociation of ionized diazomethane and diazirine (loss of atomic hydrogen).[51] The appearance energy of the [C,H,N$_2$] ions produced in these dissociations can however be used to estimate the heat of formation of the ions generated: $\Delta_f H([C,H,N_2]^+)_{diazomethane} = 1490$ kJ/mol and $\Delta_f H([C,H,N_2]^+)_{diazirine} = 1474$ kJ/mol (AE(diazomethane) = 14.8 ± 0.1 eV and AE(diazirine) = 14.2 ± 0.1 eV; $\Delta_f H$(diazomethane) = 281 kJ/mol (G2 calculation),[52] $\Delta_f H$(diazirine) = 322 kJ/mol (G2 calculation)[52] and $\Delta_f H(H^\bullet) = 217.998 \pm 0.006$).[51] These values suggest that the diazomethyl cation could be generated by both reactions but no firm conclusion can be drawn.

It is noteworthy that the diazomethyl radical (HCNN$^\bullet$) and other nitrogen containing species are of interest because of their observation in the interstellar medium and also because of their possible role in the formation of amino acids in the early stages of the formation of the solar system.[47]

3.5.11 $[C,H_2,N_2]^{+\bullet}$ – m/z 42

Isomers

Three isomers have been characterized by experiment, namely, ionized diazomethane ($CH_2N_2^{1+\bullet}$), ionized nitrilimine ($HCNNH^{1+\bullet}$) and ionized cyanamide ($NCNH_2^{1+\bullet}$) and theory provides some thermochemical data pertaining to a fourth isomer, N-isocyano-amine ($CNNH_2^{1+\bullet}$).

Identification

Gently heating a mixture of (wet) NaOH and N-nitroso-4-toluene sulfomethylamide in the ion source of the mass spectrometer produces ionized diazomethane, $CH_2N_2^{1+\bullet}$.[53] Ionized nitrilimine ($HCNNH^{1+\bullet}$) is obtained from the dissociative ionization of 1,2,4-triazol (by loss of HCN). Their respective CID mass spectra are structure-characteristic.[53]

Partial CID Mass Spectra of the $[C,H_2,N_2]^{+\bullet}$ Isomers[53]

	m/z								
Ions	12	13	14	15	16	26	27	28	29
CH_2N_2	13	22	38	—	—	19	44	100	19
HCNNH	6	24	3	2	0.6	12	40	100	93

Ionized cyanamide ($NCNH_2^{1+\bullet}$) is believed to be produced free from contamination by the direct sublimation of the molecule in the ion source at a sample temperature as low as possible.[54] The only mass spectral data available pertain to its metastable dissociation by loss of atomic nitrogen to generate the $HNCH^{1+}$ fragment ion. This dissociation is characterized by a dished-top peak ($T_{0.5} = 503 \pm 3$ meV), which could be indicative of a rearrangement before dissociation according to the following sequence:[54]

$$H_2NCN^{1+\bullet} \rightarrow HNCHN^{1+\bullet} \rightarrow HNCH^{1+} + N^\bullet$$

Precursor ions such as guanidine and 3,5-diamino-1,2,4-triazole are also reported to generate $NCNH_2^{1+\bullet}$.[54]

Thermochemistry

Combining the heat of formation and ionization energy of diazomethane (268 kJ/mol[49] and 9.0 ± 0.2 eV,[55] respectively) gives $\Delta_f H(CH_2N_2^{1+\bullet}) = 1136$ kJ/mol. Note that the heat of formation value for diazomethane listed in NIST, 206 ± 9.6 kJ/mol,[12] was obtained from ion chemistry[51] and is certainly too low. Recently, theoretical and ion chemistry values ranging from 268 to 292 kJ/mol have been reported.[45,49,52] By combination with other thermochemical data for related ions and neutrals, the lower value (268 kJ/mol) appears to be best and was therefore used here.

Calculations predict that the $HCNNH^{1+\bullet}$ ion lies higher in energy than $CH_2N_2^{1+\bullet}$ by 98 kJ/mol.[53] Using the calculated heat of formation and ionization energy values for the nitrilimine molecule, 370 kJ/mol and 9.15 ± 0.1 eV, respectively,[56] leads to $\Delta_f H(HCNNH^{1+\bullet}) = 1253$ kJ/mol, in fair agreement with the predicted relative energies of the two isomers. It is however noteworthy that there are some discrepancies between calculated $\Delta_f H(HCNNH)$ values at the different levels of theory (MP4SDTQ/ 6-31G**//MP2(full)/ 6-31G** [57] and B3LYP/6-311 + G(d,p)[53] compared with G2[56]). A calculated barrier to isomerization of 400 kJ/mol separates $CH_2N_2^{1+\bullet}$ from $HCNNH^{1+\bullet}$ (B3LYP/6-311 + G(d,p)).[53]

The cyanamide molecule is calculated to be the global minimum on the neutrals' potential energy surface, lying 129 kJ/mol below diazomethane (MP4SDTQ/6-31G**//MP2(full)/6-31G**),[57] therefore giving an estimated $\Delta_f H(NCNH_2) = 76$ kJ/mol. Combining this value with the reported ionization energy of cyanamide (10.65 eV)[58] gives $\Delta_f H(NCNH_2^{1+\bullet}) = 1180$ kJ/mol.

The fourth isomer, N-isocyano-amine ($CNNH_2^{1+\bullet}$), is calculated to lie 164 kJ/mol above $CH_2N_2^{1+\bullet}$, with a barrier of 296 kJ/mol preventing its isomerization to $HCNNH^{1+\bullet}$ (B3LYP/6-311 + G(d,p)).[53] An absolute heat of formation for this ion of 1374 kJ/mol can be obtained using an estimated $\Delta_f H(CNNH_2) = 344$ kJ/mol (average of the calculated relative energies of CH_2N_2 and $CNNH_2$ at the MP4SDTQ/6-31G**//MP2(full)/ 6-31G** [57] and B3LYP/6-311 + G(d,p)[53] levels of theory) and a calculated ionization energy of 10.68 eV (extended-2ph-TDA[59] and Green function[60]). This value is considerably higher than that predicted from the relative energies of the two isomers. Clearly, these ions are in need of further work.

3.5.12   [C,H$_3$,N$_2$]$^+$ – m/z 43

Isomers

Cyano nitrogen protonated cyanamide (H_2NCNH^{1+}) has been investigated by experiment. It is also assumed that the methyl diazonium ion ($CH_3N_2^{1+}$) can be produced but no mass spectral data are available. Finally, calculations propose that the ions H_3NCN^{1+} and cy–NC(H)N(H)$_2$–$^{1+}$ are stable species.

Identification

Protonation of cyanamide (H_2NCN) by various means generates a [C,H$_3$,N$_2$]$^+$ ion whose CID mass spectral characteristics are given below and are consistent with the H_2NCNH^{1+} connectivity.[54]

Partial CID Mass Spectra of Protonated Cyanamide[54]

				m/z					
Ion	12	13	14	15	16	17	26	27	28
H$_2$NCNH	9	18	9	11	29	10	28	100	15

The dissociative ionization of azomethane (CH$_3$NNCH$_3$, loss of CH$_3^{1\bullet}$) generates an ion whose structure has been assumed to keep the connectivity of the precursor molecule, namely the methyldiazonium ion, CH$_3$N$_2^{1+}$.[61] No mass spectral data have been presented to support this assumption.

Thermochemistry

Calculations predict that the most favored protonation site in cyanamide is the cyano nitrogen atom, yielding the H$_2$NCNH^{1+} ion as the global minimum (MP4/6-311G**//MP2/6-31G*).[54] From the PA of cyanamide, Δ_fH(H$_2$NCNH^{1+}) = 876 kJ/mol is obtained (PA(H$_2$NCN) = 805.6 kJ/mol,[12] using a calculated Δ_fH(H$_2$NCN) = 152 kJ/mol (B3LYP/6-311 + G(d,p))[53] and Δ_fH(H$^+$) = 1530 kJ/mol[12]).

An appearance energy measurement leads to Δ_fH(CH$_3$N$_2^{1+}$) = 891 ± 8 kJ/mol (using AE = 9.20 ± 0.03 eV,[61] Δ_fH(CH$_3$NNCH$_3$) = 149 ± 5 kJ/mol[62] and Δ_fH(CH$_3^{1\bullet}$) = 145.69 kJ/mol[12]). In addition, the binding energy with respect to dissociation to CH$_3^{1+}$ and N$_2$ ($\Sigma\Delta_fH_{products}$ = 1095 kJ/mol) has been determined to be 204 kJ/mol.[61] This is somewhat larger than that predicted by recent calculations (173 kJ/mol at the RB3LYP/6-311 + G(d) or G2(MP2) levels of theory).[63]

Protonation of cyanamide at the amino nitrogen and at carbon produces the stables species H$_3$NCN^{1+} and cy–NC(H)N(H)$_2^{-1+}$, respectively, lying 94 and 226 kJ/mol higher in energy than H$_2$NCNH^{1+} (MP4/6-311G**//MP2/6-31G*).[54]

3.5.13 [C,H$_4$,N$_2$]$^{+\bullet}$ – m/z 44

Isomers

There are two isomers: the diaminocarbene radical cation (H$_2$NCNH$_2^{1+\bullet}$) and its tautomer, the formamidine radical cation (H$_2$NC(H)NH$^{1+\bullet}$).

Identification[64]

These ions are generated by the reactions shown below.

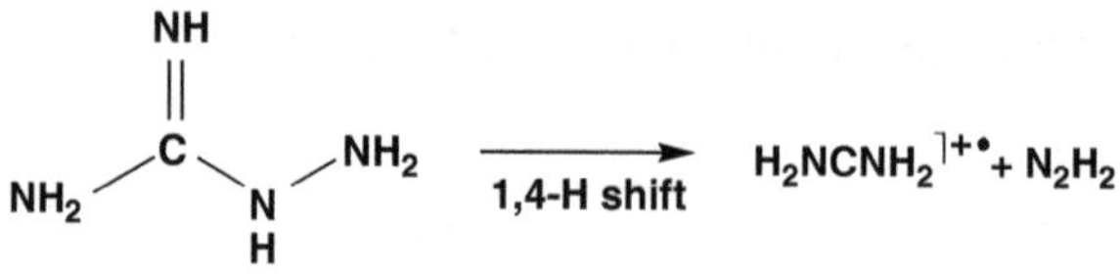

Both CID mass spectra contain the same fragment peaks but their relative intensities suffice to differentiate them.

Partial CID Mass Spectra of the $[C,H_4,N_2]^{+\bullet}$ Isomers

					m/z					
Ions	12	13	14	15	16	17	26	27	28	29
H_2NCNH_2	0.62	1.9	1.4	2.1	7.3	(85)	4.2	21	100	27
$H_2NC(H)NH$	1.2	2.6	1.7	2.6	17	30	6.3	32	100	13

Value in parenthesis contains a contribution from metastable ion dissociation.

Also, the dominant metastable dissociation for these ions is different: $H_2NCNH_2^{+\bullet}$ loses $H^\bullet$ whereas $H_2NC(H)NH^{+\bullet}$ yields $NH_3^{+\bullet}$. It is noteworthy that both isomers produce a significant doubly charged ion, albeit more intense for the carbene radical cation.

Thermochemistry

The latest calculations give $\Delta_f H(H_2NCNH_2^{+\bullet}) = 910$ kJ/mol (G2 and G2MP2)[28] somewhat lower than an earlier G2 value of 925 kJ/mol.[64] The tautomer $H_2NC(H)NH^{+\bullet}$, is predicted to be 27 kJ/mol higher in energy.[28] The energy barrier for the 1,2-hydrogen shift from nitrogen to carbon is calculated to be 247 kJ/mol (G2 and G2MP2).[28] An exhaustive computational study on the corresponding neutral species also predicts a high 1,2-hydrogen shift energy barrier (188–222 kJ/mol).[65] There are no experimental data available.

3.5.14 $[C,H_5,N_2]^+ - m/z$ 45

Isomers

Calculations yield six equilibrium structures: CH_3NHNH^{+}, $CH_3NNH_2^{+}$, $CH_2NHNH_2^{+}$, cy–$C(H_2)N(H)N(H)_2$–$^{+}$, $CH_2NNH_3^{+}$ and $[HNC \cdots NH_4]^+$.

Of these, the diazapropylium ions CH_3NHNH^{1+}, $CH_3NNH_2^{1+}$ and $CH_2NHNH_2^{1+}$ have been investigated by experiment.

Identification[66]

The dissociative ionization of 1,2- and 1,1-dimethylhydrazine ($CH_3NHNHCH_3$ and $(CH_3)_2NNH_2$, loss of $CH_3^{1\bullet}$) generates CH_3NHNH^{1+} and $CH_3NNH_2^{1+}$ ions, respectively. $CH_2NHNH_2^{1+}$ ions are obtained from the dissociative ionization of 2-hydrazinoethanol ($H_2NNHCH_2CH_2OH$, loss of $^\bullet CH_2OH$) and methylhydrazine (CH_3NHNH_2, loss of $H^\bullet$). The CID mass spectra of CH_3NHNH^{1+} and $CH_3NNH_2^{1+}$ ions are somewhat similar but differ significantly from that of CH_3NHNH^{1+} ions.

Partial CID Mass Spectra of the $[C,H_5,N_2]^+$ Isomers

Ions	m/z							
	15	16	17	22.5[a]	27	28	29	30
CH_3NHNH	15	10	5	—	10	80	100	1
CH_3NNH_2	20	20	20	9	30	100	80	5
CH_2NHNH_2	5	10	60	100	20	100	15	10

[a]Corresponds to the doubly charged ion $[C,H_5,N_2]^{+2}$.

It is noteworthy that the three species show metastable loss of HCN (m/z 18, NH_4^{1+}) and NH_3 (m/z 28, $HCNH^{1+\bullet}$) and it was suggested that the reacting configurations for the HCN and NH_3 losses were the $CH_2NNH_3^{1+}$ and $CH_2NHNH_2^{1+}$ ions, respectively.[67]

Thermochemistry

Calculations find the complex $[HCN\cdots NH_4]^+$ to be the global minimum.[68] Among the covalently bonded species, the $CH_2NHNH_2^{1+}$ ion is calculated to be the lowest energy isomer with CH_3NHNH^{1+} and $CH_3NNH_2^{1+}$ ions lying 43 and 44 kJ/mol or 37 and 33 kJ/mol higher in energy (depending on the level of theory: MP4/6-311++G(2d,2p)[68] and SDCI/ 6-31G**//6-31G** with ZPVE,[67] respectively. Several appearance energy measurements for the $CH_2NHNH_2^{1+}$ ion have yielded heat of formation values scattered in the range of 809–861 kJ/mol.[66,67]

Using the proton affinity values listed in NIST for methyldiazene (CH_3NNH) at the interior and terminal nitrogen atoms, 841 and 845 kJ/mol respectively,[12] leads to $\Delta_f H(CH_3NHNH^{1+}) = 877$ kJ/mol and $\Delta_f H (CH_3NNH_2^{1+}) = 873$ kJ/mol, ($\Delta_f H(CH_3NNH) = 188 \pm 8$ kJ/mol[40] and $\Delta_f H(H^+) = 1530$ kJ/mol).[12] However, it is noteworthy that the most recent calculations predict the proton affinity of methyldiazene to be significantly lower (813 $\pm$ 12 kJ/mol),[68] indicating that the absolute heats of formation for these ions are still uncertain.

The three ions are certainly discrete species occupying deep potential energy wells with isomerization barriers calculated to be 246 or 312 kJ/mol for $CH_3NHNH^{1+} \longrightarrow CH_3NNH_2^{1+}$ and 165 or 210 kJ/mol for $CH_3NNH_2^{1+} \longrightarrow CH_2NHNH_2^{1+}$ depending on the level of theory (MP4/6-311++G(d,p) //MP2/6-31G(d,p)+ZPE[68] and SDCI/6-31G**//6-31G** with ZPVE,[67] respectively). No reliable experimental values for the heat of formation of these ions are available.

The isomers $cy–C(H)_2N(H)N(H)_2–^{1+}$ and $CH_2NNH_3^{1+}$ are computed to be higher in energy, lying respectively, 84 or 70 kJ/mol and 14 or 10 kJ/mol above $CH_2NHNH_2^{1+}$ depending on the level of theory used (MP4/ 6-311++G(2d,2p)[68] and SDCI/6-31G**// 6-31G** with ZPVE,[67] respectively). They both occupy deep potential energy wells with high barriers preventing their isomerization to $CH_2NHNH_2^{1+}$ ($cy–C(H)_2N(H)N(H)_2–^{1+} \longrightarrow CH_2NHNH_2^{1+}$, 111 kJ/mol and $CH_2NNH_3^{1+} \longrightarrow CH_2NHNH_2^{1+}$, 210 kJ/mol).[68]

3.5.15 $[C,N_3]^+ – m/z\ 54$

Isomers

One stable species, namely the cyanodiazonium ion (NCN_2^{1+}), has been generated and characterized. Its isomer, the CN_3^{1+} ion, has also been reported by theory to be stable.

Identification[69]

A 2:1 mixture of NF_3 and H_2NCN in a chemical ionization source yields an m/z 54 ion whose CID mass spectral characteristics are consistent with the NCN_2^{1+} connectivity: m/z 26 (CN^+, 100%), 28 ($N_2^{+\bullet}$, 15%), 40 (CN_2^+, 13%), 12 ($C^{+\bullet}$, 1%) and 14 (N^+, 1%).

Thermochemistry

Only the relative energy of the two isomeric forms has been computed. The NCN_2^{1+} ion is predicted to be more stable than the CN_3^{1+} ion by 121 or 188 kJ/mol at the HF and MP2 levels of theory, respectively.[70]

3.5.16 $[C,O]^{+\bullet} – m/z\ 28$

This ion dissociates by loss of atomic oxygen to produce m/z 12 ($C^{+\bullet}$) in its CID mass spectrum. From the ionization energy and the heat of formation of the neutral (14.014$\pm$0.003 eV and -110.53 ± 0.17 kJ/mol, respectively),[12] $\Delta_f H(CO^{+\bullet}) = 1242$ kJ/mol is obtained.

3.5.17 $[C,H,O]^+ - m/z\ 29$

Isomers

The two isomers are HCO^{1+} and HOC^{1+}, the formyl and isoformyl cations, respectively. Particular interest in this pair of ions was prompted by their detection in interstellar clouds. HCO^{1+} was the first to be identified[71] and is the more abundant of the two. Laboratory identification of HOC^{1+} came shortly after.[72] The latter ion has also been shown to play an important role in the chemistry of flames and plasma.[73]

Identification

The dissociative ionization of molecules such as glyoxal $((HCO)_2$, loss of $HCO^{1\bullet})$, CH_3CHO (loss of $CH_3^{1\bullet}$) and $CH_3OC(O)H$ (loss of $CH_3O^{\bullet}$) can be used to generate essentially pure HCO^{1+} ions while dissociative ionization of CD_3OH (consecutive losses of D_2 and $D^{\bullet}$)[74] mainly produces HOC^{1+}, with HCO^{1+} being cogenerated in small amounts.[74,75] Several ion–molecule reactions have also been used to generate mixtures of the two isomers. For example, $H_3^{1+} + CO$,[75] $CO^{1+\bullet} + H_2$ and $C^+ + H_2O$.[76]

Mass spectrometric characterization of the two species relies on the fragment peak abundance ratio m/z 12:13 of the peaks in the CID mass spectra of the ions. A large ratio (≥ 3.0) is indicative of the $H-OC^{1+}$ connectivity whereas a smaller ratio (≈ 1.0) characterizes the $H-CO^{1+}$ connectivity.[74,75] It should be noted that the relative abundance of these two peaks depends on the translational kinetic energy of the ions,[77] probably because the isomers can interconvert below their first dissociation limit (see below).

The different reactivities of the isomers can also be used to characterize mixtures of them.[76,78,79] As a result of the low proton affinity of oxygen vs carbon in CO, 456.3 vs 594 kJ/mol,[12] HOC^{1+} will more readily transfer a proton to a wider variety of molecules than will HCO^{1+}. For example, the rate of proton transfer to N_2O (PA at O and N is 575 and 550 kJ/mol,[12] respectively) is almost three orders of magnitude faster for HOC^{1+} than for HCO^{1+}.[76] In the case of proton transfer to C_2H_2 (PA = 641 kJ/mol),[12] the reaction cross-section with HOC^{1+} is three times larger than for the same reaction with HCO^{1+}.[78] The reaction with H_2 is of particular interest because, in addition to proton transfer, the isomerization of HOC^{1+} to HCO^{1+} was observed to take place.[76,79] This process, where the high energy isomer is converted to the lower energy isomer in the presence of a small neutral molecule, is now known as "proton-transport catalysis" and is not uncommon.[80] Recent theoretical calculations have extensively investigated this phenomenon for the HOC^{1+} and HCO^{1+} isomers and they show that in the presence of the neutral molecule, the barrier to interconversion is substantially reduced or even reversed.[81,82]

Thermochemistry

From the proton affinity of the CO molecule at carbon and at oxygen, 594 and 426.3 kJ/mol, respectively,[12] $\Delta_f H(HCO^{1+}) = 825$ kJ/mol and $\Delta_f H(HOC^{1+}) = 993$ kJ/mol (using $\Delta_f H(CO) = -110.55 \pm 0.17$ kJ/mol and $\Delta_f H(H^+) = 1530$ kJ/mol).[12] These values are in agreement with high level ab initio calculations[83] and several experimental determinations.[76,84–86] The energy barrier for the isomerization of HCO^{1+} to HOC^{1+} is 316 kJ/mol (calculated at the B3LYP/6-311++G(d,p) and G2* level of theory).[81,82] This allows the isomers to interconvert well below the threshold for the lowest energy dissociation: $H^+ + CO$ ($\Sigma\Delta_f H_{products} = 1419$ kJ/mol).[12]

Note that because the barrier to the isomerization of these two ions (at 1141 kJ/mol) lies significantly below their lowest dissociation limit (1419 kJ/mol), an experiment that mass selects those ions with insufficient energy to decompose, such as a collision-induced dissociation study, must access both structures. This is well exemplified by the CID mass spectrum of HOC^{1+} ions which contains a signal at m/z 13.

3.5.18 $[C,H_2,O]^{+\bullet}$ – m/z 30

Isomers

Three isomers are predicted to be stable.[87] Ionized formaldehyde ($H_2CO^{1+\bullet}$), ionized hydroxymethylene ($HCOH^{1+\bullet}$) and the oxonium ion $COH_2^{1+\bullet}$. To date, only the first two ions have been generated and characterized by experiment.

Identification

$H_2CO^{1+\bullet}$ and $HCOH^{1+\bullet}$ radical cations are generated by the ionization of formaldehyde and the dissociative ionization of cyclopropanol (loss of C_2H_4) or methanol (loss of H_2), respectively. The m/z 12 to m/z 18 region of their CID mass spectra differ sufficiently to allow identification of the isomers.[88,89]

Partial CID Mass Spectra of the $[C,H_2,O]^{+\bullet}$ Isomers[89]

Ions	m/z					
	12	13	14	16	17	18
H_2CO	42	53	100	23	12	2.3
HCOH	40	100	39	8.9	40	18

Note that when O_2 is used as the CID collision gas, a very narrow but intense peak appears at m/z 15 only in the CID mass spectrum of the $HCOH^{1+\bullet}$ ion.[90] This peak corresponds to the doubly charged ion, $HCOH^{12+}$, which is predicted to be stable.[91]

The lowest energy dissociation for both isomers is the loss of $H^\bullet$ and the product ion has been identified as HCO^{1+} in both cases.[92] The metastable ion peak for the loss $H^\bullet$ from $HCOH^{1+\bullet}$ is composite and it has been proposed that two dissociation routes are responsible for this observation: a direct bond cleavage and a rearrangement to $H_2CO^{1+\bullet}$ before dissociation.[92]

Thermochemistry

$\Delta_f H(H_2CO^{1+\bullet}) = 941$ kJ/mol is obtained from the experimental IE and $\Delta_f H(H_2CO)$, 10.88 ± 0.01 eV[12] and -108.6 ± 0.46 kJ/mol,[12] respectively. Note that a second value for $\Delta_f H(H_2CO)$ of -115.90 kJ/mol is suggested by Chase in NIST.[12] On the basis of the current data for this molecule and related ions, we believe that the experimental value, -108.6 ± 0.46 kJ/mol, is to be preferred.

Calculations at the B3LYP6-311++G(3df,2pd) level of theory predict that $HCOH^{1+\bullet}$ is 26 ± 2 kJ/mol higher in energy than $H_2CO^{1+\bullet}$,[93] in agreement with earlier calculations at the QCISD(T)/6-311+G (3df,3pd)//QCISD(T)/6-31G(d,p) level of theory.[94] These computations are in agreement with recent calculations that give $\Delta_f H(HCOH^{1+\bullet}) = 965$ and 961 kJ/mol (from G2 and G2MP2 atomization energies, respectively).[28] The value for $\Delta_f H(HCOH^{1+\bullet})$ obtained from its appearance energy from methanol (H_2 loss) gives 996 kJ/mol.[89] However, calculations (G1) have shown that the reverse energy barrier for this process is not negligible and is estimated to be 22 kJ/mol.[95] Taking this into consideration, the revised experimental value (974 kJ/mol) also gives support to the above values from theory. The calculated barrier to the rearrangement of $H_2CO^{1+\bullet}$ to $HCOH^{1+\bullet}$ is 193 kJ/mol (MP3/6-31G**//6-31* with ZPVE correction).[92] The third isomer, $COH_2^{1+\bullet}$, is much higher in energy, 226 kJ/mol above $H_2CO^{1+\bullet}$ (MP3/6-31G**//6-31* with ZPVE correction), but lies in a deep potential energy well.[92]

3.5.19 $[C,H_3,O]^+ - m/z\ 31$

Isomers

There are two covalently bonded isomeric species: protonated formaldehyde ($^+CH_2OH$) and the (triplet) methoxy cation ($^3CH_3O^+$). A third isomer, a weakly bonded complex of H_2 and a formyl cation, $[H_2 \cdots HCO]^+$, has been proposed to be produced by the reaction of H_2 and HCO^{1+} at low temperature.[96]

Identification

The CID mass spectra of m/z 31 ions produced by the dissociative ionization of a wide variety of oxygen containing precursor molecules, mostly containing the functional group CH_2OH, are essentially identical, suggesting a single ion structure. The m/z 15:14 peak abundance ratio in the CID mass spectra is approximately 0.03, consistent with the $^+CH_2OH$ connectivity.[97] A weak signal for methoxy cations can however be generated by charge reversal of the corresponding anion[98,99] or by reionization of the neutral $[C,H_3,O]^\bullet$ fragment produced by metastable decomposition (a CIDI experiment, see Chapter 1.4.2.8.1) of molecular ions containing a CH_3O moiety.[15,100] The CID mass spectra of the cations so produced exhibit an intense m/z 15 peak consistent with the H_3C-O connectivity. Note that the methoxy cation has only one stable electronic state, the triplet, see below in Thermochemistry.

Both isomers lose H_2 in the metastable ion time frame to produce HCO^{1+} with a large kinetic energy release (a flat-topped metastable ion peak, see Chapter 1.4.1.4). The two peaks are superimposable, possibly resulting from the rearrangement of $^3CH_3O^+$ to $^+CH_2OH$ before dissociation.[99] There is however some evidence supporting a direct 1,1-elimination of H_2 from the methoxy cation.[101] A full reaction profile for the H_2 loss from both $^3CH_3O^+$ and $^+CH_2OH$ has been calculated.[101]

In addition to showing mass spectral differences, $^+CH_2OH$ and CH_3O^+ show different reactivities. In an experiment using the perdeuterated analogs, $^+CD_2OD$ and CD_3O^+, it was shown that protonated formaldehyde reacted by proton transfer whereas methoxy cations underwent electron transfer.[102]

Thermochemistry

Protonated formaldehyde is the lowest energy isomer. Its heat of formation, $\Delta_f H(^+CH_2OH) = 711$ kJ/mol, is calculated using $\Delta_f H(^\bullet CH_2OH) = -19 \pm 1$ kJ/mol[103] (in agreement with the computed value of -16.41 kJ/mol (G3MP2B3))[13] and IE $= 7.562 \pm 0.004$ eV.[104,105] This value is in good agreement with those obtained from appearance energy measurements.[106-108] For the methoxy cation, theoretical calculations have found that the singlet state is not a minimum on the potential energy surface, collapsing without activation to the $^+CH_2OH$ isomer.[109-111] However, the triplet state, $^3CH_3O^+$, is stable and is calculated to lie 358 kJ/mol above $^+CH_2OH$ (CCSD(T)/BSII//B3LYP/BSI).[101]

Because protonated formaldehyde is produced in all dissociative ionizations, no appearance energy measurement has allowed the direct determination of $\Delta_f H(^3CH_3O^+)$. However, some authors have estimated $\Delta_f H(^3CH_3O^+)$ using ancillary experimental thermochemical data. The values obtained were 1034 ± 20 kJ/mol[99] and 1025 ± 20 kJ/mol.[112] In addition, combining the IE and the $\Delta_f H(CH_3O^\bullet)$ (10.726 ± 0.008 eV[113] and 17 ± 4 kJ/mol,[12] respectively) gives $\Delta_f H(^3CH_3O^+) = 1052$ kJ/mol. The third isomer, the complex $[H_2\cdots HCO]^+$, is only weakly bonded (16 kJ/mol) with respect to the dissociation products $H_2 + HCO^{1+}$ ($\Sigma\Delta_f H_{products} = 825$ kJ/mol).[114]

3.5.20 $[C,H_4,O]^{+\bullet} - m/z$ 32

Isomers

Two isomers have been identified: the conventional methanol radical cation $(CH_3OH^{1+\bullet})$ and its distonic isomer $CH_2OH_2^{1+\bullet}$ (the methyleneoxonium radical cation).

Identification

Dissociative ionization of compounds such as $HOCH_2CH_2OH$, $HOCH_2$-$C(O)H$ and $HOCH_2C(O)OH$ yields the distonic ion (loss of H_2CO, CO and CO_2, respectively) whose CID mass spectral characteristics in the m/z 12 to m/z 19 region are clearly distinct from those of ionized methanol.[115,116]

In addition, the distonic ion exhibits a very intense and characteristic doubly charged ion peak at m/z 16.[116]

Partial CID Mass Spectra of the $[C,H_4,O]^{+\bullet}$ Isomers[116]

Ions	m/z						
	12	13	14	15	17	18	19
CH_3OH	8.3	15	53	100	14	7.2	3.7
CH_2OH_2	24	44	100	5.1	18	5.4	23

The lowest energy dissociation of both isomers produces $^+CH_2OH + H^\bullet$ ($\Sigma\Delta_fH_{products} = 929$ kJ/mol). The kinetic energy release values accompanying these dissociations are considerably different (9 and 34 meV for $CH_3OH^{1+\bullet}$ and $CH_2OH_2^{1+\bullet}$, respectively) and it has been shown by deuterium labeling experiments and calculations that $CH_2OH_2^{1+\bullet}$ must isomerize over an energy barrier to $CH_3OH^{1+\bullet}$ before this dissociation.[38]

Thermochemistry

$\Delta_fH(CH_3OH^{1+\bullet}) = 845$ kJ/mol is obtained from $\Delta_fH(CH_3OH) = -201$ kJ/mol and IE $= 10.84 \pm 0.01$ eV.[12] The distonic ion is the lower energy isomer with a $\Delta_fH(CH_2OH_2^{1+\bullet}) = 816$ kJ/mol (from an appearance energy measurement[116] or using the experimental proton affinity of $^\bullet CH_2OH$ (695 kJ/mol)[12] in combination with $\Delta_fH(^\bullet CH_2OH) = -19 \pm 1$ kJ/mol[103] and $\Delta_fH(H^+) = 1530$ kJ/mol).[12] Ab initio calculations (G2**) also support this value and have determined the energy barrier for the isomerization of $CH_3OH^{1+\bullet}$ to $CH_2OH_2^{1+\bullet}$ to be 108 kJ/mol.[117,118] It is interesting to note that the interconversion of $CH_3OH^{1+\bullet}$ to $CH_2OH_2^{1+\bullet}$ is catalyzed by the presence of water[119,120] and other molecules such as noble gases[121] and methanol itself.[122]

3.5.21 $[C,H_5,O]^+$ – m/z 33

Isomers

Only protonated methanol, $CH_3OH_2^{1+}$, has been characterized.

Identification

The $CH_3OH_2^{1+}$ ion can be generated by protonation of methanol,[123] from ion–molecule reactions such as $CH_4^{1+\bullet} + H_2O$[124] and $C_3H_6^{1+\bullet} + CH_3OH$[125] or by the dissociative ionization of 2-methylpropanol (loss of $C_3H_5^\bullet$).[126]

The m/z 12 to m/z 19 portion of its CID mass spectrum are dominated by the m/z 15 fragment peak. Two narrow peaks at m/z 16 and 16.5 corresponding to the doubly charged ions $CH_4O^{]2+}$ and $CH_5O^{]2+}$ are also present.[124]

Thermochemistry

From the proton affinity of methanol, $\Delta_f H$ $(CH_3OH_2^{]+}) = 575$ kJ/mol is obtained $(PA = 754.3$ kJ/mol, $\Delta_f H(CH_3OH) = -201$ kJ/mol and $\Delta_f H(H^+) = 1530$ kJ/mol).[12]

3.5.22 $[C,H_6,O]^{+\bullet}$ – m/z 34

Isomers

An electrostatically bound complex of a methyl radical with protonated water, $[CH_3 \cdots H_3O]^{+\bullet}$, has been identified.

Identification

The following reaction scheme shows how the complex $[CH_3 \cdots H_3O]^{+\bullet}$ can be generated.[127]

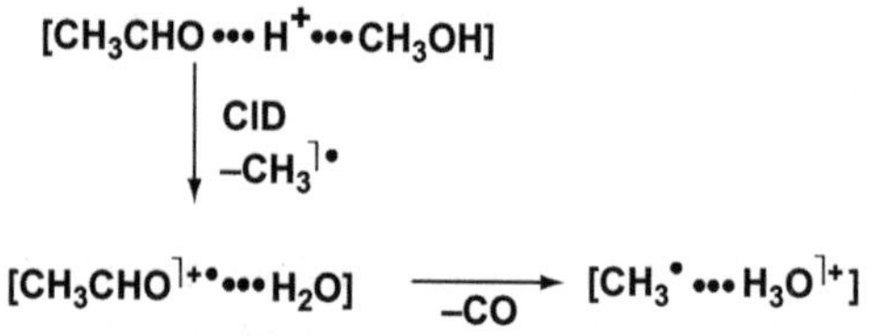

$$[CH_3CHO \cdots H^+ \cdots CH_3OH]$$
$$\downarrow \text{CID} \quad -CH_3]^{\bullet}$$
$$[CH_3CHO^{]+\bullet} \cdots H_2O] \quad \xrightarrow{\quad -CO \quad} \quad [CH_3^{\bullet} \cdots H_3O^{]+}]$$

The metastable ion dissociates solely by the loss of $CH_3^{]\bullet}$ producing H_3O^+ fragment ions, in keeping with its structure.

Thermochemistry

Calculations (G3) give a $\Delta_f H([CH_3 \cdots H_3O]^{+\bullet}) = 690$ kJ/mol with a binding energy of 58 kJ/mol.[127] No experimental value is available.

3.5.23 $[C,H_7,O]^+$ – m/z 35

Isomers

The adduct ion $[CH_4 \cdots H_3O]^+$ is the only ion that has been identified.

Identification

The electrostatically bound pair, $[CH_4 \cdots H_3O]^+$, is one of the product ions formed in the reversible reaction of methane with water at low temperatures in the ion source of a mass spectrometer.[128] The possibility that the isomeric pair ($[CH_5\cdots H_2O]^+$) may exist was rejected based on the much greater proton affinity of H_2O relative to CH_4 (691 vs 543.5 kJ/mol, respectively).[12]

Thermochemistry

Calculations (G3) give a $\Delta_f H[CH_4\cdots H_3O]^+ = 494$ kJ/mol with a binding energy of 35 kJ/mol,[127] the latter value is in good agreement with a revised value of 42 kJ/mol from experiment.[127]

3.5.24 $[C,O_2]^{+\bullet} - m/z$ 44

Ionized carbon dioxide dissociates chiefly by the loss of atomic oxygen to produce the m/z 28 ($CO^{1+\bullet}$) fragment ion peak in its CID mass spectrum. From the well established ionization energy and heat of formation of the neutral, $\Delta_f H(CO_2^{1+\bullet}) = 936$ kJ/mol (13.777 ± 0.001 eV and -393.51 ± 0.13 kJ/mol).[12]

3.5.25 $[C,H,O_2]^+ - m/z$ 45

Isomers

There are two possible isomers: $HOCO^{1+}$ and HCO_2^{1+}, the hydroxyformyl and formyloxy cations, respectively. The former ion, oxygen-protonated carbon dioxide, is a known interstellar species[129] and participates in the chemistry of interstellar clouds.[130]

Identification

The hydroxyformyl cation, $HOCO^{1+}$, is generated by protonation of CO_2 as well as by the dissociative ionization of such simple carboxylic acids as formic, acetic and oxalic acids.[89] The second isomer is more elusive but it is proposed to be generated, at least partially, in the dissociative ionization of $HC(O)OCH_3$ (loss of $CH_3^{1\bullet}$)[89] and by the collision-induced charge reversal of the corresponding anion, HCO_2^{1-}.[131,132] The structure-characteristic feature of the CID mass spectra of the two isomers lies in the small but significant difference in the m/z 16:17 ion peak abundance ratio, 0.89 and 1.0 for $HOCO^{1+}$ and HCO_2^{1+}, respectively.[89] However, it should be noted that the nondissociating HCO_2^{1+} ions produced by charge reversal of the anion were shown to have partly isomerized to $HOCO^{1+}$ within the timescale of the experiment (9 µs).[132] It is therefore likely that a flux of pure HCO_2^{1+} ions has not yet been generated.

Thermochemistry

The appearance energy of $HOCO^{1+}$ from formic acid gives $\Delta_f H(HOCO^{1+}) = 594 \pm 8$ kJ/mol[133] in agreement with 596 kJ/mol, the value obtained from the proton affinity of CO_2 ($PA(CO) = 540.5$ kJ/mol,

$\Delta_f H(CO_2) = -393.51 \pm 0.13$ kJ/mol and $\Delta_f H(H^+) = 1530$ kJ/mol.[12] There is no experimental value for $\Delta_f H(HCO_2^{1+})$. However, a recent computational study employing a selected combination of levels of theory predicts that the stable HCO_2^{1+} species produced in the charge reversal experiment are most likely a triplet state that lies some 423 kJ/mol above $HOCO^{1+}$ but below the calculated threshold for dissociation to $HCO^{1+} + O$ ($\Sigma\Delta_f H_{products} = 479$ kJ/mol[12]).[133]

3.5.26 $[C,H_2,O_2]^{+\bullet} - m/z$ 46

Isomers

Theoretical calculations conclude that at least three isomers are stable in the gas phase and therefore should be observable.[134] These ions are $HC(O)OH^{1+\bullet}$, ionized formic acid, the dihydroxycarbene radical cation ($C(OH)_2^{1+\bullet}$) and $[H_2O\cdots CO]^{+\bullet}$, an ion–neutral complex. The first two have been observed by experiment. It is noteworthy that although ionized formic acid has not been observed directly in the interstellar medium, its decomposition products HCO^{1+} and $HOCO^{1+}$ have been identified by radioastronomy.[135]

Identification

$HC(O)OH^{1+\bullet}$ is produced by the ionization of formic acid whereas $C(OH)_2^{1+\bullet}$ is obtained by the dissociative ionization of dihydroxyfumaric acid according to the following scheme:[136]

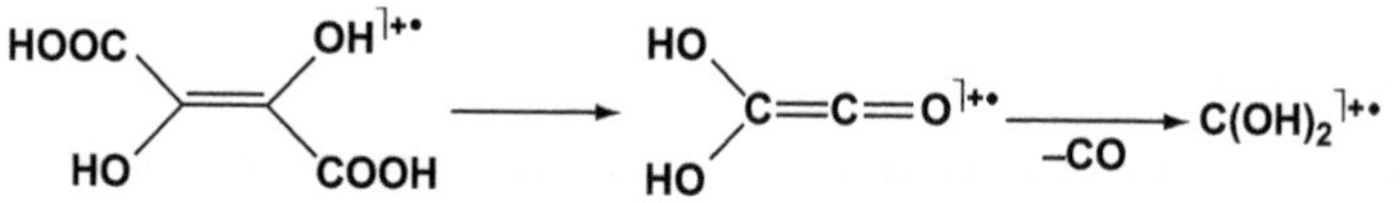

$C(OH)_2^{1+\bullet}$ ions can also be produced by the decomposition of metastable oxalic acid radical cations (($C(O)OH)_2^{1+\bullet}$, loss of CO_2).[89]

The m/z 18:17 fragment ion peak intensity ratios in the CID mass spectra of the isomers are sufficiently different to distinguish them.[89] This ratio is 4.0 for $HC(O)OH^{1+\bullet}$ ions and 1.3 for the carbene ion, $C(OH)_2^{1+\bullet}$. It is of interest to note that both ions produce, upon collisional activation, intense m/z 29 ion peaks. These have been shown (by their CID mass spectra) to be predominantly HOC^{1+} for $C(OH)_2^{1+\bullet}$ and HCO^{1+} for $HC(O)OH^{1+\bullet}$, consistent with each ions' connectivities.[136] The metastable ion mass spectra of the two isomers are different, indicating that they do not interconvert up to energies corresponding to the formation of $H_2O^{1+\bullet} + CO$ ($\Sigma\Delta H_{products} = 865$ kJ/mol[12]).[89]

Thermochemistry

From the ionization energy and the heat of formation of formic acid, $\Delta_f H(HC(O)OH^{1+\bullet}) = 714$ kJ/mol is obtained (IE = 11.33 ± 0.01 eV and

$\Delta_f H(HC(O)OH) = -378.6$ kJ/mol, respectively).[12] Theory has consistently predicted $C(OH)_2^{1+\bullet}$ to be lower in energy than its isomer and the most recent calculations (G2MP2) give $\Delta_f H(C(OH)_2^{1+\bullet}) = 662$ kJ/mol.[28,134] The only value from experiment, 732 ± 10 kJ/mol,[89] is considerably higher; it was obtained from the appearance energy of the m/z 46 fragment peak of ionized oxalic acid. A calculated energy barrier (MP3/6-31G**//4-31G including ZPVE correction) of 179 kJ/mol separates $HC(O)OH^{1+\bullet}$ from $C(OH)_2^{1+\bullet}$.[134]

3.5.27 [C,H₃,O₂]⁺ – m/z 47

Isomers

Theory predicts that there are as many as nine isomers.[137,138] Experimental attempts have been made to generate five of these, namely carbonyl-protonated formic acid ($HC(OH)_2^{1+}$), the methylenehydroperoxy ion (CH_2OOH^{1+}), the methylperoxy ion (CH_3OO^{1+}), carbon-protonated formic acid ($HOCH_2O^{1+}$) and the complexes $[H_2O–H\cdots CO]^+$ and $[H_2O–H\cdots OC]^+$. The other structures are $[CH_3\cdots O_2]^+$, cy–$OC(H)_2O(H)–^{1+}$ and cy–$O(H)C(H)$ $O(H)–^{1+}$.[139–141]

Identification

Protonation of formic acid as well as the dissociative ionization of $HC(O)OCH_2CH_3$ (loss of C_2H_4) yield the $HC(OH)_2^{1+}$ isomer.[139,142]

CH_2OOH^{1+} ions are obtained from the dissociative ionization of CH_3OOH and CH_3CH_2OOH (loss of $H^\bullet$ and $CH_3^{1\bullet}$, respectively).[139,142] Although there has been some controversy surrounding the structure of the $[C,H_3,O_2]^+$ product ion generated in the reaction of $CH_4^{1+\bullet}$ and O_2, it is now clear that the CH_2OOH^{1+} cation is produced.[143–145] The CID mass spectra of these two isomers ($HC(OH)_2^{1+}$ and CH_2OOH^{1+}) are similar, each with a dominant m/z 29 (H_2O loss) peak. This dissociation, as well as the loss of CO (generating m/z 19), is also observed in both metastable ion mass spectra. However, the presence of the structure-characteristic m/z 14 ($CH_2^{1+\bullet}$), 32 ($O_2^{+\bullet}$) and 33 (OOH^{1+}) fragment ion peaks in the CID mass spectrum of *only* the CH_2OOH^{1+} isomer clearly distinguishes between the two.[139,142] In addition, only the $HC(OH)_2^{1+}$ isomer displays charge stripping peaks at m/z 23 and 23.5.[139]

The CH_3OO^{1+} and $HOCH_2O^{1+}$ ions have both been generated by the charge reversal of the corresponding anions.[139,142] The former ion was also produced in the chemical ionization of CH_3F and O_2.[142] In both charge reversal experiments, surviving m/z 47 ions were observed indicating that they are stable (i.e., that they lie in a potential energy well). Although there are some discrepancies in the reproducibility of the CID mass spectra for the CH_3OO^{1+} ions generated as described above,[139,142] the intense m/z 15 fragment ion peak differentiates this isomer from the others.

The calculated structure for this ion shows a lengthened $C-O$ bond (1.678 Å) indicating that this ion could be regarded as a complex between CH_3^{1+} and O_2.[137,138]

The CID mass spectrum of the $HOCH_2O^{1+}$ isomer is similar to those of its isomers $HC(OH)_2^{1+}$ and CH_2OOH^{1+} but as opposed to these ions, $HOCH_2O^{1+}$ shows the structure-characteristic m/z 31 peak ($^+CH_2OH$). However, it is noteworthy that calculations find the *acyclic* $HOCH_2O^{1+}$ ion to be unstable indicating that when the corresponding radical ($HOCH_2O^{1•}$) is ionized, an oxygen–oxygen bond results, leading to the formation of a *cyclic* ion.[137,138] The cyclic ion is almost certainly the ion generated in the charge reversal experiment.

The last isomer, the complex $[H_3O\cdots CO]^+$, is coproduced with $HC(OH)_2^{1+}$ in the reaction of ionized H_2O and CO. The intense m/z 19 peak in the CID mass spectrum of the mixture clearly indicates the production of this isomer.[142]

Thermochemistry

$\Delta_f H(HC(OH)_2^{1+}) = 409$ kJ/mol is obtained from the proton affinity of formic acid $(PA(HC(O)OH) = 742$ kJ/mol, $\Delta_f H(HC(O)OH) = -378.6$ kJ/mol and $\Delta_f H(H^+) = 1530$ kJ/mol).[12] Calculations (G2) show this ion to be the global minimum.[137,138]

$\Delta_f H(CH_2OOH^{1+}) = 779$ kJ/mol is obtained from the appearance energy of the ion generated from the dissociation of CH_3CH_2OOH[139] when using $\Delta_f H(CH_3CH_2OOH) = -175$ kJ/mol (and not -166 kJ/mol which was originally used by Holmes et al.[139] or the -210 kJ/mol listed in the NIST WebBook[12] which cannot, on the basis of thermochemical arguments, be correct (for justification see Section 2.6)). This value (779 kJ/mol) is in good agreement with that obtained from reactivity experiments (774 ± 10 kJ/mol)[143] and also with theory (G2), which predicts this isomer to be 373 kJ/mol higher in energy than $HC(OH)_2^{1+}$ i.e., 782 kJ/mol.[137,138]

For the other ions, no experimental data are available. However, theory (G2) predicts that the complexes $[H_3O\cdots CO]^+$ and $[CH_3\cdots O_2]^+$ and the cyclic species cy–OC(H)$_2$O(H)–$^{1+}$ and cy–O(H)C(H)O(H)–$^{1+}$ occupy potential energy wells of significant depth and are, respectively, 19, 636, 465 and 776 kJ/mol higher in energy than $HC(OH)_2^{1+}$.[137,138] The complex $[H_2O\cdots HCO]^+$ and the ion $HOCH_2O^{1+}$ lie, respectively, 83 and 742 kJ/mol above protonated formic acid.[137,138] However, the $[H_2O\cdots HCO]^+$ ion collapses with little activation (4 kJ/mol) to $[H_3O\cdots CO]^+$ whereas the $HOCH_2O^{1+}$ ion rearranges spontaneously to CH_2OOH^{1+}.[137,138] Note that in an independent computational study (G3), the complex $[H_3O^+\cdots OC]$ was also found to be stable and estimated to lie 43 kJ/mol higher in energy than protonated formic acid.[141]

3.5.28 $[C,H_4,O_2]^{+\bullet}$ – m/z 48

Isomers

Calculations predict two peroxy species $CH_3OOH^{\rceil+\bullet}$ (methyl hydroperoxide) and $CH_3O(H)O^{\rceil+\bullet}$ (methanol oxide) and the complex $[HCO\cdots H_3O]^{+\bullet}$ to be stable.[140,146] However, only ionized methyl hydroperoxide has been characterized by experiment.[142]

Identification

The CID mass spectrum of ionized methyl peroxide ($CH_3OOH^{\rceil+\bullet}$) is dominated by a fragment peak at m/z 47 corresponding to a hydrogen atom loss. Other significant peaks are m/z 15 ($CH_3^{\rceil+}$), 29 ($[C,H,O]^+$) and 30 ($H_2CO^{\rceil+\bullet}$).[142]

Thermochemistry

The calculated (B3LYP/6-311++G(d,p)) $\Delta_f H(CH_3OOH^{\rceil+\bullet}) = 831 \pm 20$ kJ/mol[140] is in fair agreement with an estimated value of 841 based on $\Delta_f H(CH_3OOH) = -143$ kJ/mol and a calculated adiabatic IE $= 10.2 \pm 0.3$ eV (MNDO/2').[147] Note that a recent calculated $\Delta_f H(CH_3OOH) = -129 \pm 3$ kJ/mol mol (CBS/APNO)[148] seems too low, not in keeping with the CH_3O-OH bond strength of 195 ± 6 kJ/mol[44] (for justification refer to Section 2.6).

The methanol oxide radical cation is higher in energy, with a calculated $\Delta_f H(CH_3O(H)O^{\rceil+\bullet}) = 930$ kJ/mol and it is separated from $CH_3OOH^{\rceil+\bullet}$ by an interconversion barrier of approximately 105 kJ/mol (B3LYP/6-311++G(d,p)).[140] It is noteworthy that the lowest energy dissociation channel for $CH_3O(H)O^{\rceil+\bullet}$ ions is the loss of a triplet oxygen atom and this reaction is calculated to lie approximately 39 kJ/mol above the energy of the transition state for the isomerization.[140]

Finally, the complex $[HCO\cdots H_3O]^{+\bullet}$ has been calculated to be stabilized with respect to its dissociation products, ionized formaldehyde and water ($\Sigma \Delta_f H_{products} = 699$ kJ/mol), by 187 kJ/mol (ZPE/MP2/6-311+G(d,p)).[146]

3.5.29 $[C,H_5,O_2]^+$ – m/z 49

Isomers

The calculated potential energy surface of these ions shows that four isomeric species occupy potential energy wells of different depths. These are methyl hydroperoxide protonated at the α and β positions, $CH_3O(H)OH^{\rceil+}$ and $CH_3OOH_2^{\rceil+}$, respectively, the complex $[CH_2OH\cdots H_2O]^+$ and the proton-bound dimer $[H_2CO\cdots H^+\cdots H_2O]$. The ion resulting from the protonation of methyl hydroperoxide and the proton-bound dimer have been investigated by experiment.

Identification

The calculated proton affinities of CH_3OOH at the α and β positions differ by approximately 8 kJ/mol and so it could be expected that protonation yields both isomers. The CID mass spectrum of protonated methyl hydroperoxide is dominated by the m/z 31 peak ($^+CH_2OH$) but also shows structure-characteristic fragments at m/z 15 (indicative of an intact methyl group) and 32 (loss of $^\bullet OH$).[149,150] These observations can only be consistent with the α-protonated species and therefore $CH_3O(H)OH^{]+}$ ions may make up the major portion of the ion beam. However, it is very likely that $CH_3OOH_2^{]+}$ ions are also formed but that their exothermic isomerization to the lower energy isomer may prevent their observation.[149]

The CID mass spectrum of the proton-bound dimer $[H_2CO\cdots H^+\cdots H_2O]$, obtained by the protonation of formaldehyde with H_3O^+, is straightforward showing fragment peaks only in the m/z 28–31 and m/z 18–19 regions[149,150] i.e., the simple cleavage products predominate.

Thermochemistry

The global minimum on the potential energy surface of these ions is $[H_2CO\cdots H^+\cdots H_2O]$.[149,150] Its heat of formation was calculated to be 333 ± 20 kJ/mol (B3LYP/6-311++G**).[149] At the same level of theory, the other unconventional species, $[CH_2OH\cdots H_2O]^+$, is only slightly higher in energy, with a $\Delta_f H([CH_2OH\cdots H_2O]^+)=374\pm20$ kJ/mol and is separated from the proton-bound dimer by a barrier of 21 kJ/mol.

The measured proton affinity of methylhydroperoxide (adjusted to the current proton affinity scale)[151] combined with $\Delta_f H(CH_3OOH)$ and $\Delta_f H(H^+)$ (723 ± 8 kJ/mol,[150] -143 kJ/mol (see above), and 1530 kJ/mol,[12] respectively) gives $\Delta_f H(CH_3O(H)OH^{]+})=664$ kJ/mol. This is in reasonable agreement with the calculated value of 680 ± 20 kJ/mol (B3LYP/6-311++G**).[149] These computations also predict the β-protonated isomer, $CH_3OOH_2^{]+}$, to lie only 8 kJ/mol above $CH_3O(H)OH^{]+}$ and separated from $CH_3O(H)OH^{]+}$ and $[CH_2OH\cdots H_2O]^+$ by barriers of 119 and 17 kJ/mol, respectively.[149] Note that these two ions are considerably higher in energy than their dissociation products $^+CH_2OH+H_2O$ ($\Sigma\Delta_f H_{products}=469$ kJ/mol) and $H_3O^++H_2CO$ ($\Sigma\Delta_f H_{products}=486$ kJ/mol), reactions that are observed for metastable $CH_3O(H)OH^{]+}$ ions. Therefore, it is suggested that dissociation proceeds via the proton-bound dimer $[H_2CO\cdots H^+\cdots H_2O]$.[149]

3.5.30　$[C,H_6,O_2]^{+\bullet}$ – m/z 50

Isomers

Of the five species predicted by theory to be stable $[^\bullet CH_2O(H)\cdots H^+\cdots OH_2]$ (**1**) ($O\cdots H\cdots O$ bond), $[CH_3O^\bullet\cdots H^+\cdots OH_2]$ (**2**), $[H_2O\cdots H^+\cdots {}^\bullet CH_2OH]$ (**3**) ($O\cdots H\cdots C$ bond), $[CH_3O(H)\cdots H^+\cdots {}^\bullet OH]$ (**4**) and the O–O electrostatically bound complex $[CH_3O(H)\cdots OH_2]^{+\bullet}$ (**5**), isomers (**1**), (**2**) and (**4**) have been generated and investigated by experiments. It is noteworthy that isomers (**1**), (**2**) and (**3**) are proposed as intermediates in the water-catalyzed isomerization of $CH_3OH^{]+\bullet}$ to its distonic ion $CH_2OH_2^{]+\bullet}$.[117]

Identification[152]

Collision-induced loss of $H^\bullet$ from the dimer $CH_3OH_2^+/H_2O$ leads to $[^\bullet CH_2O(H)\cdots H^+\cdots OH_2]$ (**1**). This ion can also be prepared by collision-induced loss of $CH_3^{]\bullet}$ from the dimer $CH_3CH_2OH_2^+/H_2O$. Isomer (**2**), $[CH_3O^\bullet\cdots H^+\cdots OH_2]$, was produced similarly by the loss of $CH_3^{]\bullet}$ from the $(CH_3)_2OH^+/H_2O$ dimer. Loss of $CH_3^{]\bullet}$ from the proton-bound pair $(CH_3OH)_2H^+$ yields isomer (**4**), namely $[CH_3O(H)\cdots H^+\cdots^\bullet OH]$.

The CID mass spectra of these three isomers are similar but different enough to allow identification.

CID Mass Spectra of the $[C,H_6,O_2]^{+\bullet}$ Isomers

			m/z		
Ions	19	31	32	33	49
$[CH_2O(H)\cdots H\cdots OH_2]$	17	19	100	40	68
$[CH_3O\cdots H\cdots OH_2]$	17	20	100	—	70
$[CH_3O(H)\cdots H\cdots OH]$	10	—	100	82	3

Thermochemistry

Isomer (**1**) is calculated to be the global minimum.[117,152]

Calculated (G3) Heat Of Formation of the $[C,H_6,O_2]^{+\bullet}$ Isomers

Ions	Calculated $\Delta_f H^a$ (kJ/mol)
(**1**)	448
(**2**)	482
(**3**)	486[b]
(**4**)	538
(**5**)	552

[a]From Ref. [152] or otherwise indicated.
[b]Taken from Ref. [117].

3.5.31 $[C,H_2,O_3]^{+\bullet}$ – m/z 62

Isomers

Two isomeric forms have been investigated: ionized carbonic acid $(HOC(O)OH^{]+\bullet})$ and the hydrogen-bridged species $[OCO\cdots H^+\cdots {}^\bullet OH]$.

Identification

Free carbonic acid has traditionally been described as an elusive molecule. Only relatively recently, using mass spectrometric methods, has it been demonstrated to exist as a stable, discrete species in the gas phase. Thermolysis of NH_4HCO_3 in the ion source of a mass spectrometer produces an m/z 62 ion $[C,H_2,O_3]^{+\bullet}$.[153] The ion's CID mass spectrum is slightly but significantly different from the CID mass spectrum of the m/z 62 ion obtained from ionization of a mixture of H_2O and CO_2,[154] the major difference being the relative peak abundance ratio m/z 44:45 (loss of H_2O vs loss of ${}^\bullet OH$). The larger ratio for the proton-bound dimer indicates that it more readily loses water, consistent with the assigned connectivity.

Partial CID Mass Spectra of the $[C,H_2,O_3]^{+\bullet}$ Isomers

	m/z				
Ions	18	28	29	44	45
HOC(O)OH	50	3.4	6.8	13	100
$[OCO\cdots H\cdots OH]$	72	7.0	4.7	53	100

It is noteworthy that neutralization–reionization mass spectra of the two isomers are very different and can also be used to distinguish them; ionized carbonic acid produces an intense recovery signal and a fragmentation pattern very similar to its CID mass spectrum, whereas the hydrogen-bridged species shows no trace of a recovery signal and major fragment peaks at m/z 44 and 28 while m/z 18 is minor, in keeping with its unconventional structure.[153,154]

Thermochemistry

Using calculated values for the heat of formation and ionization energy of H_2CO_3 (-614 kJ/mol (CCSD(T)/aug-cc-pvtz) and 11.29 eV (G2), respectively)[155] gives $\Delta_f H(H_2CO_3^{]+\bullet}) = 475$ kJ/mol. This calculation is in good agreement with an estimated value of 460–481 kJ/mol.[153] Its isomer, $[OCO\cdots H^+\cdots {}^\bullet OH]$, is calculated to lie 36 kJ/mol higher in energy (HF/6-31G** + ZPVE).[154]

3.5.32 [C,H₃,O₃]⁺ – m/z 63

Isomers

Two ions have been characterized, namely $C(OH)_3^{1+}$, trihydroxymethyl, and the proton-bound dimer, $[H_2O \cdots H^+ \cdots OCO]$.[155,156] A third isomer, an adduct ion, $[H_2O \cdots CO_2H]^+$, is predicted by theory to occupy a potential energy well of significant depth.[155]

Identification[155,156]

Dissociative ionization of dialkylcarbonates produces $C(OH)_3^{1+}$ cations by consecutive losses of an alkenyl radical and an alkene.[157,158] The proton-bound dimer, $[H_2O \cdots H^+ \cdots OCO]$, is produced in the dissociative ionization of dihydroxyfumaric acid.[156]

The CID mass spectrum of $C(OH)_3^{1+}$ ions is dominated by the fragment ion at m/z 45 (loss of H_2O) with other significant product ions at m/z 29 (loss of CO) and 18 (loss of HOCO•) in addition to the doubly charged ion $C(OH)_3^{12+}$ at m/z 31.5. The major fragment peaks in the proton-bound dimer's CID mass spectrum are at m/z 19 (loss of CO_2) and 45 (loss of H_2O), consistent with the structure $[H_2O \cdots H^+ \cdots CO_2]$.

Thermochemistry

$\Delta_f H(C(OH)_3^{1+}) = 140 \pm 8$ kJ/mol has been determined from an appearance energy measurement using diethylcarbonate (AE = 11.72,[159] $\Delta_f H$(OC(OC$_2$H$_5$)$_2$) = 637.9 ± 0.8 kJ/mol, $\Delta_f H(C_2H_3^{1•})$ = 299 ± 5 kJ/mol and $\Delta_f H$(C$_2$H$_4$) = 52.47 kJ/mol).[12] This value is in satisfactory agreement with the computed result of 151 kJ/mol (G2(MP2), G2, QCISD(T) and B3-MP2).[155]

There are no experimental data for the other two isomers, but the proton-bound dimer $[H_2O \cdots H \cdots OCO]^+$ and the adduct ion $[H_2O \cdots CO_2H]^+$ have been calculated to have energies of −15 kJ/mol and +106 kJ/mol relative to $C(OH)_3^{1+}$ (G2MP2, G2, and QCISD(T)).[155] In addition, $\Delta_f H([H_2O \cdots H \cdots OCO]^+)$ can be estimated[160] using the empirical equation described in Section 1.3.9 to be 146 ± 8 kJ/mol. Substantial barriers to their interconversion separate the isomers.[155]

3.5.33 [C,H₄,O₃]⁺• – m/z 64

Isomers

Calculations predict that there are four possible isomers, namely $[H_2O \cdots H^+ \cdots •OCOH]$, $[H_2O \cdots HO_2CH]^{+•}$, $[H_2O \cdots HCO_2H]^{+•}$ and $H_2OC(OH)_2^{1+•}$.[156] Only the first and possibly the last species have been identified by experiment. Note that these computations also found the trihydroxymethane radical cation, $HC(OH)_3^{1+•}$, to be unstable.

Identification

After extensive rearrangement, ionized dihydroxyfumaric acid produces an ion at m/z 64 (HO(O)CC(OH) = C(OH)C(O)OH, loss of three CO molecules). This ion has been identified as the unsymmetrical proton-bound dimer, [H$_2$O$\cdots$H$^+\cdots$$^\bullet$OCOH]. Its CID characteristics are an intense m/z 19 ion (loss of OCOH$^{1\bullet}$) and relatively minor fragment peaks at m/z 63, 46, 45 and 28.[156] It is noteworthy that the authors suggest that a small amount of the distonic ion, H$_2$OC(OH)$_2$$^{1+\bullet}$, may be cogenerated.

Thermochemistry

The heat of formation of [H$_2$O$\cdots$H$^+\cdots$(O)$^\bullet$COH] was estimated, using an empirical relationship, to be 305 ± 16 kJ/mol.[156] Calculations find this ion to be the global minimum with its isomers [H$_2$O$\cdots$HO$_2$CH]$^{+\bullet}$, [H$_2$O$\cdots$HCO$_2$H]$^{+\bullet}$ and H$_2$OC(OH)$_2$$^{1+\bullet}$ lying 42, 56 and 76 kJ/mol higher in energy (B3LYP/6-31G**).[156]

3.5.34   [C,P]$^+$ – m/z 43

Carbon monophosphide is a known interstellar species.[161] Combining its calculated ionization energy and heat of formation gives $\Delta_f H(CP^{1+}) = 1577$ kJ/mol (IE $= 10.92 \pm 0.5$ eV (G2 and RCCSD(T)/cc-pV5Z//UQCISD/ 6-311G(2d) levels of theory)[162] and $\Delta_f H(CP) = 523$ kJ/mol (G2)[163]). A heat of formation value of 1592 kJ/mol, not in disagreement with the above, is deduced from the calculated binding energy in CP^{1+} with respect to P$^+$ + C (binding energy $= 4.7$ eV (B3LYP/6-311G* method)[164] and $\Sigma\Delta_f H_{\text{products}} = 2045$ kJ/mol[12]).

Note that the $\Delta_f H(CP)$ value listed in NIST, 449.89 kJ/mol, appears to be erroneous. The G2 value listed above agrees well with 518 kJ/mol which is deduced from the measured dissociation energy of gaseous carbon monophosphide (binding energy $= 515 \pm 17$ kJ/mol[165] (in agreement with a G2 value of 502 kJ/mol[164]), $\Delta_f H(C) = 716.68 \pm 0.45$ kJ/mol, and $\Delta_f H(P) = 316.5 \pm 1.0$ kJ/mol[12]). The NIST value has also been questioned by other authors.[162]

3.5.35   [C,H,P]$^{+\bullet}$ – m/z 44

Methinophosphide (or phosphaethyne), which is isoelectronic with HCN, is an interstellar species that was found in the atmospheres of Jupiter and Saturn.[166] From its ionization energy and heat of formation (10.79 ± 0.01 eV and 149.90 kJ/mol, respectively)[12] $\Delta_f H(HCP^{1+\bullet}) = 1191$ kJ/mol is deduced. This radical cation, as well as its deuterated analog, has been studied spectroscopically[167,168] but no mass spectral data are available. It is noteworthy that the isomerization HCP $\leftrightarrow$ CPH in the neutral molecule has been investigated.[169]

3.5.36 $[C,H_2,P]^+$ – m/z 45

Protonation of methinophosphide yields a $[C,H_2,P]^+$ species. Calculations predict that only H_2CP^{1+}, the carbon-protonated species, is stable.[170,171] From the proton affinity of HCP, $\Delta_f H(H_2CP^{1+}) = 981$ kJ/mol is obtained (PA $= 699$ kJ/mol[12] (not in disagreement with 684 kJ/mol obtained at the B3LYP/ 6-311++G** level of theory[171]), $\Delta_f H(HCP) = 149.91$ kJ/mol and $\Delta_f H(H^+) = 1530$ kJ/mol[12]). No mass spectral data are available. This species could be formed in interstellar clouds.[172]

3.5.37 $[C,H_3,P]^{+\bullet}$ – m/z 46

Although calculations predict four equilibrium structures for the neutral molecules, only phosphaethene (CH_2PH) has been generated.[173] From its calculated heat of formation and ionization energy (120 ± 8 kJ/mol and 9.9 ± 0.1 eV, respectively,[173] using CCSD(T) method) an estimated $\Delta_f H(CH_2PH^{1+\bullet}) = 1075$ kJ/mol is obtained. An upper limit for $\Delta_f H(CH_2PH^{1+\bullet}) = 1110 \pm 20$ kJ/mol, not in disagreement with the above, was also obtained from the measured appearance energy for the reaction: $ClCH_2PH_2 \longrightarrow CH_2PH^{1+\bullet} + HCl$ (AE $= 11.0 \pm 0.2$ eV,[174] $\Delta_f H(ClCH_2PH_2) = -43.5 \pm 4$ kJ/mol,[175] $\Delta_f H(HCl) = -92.31 \pm 0.1$ kJ/mol[12]). There are no mass spectral data for this ion.

3.5.38 $[C,H_4,P]^+$ – m/z 47

The only information available on species with a $[C,H_4,P]$ empirical formula is the calculated $\Delta_f H(CH_2PH_2^\bullet) = 179.6$ kJ/mol (G2,[41] in agreement with previous calculations at the UMP2-Full/6-31 + G(2df,p) + ZVE[176]).

3.5.39 $[C,H_5,P]^{+\bullet}$ – m/z 48

Isomers

Two isomers, namely ionized methylphosphine ($CH_3PH_2^{1+\bullet}$) and the distonic ion $CH_2PH_3^{1+\bullet}$, have been characterized.

Identification

The ion generated by dissociative ionization of *n*-hexylphosphine (loss of C_5H_{10}) exhibits a CID mass spectrum slightly different from that of the ion obtained by direct ionization of methylphosphine.[177,178]

Partial CID Mass Spectra of the $[C,H_5,P]^{+\bullet}$ Isomers[178]

	m/z							
Ions	12	13	14	15	31	32	33	34
CH_2PH_3	5.4	11	18	4.5	100	99	53	47
CH_3PH_2	3.9	6.9	14	20	100	94	53	5.9

Consistent with their respective connectivities, $CH_3PH_2^{1+\bullet}$ ions show intense fragment peaks at m/z 15 (CH_3^{1+}) and 33 (PH_2^{1+}) whereas the distonic ion $CH_2PH_3^{1+\bullet}$ produces a greater abundance of m/z 14 ($CH_2^{1+\bullet}$) and 34 ($PH_3^{1+\bullet}$).

In addition, the ions have clearly distinct reactivities. The distonic ion reacts by radical-type abstraction and proton transfer in contrast with the ionized molecule which reacts mainly by electron abstraction.[176]

Thermochemistry

Combining the ionization energy and heat of formation of CH_3PH_2 (9.12 eV[12] and -18 kJ/mol,[40] respectively) gives $\Delta_fH(CH_3PH_2^{1+\bullet}) = 862$ kJ/mol. From the calculated proton affinity at phosphorus for the $CH_2PH_2^{1\bullet}$ radical, $\Delta_fH(CH_2PH_3^{1+\bullet}) = 886$ kJ/mol is deduced (PA = 823.8 kJ/mol (G2),[41] $\Delta_fH(CH_2PH_2^{1\bullet}) = 179.6$ kJ/mol (G2)[41] and $\Delta_fH(H^+) = 1530$ kJ/mol[12]). $CH_3PH_2^{1+\bullet}$ has been predicted to be more stable than its isomer $CH_2PH_3^{1+\bullet}$ by 33 kJ/mol (G2 and G2' levels of theory),[179] not in disagreement with the above heat of formation values. The barrier for the $CH_2PH_3^{1+\bullet} \longrightarrow CH_3PH_2^{1+\bullet}$ isomerization has been computed to be 220 kJ/mol (MP3/6-31G**).[180] Note that the measured PA value for the $CH_2PH_2^{1\bullet}$ radical, 789 ± 13 kJ/mol[176] (after adjusting to the latest proton affinity scale),[151] is considerably lower than the computed G2 value listed above.

3.5.40   [C,H₆,P]⁺ – m/z 49

Protonation of methylphosphine produces the $CH_3PH_3^{1+}$ ion. Its heat of formation has been calculated to be 656.4 kJ/mol (G2),[41] in agreement with 661 kJ/mol obtained using the PA of methylphosphine and ancillary thermochemical data (PA = 851.5 kJ/mol,[12] $\Delta_fH(CH_3PH_2) = -18$ kJ/mol[40] and $\Delta_fH(H^+) = 1530$ kJ/mol[12]). No mass spectral data are available.

3.5.41   [C,S]⁺• – m/z 44

This ion dissociates by loss of atomic carbon to produce m/z 32 ($S^{+\bullet}$). Combining the ionization energy and the heat of formation of the neutral (11.33 ± 0.01 eV and 280.33 kJ/mol, respectively)[12] leads to $\Delta_fH(CS^{+\bullet}) = 1374$ kJ/mol, in agreement with calculations at a variety of levels of theory.[181] Note that the *2002 CRC Handbook of Chemistry and Physics*[182] and the Lias et al. compendium[40] give alternative values for the heat of formation of CS. It is the opinion of the authors that the 280 kJ/mol value for the heat of formation of neutral CS is the better value as it is in agreement with an experimental value (279 ± 4 kJ/mol at 298 K)[183] and also consistent with the experimentally determined bond dissociation energy of CS (D(C–S) = $\Delta_fH(C) + \Delta_fH(S) - \Delta_fH(CS)$; D(C–S at 0 K)$_{exp}$ = 709.6 ± 1.2 kJ/mol[184] or D(C–S at 298 K)$_{exp}$ = 713.3 ± 1.2 kJ/mol, $\Delta_fH(C) = 716.68 \pm 0.45$ kJ/mol[12] and $\Delta_fH(S) = 277.17 \pm 0.15$ kJ/mol).[12]

3.5.42 $[C,H,S]^+$ – m/z 45

Isomers

The two possible isomers are the thioformyl cation (HCS^{1+}), which is a known interstellar species,[129,185] and CSH^{1+}. Experimentally, only the former is observed.

Identification

The thioformyl cation is obtained by protonation of CS. This ion is also a very common fragment ion in the mass spectra of organosulfur compounds.

Thermochemistry

From the proton affinity of CS, $\Delta_f H(HCS^{1+}) = 1019$ kJ/mol is deduced (PA $= 791.5$ kJ/mol, $\Delta_f H(CS) = 280$ kJ/mol and $\Delta_f H(H^+) = 1530$ kJ/mol).[12] This value is consistent with the value obtained when combining the heat of formation and ionization energy of HCS ($\Delta_f H(HCS) = 300.4 \pm 8.4$ kJ/mol and $\leq 7.499 \pm 0.005$ eV)[186] but is considerably larger than an experimental value of ≤ 975 kJ/mol obtained by measuring the photoionization appearance energies of the HCS^{1+} ion generated from thiirane (cy–C_2H_4S–), thietane (cy–C_3H_6S–) and tetrahydrothiophene (cy–C_4H_8S–).[187]

The isomeric form CSH^{1+} is calculated to lie 298–317 kJ/mol higher in energy (at a variety of levels of theory).[188–191] On the singlet potential energy surface, it is expected to collapse with very little activation to HCS^{1+}.[188,189,191]

3.5.43 $[C,H_2,S]^{+\bullet}$ – m/z 46

Isomers

Calculations predict that there are three isomeric forms, namely, ionized thioformaldehyde ($CH_2S^{1+\bullet}$), the carbene radical cations $HCSH^{1+\bullet}$ and $H_2SC^{1+\bullet}$.[191] $CH_2S^{1+\bullet}$ is found in interstellar clouds and could be formed by the reaction of HCS^{1+} and H_2.[191] No mass spectral characteristics have been published.

Thermochemistry

Combining the heat of formation and ionization energy of CH_2S (118 ± 8.4 kJ/mol[12] (in agreement with G2 calculation)[192] and 9.376 ± 0.003 eV,[12] respectively) gives $\Delta_f H(CH_2S^{1+\bullet}) = 1023$ kJ/mol. The latest calculations find that the thiohydroxycarbene radical cation ($HCSH^{1+\bullet}$) lies 121 kJ/mol higher in energy than $CH_2S^{1+\bullet}$ (in agreement with earlier calculations at the MRDCI + ZPVE level)[188] and is separated from this ion by a barrier to isomerization of 96 kJ/mol (QCIS(T)/6-311++G(2df,2pd)//MP2/6-311

++G(d,p) + ZPE).[191] This is supported by a G2 absolute value of 1146 ± 5 kJ/mol.[28] The third isomer, $H_2SC^{1+\bullet}$, is considerably higher in energy lying 377 kJ/mol above $CH_2S^{1+\bullet}$.[191]

3.5.44 $[C,H_3,S]^+$ – m/z 47

Isomers

The mercapto-methyl and thiomethoxy cations, $^+CH_2SH$ and CH_3S^+, respectively, have been studied both by experiment and by theory. Two other isomers, $^+CHSH_2$ and $[H_2\cdots HCS]^+$ have been investigated by theory.

Identification

Fragment ions of elemental composition $[C,H_3,S]$ are very common in the mass spectra of organosulfur compounds and at low ionizing electron energy (ca 12 eV), all show very similar CID mass spectra indicating the likelihood of a single ion structure. From its dissociation characteristics, and by analogy with its oxygen counterpart, this ion has been assigned the connectivity $^+CH_2SH$.[193,194] This ion is also specifically produced by protonation of thioformaldehyde.[195]

It was however noted that at higher ionizing energy (70 eV), CH_3SSCH_3 and to a lesser extent other precursor molecules, produced an m/z 47 fragment ion with a slightly but significantly different CID mass spectrum (e.g., an increased relative intensity of the m/z 15 fragment ion) suggesting that the thiomethoxy cation (CH_3S^+) is also present in the ion flux under such conditions.[193,194]

Partial CID Mass Spectra of the Fragment $[C,H_3,S]^+$ Ion from a Variety of Precursors[194]

Precursor Molecule	m/z						Ion Structure
	12	13	14	15	32	33	
CH_3SH							
70 eV	2.8	6.2	18	2.2	70	100	CH_2SH
12 eV	3.8	9.8	19	2.5	75	100	CH_2SH
CH_3SCH_3							
70 eV	4.3	6.7	18	3.3	73	100	CH_2SH
12 eV	5.1	8.5	24	3.0	72	100	CH_2SH
CH_3SSCH_3							
70 eV	4.4	8.3	21	15	95	100	CH_3S
12 eV	4.4	7.3	23	4.7	82	100	CH_2SH

Note that the two *metastable* ions, $^+CH_2SH$ and CH_3S^+, dissociate by loss of H_2 to generate HCS^{1+} ions. The large kinetic energy release $(T_{0.5} \approx 0.9\,eV)$[196]

accompanying this dissociation is within experimental error the same for both ions, a similarity with the behavior of the $^+CH_2OH/CH_3O^+$ pair.

Thermochemistry

The heat of formation of $^+CH_2SH$ can be obtained from the proton affinity of thioformaldehyde, $\Delta_fH(^+CH_2SH) = 888$ kJ/mol (PA(CH_2S) = 759.7 kJ/mol, $\Delta_fH(CH_2S) = 118 \pm 8.4$ kJ/mol and $\Delta_fH(H^+) = 1530$ kJ/mol).[12]

From the latest calculations (CCSD(T)/B2[197] and QCISD(T)/6-31++G (2df,2pd)//MP2/6-311++G(d,p) + ZPE[191] levels of theory) the following observations can be made: The global minimum on the combined singlet and triplet potential energy surfaces for [C,H$_3$,S]$^+$ ions is the mercapto-methyl cation ($^+CH_2SH$) in its singlet state. It lies approximately 120 kJ/mol lower in energy than the triplet thiomethoxy cation, $^3CH_3S^+$. These observations are consistent with previous calculations.[198,199] In addition, the $^1CH_3S^+$ ion was found not to be a minimum on the potential energy surface, indicating that ions formed in the dissociative ionization of CH_3SSCH_3 are $^3CH_3S^+$. The barrier for the isomerization of $^3CH_3S^+$ to $^3CH_2SH^{1+}$ is predicted to be 192 kJ/mol.

A detailed computational study[197] on the mechanism of H_2 elimination from these cations indicates that in the case of the $^1CH_2SH^{1+}$ ion there is a significant reverse energy barrier (134 kJ/mol) associated with the formation of HSC^{1+} which is in keeping with the observed kinetic energy release. As for the $^3CH_3S^+$ ion, a concerted 1,1-H_2 elimination or isomerization to $^1CH_2SH^{1+}$, followed by a 1,2 H_2 elimination, could both occur in the metastable ion time frame and involve energies again consistent with the observed kinetic energy release. This is remarkably similar to the behavior of the oxygen analogs, $^+CH_2OH$ and CH_3O^+.

Other isomers investigated by theory are the carbene ion $CHSH_2^{1+}$ and the complex $[H_2 \cdots HCS]^+$. The former ion in its singlet state is calculated to lie 336 kJ/mol above $^1CH_2SH^{1+}$ (at the QCISD(T)/6-31++ G(2df,2pd)//MP2/6-311++G(d,p) + ZPE level of theory)[191] whereas the latter, with a binding energy of only 5 kJ/mol, is 179 kJ/mol higher in energy than $^1CH_2SH^{1+}$ (at the HF/TZ + 2d1p//TZ + d level of theory).[198]

It is noteworthy that there is some uncertainty related to the heat of formation of the neutral $^\bullet CH_2SH$. The value obtained from ion chemistry is 150 ± 8.4 kJ/mol[200,201] whereas G2 calculations give 166.1 kJ/mol.[41] When considering the thermochemical data of related species, the latter value appears to be better.

3.5.45 [C,H$_4$,S]$^{+\bullet}$ – m/z 48

Isomers

There are two isomers: the conventional methanethiol radical cation ($CH_3SH^{1+\bullet}$) and the distonic radical cation ($CH_2SH_2^{1+\bullet}$).

Identification[38,39]

Direct ionization of methanethiol and dissociative ionization of 2-mercapto-ethanol (HOCH$_2$CH$_2$SH, loss of H$_2$CO) generate ions with very similar MI (both ions lose H$^•$ producing a Gaussian peak with a kinetic energy release of 28 meV) and CID mass spectral characteristics. The observed fragment peaks can be rationalized based on the connectivities of both the CH$_3$SH$^{]+•}$ and the CH$_2$SH$_2$$^{]+•}$ ions and so both precursor molecules likely yield a mixture of product ions. However, the more intense m/z 15 fragment peak for the ion generated directly from methanethiol suggests the presence of a significant amount of CH$_3$SH$^{]+•}$ in the ion flux. Also, the greater relative abundance of m/z 34 (SH$_2$$^+$) and 14 (CH$_2$$^{+•}$) and the more intense doubly charged ion, [C,H$_4$,S]$^{2+}$ (m/z 24) for the product ion obtained from 2-mercapto-ethanol, are consistent with the distonic ion (CH$_2$SH$_2$$^{]+•}$) being predominantly formed.

Partial CID Mass Spectra of the [C,H$_4$,S]$^{+•}$ Isomers

					m/z				
Ions	14	15	22.5	23	23.5	24	32	33	34
CH$_3$SH	8.3	42	0.37	3.0	5.0	7.2	61	100	16
CH$_2$SH$_2$	13	37	0.50	2.7	2.5	9.7	72	100	29

In contrast, reactivity experiments[202] with neutral reagents with ionization energies ≤9.5 eV suggest that ionization of methanethiol and dissociative ionization of 2-mercapto-ethanol generate discrete ion fluxes of CH$_3$SH$^{]+•}$ and CH$_2$SH$_2$$^{]+•}$, respectively. However, in the presence of reagents with ionization energies ≥9.9 eV, some isomerization before reaction is expected to take place at longer reaction times. That the ion fluxes produced could be different mixtures of the two isomers was discussed by the authors.

In addition, this study sets an upper limit for the proton affinity at sulfur in the $^•$CH$_2$SH radical at 724 kJ/mol (this value was adjusted to reflect the current proton affinity scale),[151] in excellent agreement with a calculated value of 725.8 kJ/mol[41] but somewhat lower than that quoted in NIST (733.9 kJ/mol).[12]

Thermochemistry

Combining the heat of formation and the ionization energy of CH$_3$SH, -22.8 ± 0.6 kJ/mol[12] and 9.45458 ± 0.00036 eV[203] (in agreement with earlier experimental measurement of 9.4553 ± 0.0006 eV),[204] gives a $\Delta_f H(\text{CH}_3\text{SH}^{]+•}) = 889$ kJ/mol. The latest theoretical calculations predict the distonic ion to be 90 kJ/mol higher in energy than CH$_3$SH$^{]+•}$ (QCISD(T) method without ZPE taken into account),[205] about 14 kJ/mol higher than previously calculated[190,199] but in good agreement with a G2 value of 971 kJ/mol.[41] The value obtained from the appearance energy measurement, $\Delta_f H(CH_2SH_2^{]+\bullet}) = 916$ kJ/mol (loss of H_2CO from $HOCH_2CH_2SH$: $AE = 10.42 \pm 0.05$ eV,[38] $\Delta_f H(H_2CO) = -108.6 \pm 0.46$ kJ/mol[12] and $\Delta_f H$ $(HOCH_2CH_2SH) = -198$ kJ/mol[36]), is certainly too low.

The barrier for the isomerization of $CH_3SH^{]+\bullet}$ to $CH_2SH_2^{]+\bullet}$ is calculated to be 190 kJ/mol (QCISD(T) method without the ZPE being taken into account[205] and MP3/6-31G**//4-31G[199]). The transition state for isomerization therefore lies below the dissociation limit to $^+CH_2SH + H^\bullet$ ($\Sigma\Delta_f H_{products} = 1106$ kJ/mol) indicating that the ions can interconvert before dissociation. This is in keeping with the MI and CID mass spectral characteristics of the two ions.

3.5.46 [C,H$_5$,S]$^+$ – m/z 49

$CH_3SH_2^{]+}$ ions are proposed to be generated by dissociative photoionization of dimethyldisulfide (CH_3SSCH_3) and by protonation of methanethiol (CH_3SH). From the proton affinity of methanethiol, $\Delta_f H(CH_3SH_2^{]+}) = 734$ kJ/mol is deduced (PA(CH_3SH) = 773.4 kJ/mol, $\Delta_f H(H^+) = 1530$ kJ/mol and $\Delta_f H(CH_3SH) = -22.8 \pm 0.59$ kJ/mol).[12] In agreement with this, the latest appearance energy measurement for the ion sets an upper limit of 728 kJ/mol for its heat of formation.[206]

3.5.47 [C,S$_2$]$^{+\bullet}$ – m/z 76

The CID mass spectrum of ionized carbon disulfide is dominated by the fragment ion peaks corresponding to the production of $CS^{]+\bullet}$ (m/z 44) and $S^{+\bullet}$ (m/z 32) ions.[207] Interestingly, it also generates $S_2^{]+\bullet}$ ions[207] by a mechanism that must involve geometric changes in the molecular ion before dissociation.[208]

Combining the heat of formation and ionization energy of carbon disulfide (116.94 kJ/mol and 10.073 ± 0.005 eV, respectively)[12] gives $\Delta_f H(CS_2^{]+\bullet}) = 1089$ kJ/mol.

3.5.48 [C,H,S$_2$]$^+$ – m/z 77

Isomers

In principle, protonation of carbon disulfide can take place at either carbon or sulfur atoms. By experiment, only the latter ($HSCS^{]+}$) has been observed, although theory predicts $HCS_2^{]+}$ to be stable.

Identification[207]

Protonation of CS_2 using $CH_5^{]+}$ in the ion source of the mass spectrometer gives an ion that dissociates mainly to m/z 45 ([H,C,S]$^+$), 44 ($CS^{]+}$), 33 ($HS^{]+}$) and 32 ($S^{+\bullet}$), consistent with the $HSCS^{]+}$ connectivity.

Thermochemistry

From the proton affinity of carbon disulfide, $\Delta_f H(HSCS^{1+}) = 965$ kJ/mol is obtained (PA = 681.9 kJ/mol (in agreement with a calculated value, 686 kJ/mol (QCISD(T)/6-311 + G(2d,p) + ZPVE)[207]), $\Delta_f H(CS_2) = 116.94$ kJ/mol and $\Delta_f H(H^+) = 1530$ kJ/mol[12]). Calculations predict that the carbon-protonated species (HCS_2^{1+}) is 102 kJ/mol higher in energy.[207]

3.5.49 $[C,H_2,S_2]^{+\bullet}$ – m/z 78

Isomers

Three isomers have been investigated by experiment: ionized dithioformic acid $(HC(S)SH^{1+\bullet})$, ionized dimercaptocarbene $(HSCSH^{1+\bullet})$ and ionized dithiirane $(cy–CH_2SS–^{1+\bullet})$. Ionized thiosulfine $(CH_2SS^{1+\bullet})$ and $SH_2CS^{1+\bullet}$ have also been found by computations to be stable.

Identification[209]

The following reactions generate the ions of interest.

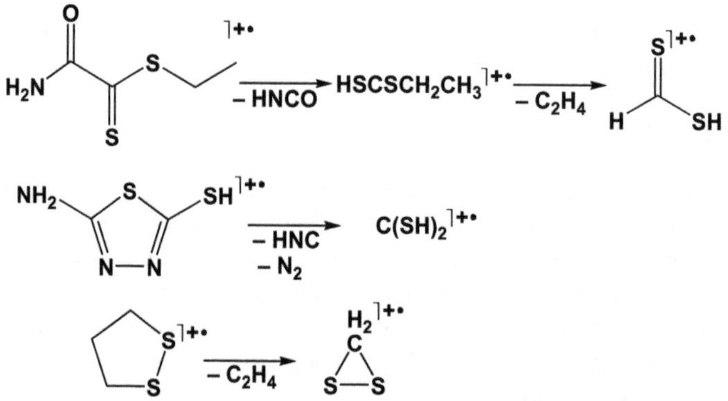

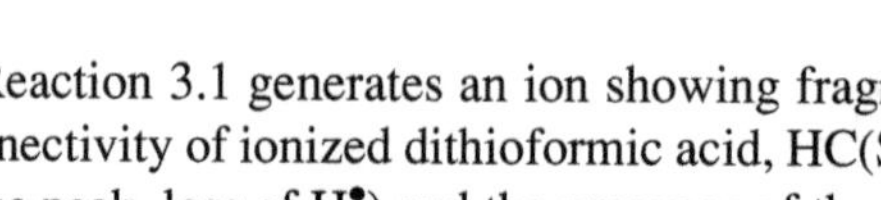

Reaction 3.1 generates an ion showing fragment peaks consistent with the connectivity of ionized dithioformic acid, $HC(S)SH^{1+\bullet}$: the large m/z 77 peak (base peak, loss of $H^\bullet$) and the presence of the fragment peaks at m/z 45 (loss of $SH^\bullet$), 33 (SH^{1+}) and 32 (S^+).

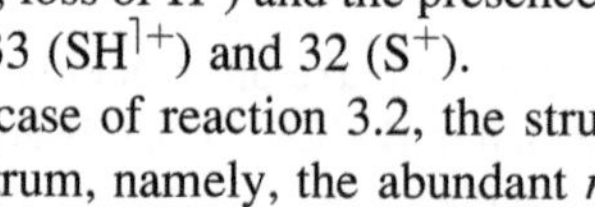

In the case of reaction 3.2, the structure-identifying features of the CID mass spectrum, namely, the abundant m/z 44 $(CS^{1+\bullet})$ and 45 $([C,H,S]^+)$ as well as the m/z 66 $(H_2S_2^{1+\bullet})$ and the doubly charged peak at m/z 39, suggest the formation of the dimercaptocarbene radical cation, $HSCSH^{1+\bullet}$.

The CID mass spectrum of the ion formed in reaction 3.3 shows an important fragment peak at m/z 64 $(S_2^{1+\bullet})$ which suggests two possible structures, namely, $CH_2SS^{1+\bullet}$ (ionized thiosulfine) or $cy–CH_2SS–^{1+\bullet}$ (ionized dithiirane). Considering the relative energy of these two species (see below), the formation of $CH_2SS^{1+\bullet}$ can be ruled out. The CID mass spectra of these three ions are sufficiently different to allow their differentiation.

Partial CID Mass Spectra of the $[C,H_2,S_2]^{+\bullet}$ Isomers

Ions	m/z					
	12	13	14	32	33	34
HC(S)SH	0.74	0.66	—	100	66	26
HSCSH	0.98	—	—	100	30	90
cy–CH$_2$SS–	0.74	1.1	1.7	100	a	33

aShoulder along side a broad and intense m/z 32.

Thermochemistry

No experimental data are available for these ions but theoretical calculations predict that the cy–CH$_2$SS–$^{1+\bullet}$ is the lowest energy isomer. The other ions, HC(S)SH$^{1+\bullet}$, CH$_2$SS$^{1+\bullet}$ and HSCSH$^{1+\bullet}$ lie respectively, 10, 74, and 93 kJ/mol higher in energy (B3LYP/6-31G**).[209] These values are in agreement with another computational study that further predicts that the SH$_2$CS$^{1+\bullet}$ ion is also stable, being higher in energy than HC(S)SH$^{1+\bullet}$ by 153 or 185 kJ/mol, depending on the level of theory (UMP2/6-31G** and UQCISD(T)/6-311 ++ G**).[210] The absolute heat of formation of the carbene ion has been calculated to be 1064 kJ/mol (G2).[28] The barrier for the isomerization for HC(S)SH$^{1+\bullet}$ $\longrightarrow$ HSCSH$^{1+\bullet}$ has been estimated to be approximately 192 kJ/mol (UQCISD(T)/6-311 ++ G(d,p) + ZPE).[210]

3.5.50 $[C,H_3,S_2]^+$ – m/z 79

Isomers

Six isomers have been proposed by theory to be stable. These are CH$_3$SS^{1+} (1), CH$_2$S(H)S^{1+} (2), CH$_2$SSH^{1+} (3), cy–S(H)CH$_2$S–$^{1+}$ (4), HSC(H)SH^{1+} (5) and CHS(H)SH^{1+} (6). None of these have been characterized by mass spectrometry.

Identification

Dissociative ionization of dimethyl disulfide (CH$_3$SSCH$_3$, loss of CH$_3$$^{1\bullet}$) yields a $[C,H_3,S_2]^+$ ion. On the basis of thermochemical considerations (comparison of experimental and calculated heats of formation) it is suggested that the ion produced at threshold from dimethyl disulfide is most likely CH$_2$SSH^{1+} and not its isomer CH$_3$SS^{1+}.[211–213]

Thermochemistry

Theory predicts that the global minimum is $HSC(H)SH^{1+}$. The absolute heat of formation of the ions, calculated at the G2 level of theory, is $\Delta_f H(1) = 910.4$ kJ/mol, $\Delta_f H(2) = 1120.1$ kJ/mol, $\Delta_f H(3) = 877.8$ kJ/mol, $\Delta_f H(4) = 895.8$ kJ/mol, $\Delta_f H(5) = 834.3$ kJ/mol and $\Delta_f H(6) = 1225.9$ kJ/mol, in agreement with other calculations.[211,212] The transition state energies corresponding to isomerization have also been calculated (G2): $1 \longrightarrow 2 = 624$ kJ/mol, $2 \longrightarrow 3 = 29$ kJ/mol, $3 \longrightarrow 4 = 109$ kJ/mol, $3 \longrightarrow 6 = 368$ kJ/mol, $4 \longrightarrow 5 = 184$ kJ/mol.[212]

3.5.51 $[C,S_3]^{+\bullet} - m/z\ 108$

Isomers

Computations find five different minima on the potential energy surface of these ions: ionized carbondisulfide S-sulfide ($SCSS^{1+\bullet}$), three different states for the C_{2v} carbon trisulfide ion ($SC(S)S^{1+\bullet}$) and a four-membered ring cy-$CSSS-^{1+\bullet}$, $SCSS^{1+\bullet}$ and $SC(S)S^{1+\bullet}$ have been generated and characterized.

Identification

Self-chemical ionization of carbon disulfide produces the $(CS_2)_2^{1+\bullet}$ dimer which dissociates to yield an m/z 108 fragment ion (loss of CS).[214] This ion's CID mass spectrum comprises the following fragment ions: m/z 76 (85%, loss of S), 64 (100%, loss of CS), 54 (3%, CS_3^{12+}), 44 (12%, $CS^{1+\bullet}$) and 32 (5%, $S^{+\bullet}$). $CS_3^{1+\bullet}$ ions can also be obtained by the dissociative ionization of isomeric dithiolethiones ((1) and (2))[214,215] or by charge reversal of the anion generated from the dissociative ionization of 4,5-dioxo-2-thioxo-1,3-diithiolan (3)[215] according to following reactions:

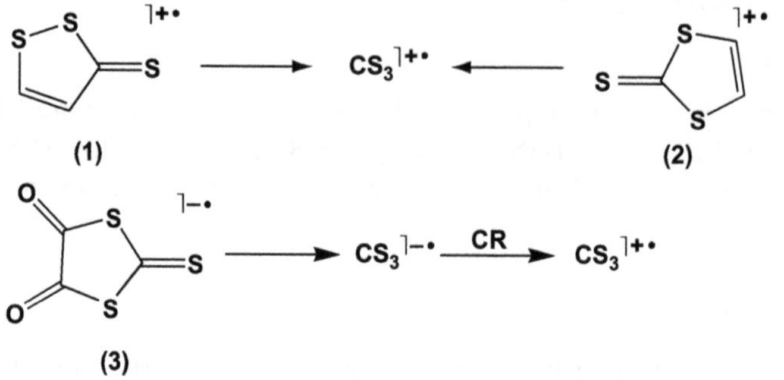

(1)   (2)

(3)

In all cases, the CID mass spectra are very similar. ^{34}S-labeling experiments[215] on the isomeric dithiolethiones ((1) and (2)) suggest that mixtures of ions, most likely comprising $SCSS^{1+\bullet}$ and $SC(S)S^{1+\bullet}$ are formed in the dissociations, supporting the observed similarities of the CID mass spectra.[214,215]

Thermochemistry[214]

Theoretical calculations find $SCSS^{1+\bullet}$ to be the global minimum on the potential energy surface. Carbon trisulfide ions $(SC(S)S^{1+\bullet})$ in the $^2A2$, $^2B2$ and $^2B1$ states lie 42.4, 46.3, and 125.4 kJ/mol, respectively, above $SCSS^{1+\bullet}$ while the cyclic ion is predicted to be even higher in energy at 262.8 kJ/mol above $SCSS^{1+\bullet}$ (G2(MP2,SVP)).

3.5.52 $[C,S_4]^{+\bullet}$ – m/z 140

In addition to $CS_3^{1+\bullet}$ ions, self-chemical ionization of carbon disulfide leads to the generation of $CS_4^{1+\bullet}$. This ion can also be obtained from the ion-molecule reaction of $CS_3^{1+\bullet} + CS_2$. The ion formed dissociates to yield fragment ions at m/z 96 $(S_3^{1+\bullet})$ and 108 $CS_3^{1+\bullet}$.[214] No thermochemical data are available.

3.5.53 $[C,H_n,X_m]^{+\bullet/+}$ – where $n \leq 2$, $m \leq 4$ and $X = F$, Cl, Br and/or I

These ions do not have isomeric forms. The thermochemical data pertaining to these species are given below. The footnotes indicate where the data are equivocal.

Selected Values for Ions of Empirical Formula $[C,H_n,X_m]$ where $n \leq 2$, $m \leq 4$ and $X = F$, Cl, Br and/or I

Species	$\Delta_f H$(neutral)[44] (kJ/mol)	Ionization energy[12] (eV)	$\Delta_f H$(ion)[a] (kJ/mol)
CF	240.6±10	9.11±0.01	1120
	(240–249)[216–218]	(8.95–9.43)[216,218,219]	(1133, 1135)[216,218]
CCl	443.1±13.0[b]	8.9±0.2	1302
	(433–439)[217,220,221]	(8.63–8.71)[220,221]	(1271–1276)[220,221]
CBr	510±63	10.43±0.02[c]	1516
	(497–500)[217,222,223]	≤9.11±0.78[223]	
		(8.55, 8.65)[223]	(1324, 1325)[223]
CI	570±35[b]		
	(557.6)[217]		
CHF	143.0±12.6	10.06±0.05	1114
	(143–147)[218,224–226]	(10.2)[218]	(1129)[218]
CHCl	326.4±8.4	9.84[d]	1205[e]
	(315–326)[220,221,224–226]	(9.10–9.20)[220,221]	(1200–1208)[220,221]
CHBr	373±18	9.17±0.23[223]	1257
	(371–380)[222,223,225]	(9.00, 9.02)[223]	(1229, 1239)[223]
CHI	428±21[227]		
	(425, 427)[225]		

(*continued*)

(Continued)

Species	$\Delta_f H$(neutral)[44] (kJ/mol)	Ionization energy[12] (eV)	$\Delta_f H$(ion)[a] (kJ/mol)
CH_2F	-31.8 ± 4.2 $(-29, -32.6)$[41,228]	9.04 ± 0.01	840
CH_2Cl	117.2 ± 2.9 $(117-119)$[41,220,221,228]	8.75 ± 0.01 $(8.61-8.65)$[220,221]	961
CH_2Br	171.1 ± 2.7 $(165-175)$[41,223,229]	8.61 ± 0.01 $(8.37-8.68)$[229]	1002
CH_2I	229.7 ± 8.4	8.40 ± 0.03	1040
CF_2	-182.0 ± 6.3 $(-187$ to $-206)$[216,218,225,230]	11.44 ± 0.03 (11.54)[218]	922
CCl_2	226 $(214-231)$[220,224,225,230]	9.27 ± 0.04 $(9.19, 9.21)$[220]	1120
CBr_2	343.5 $(336-350)$[222,225,226]	10.11 ± 0.09	1319
CI_2	468 ± 60 (450)[225]		
$CClF$	31.0 ± 13.4 $(14-32)$[224-226,230]	10.62	1058
$CClBr$	267	10.4 ± 1[f]	1232
CHF_2	-238.9 ± 4.2 $(-241, -245)$[218,228]	$8.74 - \leq 8.90$ (8.88)[218]	604–620 (612)[218]
$CHBr_2$	188.3 ± 9.2 $(176-188)$[223]	$8.13 \pm 0.16, 8.3 \pm 0.3,$ 8.61 ± 0.01[231] $(8.47, 8.57)$[223]	973–1019
$CHCl_2$	93.3 ± 4.2 $(91-94)$[220,221,228]	$8.32 \pm 0.01, 8.45$ $(8.14, 8.17)$[220,221]	896, 909 (879)[220,221]
CHI_2	314.4 ± 3.3		
$CHBrCl$	143 ± 6		
CH_2F_2	-450.66[g] (-451.6)[232]	12.71 $(12.44, 12.78)$[232,233]	776
CH_2Br_2	$-14.8-10.0$[234] $(-0.4, -19)$[223]	10.41 ± 0.13 $(10.26, 10.25)$[223]	987–1014
CH_2I_2	$118 \pm 4.2, 122 \pm 4.2$[g]	9.46 ± 0.2	1031, 1035
CF_3	-465.7 ± 2.1 $(-465$ to $-483)$[216,218,228,230]	$8.5 \pm 0.8-9.3 \pm 0.2$ $(8.77-9.25)$[216,218,219]	354–431 $(412, 424)$[216,218]
CCl_3	71.1 ± 2.5 $(40-75)$[220,228,230]	8.109 ± 0.005 (7.90)[220]	853 (838)[220]
CBr_3	205.0 ± 8.4 (214.76)[37]	7.5 ± 0.2	929
CI_3	424.9 ± 2.8	—	—
$CClF_2$	-279.0 ± 8.4 (-274.7)[228]	$9.0 \pm 0.5, 13.4 \pm 1$[f] 8.7[h]	556[g]
CCl_2F	-89.0 ± 8.4 (-94.3)[228]	$8.5 \pm 0.5, 12.4 \pm 1$[f] 8.2[h]	703[g]
$CBrClF$	-35.5 ± 6.3	—	—
$CBrF_2$	-224.7 ± 12.6	12.5 ± 1[f]	1479
CBr_2Cl	163 ± 8	13.8 ± 1[f]	—
$CBrCl_2$	124 ± 8	—	—

(Continued)

Species	$\Delta_f H$(neutral)[44] (kJ/mol)	Ionization energy[12] (eV)	$\Delta_f H$(ion)[a] (kJ/mol)
CHF_3	-697.05[g]	13.86	640
	(-698.5)[232]	(14.03)[232]	
$CHCl_3$	-103.18[g]	11.37 ± 0.02	994
	(-101.9)[232]	(11.49)[232]	
$CHBr_3$	55.4 ± 3.3[g]	10.50 ± 0.02	1068
CHI_3	—	9.23 ± 0.02	—
CF_4	-930 ± 20[g]	≤ 14.7	≤ 488
	(-936.9)[232]	(14.49)[232]	
CBr_4	50.21[g]	10.31	1045
CI_4	267.94[g]	8.95, 9.10	1131, 1146
FCO	-171.5[g]	9.3 ± 0.1[235]	745.3 ± 9.6[235]
	-152.1 ± 12[235]	8.76 ± 0.32[237]	726 ± 6[238]
	-142 ± 21[236]		
	(-185 ± 2)[i]		
ClCO	22 ± 4[239,j]	$(8.28, 8.27)$[240]	$(780.0, 772.4)$[240]
	$(-19.4, -24.1,$[240]		
	-19.2 ± 2[k]$)$		
BrCO	(2.9 ± 6)[k]		$(764, 664)$[l]
F_2CO[m]	-638.90[12]	13.04 ± 0.03	633 ± 6[238]
	$\geq -624 \pm 6$[238]	13.62[241]	
Cl_2CO[m]	-220.08[12]	≈ 11.2	≈ 861
	(-216 ± 2)[k]	11.90[241]	
Br_2CO[m]	-113.5 ± 0.5[12]	10.8	929
	(-95 ± 4)[k]	10.98[241]	
$FClCO$[m]	-426.77[12]	12.58[241]	787
$FBrCO$[m]		11.87[241]	
$BrClCO$[m]		11.30[241]	
FCO_2	-356 ± 12[242]		
	(-360)[243]		

Parentheses indicate the values or range of values obtained from recent computations.

[a]Deduced from $\Delta_f H$(ion) $= $ IE $+ \Delta_f H$(neutral) or otherwise stated.
[b]Revised value from Y.-R. Luo, private communication.
[c]Considering how the ionization energy varies in the series CX, CHX and CH_2X for X=F and Cl, and in addition to the more recent IE measurement, the NIST value must certainly be too high.
[d]Comparing with other experimental and calculated values and considering the value obtained from IE $= \Delta_f H$(ion) $- \Delta_f H$(neutral), the NIST value is certainly too high.
[e]$\Delta_f H$(0 K) from Ref. [221].
[f]From Ref. [244]. These values are not very accurate.
[g]From NIST.[12]
[h]Deduced from the heat of formation of the ion and neutral.
[i]$\Delta_f H$(0 K) from Ref. [245].
[j]Note that the NIST value, -62.76 kJ/mol,[12] must certainly be to high.
[k]$\Delta_f H$(0 K) from Ref. [246].
[l]Deduced from the calculated dissociation energies to $CO^{+\bullet} + Br$ and $CO + Br^+$ of 590 and 467 kJ/mol,[247] respectively, and ancillary data from NIST.[12] Note that the authors also predict the isomer $COBr^+$ to be stable and to lie 312 kJ/mol lower in energy than $BrCO^+$.
[m]The electron-impact mass spectrum of the compound is provided in Ref. [248].

3.5.54 $[CCl_4]^{+\bullet}$ – m/z 152

Isomers

Two isomers have been characterized, ionized carbon tetrachloride ($CCl_4^{1+\bullet}$) and the $Cl_2C-Cl-Cl^{1+\bullet}$ radical cation. Theoretical calculations predict a total of at least six bonded species of various geometries to be stable.

Identification

It was only in 1985, in a carefully conducted experiment, that Drewello et al. were able to produce for the first time a weak but recognizable signal (intensity <0.01% of the base peak $CCl_3^{1+\bullet}$) corresponding to $CCl_4^{1+\bullet}$ by direct ionization of carbon tetrachloride.[249] Until then, electron-impact ionization of this molecule was believed to generate no molecular ion although theoretical calculations at the modest MINDO/3 level predicted this ion to be stable.[250] Indeed, upon ionization, carbon tetrachloride is subject to Jahn–Teller effects which distort its T_d geometry. Because of the different geometries of the molecule and molecular ion, the ion production efficiency near the ionization threshold is therefore very low. The distonic ion, $Cl_2C-Cl-Cl^{1+\bullet}$, is obtained from the decarbonylation of ionized $Cl_3CC(O)Cl$.[249]

Both metastable ions lose $Cl^\bullet$ to give CCl_3^{1+} but the corresponding peak shapes are different. In the case of $CCl_4^{1+\bullet}$ the peak is dished ($T = 155 \pm 4$ meV, measured from the width across the summit of the peak) whereas the distonic ion produces a narrow Gaussian peak ($T_{0.5} = 6.2 \pm 0.5$ meV).[249,251] These observations are consistent with the proposed reaction path for the decomposition of $CCl_4^{1+\bullet}$ which requires this ion to isomerize (over a calculated barrier of 57 kJ/mol (RHF-MNDO)) to the distonic species before dissociation.[249] Their CID mass spectra (O_2) are also slightly different, $Cl_2C-Cl-Cl^{1+\bullet}$ generating more abundant Cl^+ and $CCl_2^{1+\bullet}$ fragment peaks, consistent with its connectivity.[251]

Thermochemistry

The heat of formation of ionized carbon tetrachloride can be deduced from the $\Delta_f H$ and IE of the molecule. NIST list values for the former ranging from −94 to −125 kJ/mol, their recommended value being −95.98 kJ/mol.[12] The latest calculated value (G3) is −103.0 kJ/mol,[232] in agreement with the MINDO/3 study.[250] The G3 computations[232] also give $\Delta_f H(CCl_4^{1+\bullet}) = 999.7$ kJ/mol leading to an $IE(CCl_4) = 11.43$ eV, in good agreement with experimental and G2(MP2) computed values.[252] Therefore, the value recommended in NIST may be slightly too high. In the MINDO/3 study,[250] the potential energy surface showed seven bonded states for $CCl_4^{1+\bullet}$, the global minimum having the C_{2v} geometry. The heat of formation of $Cl_2C-Cl-Cl^{1+\bullet}$ has been calculated at a modest level of theory (MNDO/3) to be 1069 kJ/mol,[249] 16 kJ/mol higher in energy than the conventional molecular ion. Note however that in this study the $\Delta_f H(CCl_4^{]+\bullet}) = 1053$ kJ/mol, is in disagreement with the more recent value.

3.5.55 [C,Br$_3$,Cl]$^{+\bullet}$ – m/z 284^{249}

By analogy with the distonic ion $Cl_2C-Cl-Cl^{]+\bullet}$, the distonic ion $Br_2C-Br-Cl^{]+\bullet}$ is believed to be generated from the decarbonylation of $CBr_3C(O)Cl$. The ion produced by CO loss from this molecule exclusively loses $Cl^\bullet$ upon collisional activation, consistent with the proposed connectivity.

3.5.56 [C,H,Cl,F]$^{+}$ – m/z 67^{253}

Isomers

Three isomeric forms are calculated to be stable, namely the complexes $[HF\cdots CCl]^+$ and $[HCl\cdots CF]^+$ and the $CHClF^{]+}$ ion. Only the latter ion has been observed by experiment.

Identification

$CHClF^{]+}$ ions can be generated by the dissociative ionization of $CHClF_2$ (loss of $F^\bullet$) or by charge exchange ionization with $N_2^{]+\bullet}$ followed by loss of $F^\bullet$. The ion's CID mass spectrum is dominated by fragment ions at m/z 32 ($CHF^{]+\bullet}$) and 31 ($CF^{]+}$) with minor peaks at m/z 48 ($CHCl^{]+\bullet}$) and 35 (Cl^+), consistent with the proposed structure.

Thermochemistry

The $CHClF^{]+}$ ion is calculated to be the global minimum. Combining the heat of formation and the calculated IE of the $CHClF^{]\bullet}$ radical (-60.7 ± 10.0 kJ/mol^{44} (in agreement with a computed value of -63.8 ± 4 kJ/mol)^{228} and 8.44 eV (G2(MP2),^{253} respectively) gives $\Delta_f H(CHClF^{]+}) = 754$ kJ/mol. The complexes $[HCl\cdots CF]^+$ and $[HF\cdots CCl]^+$ are calculated to lie 879 and 895 kJ/mol, respectively, both higher in energy and separated from $CHClF^{]+}$ by isomerization barriers of 569 and 632 kJ/mol. These isomerization barriers lie above the threshold for dissociation to $CCl^{]+} + HF$ and $CF^{]+} + HCl$ and therefore prevent the isomers from interconverting before dissociation.

3.5.57 [C,H$_2$,Cl$_2$]$^{+\bullet}$ – m/z 84

Isomers

Theory predicts five stable species: the classical radical cation ($CH_2Cl_2^{]+\bullet}$), its distonic isomer $Cl(H)C-Cl-H^{]+\bullet}$, $[CH_2-Cl-Cl]^{+\bullet}$ (isodichloromethane) and two loosely bonded species $[CCl_2\cdots H-H]^{+\bullet}$ and $[HC(Cl)\cdots H\cdots Cl]^{+\bullet}$. The first two isomers have been characterized by mass spectrometry.

Identification

Direct ionization of dichloromethane leads to an ion with mass spectral characteristics different from those of the ion generated by dissociative ionization of dichloroacetic acid (loss of CO_2).[116,254,255] On the basis of its CID mass spectrum, the latter ion has been assigned the connectivity of the distonic ion $Cl(H)C–Cl–H]^{+•}$.

Partial CID Mass Spectra of the $[C,H_2,Cl_2]^{+•}$ Isomers[255]

						m/z			
Ions	41	41.5	42	47	48	49	70	82	83
CH_2Cl_2	2	0.05	0.2	(4)	(8)	(100)	0.5	1	(10)
$Cl(H)C–Cl–H$	—	—	5	11	23	(100)	—	2	9

Values in parentheses contain contributions from metastable ion dissociation. Note that m/z 41, 41.5 and 42 correspond to doubly charged ions.

Note that the classical radical cation and two other isomers, namely the distonic ion $Cl(H)C–Cl–H]^{+•}$ and $[CH_2–Cl–Cl]^{+•}$, (best viewed as a chloro-methyl cation electrostatically bonded to a chlorine atom), have also been identified in the electron bombardment matrix-isolation (EBMI) Fourier-transform infrared spectroscopic study of dichloromethane.[256]

Thermochemistry

Combining the heat of formation and ionization energy of methylene chloride (-95.52 kJ/mol and 11.33 ± 0.04 eV, respectively)[12] gives $\Delta_f H(CH_2Cl_2^{+•}) = 998$ kJ/mol. Calculations (MP2(full)/6-311++g(2d,2p)) predict that the distonic isomer $Cl(H)C–Cl–H]^{+•}$ lies 47 kJ/mol above $CH_2Cl_2^{+•}$,[255] significantly higher than predicted previously (34 kJ/mol at MP4SDTQ/6-31++G**//MP2(FC)/6-31++G**).[257] The transition state for the rearrangement into the distonic ion is calculated to lie 33 kJ/mol above the dissociation to CH_2Cl^{+} and $Cl^{•}$.[255] The ion $[CH_2–Cl–Cl]^{+•}$ is calculated to lie 81 kJ/mol above the classical ion with a very small barrier to isomerization of 2 kJ/mol.[255] The magnitude of the calculated barrier is significantly smaller than that obtained from the EBMI investigation[256] (43 kJ/mol).

Note that the species $[CH_2–Cl–Cl]^{+•}$, $[CCl_2\cdots H–H]^{+•}$ and $[HC(Cl)\cdots H\cdots Cl]^{+•}$ were also investigated at the MP4SDTQ/6-31++G** //MP2(FC)/6-31++G** level and predicted to be higher in energy than the classical cation by 34.5, 103 and 54.1 kJ/mol, respectively.[257] Note also that the magnitude of the relative energy of the $[CH_2–Cl–Cl]^{+•}$ ion obtained at the different levels of theory is very different.

3.5.58 $[C,H_2,F_3]^+ - m/z\ 71^{258}$

Isomers

These species have been investigated solely by theory which predicts three stable complexes: $[CF_2H\cdots FH]^+$, $[CF_3\cdots H_2]^+$ and $[CFH_2\cdots F_2]^+$.

Thermochemistry

The complex $[CF_2H\cdots FH]^+$ is calculated to be the global minimum with the isomeric ions $[CF_3\cdots H_2]^+$ and $[CFH_2\cdots F_2]^+$ lying, respectively, 124 and 564 kJ/mol higher in energy (G2). Assuming that the lowest energy isomer is produced by protonation of trifluoromethane, an absolute heat of formation for the $[CF_2H\cdots FH]^+$ complex could be obtained. From experiment, the proton affinity of CHF_3 has been determined to be 615 ± 20 kJ/mol[259] and when adjusted to the current PA scale is listed as 619.5 kJ/mol.[12] However, G2 calculations predict this quantity to be appreciably lower, 582 kJ/mol. Therefore, the heat of formation of the $[CF_2H\cdots FH]^+$ is estimated to lie in the range of 213–251 kJ/mol ($\Delta_fH(CHF_3) = -697.05$ kJ/mol and $\Delta_fH(H^+) = 1530$ kJ/mol).[12]

3.5.59 $[C,H_3,F]^{+\bullet} - m/z\ 34$

Isomers

Two isomers have been identified, ionized methyl fluoride ($CH_3F^{]+\bullet}$) and the distonic radical cation ($CH_2FH^{]+\bullet}$).

Identification

Dissociative ionization of fluoroacetic acid (loss of CO_2) produces an ion that is markedly different from that obtained by ionization of fluoromethane. The CID mass spectrum of the former ion has no significant peak at m/z 15 and also differs from $CH_3F^{]+\bullet}$ by having a much larger m/z 20:19 fragment peak intensity ratio (~4.5 for $CH_2FH^{]+\bullet}$ and ~2.0 for $CH_3F^{]+\bullet}$).[38,260] Both these observations are consistent with the connectivity of the distonic ion, $CH_2FH^{]+\bullet}$. In addition, a small peak at m/z 17, corresponding to a doubly charged $[C,H_3,F]^{2+}$ ion, is only observed in the CID mass spectrum of the distonic ion, an observation characteristic of such species.[38]

Thermochemistry

Combining the heat of formation and ionization energy of methyl fluoride (-247 kJ/mol and 12.50 ± 0.04 eV, respectively)[12] gives Δ_fH ($CH_3F^{]+\bullet}$) $= 959$ kJ/mol. This value is significantly lower than ca 975 kJ/mol predicted by theory (G2 and G2').[179] Note that the NIST value for the IE (12.50 ± 0.04 eV) is slightly lower than that obtained from computations (12.65 or 12.60 eV, G2 and G2', respectively).[179] In addition, Chase in the

NIST WebBook lists two values for $\Delta_f H(CH_3F)$: -247 kJ/mol from experiment and -234.30 kJ/mol from his review.[12] Calculations at the G2 and G2' levels of theory[179] are in agreement with the experimentally derived -247 kJ/mol[261] and it was therefore selected by the authors to be the better value.

Computed relative[180] and absolute[179] energies place the distonic ion lower in energy than its conventional isomer. The heat of formation of the distonic isomer was calculated to be 965 kJ/mol (G2 and G2'),[41,179] in good agreement with a revised (from a review of ancillary thermochemical data) appearance energy measurement (965 ± 15 kJ/mol).[179] Calculations predict a barrier of 95 kJ/mol for the isomerization of $CH_3F^{]+\bullet}$ to $CH_2FH^{]+\bullet}$ (MP4/6-311G(df,p) + ZPVE correction).[180]

3.5.60 $[C,H_3,Cl]^{+\bullet}$ – m/z 50

Isomers

Two isomers have been identified, ionized methyl chloride ($CH_3Cl^{]+\bullet}$) and the distonic methylenechloronium radical cation ($CH_2ClH^{]+\bullet}$).

Identification

Dissociative ionization of $ClCH_2C(O)OH$ and $ClCH_2C(O)H$ (loss of CO_2 and CO, respectively) both yield ions whose CID mass spectra are the same but distinct from that of ionized methyl chloride. These ions are assigned the $CH_2ClH^{]+\bullet}$ connectivity. It is noteworthy that both of these reactions also lead to the cogeneration of some $CH_3Cl^{]+\bullet}$ ions,[38,262] a feature uncovered by neutralization–reionization mass spectrometry.

The structure-indicative fragment intensity ratios Cl^+:$HCl^{]+\bullet}$ and $CH_2^{]+\bullet}$:$CH_3^{]+}$ are used to distinguish between the two. For $CH_3Cl^{]+\bullet}$, Cl^+:$HCl^{]+\bullet} > 1$ and $CH_2^{]+\bullet}$:$CH_3^{]+} < 1$ whereas the distonic ion is characterized by Cl^+:$HCl^{]+\bullet} < 1$ and $CH_2^{]+\bullet}$:$CH_3^{]+} > 1$.[38,262] In addition, the distonic ion shows a very prominent doubly charged ion, $CH_2ClH^{]2+}$ (m/z 26),[38,262] consistent with calculations that predict this ion to reside in a potential energy well sufficiently deep for it to be observable.[263,264]

Thermochemistry

Combining the heat of formation and the ionization energy of methyl chloride (-83.68 kJ/mol and 11.26 ± 0.03 eV, respectively)[12] gives $\Delta_f H$ ($CH_3Cl^{]+\bullet}$) = 1003 kJ/mol, in agreement with the calculated value, 1006 kJ/mol (G2 and G2').[179] Theory predicts that the distonic ion atypically lies at a higher energy than its conventional counterpart, $\Delta_f H(CH_2ClH^{]+\bullet})$ = 1048 kJ/mol.[41,179] Because the dissociative ionization of $ClCH_2C(O)OH$ generates some $CH_3Cl^{]+\bullet}$ in addition to the distonic ion, the reported value for $\Delta_f H(CH_2ClH^{]+\bullet})$ obtained from an early appearance energy measurement must likely be too low.[38]

A barrier of 133 kJ/mol is calculated for the isomerization of $CH_3Cl^{\rceil+\bullet}$ to $CH_2ClH^{\rceil+\bullet}$ (MP3/6-31G** with ZPVE correction).[180] This means that the isomers probably cannot interconvert below the dissociation limit to $CH_2Cl^{\rceil+} + H^\bullet$ ($\Sigma\Delta_f H_{products} = 1168$ kJ/mol).

3.5.61 $[C,H_3,Br]^{+\bullet} - m/z\ 94$

Isomers

Two isomers have been identified, ionized methyl bromide ($CH_3Br^{\rceil+\bullet}$) and the distonic methylenebromonium radical cation ($CH_2BrH^{\rceil+\bullet}$).

Identification[38]

The same ion is generated by the dissociative ionization of both $BrCH_2C(O)OH$ and $BrCH_2CH_2OH$ (loss of CO_2 and H_2CO, respectively) and shows a CID mass spectrum different from that of ionized methyl bromide. The former ion is assigned the connectivity of the distonic ion $CH_2BrH^{\rceil+\bullet}$.

The structure-indicative fragment peak abundance ratio used to distinguish the conventional ion from its distonic isomer is Br^+:$HBr^{\rceil+\bullet}$. For $CH_3Br^{\rceil+\bullet}$, this ratio is greater than one whereas for $CH_2BrH^{\rceil+\bullet}$ it is less than one. It is again noteworthy that as for the chloro ions, the dissociative ionization of $BrCH_2CHO$ generates a mixture of the isomers.

Thermochemistry

Combining the heat of formation and the ionization energy of methyl bromide (-34.3 ± 0.8 kJ/mol and 10.541 ± 0.003 eV, respectively)[12] gives $\Delta_f H(CH_3Br^{\rceil+\bullet}) = 983$ kJ/mol. Calculations (G2) predict the heat of formation of the distonic ion to be 1083.6 kJ/mol.[41] There is however a large discrepancy between this value and that from an appearance energy measurement (992 ± 12 kJ/mol), very likely because the two isomers are cogenerated.[38]

3.5.62 $[C,H_3,I]^{+\bullet} - m/z\ 142$

Isomers

Again, like the other halo-analogs, two isomers have been identified, namely ionized methyl iodide ($CH_3I^{\rceil+\bullet}$) and the distonic radical cation ($CH_2IH^{\rceil+\bullet}$).

Identification

The CID mass spectrum of the product ion generated in the dissociative ionization of $ICH_2C(O)OH$ (loss of CO_2) is only slightly different from that of ionized methyl iodide. The former ion is proposed to be $CH_2IH^{\rceil+\bullet}$. The m/z 15:14 peak abundance ratio is greater than one for the conventional isomer but less than one for the distonic ion. In addition, the peak shapes of the m/z 127 ($M^{+\bullet} - CH_3^{1\bullet}$) fragment ion are complex but are different for the two isomers.[38]

Thermochemistry

Combining the heat of formation and the ionization energy of methyl iodide (14.3 ± 1.4 kJ/mol and 9.54 ± 0.02 eV, respectively)[12] gives $\Delta_f H(CH_3I^{1+\bullet}) = 906$ kJ/mol. There are no reported values, from theory or experiment, for the heat of formation of the distonic ion. However, by analogy with the other halo ions, it is likely to be higher by at least 100 kJ/mol.

3.5.63 $[C,H_4,X]^+$ – WHERE X = F, Cl, Br AND I

Protonation of the methyl halides occurs at the halogen atom to generate a single isomer.

Available Thermochemical Data for CH_3XH^{1+} Ions Where X=F, Cl, Br and I

Ions	PA[12] (kJ/mol)	$\Delta_f H(CH_3X)$[12] (kJ/mol)	$\Delta_f H(CH_3XH^{1+})^a$ (kJ/mol)
CH₃FH	598.9	−247	684
CH₃ClH	647.3	−83.68	799
CH₃BrH	664.2	−34.3	832
CH₃IH	691.7	14.3	853

$^a\Delta_f H(CH_3XH^{1+}) = \Delta_f H(CH_3X) + \Delta_f H(H^+) - PA$; using $\Delta_f H(H^+) = 1530$ kJ/mol.[12]

It is worth noting that these data allow the calculation of the methyl cation affinity of the halogen hydrides.

3.5.64 $[C,N,O]^+$ – m/z 42

Dissociative ionization of HNCO (loss of H$^\bullet$) yields NCO^{1+} ions.[265,266] A photoionization appearance energy measurement yields $\Delta_f H(NCO^{1+}) = 1275$ kJ/mol,[266] in agreement with an electron-impact measurement.[265] These values also agree with the value obtained by combining the heat of formation and ionization energy of the NCO$^{1\bullet}$ radical (129 ± 10 kJ/mol and 11.759 ± 0.06 eV, respectively, giving 1264 kJ/mol).[266] Note that a lower value is obtained if the calculated $\Delta_f H(NCO^{1\bullet})$ (120.85 kJ/mol, G2MP2)[37] is used.

3.5.65 $[C,H,N,O]^{+\bullet}$ – m/z 43

Isomers

There are three isomeric ions that have been investigated by experiment: HNCO$^{1+\bullet}$, the isocyanic acid radical cation, HCNO$^{1+\bullet}$, the fulminic acid radical cation and finally, CNOH$^{1+\bullet}$.[267] Theory also predicts three other linear ions, HOCN$^{1+\bullet}$, HNOC$^{1+\bullet}$ and HCON$^{1+\bullet}$ as well as two cyclic species, cy–N(H)OC–$^{1+\bullet}$ and cy–C(H)ON–$^{1+\bullet}$, to be local minima on the potential energy surface.[268]

Identification[267]

HNCO$^{1+\bullet}$ radical cations are generated by the ionization of the products of the thermal depolymerization of cyanuric acid or by heating KNCO in the presence of KHSO$_4$.[269] Thermolysis of chlorooximino acetic acid (HONC(Cl)C(O)OH) in the ion source of the mass spectrometer and dissociative ionization of methyl chlorooximino acetate (HONC(Cl)C(O)OCH$_3$) generate two other distinct isomers. The CID mass spectral characteristics of the latter two ions are consistent with the connectivities in HCNO$^{1+\bullet}$ and CNOH$^{1+\bullet}$ ions, respectively. The CID mass spectra of these ions are sufficiently different to allow identification.

Partial CID Mass Spectra of the [C,H,N,O]$^{+\bullet}$ Isomers

						m/z						
Ions	12	13	14	15	16	17	26	27	28	29	30	31
HNCO	5.8	0.14	7.2	33	1.4	0.57	23	55	100	(77)	26	—
HCNO	1.6	5.2	1.6	3.1	0.78	0.39	9.1	18	6.5	(7.8)	100	—
CNOH	4.4	1.6	6.7	22	1.1	4.4	37	33	69	(100)	43	1.6

Values in parentheses contain contributions from metastable ion dissociation.

It is noteworthy that in the CID mass spectrum of the CNOH$^{1+\bullet}$ ion, the presence of the fragment peaks at m/z 15 and 28 suggests that HNCO$^{1+\bullet}$ ions may be cogenerated. The three ions share common metastable dissociations (loss of H$^\bullet$ and loss of N$^\bullet$) but with distinct characteristics (kinetic energy release and peak shape) suggesting that the isomers do *not* interconvert before these dissociations.

Thermochemistry

The potential energy surface of these ions has been extensively studied at the B3LYP/6-311G(d,p) level of theory.[268] Calculations show that the HNCO$^{1+\bullet}$ radical cation is the global minimum on the doublet surface.

Combining the ionization energy and the heat of formation of isocyanic acid (-101.67 kJ/mol and 11.695 ± 0.005 eV, respectively)[12] leads to $\Delta_f H(\text{HNCO}^{1+\bullet}) = 1017$ kJ/mol. Computations place the HCNO$^{1+\bullet}$ and CNOH$^{1+\bullet}$ ions, respectively, 205 and 371 kJ/mol higher in energy than HNCO$^{1+\bullet}$ and the barriers for their interconversion are high.[268] The relative energies of the other isomers, namely HOCN$^{1+\bullet}$, HNOC$^{1+\bullet}$, HCON$^{1+\bullet}$, cy–N(H)OC–$^{1+\bullet}$ and cy–C(H)ON–$^{1+\bullet}$, were also calculated to be 131, 370, 510, 417 and 277 kJ/mol, respectively.

3.5.66 $[C,H_2,N,O]^+ - m/z$ 44

Isomers

Four stable isomeric species have been identified by experiment, namely, H_2NCO^{1+} (1), $HNCOH^{1+}$ (2), H_2CNO^{1+} (3) and $HCNOH^{1+}$ (4). Several other species, namely, $NCOH_2^{1+}$ (5), $HNC(H)O^{1+}$ (6), H_2NOC^{1+} (7), cy-$N(H)_2CO-^{1+}$ (8), cy-$C(H)_2NO-^{1+}$ (9), $CN(H)OH^{1+}$ (10), $CNOH_2^{1+}$ (11), H_2CON^{1+} (12) and $HCONH^{1+}$ (13) have been found by computation to be local minima on the potential energy surface.[270,271]

Identification

Dissociative ionization of $CH_3C(O)NH_2$ (loss of $CH_3^{1\bullet}$) generates an ion whose CID mass spectrum is consistent with the H_2NCO^{1+} (1) connectivity.[272] Protonation of isocyanic acid (HNCO) yields the same ion,[272] showing that protonation takes place exclusively at N, in keeping with the N and O proton affinities of the acid (753[12] and 650 kJ/mol,[270] respectively). $HNCOH^{1+}$ (2) ions are produced in the dissociation of ionized methylamino-hydroxycarbene $(CD_3N(H)COH^{1+\bullet}$, loss of $CD_3^{1\bullet})$.[273] Dissociative ionization of nitromethane (loss of $^\bullet OH$) generates H_2CNO^{1+} (3) ions.[272] $HCNOH^{1+}$ (4) ions are obtained by the dissociative ionization of $CH_3CH = NOH$ (loss of $CH_3^{1\bullet}$) and glyoxime (HON = CHCH = NOH, loss of $HCNOH^{1\bullet}$) or by protonation of fulminic acid (HCNO).[267] The four isomers can easily be distinguished based on their distinct CID mass spectra.

Partial CID Mass Spectra of the $[C,H_2,N,O]^+$ Isomers[267,273]

Ions	m/z												
	12	13	14	15	16	17	18	26	27	28	29	30	31
H_2NCO	4	1	5	17	(95)	1	—	18	48	100	44	11	—
$HNCOH$	9	1	6	21	44	7	6	31	100	71	94	18	1
H_2CNO	1	1	4	1	6	—	—	3	10	8	6	100	—
$HCNOH$	3	7	3	6	(80)	7	7	22	100	20	33	72	8

Values in parentheses contain contributions from metastable ion dissociation.

The ion $HNC(H)O^{1+}$ (6) was expected to be generated by charge reversal of the corresponding anion but, instead, it was concluded that the resulting cation was unstable and rapidly isomerized to $H_2NCO^{1+\bullet}$.[274] It is noteworthy that $HNC(H)O^{1+}$ ions were initially proposed to be generated in the dissociative ionization of $HCON(H)CH_3$ (loss of $CH_3^{1\bullet}$)[272] but further investigations showed that the ion produced was predominantly $HNCOH^{1+}$.[273] Note also that the presence of peaks at m/z 28 and 16 in the CID mass spectrum of the HCNOH^{1+} ions indicates that this ion can isomerize to H$_2$NCO^{1+}.[273]

Thermochemistry

H$_2$NCO^{1+} (1) is the lowest energy isomer. $\Delta_f H(1) = 675$ kJ/mol can be obtained from the proton affinity of isocyanic acid (PA $= 753$ kJ/mol, $\Delta_f H(\text{HNCO}) = -101.67$ kJ/mol and $\Delta_f H(\text{H}^+) = 1530$ kJ/mol).[12] This value is in good agreement with an AE measurement (672 kJ/mol,[272] (CH$_3$C(O)NH$_2$ $\longrightarrow$ H$_2$NCO^{1+} + CH$_3^{1\bullet}$; AE $= 10.94$ eV,[272] $\Delta_f H(\text{CH}_3\text{C(O) NH}_2) = -238.33$ ± 0.78 kJ/mol[12] and $\Delta_f H(\text{CH}_3^{1+}) = 145.69$ kJ/mol).[12] Calculations (G2 and G1) predict that HNCOH^{1+} is approximately 67 kJ/mol higher in energy than H$_2$NCO^{1+} and is separated from it by a barrier to isomerization of approximately 336 kJ/mol.[270,271]

HCNOH^{1+} and H$_2$CNO^{1+} ions are calculated (G2) to lie respectively 284 and 318 kJ/mol above H$_2$NCO^{1+} (leading to $\Delta_f H(\text{HCNOH}^{1+}) = 959$ kJ/mol and $\Delta_f H(\text{H}_2\text{CNO}^{1+}) = 993$ kJ/mol) and to occupy deep potential energy wells.[271,272] The measured heat of formation for these ions, 990 kJ/mol (CH$_3$CHNOH $\longrightarrow$ HCNOH^{1+} + CH$_3^{1\bullet}$) and 1014 kJ/mol (CH$_3$NO$_2$ $\longrightarrow$ H$_2$CNO^{1+} + $^\bullet$OH) is not in disagreement with the values deduced from the relative energies. The measured heat of formation for the HCNOH^{1+} ion (990 kJ/mol) is considered to be an upper limit.[148]

Theory also predicts that HNC(H)O^{1+} lies even higher in energy than the other isomers, 383 or 418 kJ/mol (G2 and G1, respectively)[270,271] above H$_2$NCO^{1+} with a small isomerization barrier to HNCOH^{1+} (27 kJ/mol),[271] in keeping with observations.[273]

Relative Energies of the [C,H$_2$,N,O]$^+$ Isomers[271]

Ions	Relative Energies (G2) (kJ/mol)
(1)	0
(2)	69
(3)	318
(4)	284
(5)	309
(6)	383
(7)	433
(8)	438
(9)	451
(10)	516
(11)	532
(12)	585
(13)	617

3.5.67　$[C, H_3, N, O]^{+\bullet}$ – m/z 45

Isomers

Six isomers have been investigated by experiment, namely, ionized formamide ($H_2NCHO^{1+\bullet}$), its enol form ($HNC(H)OH^{1+\bullet}$), the aminohydroxycarbene ($H_2NCOH^{1+\bullet}$), ionized nitrosomethane ($CH_3NO^{1+\bullet}$), ionized formaldoxime ($CH_2NOH^{1+\bullet}$) and ionized formaldonitrone ($CH_2N(H)O^{1+\bullet}$). It is noteworthy that the most recent theoretical calculations find that the ion $NH_3CO^{1+\bullet}$ and the complex $[OC\cdots NH_3]^{+\bullet}$ are only weakly bound, occupying shallow potential energy wells.[275]

Identification

Ionization of formamide generates $H_2NCHO^{1+\bullet}$ ions.[276,277] Dissociative ionization of *N*-formylhydrazone ($CH_3C(H)NN(H)CHO$, loss of CH_3CN)[277] produces the enol form of ionized formamide ($HNC(H)OH^{1+\bullet}$) whereas methyl carbamate ($H_2NC(O)OCH_3$, loss of H_2CO)[276,277] and oximide ($H_2NC(O)C(O)NH_2$, loss of $HNCO$)[277] yield the carbene ion $H_2NCOH^{1+\bullet}$.

Because the metastable ion and collision-induced dissociation characteristics of these isomers are very similar, their respective connectivities had to be confirmed by investigating the structure of some of their dissociation product ions.[277] As anticipated and based on their connectivity, $HNC(H)OH^{1+\bullet}$ ions dissociate to $HNCH^{1+}$ (loss of $^\bullet OH$) and to $HCOH^{1+\bullet}$ (loss of $NH^\bullet$). By investigating the perdeuterated analogs of $H_2NCHO^{1+\bullet}$ and $H_2NCOH^{1+\bullet}$ it was concluded that, consistent with the ions connectivity, the former largely generated HCO^{1+} (loss of $^\bullet NH_2$) and $HNCH^{1+}$ (loss of $^\bullet OH$) whereas the latter gave COH^{1+} (loss of $^\bullet NH_2$) and H_2NC^{1+} (loss of $^\bullet OH$). To directly distinguish between the isomers, their respective m/z 27:28:29 fragment ion intensity ratio (see table below) can be used.[277]

The dissociative ionization of nitromethane (loss of atomic oxygen)[276,278] and ethyl nitrite (CH_3CH_2ONO, loss of H_2CO)[278] as well as the dissociation of protonated nitromethane (loss of $^\bullet OH$),[278] yields an ion with CID mass spectral characteristics very different from all the other ions discussed above and consistent with the structure of ionized nitrosomethane, $CH_3NO^{1+\bullet}$.[276,278,279]

Thermolysis of trimeric formaldoxime hydrochloride and ionization of the products yield yet another ion. Its CID mass spectrum is distinct from all of the others by the significant fragment ion at m/z 30 ($NO^{1+\bullet}$) and smaller peaks at m/z 14, 18 and 31, consistent with a $CH_2NOH^{1+\bullet}$ connectivity.[279]

The dissociative ionization of 1,2-oxazolidine (cy-N(H)CH$_2$CH$_2$CH$_2$O–, loss of C_2H_4) produces the formaldonitrone radical cation, $CH_2N(H)O^{1+\bullet}$. The CID mass spectrum of this ion is similar to but distinguishable from that of $CH_3NO^{1+\bullet}$ and $CH_2NOH^{1+\bullet}$ ions by the following features: an m/z 30 fragment peak significantly more abundant than for the $CH_2NOH^{]+\bullet}$ ion and the presence of a m/z 31 peak which is absent in the spectrum of $CH_3NO^{]+\bullet}$.[279]

Partial CID Mass Spectra of the $[C,H_3,N,O]^{+\bullet}$ Isomers[277,279]

Ions	m/z									
	15	16	17	18	26	27	28	29	30	31
H_2NCHO	3.0	34	(100)	—	3.7	9.4	21	53	5.3	
H_2NCOH	6	55	(55)		12	60	92	100	10	
$HNC(H)OH$	4	28	(47)		7	30	100	55	20	
CH_2NOH	—	5.4	5.7	5.9	8.6	42	(100)	5.6	(38)	3.9
CH_3NO	5.9	1.8	—	—	1.8	5.7	7.8	7.3	(100)	—
$CH_2N(H)O$	0.052	1.5	0.62	0.62	—	6.2	26	1.7	(100)	3.8

Values in parentheses contain contributions from metastable ion dissociations.

Thermochemistry

Calculations locate the carbene ion as the global minimum[275,280] with $\Delta_fH(H_2NCOH^{]+\bullet}) = 776 \pm 5$ kJ/mol (G2(MP2,ZVP))[275] somewhat lower than the measured value of ≤ 796 kJ/mol obtained from the appearance energy of the ion generated from methyl carbamate ($CH_3OC(O)NH_2$, loss of H_2CO).[276,281] The latter value must however be considered an upper limit because the dissociation leading to the carbene ion is not the lowest energy process.

Calculations (G2(MP2,ZVP)) find $\Delta_fH(H_2NCHO^{]+\bullet}) = 797 \pm 5$ kJ/mol and $\Delta_fH(HNC(H)OH^{]+\bullet}) = 847 \pm 5$ kJ/mol.[275] Combining the heat of formation and ionization energy of H_2NCHO (-186 kJ/mol and 10.16 ± 0.06 eV, respectively)[12] gives $\Delta_fH(H_2NCHO^{]+\bullet}) = 794$ kJ/mol, in agreement with the above calculations. These three ions occupy deep potential energy wells. The calculated barriers for isomerization of the carbene ion into ionized formamide and its enol form are 171 ± 5 kJ/mol and 239 ± 5 kJ/mol, respectively (G2(MP2,ZVP)).[275] From the heat of formation and ionization energy of nitrosomethane (70 ± 4 kJ/mol[282] and 9.3 eV, respectively),[12] $\Delta_fH(CH_3NO^{]+\bullet}) = 967$ kJ/mol is obtained.

An independent series of calculations predicts $CH_2NOH^{]+\bullet}$ and $CH_2N(H)O^{]+\bullet}$ ions to lie 190 and 183 kJ/mol above ionized formamide (QCISD(T)/6-311G(3df,2p)).[278] These two ions, in addition to $CH_3NO^{]+\bullet}$, were also investigated at the G2(MP2) level of theory.[279] The three ions were found to occupy deep potential energy wells with $CH_3NO^{]+\bullet}$ being the lowest energy species. $CH_2NOH^{]+\bullet}$ and $CH_2N(H)O^{]+\bullet}$ were computed to be 21 and 28 kJ/mol higher in energy. The energy barriers for isomerization were also calculated to be 175 kJ/mol for $CH_2N(H)O^{]+\bullet} \longrightarrow CH_3NO^{]+\bullet}$ and 227 kJ/mol for $CH_2N(H)O^{]+\bullet} \longrightarrow CH_2NOH^{]+\bullet}$. Note that $CH_2N(H)O^{]+\bullet}$

was calculated to be 136 kJ/mol below the dissociation products $CH_3^{1\bullet} + NO^{1+}$ ($\Sigma\Delta_f H_{products} = 1130$ kJ/mol)[12] which allows estimated values for $\Delta_f H(CH_2N(H)O^{1+\bullet}) = 994$ kJ/mol, $\Delta_f H(CH_2NOH^{1+\bullet}) = 1001$ kJ/mol and $\Delta_f H(CH_3NO^{1+\bullet}) = 973$ kJ/mol (in very good agreement with the experimental value listed above).

3.5.68   [C,H₄,N,O]⁺ – m/z 46

Isomers

In principle, the protonation of formamide (H_2NCHO) and formaldoxime (CH_2NOH) could yield several possible isomers. However, only the ion resulting from protonation at oxygen in formamide, H_2NCHOH^{1+} (**1**), has been investigated by experiment. Theory however predicts that $HNC(H)OH_2^{1+}$ (**2**), H_3NCOH^{1+} (**3**), $H_2NCOH_2^{1+}$ (**4**), [$HCNH\cdots OH_2$]⁺ (**5**), H_2CNHOH^{1+} (**6**), H_3CNHO^{1+} (**7**), $H_2CNOH_2^{1+}$ (**8**), H_3CNOH^{1+} (**9**), cy–(H)₂CN(H)O(H)–]¹⁺ (**10**) and H_3NCHO^{1+} (**11**) occupy potential energy wells of significant depth.

Identification[281]

The loss of $HCO^{1\bullet}$ from ionized methyl carbamate ($H_2NC(O)OCH_3^{1+\bullet}$) produces a [C,H₄,N,O]⁺ fragment ion whose CID mass spectrum shows intense fragment peaks at m/z 29 (HCO^{1+} and/or $HCNH_2^{1+}$), 28 ($CO^{1+\bullet}$ and/or $HNCH^{1+}$), 18 (NH_4^{1+}) and 17 ($NH_3^{1+\bullet}$). Note that the first three dissociation products are also generated by the metastable [C,H₄,N,O]⁺ ions. None of the above features allow an unambiguous identification of the connectivity of the ion produced from methyl carbamate. However, by comparing the NR mass spectrum of the fragment [C,H₄,N,O]⁺ ions with that of their deuterated analog, [C,H₂,D₂,N,O]⁺ (loss of $DCO^{1\bullet}$ from $H_2NC(O)OCD_3$), it became evident that these ions were H_2NCHOH^{1+} (**1**) and not H_3NCHO^{1+} (**11**). Results obtained from ¹³C- and ¹⁸O-labeled precursors also support the assigned connectivity.

Thermochemistry

From the proton affinity of formamide $\Delta_f H(H_2NCHOH^{1+}) = 522$ kJ/mol is obtained (PA(H_2NCHO) = 822 kJ/mol, $\Delta_f H(H_2NCHO) = -186$ kJ/mol and $\Delta_f H(H^+) = 1530$ kJ/mol).[12] Theory predicts that oxygen is the favored protonation site (G2MP2), with the nitrogen-protonated species, H_3NCHO^{1+} (**11**), calculated to lie 69 kJ/mol higher in energy than H_2NCHOH^{1+} and separated from it by a barrier of 160 kJ/mol (G2(MP2)).[283] Note that these calculations find that the H_3NCHO^{1+} ion lies in a potential energy well of significant depth (92 kJ/mol, G2(MP2)),[283] in agreement with previous calculations.[284]

Two separate theoretical studies have shown that several ions with a C–N–O backbone occupy potential energy wells of significant depth and their relative energies are as follows:

Computed Relative Energies for the $[C,H_4,N,O]^+$ Ions

Ions	Relative Energy (G2(MP2))[283] (kJ/mol)	Ions	Relative Energy (MP4 + ZPE correction)[285] (kJ/mol)
(1)	0	(5)	0
(2)	179	(6)	187
(3)	268	(7)	261
(4)	265	(8)	264
		(9)	324
		(10)	344

3.5.69 $[C,H_5,N,O]^{+\bullet}$ – m/z 47

Isomers

The O-methylhydroxylamine radical cation ($CH_3ONH_2^{]+\bullet}$) and the unconventional ion $[H_2CO\cdots H\cdots NH_2]^{+\bullet}$ have been investigated by experiment. No less than seven other isomers, namely $[HCO^\bullet\cdots H^+\cdots NH_3]$, $H_3NCH_2O^{]+\bullet}$, $H_2NCH_2OH^{]+\bullet}$, $[CH_2NH_2^+\cdots HO^\bullet]$, $CH_2NH_2OH^{]+\bullet}$, $CH_3NH_2O^{]+\bullet}$ and $H_2NO(H)CH_2^{]+\bullet}$ are predicted by theory to occupy potential energy wells of significant depth.[286]

Identification[286]

Ionized methyl carbamate ($NH_2C(O)OCH_3$, loss of CO) generates an ion whose CID mass spectral characteristics differ from those of ionized O-methylhydroxylamine, $CH_3ONH_2^{]+\bullet}$.

Partial CID Mass Spectra of the $[C,H_5,N,O]^{+\bullet}$ Isomers

Ions	m/z									
	14	15	16	17	18	28	29	30	31	32
CH_3ONH_2	1.6	8.1	3.2	15	—	9.7	40	40	32	100
$[H_2CO\cdots H\cdots NH_2]$	—	—	3.4	47	100	5.7	32	24	2.2	—

The CID mass spectrum alone is not enough to assign an atom connectivity to the ion generated from methyl carbamate. The ion does not dissociate by loss of 15 mass units to yield an m/z 32 fragment ion (suggesting that it does not contain a methyl group) and the m/z 31, 30, 29 and 17 fragment ions have been identified by isotopic labeling experiments to be $^+CH_2OH$ (and not $HNO^{]+}$), $H_2CO^{]+\bullet}$, $HCO^{]+}$ and $NH_3^{]+\bullet}$ respectively. On the basis of these observations, the ion was therefore assigned the structure $[H_2CO\cdots H\cdots NH_2]^{+\bullet}$. That the NR mass spectrum of this ion contains no recovery signal and significant fragment peaks at m/z 30, 29, 28, 17 and 16 ($H_2CO^{1+\bullet}$, HCO^{1+}, $CO^{1+\bullet}$, $NH_3^{1+\bullet}$ and $^+NH_2$) also supports the structure assignment. Note that the magnitude of the computed N−O distance (2.68 Å) suggests that $[H_2CO\cdots H\cdots NH_2]^{+\bullet}$ ions should be regarded as an ion–neutral complex, $[H_2CO\cdots NH_3]^{+\bullet}$.[286]

Thermochemistry[286]

Combining the ionization energy and heat of formation of the *O*-methyl hydroxylamine molecule (9.25 ± 0.05 eV and -26 ± 8 kJ/mol, respectively)[40] leads to $\Delta_f H(CH_3ONH_2^{1+\bullet}) = 866$ kJ/mol. Note that NIST lists 9.55 eV[12] for the ionization energy of the *O*-methyl hydroxylamine molecule, which, based on thermochemical arguments, is most likely too high. An appearance energy measurement of the ion generated following CO loss from methyl carbamate gives $\Delta_f H([H_2CO\cdots NH_3]^{+\bullet}) = 769$ kJ/mol (AE = 11.16 ± 0.01 eV, $\Delta_f H(NH_2C(O)OCH_3) = -418$ kJ/mol[40] and $\Delta_f H(CO) = -110.5$ kJ/mol[12]).

Calculations of the relative energies of these isomers find neither of these ions to be the global minimum on the potential energy surface; the proton-bridged species $[HCO^\bullet\cdots H^+\cdots NH_3]$ was identified as the thermodynamically most stable isomer (SDCI/6-31G**). At this level of theory, the relative energy of the isomers $[H_2CO\cdots NH_3]^{+\bullet}$ and $CH_3ONH_2^{1+\bullet}$ is calculated to be 90 kJ/mol, in agreement with their absolute experimental heats of formation. It is noteworthy that the distonic ion of methylhydroxylamine, $CH_2ONH_3^{1+\bullet}$, is found to collapse without activation to $[H_2CO\cdots NH_3]^{+\bullet}$.

The other ions found to occupy potential energy wells of significant depth are $H_3NCH_2O^{1+\bullet}$, $H_2NCH_2OH^{1+\bullet}$, $[CH_2NH_2^+\cdots HO^\bullet]$, $CH_2NH_2OH^{1+\bullet}$, $CH_3NH_2O^{1+\bullet}$ and $H_2NO(H)CH_2^{1+\bullet}$ with calculated relative energies of 72, 90, 102, 251, 255 and 365 kJ/mol above $[HCO^\bullet\cdots H^+\cdots NH_3]$.

3.5.70 $[C,H,N,O_2]^{+\bullet}$ – m/z 59

Isomers

There are two isomeric species, namely $HCNO_2^{1+\bullet}$ and $ONC(O)H^{1+\bullet}$, ionized nitrocarbene and the nitrosoformaldehyde radical cations, respectively.

Identification[287]

Electron-impact ionization of ethyl nitrodiazoacetate ($C_2H_5O_2CC(N_2)NO_2$) in the presence of ammonia generates a $[C,H,N,O_2]$ radical anion (loss of N_2, CO_2 and C_2H_4). Charge reversal of this anion produces a cation with CID mass spectral characteristics consistent with the $HCNO_2^{1+\bullet}$ connectivity, i.e., m/z 46 ($NO_2^{1+\bullet}$) and 43 ($[H,C,N,O]^+$).

Heating the Diels–Alder adduct of nitrosoformaldehyde and 9,10-dimethylanthracene in the ion source of the mass spectrometer yields a $[C,H,N,O_2]^{+\bullet}$ ion with a distinctive CID mass spectrum. The fragment ion peaks at m/z 29 ($[HCO]^+$) and 42 ($[NCO]^+$) indicate that this ion is $ONC(O)H^{]+\bullet}$.

No thermochemical data are available for either ion.

Partial CID Mass Spectra of the $[C,H,N,O_2]^{+\bullet}$ Isomers

Ions	\multicolumn{10}{c}{m/z}
	26	**27**	**28**	**29**	**30**	**42**	**43**	**44**	**45**	**46**
$HCNO_2$	5.4	10	27	36	100	42	62	4	0.11	31
$ONC(O)H$	7.8	13	57	94	100	19	2.2	—	—	—

3.5.71 $[C,H_2,N,O_2]^+$ – m/z 60

Isomers

Eight isomeric forms of $[C,H_2,N,O_2]^+$ having the following structures have been investigated by theory:

$$CH_2NO_2 \quad CH_2ONO \quad HC(O)NOH \quad HC(O)N(H)O$$
$$\mathbf{1} \qquad\qquad \mathbf{2} \qquad\qquad \mathbf{3} \qquad\qquad \mathbf{4}$$

$$HCN(O)OH \quad OCN(H)OH$$
$$\mathbf{5} \qquad\qquad \mathbf{6}$$

structure 7: four-membered ring with N at top, two O atoms at sides, and CH_2 at bottom

structure 8: H_2C —— NO three-membered ring with O at bottom

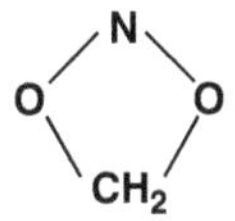

7 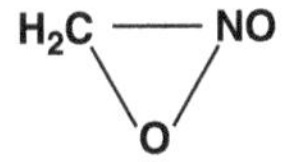 **8**

Identification[288]

Cations $CH_2ONO^{]+}$ (**2**) and $OCHN(H)OH^{]+}$ (**6**) have been proposed to be generated in the following dissociative ionization reactions:

$$CH_3CH_2ONO \rightarrow CH_2ONO^{]+} + CH_3^{]\bullet}$$

$$HON(H)C(O)OC_2H_5 \rightarrow OCHN(H)OH^{]+} + C_2H_5O^{]\bullet}$$

It is noteworthy that the latter reaction also produces an isobaric $[C_2,H_6,N,O]^+$ species by the loss of $HCO_2^{]\bullet}$. The abundance ratio $OCHN(H)OH^{]+}$: $[C_2,H_6,N,O]^+$ is reported to be 9:1. No mass spectral data on the ions have been provided.

Thermochemistry[288]

Theoretical calculations predict that $OCHN(H)OH^{1+}$ (**6**) is the global minimum, with the other ions lying higher in energy.

Computed Relative Energies of the $[C,H_2,N,O_2]^+$ Isomers

Ions	Relative Energy QCISD(T)/ 6-311 + G(3df,2p) (kJ/mol)
(**1**)	534
(**2**)	33
(**3**)	157
(**4**)	89
(**5**)	460
(**6**)	0
(**7**)	223
(**8**)	278

Note that for isomer (**1**) the singlet state was found to be unstable. The listed energy corresponds to the higher energy triplet state.

3.5.72 $[C,H_3,N,O_2]^{+\bullet}$ – m/z 61

Isomers

Six isomers have been investigated by experiment: ionized nitromethane $(CH_3NO_2^{1+\bullet})$, ionized *aci*-nitromethane $(CH_2N(O)OH^{1+\bullet})$, ionized methyl nitrite $(CH_3ONO^{1+\bullet})$, the ionized keto and enol forms of formohydroxamic acid $(HC(O)NHOH^{1+\bullet}$ and $HC(OH)NOH^{1+\bullet}$, respectively) and ionized carbamic acid $(H_2NC(O)OH^{1+\bullet})$.

Identification

$CH_3NO_2^{1+\bullet}$, $CH_3ONO^{1+\bullet}$ and $HC(O)NHOH^{1+\bullet}$ are generated by ionization of the corresponding molecules. The *aci* form of nitromethane, $CH_2N(O)OH^{1+\bullet}$, can be generated either by a specific 1,5-hydrogen shift in $CH_3CH_2CH_2NO_2^{1+\bullet}$ followed by the loss of C_2H_4[289] or by dissociative ionization of $HOCH_2CH_2NO_2$ (loss of H_2CO).[290] Ionized glyoxime (HON-CHCHNOH$^{1+\bullet}$, by loss of [H,C,N]) produces yet another ion to which the connectivity of the enol form of formohydroxamic acid, $HC(OH)NOH^{1+\bullet}$, has been assigned.[290] Ionized carbamic acid, $H_2NC(O)OH^{1+\bullet}$, was generated by ionization of the neutral products produced by the thermolysis of ammonium carbamate, $NH_4CO_2NH_2$.[291] The ions can be distinguished by their CID mass spectra.

Partial CID Mass Spectra of the $[C,H_3,N,O_2]^{+\bullet}$ Isomers[290–292]

Ions							m/z						
	15	16	17	28	29	30	31	32	33	43	44	45	46
CH_3NO_2	3.4	—	—	—	2.2	80	2.7	—	—	4.0	41	3.4	100
$CH_2N(O)OH$	—	—	—	—	2.3	5.8	2.3	—	—	15	100	1.2	—
CH_3ONO	0.5	—	—	1.4	12	100	35	—	—	—	—	—	0.5
$HC(O)NHOH$	—	1.1	0.8	2.3	15	6.8	1.5	11	100	6.8	1.5	0.08	—
$HC(OH)NOH$	—	1.0	0.7	3.2	9.7	8.9	1.5	3.7	100	32	28	—	—
$H_2NC(O)OH$	1.8	19	90	4.4	5.0	1.4	—	2.2	—	22	98	100	4.5

It is interesting however to note that the isomers $CH_3NO_2^{\,]+\bullet}$, $CH_2N(O)OH^{\,]+\bullet}$ and $CH_3ONO^{\,]+\bullet}$ and also the isomers $HC(OH)NOH^{\,]+\bullet}$ and $H_2NC(O)OH^{\,]+\bullet}$ share common metastable dissociation channels. To account for these observations, their isomerization and interconversion before dissociation have been proposed.

MI Mass Spectra of the $[C,H_3,N,O_2]^{+\bullet}$ Isomers[290]

Ions				m/z		
	30	31	33	43	44	60
CH_3NO_2	48	2.5		4.7	100	73
$CH_2N(O)OH$				15	100	1.8
CH_3ONO		100				15
$HC(O)NHOH$			100			
$HC(OH)NOH$			100	3.2	25	

From metastable ion appearance energy measurements and kinetic energy releases, $CH_3NO_2^{\,]+\bullet}$ ions are proposed to isomerize to $CH_3ONO^{\,]+\bullet}$ before $H^\bullet$ loss (producing $CH_2ONO^{\,]+}$ ions, m/z 60) and to $CH_2N(O)OH^{\,]+\bullet}$ before dissociation by the losses of $^\bullet OH$ and H_2O (yielding product ions $CH_2NO^{\,]+}$ (m/z 44) and $HCNO^{\,]+}$ (m/z 43)). There is no evidence to indicate that the reverse isomerization, $CH_2N(O)OH^{\,]+\bullet}$ to $CH_3NO_2^{\,]+\bullet}$, takes place (absence of m/z 30 in MI mass spectra of $CH_2N(O)OH^{\,]+\bullet}$) or that $CH_3ONO^{\,]+\bullet}$ communicates with $CH_3NO_2^{\,]+\bullet}$ and $CH_2N(O)OH^{\,]+\bullet}$.[290]

In order to lose CO, $HC(O)NHOH^{\,]+\bullet}$ ions must first undergo a rate-determining isomerization to the enol form, $HC(OH)NOH^{\,]+\bullet}$, which in turn is proposed to isomerize to a proton-bound pair $[HNO\cdots H^+\cdots^\bullet OCH]$ before the dissociation leading to $NH_2OH^{\,]+\bullet}$ ions, (m/z 33). There is no evidence to suggest that $CH_3NO_2^{\,]+\bullet}$, $CH_2N(O)OH^{\,]+\bullet}$ and $CH_3ONO^{\,]+\bullet}$ ions communicate with their isomers $HC(O)NHOH^{\,]+\bullet}$ and $HC(OH)NOH^{\,]+\bullet}$.[290]

Note that $CH_3NO_2^{\,]+\bullet}$, $CH_2N(O)OH^{\,]+\bullet}$ and $CH_3ONO^{\,]+\bullet}$ have also been distinguished by their neutral fragment mass spectra.[293] In dissociative neutralization, a not widely used technique, the resulting spectrum (neutral fragment mass spectra) is obtained by collisional neutralization of a mass-selected beam of kilovolt ions (usually using alkali metal vapor as neutralizing gas) followed by analysis of the neutral products formed (no reionization is required before analysis).[294,295]

Thermochemistry

Combining the heat of formation and ionization energy of nitromethane (-81 ± 1 kJ/mol and 11.08 ± 0.04 eV, respectively)[12] and methyl nitrite (-65.44 ± 0.44 kJ/mol and 10.44 eV, respectively)[12] gives $\Delta_f H(CH_3NO_2^{1+\bullet}) = 988$ kJ/mol and $\Delta_f H(CH_3ONO^{1+\bullet}) = 942$ kJ/mol. The heat of formation of the aci-form of nitromethane, $\Delta_f H(CH_2N(O)OH^{1+\bullet}) = 958$ kJ/mol, was obtained from the measurement of its appearance energy from nitropropane (AE $= 11.75$ eV, $\Delta_f H(C_3H_7NO_2) = -124$ kJ/mol and $\Delta_f H(C_2H_4) = 52$ kJ/mol).[290]

Ab initio calculations predict that the enol form of formohydroxamic acid radical cation, $HC(OH)NOH^{1+\bullet}$, lies 121 kJ/mol lower in energy than nitromethane (STO/6-31G).[296] No data are available for the formohydroxamic acid radical cation itself. MNDO calculations predict $\Delta_f H(H_2NC(O)OH^{1+\bullet}) = 628$ kJ/mol, not too far from the 653 kJ/mol obtained using a calculated (MNDO) $\Delta_f H(NH_2C(O)OH)$ and an estimated ionization energy.[291]

3.5.73 $[C,H_4,N,O_2]^+ - m/z\ 62$

Isomers

Three isomers have been proposed: $H_2NC(OH)_2^{1+}$, CH_3ONOH^{1+} and CH_3OH-NO^{1+}. However, the mass spectral characteristics of the first ion suggest that other isomeric forms, $H_2NC(O)OH_2^{1+}$, $H_3NC(O)OH^{1+}$ and $HNC(OH_2)OH^{1+}$ are accessible.

Identification

Dissociation of ionized ethyl carbamate ($H_2NC(O)OC_2H_5$, loss of $C_2H_3^{1\bullet}$) produces an abundant m/z 62 fragment ion.[158,297] The MI mass spectrum of this ion is dominated by the loss of H_2O (m/z 44) with minor peaks corresponding to m/z 18 and 19 (NH_4^+ and H_3O^{1+}, respectively). Collisional activation opens other dissociation channels, such as $^\bullet OH$ loss (m/z 45) and generation of $NH_3^{1+\bullet}$ (m/z 17, loss of $C(O)OH^{1\bullet}$). This ion has been assigned the $H_2NC(OH)_2^{1+}$ connectivity.[297] However, to account for the observed dissociations, isotopic labeling experiments indicate that protonated carbamic acid is not the reacting configuration in all cases and therefore the isomers $H_2NC(O)OH_2^{1+}$, $H_3NC(O)OH^{1+}$ and $HNC(OH_2)OH^{1+}$ have been proposed as intermediate species.[297]

Solvation of methanol with NO^{1+} ions and proton transfer from H_3O^{1+} to methyl nitrite (CH_3ONO) (note that both reactions are of low exothermicity) produce an m/z 62 product ion whose MI mass spectrum shows a Gaussian peak at m/z 30 (NO^{1+}).[298] When methane is used as the reagent for the proton transfer reaction (a reaction of greater exothermicity), the m/z 62 ion produced is also metastable yielding m/z 30 (NO^{1+}). However, the peak is now composite in shape: a Gaussian peak atop a dished or flat-topped peak,[299] possibly indicating that two species are generated. From the mass spectral characteristics of the ions generated by the above reactions, ions of connectivity CH_3OH-NO^{1+} are proposed to be produced in the solvation of methanol and by protonation reactions of low exothermicity. The exothermic protonation is however believed to generate, at least in part, another isomer, CH_3ONOH^{1+}.[298]

Partial CID Mass Spectra of the $[C,H_4,N,O_2]^+$ Isomers[299]

	m/z						
Ions	15	29	31	32	45	46	47
CH_3OH-NO	504	45	100	50	—	—	7.5
CH_3ONOH	18	55	100	16	27	23	10

The m/z 30 metastable ion peak is not included in the table.

Thermochemistry

The appearance energy of the m/z 62 fragment ion from ethyl carbamate was measured to be 10.7 eV, yielding a $\Delta_f H(H_2NC(OH)_2^{1+}) = 287$ kJ/mol $(\Delta_f H(H_2NC(O)OC_2H_5) = -446.3$ kJ/mol[12] and $\Delta_f H(C_2H_3^{1\bullet}) = 299 \pm 5$ kJ/mol[12]).[297] This value is in quite good agreement with a calculated (G3) value of 282 kJ/mol.[300] MNDO calculations predict the heats of formation of the isomers $H_3NC(O)OH^{1+}$, $H_2NC(O)OH_2^{1+}$ and $HNC(OH_2)OH^{1+}$ to be 437, 455 and 486 kJ/mol, respectively.[297] No thermochemical data are available for the CH_3ONOH^{1+} and CH_3OH-NO^{1+} ions.

3.5.74 $[C,H_3,N,O_3]^{+\bullet} - m/z$ 77

Ionization of methyl nitrate (CH_3ONO_2) yields no molecular ion. The base peak in the normal mass spectrum is m/z 46, $NO_2^{1+\bullet}$ (loss of 31, presumably $CH_3O^{\bullet}$). This is in keeping with the molecule's ionization energy $(11.53 \pm 0.01$ eV $= 1112$ kJ/mol)[12] being greater than the energy required for the dissociation $CH_3ONO_2 \longrightarrow NO_2^{1+\bullet} + CH_3O^{\bullet}$ $(\Sigma\Delta_f H_{products} = 1097$ kJ/mol).[12]

3.5.75 $[C,H_4,N,O_3]^+ - m/z$ 78

Isomers

Two isomers have been investigated by experiment: the protonated methoxy and terminal oxygen methyl nitrate, $CH_3O(H)-NO_2^{1+}$ and $CH_3ON(OH)O^{1+}$, respectively.

Identification

As with methyl nitrite (see $[C,H_3,N,O_3]^{+\bullet}$ above) protonation of methyl nitrate yields, depending on the exothermicity of the ion–molecule reaction, species with slightly different MI mass spectral characteristics. Highly exothermic protonation, such as that obtained when using H_3^{1+} as protonating agent (exothermicity $= 310$ kJ/mol), yields an ion whose MI mass spectrum shows a composite m/z 46 fragment ion peak (NO_2^{1+}).[298] The peak is dominated by its broad component. Less exothermic protonation, resulting from the use of $C_2H_5^{1+}$ (exothermicity $= 54$ kJ/mol), also yields a metastable product ion that decomposes to give m/z 46 (NO_2^{1+}) with the difference being that the composite peak is now dominated by the narrow component.[298] (Refer to Physicochemical aspects of metastable ion dissociations, in Section 1.4.1 for a discussion on composite metastable ion peaks.)

The ion–molecule reaction of NO_2^{1+} with CH_3OH in an ion source also yields a metastable m/z 78 ion which loses CH_3OH to produce a very narrow m/z 46 Gaussian peak $(T_{0.22} = 0.95$ meV).[298] It is noteworthy that the reaction using $CH_3^{18}OH$ yields exclusively unlabeled NO_2^{1+} ions, strongly suggesting the retention of an intact methanol moiety and the generation of an electrostatically bound pair (see also below).[298] The CID mass spectra (collision gas, air) of the ions generated by the above ion–molecule reaction and by the exothermic protonation of CH_3ONO_2 are significantly different.[298,301]

Partial CID Mass Spectra of the $[C,H_4,N,O_3]^+$ Isomers[301]

	m/z					
Ions	15	28	29	30	31	32
$CH_3O(H)-NO_2$	2.5	3.3	32	48	100	38
$CH_3ON(OH)O$	3.9	—	28	100	35	11

On the basis of these CID mass spectra, it was proposed that the ion–molecule reaction yields a single isomer, $CH_3O(H)-NO_2^{1+}$, whereas the protonation reactions yield mixtures of both isomeric forms, $CH_3O(H)-NO_2^{1+}$ and $CH_3ON(OH)O^{1+}$, the composition of which depends on the exothermicity of the reaction that produces them.[298,301]

Thermochemistry

Calculations predict that the $CH_3O(H)-NO_2^{1+}$ ion is the most stable isomer (G2MP2).[301] On the basis of the proton affinity of the methoxy oxygen in methyl nitrate, 733.6 kJ/mol,[12] $\Delta_fH(CH_3O(H)-NO_2^{1+}) = 674$ kJ/mol (using $\Delta_fH(CH_3NO_3) = -122 \pm 1$ kJ/mol and $\Delta_fH(H^+) = 1530$ kJ/mol).[12] Its optimized geometry shows an elongated O–N bond (2.365 Å at MP2/6-311 + G(2df,p)), consistent with the ion being only weakly bound with respect to its dissociation to $CH_3OH + NO_2^{]+}$ ($\Sigma\Delta_f H_{products} = 757$ kJ/mol^{12}).[301] The calculated proton affinity of the terminal oxygen varies from 712 to 725 kJ/mol, depending on which rotomer of the $CH_3ON(OH)O^{]+}$ ion is produced (G2MP2).[301]

3.5.76 $[C,N_2,O]^{+\bullet} - m/z\ 56$[302]

Chemical ionization of a CO/N_2 mixture yields a $[C,N_2,O]^{+\bullet}$ product ion which can be isolated from isobaric $N_4^{]+\bullet}$ and $C_2O_2^{]+\bullet}$ ions by proper isotopic labeling. The CID mass spectrum of this ion is dominated by fragment peaks corresponding to $CO^{]+\bullet}$ and $N_2^{]+\bullet}$ and minor peaks assigned to $NCO^{]+}$ and $CNN^{]+}$ ions. These observations are consistent with an ion having an $N-N-C-O$ connectivity. Computations (B3LYP) find a single minimum on the doublet surface, a transplanar ion ($^2A'$). On the quartet surface, two ions of C_s geometry are found to be stable, lying high above the doublet by 388 ± 20 kJ/mol ($^4A'$) and 420 ± 30 kJ/mol ($^4A''$), depending on the level of theory used.

It is noteworthy that neutralization–reionization of this ion generates a recovery signal suggesting that a neutral $[C,N_2,O]$ species was produced. On the basis of experimental results, the neutral generated is believed to be the bent $N-N-C-O$ ($^3A''$) species originating from the reionization of one of the quartet state ion.

3.5.77 $[C,H_2,N_2,O]^{+\bullet} - m/z\ 58$[66]

The dissociative ionization of methyl carbazate ($H_2NNHC(O)OCH_3$) generates an m/z 58 ion by the loss of methanol. On the basis of its mass spectral characteristics, the product ion has been assigned the connectivity of the aminyl isocyanate radical cation ($H_2NNCO^{]+\bullet}$). In keeping with the above structure assignment, the $H_2NNCO^{]+\bullet}$ ion loses CO to generate the $H_2NN^{]+\bullet}$ ion and not the isomer $HNNH^{]+\bullet}$. When neutralized and reionized, $H_2NNCO^{]+\bullet}$ produces a recovery signal and fragment peaks indicating that neutral aminyl isocyanate is a stable species, as predicted by theory.[303] There are no reported thermochemical data for this ion.

3.5.78 $[C,N,S]^+ - m/z\ 58$

Isomers

The two isomeric forms $NCS^{]+}$ and $CNS^{]+}$ have been characterized.

Identification[304]

Collision-induced demethylation of CH_3NCS (methylthiocyanate) and CH_3CNS (acetonitrile-N-sulfide) produces the $NCS^{]+}$ and $CNS^{]+}$ ions, respectively. The distinctive feature in the ions' CID mass spectra is the m/z 56:44 ($NS^{]+}$:$CS^{]+\bullet}$) peak abundance ratio. Consistent with their respective connectivities, this ratio is 0.63 for $NCS^{]+}$ and 1.56 for $CNS^{]+}$.

There are no available thermochemical data.

3.5.79 [C,H,N,S]$^{+\bullet}$ – m/z 59

Isomers

Two isomers have been characterized: HNCS$^{1+\bullet}$ and HCNS$^{1+\bullet}$, ionized isothiocyanic and thiofulminic acids. The former molecule has been identified in interstellar clouds.[305]

Identification[306]

Direct ionization of isothiocyanic acid produces an ion with a CID mass spectrum different from that of the ion obtained from the dissociative ionization of 1,3,4-oxathiazole-2-one (loss of CO_2). Given the structure of the precursor molecule and the presence of the intense m/z 46 (NS^{1+}) fragment peak, the latter ion was assigned the HCNS$^{1+\bullet}$ connectivity.

Partial CID Mass Spectra of the [C,H,N,S]$^{+\bullet}$ Isomers

	m/z						
Ions	26	26	32	33	44	45	46
HNCS	3.0	17	100	13	23	3.0	4.5
HCNS	2.3	9.0	100	9.8	3.0	1.5	26

Other precursor molecules such as 3-amino- and 3-methyl-1,2,4-triazole-5-thiol, trithiocyanuric acid, ethyl thiocyanate and ethyl isothiocyanate were shown to produce the isomer HNCS^{1+}. Sulfuration of HCN in the ion source of a mass spectrometer using CS$_3$$^{1+\bullet}$ as a S$^{+\bullet}$ transfer agent produces ionized thiofulminic acid.[307]

Thermochemistry

The thiofulminic acid radical cation (HCNS$^{1+\bullet}$) has been calculated to be stable with respect to unimolecular dissociation to HCN + S$^{+\bullet}$ by 419.1 kJ/mol (G2(MP2,SVP)).[307] Therefore its heat of formation can be estimated to be 993 kJ/mol ($\Sigma\Delta_f H_{products} = 1412$ kJ/mol).[12] No data are available for the isomeric ion. However, NIST lists an ionization energy for HNCS of ≤ 9.92 eV.[12] Calculations predict that isothiocyanic acid (HNCS) is the global minimum on the [C,H,N,S] potential energy surface with thiofulminic acid (HCNS) lying 144 kJ/mol above it (B3LYP//B3LYP and CCSD(T)//MP2).[308]

3.5.80   [C,N₂,S]⁺• – *m/z* 72

Isomers

Ionized thionitrosyl cyanide (NCNS¹⁺•) and its isomer thiazyl cyanide (NCSN¹⁺•) have been generated and characterized by experiment.

Identification[304]

Electron-impact of (1,2,5)-thiadiazolo[3,4-c](1,2,5)-thiadiazole produces a *m/z* 72 ([C,N₂,S]⁺•) fragment ion.

The CID mass spectrum of this ion is dominated by an *m/z* 46 fragment peak corresponding to NS¹⁺ cations with other significant and structure-characteristic peaks at *m/z* 40 (CN₂¹⁺•) and 32 (S⁺•). These peaks suggest that the ion generated possesses the connectivity of thionitrosyl cyanide (NCNS¹⁺•). It is noteworthy that the molecular species is obtained by NRMS of the NCNS¹⁺• ion.

The reaction of thiazyl chloride (ClNS, generated from the flash-vacuum pyrolysis of N₃S₃Cl₃) with silver cyanide in the ion source of the mass spectrometer also produces an *m/z* 72 fragment peak. The CID mass spectrum of this ion is similar to that of NCNS¹⁺• but differs by having a significantly more intense *m/z* 44 fragment peak (CS¹⁺). On the basis of this observation and other spectroscopic evidence,[309,310] this ion was assigned the NCSN¹⁺• connectivity. It is noteworthy that the similarities of the two CID mass spectra are suggested to be due to postcollisional isomerization, a phenomenon observed for other sulfur-containing ions.[311–313]

Thermochemistry[304]

From the calculated heat of formation and ionization energy of the thionitrosyl cyanide molecule, $\Delta H_f(\text{NCNS}^{1+\bullet}) = 1356$ kJ/mol ($\Delta H_f^{0K}(\text{NCNS}) = 410 \pm 12$ kJ/mol and IE $= 9.8 \pm 0.3$ eV at the QCISD(T)/6-311 + G(d) level of theory). Molecular thiazyl cyanide (NCSN) is calculated to lie 59 kJ/mol (0 K) above NCNS and separated from it by a large energy barrier (276 kJ/mol at the QCISD(T)/6-311 + G(3df) level of theory).

3.5.81   [C,H₂,N₂,S]⁺• – *m/z* 74[314]

Isomers

Radical cations NH₂CNS¹⁺• and HNCNSH¹⁺• are both computed to be stable and may have been generated.

Identification

4-Carboxamido-3-cyano-1,2,5-thiadiazole   (cy–SNC(C(O)NH$_2$)C(CN)N–)
lose a CO molecule to generate what is believed to be the 4-amino-2-
cyanothiadiazole (cy–SNC(NH$_2$)C(CN)N–) radical cation. This ion undergoes
further dissociation and produces an m/z 74 ion. The fragment ion's CID mass
spectrum shows peaks consistent with the connectivity of NH$_2$CNS$^{1+\bullet}$ or
HNCNSH$^{1+\bullet}$ radical cations. CI of cyanamide (NCNH$_2$) using CS$_2$ as reagent,
also yields this ion. Dissociative ionization of 2-amino-1,3,4-thiadiazole (cy–
SC(H)NNC(NH$_2$)–) also generates an m/z 74 ion. The CID mass spectrum of
the latter ion is very similar to that of NH$_2$CNS$^{1+\bullet}$/HNCNSH$^{1+\bullet}$, with the
difference being that the intensity of the charge-stripping peak is very much
reduced. To account for this observation, Flammang et al. studied the dissociation characteristic of the m/z 74 ion generated from 2-amino-1,3,4,5-thiatria-
zole (cy–SNNNC(NH$_2$)–) and concluded that, in this case, the m/z 74 ion could
be the isomeric ion cy–SNC(NH$_2$)–$^{1+\bullet}$. It was thereafter concluded that dissociative ionization of 2-amino-1,3,4-thiadiazole generated a mixture of
NH$_2$CNS$^{1+\bullet}$/HNCNSH$^{1+\bullet}$ and cy–SNC(NH$_2$)–$^{1+\bullet}$.

Thermochemistry

Computations (G2(MP2,SVP)) find the NH$_2$CNS$^{1+\bullet}$ to be lower in energy
than HNCNSH$^{1+\bullet}$ by 24.4 kJ/mol. These ions are separated by an isomerization energy barrier of 259.2 kJ/mol above NH$_2$CNS$^{1+\bullet}$. This is significantly
lower in energy than the calculated lowest energy dissociation
(H$^{\bullet}$ + HNCNS^{1+}, 361.2 kJ/mol above NH$_2$CNS$^{1+\bullet}$) and is therefore consistent with the ions being indistinguishable.

3.5.82   [C,H$_2$,O,P]$^+$ – m/z 61[163]

Calculations (G2) show that 19 and 15 local minima are found on the triplet
and singlet potential energy surfaces, respectively. The global minimum on
the combined surfaces is ^{1}PH$_2$CO^{1+}. It lies 582 kJ/mol below the dissociation
products P$^+$ + H$_2$CO (G2). The lowest energy triplet, ^{3}PC(H)OH^{1+}, is 197
kJ/mol higher in energy (G2) than ^{1}PH$_2$CO^{1+}. Note that other singlet species
such as cy–C(H)$_2$OP–$^{1+}$, HPCOH^{1+}, cy–C(H)OP(H)–$^{1+}$ and PC(H)OH^{1+} are
lower in energy than ^{3}PC(H)OH^{1+}, lying 114, 134, 145 and 184 kJ/mol,
respectively, above ^{1}PH$_2$CO^{1+}. On the singlet potential energy surface, all the
isomers are predicted to interconvert below dissociation to P$^+$ + H$_2$CO. No
experimental data are available for these ions.

3.5.83   [C,H$_3$,O,P]$^+$ – m/z 62[315]

Computations of the relative stability of the CH$_3$PO^{1+}, CH$_3$OP^{1+} and
CH$_2$POH^{1+} isomers have found that among these species, the latter ion is

the most stable with the former two very close in energy, lying 17 kJ/mol above CH_2POH^{1+}. No attempts to generate these have been published.

3.5.84 $[C,H_4,O,P]^+ - m/z\ 63^{315}$

Calculations find 16 minima on the triplet potential energy surface (G1 and G2). Of these, CH_3POH^{1+} is found to be the global minimum, lying 333 kJ/mol below the products $CH_3OH + P^+$. All other isomers also lie below these products.

No experimental data are available for these ions.

Relative Energies of the $[C,H_4,O,P]^+$ Isomers

Ions	Relative Energy with Respect to $CH_3OH + P^+$ (kJ/mol)	Ions	Relative Energy with Respect to $CH_3OH + P^+$ (kJ/mol)
CH_3POH	−333	$PH_2C(H)OH$	−214
$CH_2P(H)OH$	−302	PH_2OCH_2	−198
$CH_3O(H)P$	−301	PH_3COH	−172
CH_2POH_2	−266	$CH_2O(H)PH$	−167
$PC(H)_2OH_2$	−261	$CHP(H)OH_2$	−158
CH_3OPH	−237	$PH_3C(H)O$	−151
$PHC(H)_2OH$	−224	$PHC(H)OH_2$	−146
$CH_3P(H)O$	−220	$PH_2O(H)CH$	−32

3.5.85 $[C,O_2,P]^+ - m/z\ 75^{316}$

These ions have been investigated solely by theory. Five minima, on both the singlet and triplet potential energy surfaces, are found. Of these, four are complexes $[P \cdots OCO]^+$ (**1**), $[P \cdots CO_2]^+$ (**2**), $[OP \cdots CO]^+$ (**3**) and $[OP \cdots OC]^+$ (**4**) and one is cyclic, cy–POCO–$^{1+}$ (**5**). The most stable species on the singlet and triplet surfaces are the complexes $[OP \cdots CO]^+$ and $[P \cdots OCO]^+$, respectively (MP4//MP2). Isomer (**2**), although stable, lies above the products $P^+ + CO_2$ and therefore should not be formed spontaneously.

Binding Energies for the $[C,O_2,P]^+$ Isomers

Ions	Binding Energy with Respect to Products (kJ/mol)		Products
	Singlet	Triplet	
(**1**)	151	133	$P^+ + CO_2$
(**2**)	−188	−277	$P^+ + CO_2$
(**3**)	50	226	$OP^+ + CO$
(**4**)	29	87	$OP^+ + CO$

3.5.86 $[C,H_3,O_2,P]^{+\bullet} - m/z\,78$

Isomers

Four ions have been characterized by mass spectrometry: $CH_3OP=O^{]+\bullet}$ (1), $CH_2OPOH^{]+\bullet}$ (2), $CH_2=P(O)(OH)^{]+\bullet}$ (3) and $CH_3P(O)_2^{]+\bullet}$ (4). Some 16 other structures including $[HOP\cdots HCHO]^{+\bullet}$ (5), $CH_2O(H)PO^{]+\bullet}$ (6), $[HOCH_2\cdots OP]^{+\bullet}$ (7), $[CH_2OH\cdots OP]^{+\bullet}$ (8), $CH_2OP(H)O^{]+\bullet}$ (9), $[-OCH_2O-]PH^{]+\bullet}$ (10), $[-OCH_2O(H)-]P^{]+\bullet}$ (11), $HC(O)P(H)OH^{]+\bullet}$ (12), $[HOPH\cdots OCH]^{+\bullet}$ (13), $[-CH_2O-]POH^{]+\bullet}$ (14), $HOCH_2PO^{]+\bullet}$ (15), $HOCH=P(H)O^{]+\bullet}$ (16), $[HOP(H)H\cdots CO]^{+\bullet}$ (17), $[HPO(H)H\cdots CO]^{+\bullet}$ (18), $[OCPH\cdots OH_2]^{+\bullet}$ (19) and $[HOPH_2\cdots OC]^{+\bullet}$ (20) have been calculated to be stable.

Identification

Dissociative ionization of methyl dichlorophosphate ($CH_3OP(O)Cl_2$, two consecutive losses of $Cl^\bullet$) and collisional activation of ionized ethyl ethylene phosphite (see below) produce $m/z\,78$ fragment ions with similar but distinguishable CID mass spectra.[317,318]

The structure-indicative m/z 62 (loss of atomic oxygen) and 64 (loss of methylene) fragment peaks in the ions' respective CID mass spectra suggest that methyl dichlorophosphate produces the $CH_3OPO^{]+\bullet}$ ion whereas the ethyl ethylene phosphite precursor molecule generates the $CH_2OPOH^{]+\bullet}$ distonic isomer.[317,318]

The $CH_2=P(O)(OH)^{]+\bullet}$ (3) ion is produced following water loss from ionized methyl phosphonic acid ($CH_3P(O)(OH)_2$) and is characterized by intense fragment peaks corresponding to the loss of atomic oxygen (m/z 62) and methylene (m/z 64).[319] The more elusive keto tautomer $CH_3P(O)_2^{]+\bullet}$ (4) can be generated by ionization of the *neutral* fragments formed in the following dissociations[319] and shows a CID mass spectrum dominated by the fragment peak at m/z 47 ($PO^{]+}$) and so is clearly distinct from the others.

Partial CID Mass Spectra of the $[C,H_3,O_2,P]^{+\bullet}$ Isomers[317–319]

					m/z					
Ions	14	15	31	47	48	49	50	62	63	64
CH_3OPO	—	8.6	100	43	41	8.6	49	5.7	—	—
CH_2OPOH	0.63	—	100	59	81	19	16	—	—	3.1
$CH_2=P(O)(OH)$	2.9	—	100	87	84	13	10	8.8	—	34
$CH_3P(O)_2$	—	18	15	100	—	—	—	10	5.1	—

It is noteworthy that the MI mass spectra of $CH_3OP=O^{]+\bullet}$ and $CH_2OPOH^{]+\bullet}$ are almost identical, showing dissociations by loss of H_2O (m/z 60), CO (m/z 50) and $PO^{]\bullet}$ (m/z 31).[317] The ion $CH_2=P(O)(OH)^{]+\bullet}$ also exhibits the same metastable dissociations with, however, noticeable differences in the relative abundance of each peak. The kinetic energy release associated with CO loss is the same for the three isomers ($T_{0.5}=40$ meV),[317] strongly suggesting that these ions can interconvert before dissociation. For isomers $CH_3OPO^{]+\bullet}$ and $CH_2OPOH^{]+\bullet}$, ^{18}O-labeling experiments further support interconversion before loss of CO.[317] That the ions can isomerize before dissociation is also in keeping with their CID mass spectra being very similar.

Thermochemistry[317]

Calculated Heat of Formation for the $[C,H_3,O_2,P]^{+\bullet}$ Isomers

Ions	Calculated $\Delta_f H$ (CBS-QB3) (kJ/mol)	Ions	Calculated $\Delta_f H$ (CBS-QB3) (kJ/mol)
(1)	567	(11)	597
(2)	508	(12)	569
(3)	500	(13)	644
(4)	628	(14)	562
(5)	643	(15)	581
(6)	613	(16)	593
(7)	582	(17)	552
(8)	549	(18)	560
(9)	586	(19)	546
(10)	597	(20)	548

The computed barriers for the interconversion of the ions are 90 kJ/mol for $CH_3OP=O^{]+\bullet} \longrightarrow CH_2OPOH^{]+\bullet}$, 195 kJ/mol for $CH_3OP=O^{]+\bullet} \longrightarrow CH_2=P(O)(OH)^{]+\bullet}$ and 163 kJ/mol for $CH_3OP=O^{]+\bullet} \longrightarrow CH_3P(O)_2^{]+\bullet}$.

3.5.87 $[C,H_5,O_3,P]^{+\bullet}$ – m/z 96[318]

The insecticide acephate $(CH_3S(CH_3O)P(O)N(H)C(O)CH_3)$ shows an m/z 96 fragment ion produced by the consecutive losses of $[C_2,H_3,N]$ and CH_2S following considerable skeletal rearrangement. On the basis of its metastable ion dissociations (losses of H_2O (m/z 78) and H_2CO (m/z 66)) and on its CID mass spectrum showing prominent peaks at m/z 81 (loss of $CH_3^{\bullet}$) and 65 (loss of $CH_3O^{\bullet}$), this ion has been assigned the $CH_3OP(OH)_2^{+\bullet}$ connectivity. No thermochemical data are available for this ion.

3.5.88 $[C,O,S]^{+\bullet}$ – m/z 60

Carbonyl sulfide (OCS) is an important molecule in the global atmospheric cycling of sulfur and has been detected in the atmospheres of Jupiter and Venus and in dense molecular clouds.[320] Combining the heat of formation and ionization energy of the molecule, -138.41 kJ/mol and 11.18 ± 0.01 eV, respectively,[12] gives $\Delta_f H(OCS^{1+\bullet}) = 940$ kJ/mol. Its lowest energy dissociation channel leads to the formation of $CO + S^{+\bullet}$.[321,322]

3.5.89 $[C,H,O,S]^{+}$ – m/z 61

Isomers

In principle, protonation of OCS could take place at carbon, oxygen or sulfur, yielding three isomers, namely, $OC(H)S^{1+}$, $HOCS^{1+}$ and $HSCO^{1+}$, respectively. $[H,C,O,S]^{+}$ ions have been proposed as interstellar species.[323–325] None of the above ions have been investigated mass spectrometrically.

Thermochemistry

Theory predicts that all three isomers are stable, the $HSCO^{1+}$ ion being the global minimum. The most recent calculations[326] predict that $HOCS^{1+}$ and $OC(H)S^{1+}$ are higher in energy, lying 14–24 kJ/mol (depending on the level of theory, B3LYP/6-311++G**, CCSD(T)/6-311++G** or G3 (0 K)) and 241 kJ/mol (B3LYP/6-311++G** including zero point energies) respectively, above $HSCO^{1+}$, in agreement with earlier calculations.[327,328] The proton affinity of OCS is listed to be 628.5 kJ/mol,[12] but the connectivity of the resulting [H,C,O,S] ion has not been identified mass spectrometrically.

At present, only the O-protonated species has been identified spectroscopically.[328–330] It is interesting to note that Saebo et al.[328] argue that a priori there is no reason why protonation, which in their study is achieved in a low temperature hollow cathode by the reaction of H_3^{1+} ions with OCS, could not take place at both terminal atoms (i.e., oxygen and sulfur). However, because the exothermicity of the protonation reaction is greater at S than at O, the $HSCO^{1+}$ ions are generated at a higher rotational temperature than the

HOCS^{1+} ions. This results in the infrared bands being composed of more lines of lower intensity, which may explain the lack of spectral evidence supporting the production of HSCO^{1+} ions. In addition, in a combined experimental and theoretical analysis of the effects of vibrational mode and collision energy on the protonation reaction, there is some evidence to suggest that the S-protonated isomer is produced at least at low collision energy.[326]

3.5.90 [C,H$_2$,O,S]$^{+\bullet}$ – m/z 62

Isomers

Of the three sought-after isomers, HC(S)OH$^{1+\bullet}$, HC(O)SH$^{1+\bullet}$ and HOCSH$^{1+\bullet}$, ionized thioformic and thionformic acids and the thioxy-hydroxy-carbene radical cations, respectively, only the latter has been observed by experiment. Note that no attempts to generate and characterize ionized sulfine (CH$_2$SO$^{1+\bullet}$) have been reported. This is most likely due to the relative short lifetime of the sulfine molecule in the gas phase.[331]

Identification[332]

Dissociative ionization of O-ethyl thioformate and S-ethyl thioformate, (HC(S)OC$_2$H$_5$ and HC(O)SC$_2$H$_5$, respectively, loss of C$_2$H$_4$ in both cases) produces species with identical CID mass spectral characteristics: m/z 61 (loss of H$^\bullet$), 45 ([H,C,S]$^+$, loss of $^\bullet$OH), 34 (SH$_2$$^{1+\bullet}$, loss of CO) and 29 ([H,C,O]$^+$, loss of $^\bullet$SH). These observations alone could imply either that a single species is generated from both precursor molecules or that a mixture of isomers is produced. However, further mass spectrometric experiments combined with the species' reactivity toward dimethyldisulfide strongly suggest that a single ion, the ionized carbene (HOCSH$^{1+\bullet}$), is generated.

Thermochemistry

Calculations (UQCISD(T)/6-311++G**)[333] predict that ionized thioformic acid (HC(S)OH$^{1+\bullet}$) is lower in energy than both HC(O)SH$^{1+\bullet}$ and HOCSH$^{1+\bullet}$ ions by 88 and 77 kJ/mol, respectively, in agreement with earlier calculations (MP4/6-31G**).[334] The absolute heat of formation of the carbene has been calculated to be 873 or 864 kJ/mol, depending on the level of theory used (G2 and G2MP2).[28] Barriers to isomerization of 118 and 165 kJ/mol prevent the carbene ion from freely interconverting with HC(S)OH$^{1+\bullet}$ and HC(O)SH$^{1+\bullet}$, respectively.[333]

It is worth adding that an approximate $\Delta_f H$(CH$_2$SO$^{1+\bullet}$) $\approx$ 957 kJ/mol is obtained by combining its calculated heat of formation (-30 ± 6 kJ/mol, at the CBS-QB3 level of theory,[335] in agreement with -38 ± 10 kJ/mol from DFT calculations[336]) with the measured ionization energy (10.23 eV[337]).

3.5.91 $[C,H_3,O,S]^+$ – m/z 63

Isomers

Protonation of sulfine at carbon and sulfur yields two isomers, CH_3SO^{1+} (**1**) and CH_2SOH^{1+} (**2**), respectively. Theory finds no less than seven other stable isomeric forms:

Identification

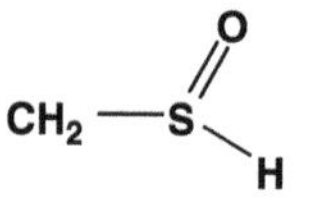

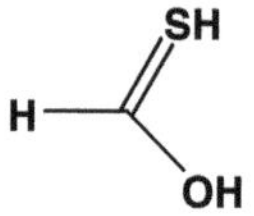

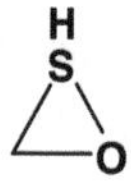

Dissociative ionization of di-n-butyl sulfoxide ((n-C_4H_9)$_2$SO, consecutive losses of C_4H_8 and $C_3H_7{}^{1\bullet}$) produces an m/z 63 ion whose CID mass spectral characteristics are different from those of the ion generated from dimethylsulfoxide ($CH_3S(O)CH_3$, loss of $CH_3{}^{1\bullet}$).[338,339]

Partial CID Mass Spectra of the $[C,H_3,S,O]^+$ Isomers[338]

Ions	13	14	15	32	33	34	44	45	46	47	48	49
CH_3SO	2	3.3	16	23	17	11	15	(123)	26		100	6.5
CH_2SOH	2.7	6.2	0.3	78	21	4	100	(446)	(84)	6	54	53

Values in parentheses contain contributions from metastable ion dissociation.

From their respective dissociation characteristics, it was concluded that di-n-butyl sulfoxide generates oxygen-protonated sulfine (CH_2SOH^{1+}) while dimethylsulfoxide gave CH_3SO^{1+} ions.

However, it was shown that depending on the energy content of the dimethylsulfoxide precursor molecule, the product ion could be either isomer.[339–341] Indeed, source-generated m/z 63 ions (generated from precursors dissociating with rate constants $>10^6$ s^{-1}) were identified as mainly CH_3SO^{1+} ions but when nondissociating precursor molecules are sampled (dissociation rate constants $<10^6$ s^{-1}), the ion produced by loss of methyl is the sulfur-protonated species (CH_2SOH^{1+}). This was rationalized by the isomerization of ionized dimethylsulfoxide to its aci-form ($CH_2S(OH)CH_3$) before dissociation (see below).

Thermochemistry

Of all nine isomers, $HC(SH)OH^{1+}$ is predicted to be the global minimum. Recent calculations predict that the CH_2SOH^{1+} ion is lower in energy than its isomer CH_3SO^{1+} by about 92 kJ/mol (G2(MP2,SVP)),[341] in agreement with previous calculations at a variety of levels of theory.[342,343] The absolute heat of formation of these ions was also calculated to be $\Delta_f H(CH_2SOH^{1+}) = 707$ and $\Delta_f H(CH_3SO^{1+}) = 793$ kJ/mol (CBS-QB3),[335] in agreement with the above relative energies. The isomerization barrier separating CH_3SO^{1+} from CH_2SOH^{1+} is estimated to lie somewhere in the range of 292–354 kJ/mol depending on the level of theory.[343]

Considering that the measured appearance energy for methyl loss from dimethylsulfoxide actually corresponds to the formation of the CH_2SOH^{1+} ion and not, as initially proposed, to CH_3SO^{1+}, the experimental $\Delta_f H(CH_2SOH^{1+}) = 730$ kJ/mol (using AE = 10.64 ± 0.07 eV,[344] $\Delta_f H((CH_3)_2SO) = -150.5 \pm 1.5$ kJ/mol[12] and $\Delta_f H(CH_3^{1\bullet}) = 145.69$ kJ/mol[12]) is somewhat higher than that predicted by theory. In order for $(CH_3)_2SO^{1+\bullet}$ to dissociate to $CH_2SOH^{1+} + CH_3^{1\bullet}$, it must first isomerize to $CH_2S(OH)CH_3^{1+\bullet}$. Calculations have shown that the barrier for this isomerization lies some 10–30 kJ/mol above the dissociation limit (depending on the level of theory)[335,341] and therefore the experimental result (730 kJ/mol) sets only an upper limit for the heat of formation of CH_2SOH^{1+}.

Computed Relative Energies of the Isomers [C,H₃,S,O]+ [339]

Ions	Relative Energies RHF/6-31G(d,p) (kJ/mol)
(3)	57
(4)	309
(5)	0
(6)	273
(7)	190
(8)	178
(9)	62

It is noteworthy that by combining the calculated[335] $\Delta_f H(CH_2SO)$ and $\Delta_f H(CH_2SOH^{1+})$, -30 ± 6 kJ/mol and 707 kJ/mol, respectively, with $\Delta_f H(H^+) = 1530$ kJ/mol yields $PA(CH_2SO) = 793$ kJ/mol, in agreement with the reevaluated (new proton affinity scale) experimental value 787.6 ± 2.6 kJ/mol;[345] the value listed by NIST, 798.9 kJ/mol,[12] is not preferred.

3.5.92 $[C,H_4,O,S]^{+\bullet}$ – m/z 64

Isomers

Two isomers have been generated: ionized methanesulfenic acid $(CH_3SOH^{]+\bullet})$ and the distonic ion $CH_2S(H)OH^{]+\bullet}$. A third species, an ion–neutral complex of ionized thioformaldehyde and water $[CH_2S \cdots OH_2]^{+\bullet}$, has been proposed as an intermediate. Methanesulfenic acid itself presumably plays a significant role in the photochemical degradation of methanethiol and dimethylsulfide and major natural and man-made pollutants.[346–349] In addition, sulfenic acids in general are also implicated as intermediates in the formation of the lachrymatory substances found in freshly cut garlic and onion.[350]

Identification[338]

Ionization of the pyrolysis products of methyl t-butyl sulfoxide $(CH_3S(O)C_4H_9)$ and dissociative ionization of methyl n-propyl sulfoxide $(CH_3S(O)C_3H_7$, loss of $C_3H_6)$, generate an ion to which the $CH_3SOH^{]+\bullet}$ connectivity was assigned. Dissociative ionization of di-n-butyl sulfoxide generates (consecutive losses of C_4H_8 and C_3H_6) an ion with slightly different CID mass spectral characteristics. On the basis of the more prominent m/z 50 $(HSOH^{]+\bullet})$ and 46 (loss of H_2O) fragment ions, the latter ion was assigned the connectivity of the distonic ion $(CH_2S(H)OH^{]+\bullet})$. The ions have essentially identical MI mass spectra.

Partial CID Mass Spectra of the $[C,H_4,O,S]^{+\bullet}$ Isomers[338]

Ions	m/z										
	14	15	32	33	44	45	46	47	48	49	50
CH₃SOH	1.7	4.1	15	5	16	100	(29)	(69)	18	33	1.1
CH₂S(H)OH	2.2	2.6	26	4	9.4	100	(92)	(72)	17	32	10

Values in parentheses contain contributions from metastable ion dissociation.

MI Mass Spectra of the $[C,H_4,O,S]^{+\bullet}$ Isomers

Ions	m/z				
	31	45	46	47	49
CH₃SOH	12	7	100 (29)	79 (19)	4
CH₂S(H)OH	11	6	100 (27)	80 (20)	3

Values in parentheses are the kinetic energy release $(T_{0.5})$ in meV.

These observations, and results of isotopic labeling, suggest that before dissociation by loss of $^\bullet$OH (producing CH_2SH^{1+} fragment ions (m/z 47)), the methanesulfenic acid radical cation isomerizes to the distonic isomer $CH_2S(H)OH^{1+\bullet}$. In addition, a further hydrogen shift produces a third isomer, $[CH_2S\cdots OH_2]^{+\bullet}$, the reacting configuration for the water loss (producing $CH_2S^{1+\bullet}$, m/z 47).

Thermochemistry

From the latest theoretical calculations (G3/MP2), $\Delta_f H(CH_3SOH) = -139.4$ kJ/mol,[351] in agreement with earlier calculations at the G2(MP2) level of theory $(-141.8$ kJ/mol).[352] Combining this value with the experimental ionization energy, 8.67 ± 0.03 eV (in agreement with values computed at the G2, G1, and G2MP2 levels of theory),[353] leads to $\Delta_f H(CH_3SOH^{1+\bullet}) = 697$ kJ/mol, not in disagreement with 685 kJ/mol obtained from an appearance energy measurement ($CH_3CH_2CH_2S(O)CH_3 \longrightarrow CH_3SOH^{1+\bullet} + C_3H_6$; AE $= 9.36 \pm 0.05$ eV,[354] $\Delta_f H(CH_3CH_2CH_2S(O)CH_3) = -198$ kJ/mol (by additivity)[58,355,356] and $\Delta_f H(C_3H_6) = 20.41$ kJ/mol[12]). The heats of formation of the distonic ions $CH_2S(H)OH^{1+\bullet}$ and $CH_2SOH_2^{1+\bullet}$ have been calculated to be 795 and 709 kJ/mol, respectively (MNDO).[338]

3.5.93 $[C,H,O,F]^{+\bullet} - m/z$ 48

Isomers

Two isomers have been identified by experiment, ionized formyl fluoride ($HC(O)F^{1+\bullet}$) and ionized fluorohydroxycarbene ($FCOH^{1+\bullet}$). A third isomer, the complex $[HF\cdots CO]^{+\bullet}$, is calculated to be a stable species.

Identification[357]

Direct ionization of formyl fluoride yields an ion with a CID mass spectrum distinct from that of the ion obtained by the dissociative ionization of methyl fluoroformate ($FC(O)OCH_3$, loss of H_2CO).

Partial CID Mass Spectra of the $[C,H,O,F]^{+\bullet}$ Isomers

	m/z			
Ions	28	29	31	32
HC(O)F	100	73	100	18
FCOH	100	85	47	44

In addition to the relative abundance of the fragment peaks, the CID mass spectra of the $HC(O)F^{1+\bullet}$ and $FCOH^{1+\bullet}$ isomers differ in the following ways:

(1) the peak shape corresponding to fluorine loss (m/z 29) is Gaussian for the ionized carbene and dished for the formyl fluoride radical cation and (2) only FCOH$^{1+\bullet}$ exhibits a structure-indicative m/z 17 ($^{+}$OH) fragment ion peak.

Thermochemistry

Combining the heat of formation and ionization energy of formyl fluoride (-376.56 kJ/mol and 12.37 ± 0.02 eV,[12] the latter value being in agreement with calculations[93]) gives $\Delta_f H(HC(O)F^{1+\bullet}) = 817$ kJ/mol. (However, note that calculations at the G2 and MP4 levels of theory suggest a higher value for $\Delta_f H(HC(O)F) = -393.7$ kJ/mol[358] or -392 ± 6 kJ/mol,[359] respectively, which would lead to $\Delta_f H(HC(O)F^{1+\bullet}) \approx 801$ kJ/mol.) Calculations predict that the ionized carbene (FCOH$^{1+\bullet}$) is the global minimum on the potential energy surface and, depending on the level of theory, lies 57, 35 or 37 kJ/mol lower in energy than its isomer (at the MP2/6-31G**//MP2/6-31G**[357] and G1[360] and B3LYP6-311++G(3df,2pd)[93] levels of theory, respectively). G2(MP2) calculations also predict that the absolute $\Delta_f H(FCOH^{1+\bullet}) = 760$ kJ/mol,[28] in agreement with the MP2/6-31G**//MP2/6-31G**[357] relative energy and $\Delta_f H(HC(O)F^{1+\bullet}) = 817$ kJ/mol. FCOH$^{1+\bullet}$ is separated from its isomer by a large barrier to interconversion of about 200 kJ/mol.[357,360] In addition, calculations at the MP2/6-31G**//MP2/6-31G** level of theory[357] predict that the ion–neutral complex [HF$\cdots$CO]$^{+\bullet}$ is stable, lying 77 kJ/mol higher in energy than FCOH$^{1+\bullet}$ and separated from it by an isomerization barrier of 261 kJ/mol.

3.5.94   [C,H,O,Cl]$^{+\bullet}$ – m/z 64

Isomers

Ionized formyl chloride (HC(O)Cl$^{1+\bullet}$) and its ionized carbene isomer ClCOH$^{1+\bullet}$ have been investigated solely by theory.

Thermochemistry

$\Delta_f H(HC(O)Cl^{1+\bullet}) = 918$ kJ/mol is obtained by combining the calculated heat of formation (G2) and ionization energy of the molecule ($\Delta_f H(HC(O)Cl = -192.7 \pm 6$ kJ/mol[358] and 11.51 eV,[12] the latter value in agreement with other calculations[93]). Calculations at the B3LYP6-311++G(3df,2pd) level suggest that the carbene ion is lower in energy than its conventional isomer by 28 ± 2 kJ/mol.[93]

3.5.95   [C,H,O$_2$,F]$^{+\bullet}$ – m/z 64[361]

Ionized fluoroformic acid (FC(O)OH$^{1+\bullet}$) can be generated by the collision-induced dissociative deprotonation of the protonated acid, FC(OH)$_2$$^{1+}$. Its CID mass spectrum is dominated by the structure-characteristic fragment peaks at

m/z 47 (FCO$^{]+}$) and 45 (C(O)OH$^{]+}$), in addition to the minor peaks at m/z 31 (FC$^{]+}$) and 17 ($^+$OH). Note that the elusive fluoroformic acid molecule has been generated by NRMS of the ion. The heat of formation of the molecule is calculated to be about -615 kJ/mol at a variety of levels of theory.[243,362]

3.5.96   [C,H$_2$,O$_2$,F]$^+$ – m/z 65[361]

Dissociative ionization of ethyl fluoroformate (FC(O)OCH$_2$CH$_3$) yields an m/z 65 ion (loss of C$_2$H$_3$$^{]•}$) to which the connectivity of protonated fluoroformic acid, FC(OH)$_2$$^{]+}$, has been assigned. The CID mass spectrum of this ion exhibits structure-characteristic peaks at m/z 48 (FCOH$^{]+•}$), 31 (FC$^{]+}$) and 29 (COH$^{]+}$), in addition to the doubly charged ions at m/z 32.5 (FC(OH)$_2$$^{]2+}$) and 23 (C(OH)$_2$$^{]2+}$, loss of F$^•$), supporting this assignment.

3.5.97   [C,H$_3$,P,S$_2$]$^{+•}$ – m/z 110[363]

Isomers

Two isomers have been investigated: the (methylthio)thioxophosphane radical cation (CH$_3$SP$=$S$^{]+•}$) and the dithioxophosphorane radical cation (CH$_3$P(S)$_2$$^{]+•}$).

Identification

CH$_3$SP$=$S$^{]+•}$ and CH$_3$P(S)$_2$$^{]+•}$ can be generated by the following dissociative ionizations:

$$\text{(structure)} \xrightarrow{-H_3CSPS_2} \text{(structure)} \xrightarrow{-S} H_3CSPS^{]+•} \quad (1)$$

$$\text{(structure)} \xrightarrow{-CH_3PS_2} \text{(structure)} \quad (2)$$

The CID mass spectra of the ions are closely similar. Structure-indicative fragment peaks at m/z 63 (base peak, PS$^{]+}$) and 47 (CH$_3$S$^+$) in the CID mass spectrum of the ion generated in reaction (1) suggest the production of an ion of CH$_3$SP$=$S$^{]+•}$ connectivity. The base peak (m/z 95, PS$_2$$^{]+}$) in the mass spectrum of the ion generated in reaction (2) suggests that a second isomer, an ion of connectivity CH$_3$P(S)$_2$$^{]+•}$ is produced.

Partial CID Mass Spectra of the $[C,H_3,P,S_2]^{+\bullet}$ Isomers

	m/z			
Ions	45	47	63	95
$CH_3SP=S$	11	31	100	11
$CH_3P(=S)_2$	13	30	85	100

However, in order to account for the observed m/z 47 (CH_3S^+) in the CID mass spectrum of $CH_3P(S)_2^{+\bullet}$, a 1,2-methyl shift from P to S must take place below the dissociation limit. That both CID mass spectra are so similar strongly suggest that indeed these ions can interconvert before dissociation.

Thermochemistry

At a modest level of theory (MNDO), the heats of formation of the ions have been estimated to be $\Delta_f H(CH_3SP=S^{+\bullet}) = 916$ kJ/mol and $\Delta_f H(CH_3P(=S)_2^{+\bullet}) = 1013$ kJ/mol. No information is available concerning the energy of the barrier for isomerization.

3.5.98 $[C,H_5,N,O,P]^+ - m/z\ 78$[318]

Isomers

Two ions have been investigated by experiment: $CH_3OPNH_2^{+}$ and the methyl-shifted isomer $CH_3P(O)NH_2^{+}$. The isomeric ions $CH_2OP(H)NH_2^{+}$, $CH_3O(H)PNH_2^{+}$ and $CH_3OP(H)NH^{+}$ are predicted by theory to be stable.

Identification

$CH_3OP(O)NH_2^{+\bullet}$ and $CH_3OP(OH)NH_2^{+\bullet}$, two fragment ions of methamidophos $((CH_3O)(CH_3S)P(O)NH_2$, a pesticide extensively used for crop production and fruit tree treatment), have been identified as precursors for m/z 78 ions of elemental composition $[C,H_5,N,O,P]^+$. On the basis of the connectivity in these precursors, the m/z 78 ion generated has been assigned the $CH_3OP(NH_2)^{+}$ structure (loss of atomic oxygen and loss of $^\bullet OH$ from the above precursors, respectively). The isomeric ion, $CH_3P(O)(NH_2)^{+}$, is generated from the dissociative ionization of $CH_3P(O)(NH_2)Cl$ (loss of $Cl^\bullet$). These ions have similar but distinguishable CID mass spectra.

Partial CID Mass Spectra of the $[C,H_5,N,O,P]^+$ Isomers

	m/z				
Ions	46	47	48	62	63
CH_3OPNH_2	100	70	24	9	17
$CH_3P(O)NH_2$	58	100	13[a]	21	58

[a]Unresolved peak.

3.5.99 THERMOCHEMISTRY

Calculated Heat of Formation for the [C,H$_5$,N,O,P]$^+$ Isomers

Ions	Calculated $\Delta_f H$ (CBS-QB3) (kJ/mol)
CH$_3$P(O)NH$_2$	454
CH$_3$OPNH$_2$	466
CH$_2$OP(H)NH$_2$	577
CH$_3$O(H)PNH$_2$	598
CH$_3$OP(H)NH	619

3.5.100 [C,H$_5$,N,O$_2$,P]$^+$ – m/z 94[318]

The base peak in the normal mass spectrum of methamidophos (CH$_3$S(CH$_3$O)-P(O)NH$_2$) corresponds to an isobarically pure ion of elemental composition [C,H$_5$,N,O$_2$,P] (m/z 94). This ion is generated by a direct bond cleavage in the molecular ion (loss of CH$_3$S$^\bullet$) and based on the connectivity of the precursor was assigned the structure CH$_3$OP(O)NH$_2$$^{1+}$. It has been calculated to be lower in energy than its isomers: $\Delta_f H$(CH$_3$OP(O)NH$_2$$^{1+}$) = 249 kJ/mol, $\Delta_f H$(CH$_2$OP(O)NH$_3$$^{1+}$) = 308 kJ/mol and $\Delta_f H$(CH$_2$OP(OH)NH$_2$$^{1+}$) = 279 kJ/mol (CBS-QB3).

The insecticide acephate (CH$_3$S(CH$_3$O)P(O)N(H)C(O)CH$_3$) also shows in its normal mass spectrum (consecutive losses of CH$_3$S$^\bullet$ and CH$_2$CO) an m/z 94 fragment ion of elemental composition [C,H$_5$,N,O$_2$,P]. The CID mass spectrum of this ion and the ion produced from methamidophos were closely similar, suggesting that the ions are the same. The mass spectral data have not been published.

3.5.101 [C,H$_6$,N,O$_2$,P]$^{+\bullet}$ – m/z 95[318]

Ionized methamidophos (CH$_3$S(CH$_3$O)P(O)NH$_2$$^{1+\bullet}$) also shows a prominent isobarically pure fragment peak at m/z 95 of elemental composition [C,H$_6$,N,O$_2$,P]. This ion most likely possesses the CH$_3$OP(OH)NH$_2$$^{1+\bullet}$ connectivity and is proposed to be generated by loss of CH$_2$S following a 1,4-H shift in the molecular ion (leading to the distonic CH$_3$OP(OH)(NH$_2$)SCH$_2$$^{1+\bullet}$). This ion has been found to be stable and lower in energy than its isomer CH$_3$OP(O)(H)NH$_2$$^{1+\bullet}$ (CBS-QB3): $\Delta_f H$(CH$_3$OP(OH)NH$_2$$^{1+\bullet}$) = 253 kJ/mol and $\Delta_f H$(CH$_3$OP(O)(H)NH$_2$$^{1+\bullet}$) = 379 kJ/mol. Ionized acephate (CH$_3$S(CH$_3$O)P(O)N(H)C(O)CH$_3$) also exhibits an m/z 95 ion but in this case the fragment peak is not isobarically pure, consisting of a 5:1 mixture of [C,H$_6$,N,O$_2$,P]$^{+\bullet}$ and [C,H$_4$,O$_3$,P]$^{+\bullet}$. No mass spectral data on this ion have been published.

3.5.102   [C,N,S,Cl] – m/z 93[314]

Dissociative ionization of 3,4-dichlorothiadiazole (cy–NC(Cl)C(Cl)NS–, loss of ClCN) produces an intense m/z 93 fragment peak. On the basis of the connectivity in the molecule, as well as the CID mass spectrum of this ion (m/z 32 (S^+), 46 (NS^{1+}), 47 (CCl^{1+}), 58 (base peak, loss of $Cl^•$) and 61 (loss of S)), the assigned structure was $ClCNS^{1+•}$. This ion is also produced from the monochloro analog and by sulfuration of cyanogen chloride (ClCN, using CS_2 under CI conditions). It is computed to lie 417.2 kJ/mol (G2(MP2,SVP)) below its dissociation products $CCl^{1•} + NS^{1+}$.

3.5.103   [C,H₃,O,S,P]⁺• – m/z 94[364]

Isomers

Two isomers, $CH_3SP = O^{1+•}$ and $CH_3OP = S^{1+•}$, have been investigated by theory. Only the former has however been generated and studied experimentally.

Identification

The (methylthio)oxophosphane radical cation ($CH_3SP = O^{1+•}$) has been generated by the dissociative ionization of (dimethoxy)thiophosphoryl chloride (($CH_3O)_2P(= S)Cl$, by successive losses of $Cl^•$ and $CH_3O^•$). The CID mass spectrum of the ion consists of abundant structure-indicative fragment ions at m/z 79 ($OPS^{1+•}$), 47 (CH_3S^+), 46 ($SCH_2^{1+•}$) and 45 (SCH^{1+}). The absence of peaks at m/z 31 ([C,H₃,O]⁺), 30 ($H_2CO^{1+•}$), 29 (HCO^{1+}) and 28 ($CO^{1+•}$) rules out the presence, in the ion beam, of any substantial amount of the isomeric ion $CH_3OP = S^{1+•}$. However, the latter ion is most likely cogenerated in small amounts, as suggested by the presence of significant peaks at m/z 61 ($POCH_2^{1+}$), 63 (PS^{1+}) and 64 (PSH^{1+}). It is noteworthy that the NRMS shows a significant recovery signal indicating that the neutral counterpart of $CH_3SP = O^{1+•}$ is stable.

Thermochemistry

Computations give $\Delta_f H(CH_3SP = O^{1+•}) = 601$ kJ/mol and $\Delta_f H(CH_3OP = S^{1+•}) = 687$ kJ/mol (MOPAC with PM3 parameters).

REFERENCES

1. Bersuker, I.B. *Chem. Rev.* **2001**, *101*, 1067.
2. Ignatyev, I.S., Schaeffer III, H.F., Schleyer, P.V.R. *Chem. Phys. Lett.* **2001**, *337*, 158.
3. Smith, D., Adams, N.G. *Astrophys. J.* **1978**, *220*, L87.
4. Anicich, V.G., Huntress Jr., W.T. *Astrophys. J. Suppl. Ser.* **1986**, *62*, 553.

5. Knight Jr., L.B., Steadman, J., Feller, D., Davidson, E.R. *J. Am. Chem. Soc.* **1984**, *106*, 3700.
6. Signorell, R., Merkt, F. *J. Chem. Phys.* **1999**, *110*, 2309.
7. Wetmore, S.D., Boyd, R.J., Eriksson, L.A., Laaksonen, A. *J. Chem. Phys.* **1999**, *110*, 12059.
8. Miller, S.L., Urey, H.C. *Science* **1959**, *130*, 245.
9. Tal'roze, V.L., Lyubimova, A.K. *Dokl. Akad. Nauk SSSR* **1952**, *86*, 909.
10. Smith, D., Adams, N.G., Alge, E. *J. Chem. Phys.* **1982**, *77*, 1261.
11. Olah, G.A., Rasul, G. *Acc. Chem. Res.* **1997**, *30*, 245.
12. *NIST Chemistry WebBook, NIST Standard Reference Data Base Number 69*, National Institute of Standards and Technology: Gaithersburg, MD, **2005**.
13. Janoschek, R., Rossi, M.J. *Int. J. Chem. Kinet.* **2002**, *34*, 550.
14. Goldsmith, P.F., Langer, W.D., Ellder, J., Irvine, W., Kollberg, E. *Astrophys. J.* **1981**, *249*, 524.
15. Burgers, P.C., Holmes, J.L., Mommers, A.A., Szulejko, J.E., Terlouw, J.K. *Org. Mass Spectrom.* **1984**, *19*, 442.
16. Bieri, G., Jonsson, B.-O. *Chem. Phys. Lett.* **1978**, *56*, 446.
17. McLafferty, F.W., McGilvery, D.C. *J. Am. Chem. Soc.* **1980**, *102*, 6189.
18. Petrie, S., Freeman, C.G., Meot-Ner, M., McEwan, M.J., Ferguson, E.E. *J. Am. Chem. Soc.* **1990**, *112*, 7121.
19. Burgers, P.C., Holmes, J.L., Mommers, A.A., Terlouw, J.K. *Chem. Phys. Lett.* **1983**, *102*, 1.
20. Hansel, A., Scheiring, C., Glantschnig, M., Lindinger, W., Ferguson, E.E. *J. Chem. Phys.* **1998**, *109*, 1748.
21. Peterson, K.A., Mayrhofer, R.C., Woods, R.C. *J. Chem. Phys.* **1990**, *93*, 4946.
22. Langer, W.D., Graedel, T.E. *Astrophys. J. Suppl. Ser.* **1989**, *69*, 241.
23. Burgers, P.C., Holmes, J.L., Terlouw, J.K. *J. Am. Chem. Soc.* **1984**, *106*, 2762.
24. Burgers, P.C., Terlouw, J.K., Weiske, T., Schwarz, H. *Chem. Phys. Lett.* **1986**, *132*, 69.
25. DeFrees, D.J., Binkley, J.S., Frisch, M.J., McLean, A.D. *J. Chem. Phys.* **1986**, *85*, 5194.
26. Frisch, M.J., Raghavachari, K., Pople, J.A., Bouma, W.J., Radom, L. *Chem. Phys.* **1983**, *75*, 323.
27. Polce, M.J., Kim, Y., Wesdemiotis, C. *Int. J. Mass Spectrom. Ion Processes* **1997**, *167/168*, 309.
28. Flammang, R., Nguyen, M.T., Bouchoux, G., Gerbaux, P. *Int. J. Mass Spectrom.* **2000**, *202*, A8.
29. Nguyen, M.T., Rademakers, J., Martin, J.M.L. *Chem. Phys. Lett.* **1994**, *221*, 149.
30. Henriksen, J., Hammerum, S. *Int. J. Mass Spectrom.* **1998**, *179/180*, 301.
31. Holmes, J.L., Lossing, F.P., Mayer, P.M. *Chem. Phys. Lett.* **1992**, *198*, 211.
32. Williams, D.H., Hvistendahl, G. *J. Am. Chem. Soc.* **1974**, *96*, 6753.
33. DeOliveira, G., Martin, J.M.L., Silwal, I.K.C., Liebman, J.F. *J. Comp. Chem.* **2001**, *22*, 1297.
34. Sana, M., Leroy, G., Hilali, M., Nguyen, M.T., Vanquickenborne, L.G. *Chem. Phys. Lett.* **1992**, *190*, 551.
35. Hammerum, S., Solling, T.I. *J. Am. Chem. Soc.* **1999**, *121*, 6002.
36. Lossing, F.P., Lam, Y.-T., Maccoll, A. *Can. J. Chem.* **1981**, *59*, 2228.

37. Janoschek, R., Rossi, M.J. *Thermochemical Properties from G3MP2B3 Calculations*, Wiley InterScience, **2004**.

38. Holmes, J.L., Lossing, F.P., Terlouw, J.K., Burgers, P.C. *Can. J. Chem.* **1983**, *61*, 2305.

39. Terlouw, J.K., Heerma, W., Dijkstra, G., Holmes, J.L., Burgers, P.C. *Int. J. Mass Spectrom. Ion Phys.* **1983**, *47*, 147.

40. Lias, S.G., Bartmess, J.E., Liebmann, J.F., Holmes, J.L., Levin, R.D., Mallard, W.G. *J. Phys. Chem. Ref. Data* **1988**, *17*.

41. Mayer, P.M., Glukhovtsev, M.N., Gauld, J.W., Radom, L. *J. Am. Chem. Soc.* **1997**, *119*, 12889.

42. Bouma, W.J., Dawes, J.M., Radom, L. *Org. Mass Spectrom.* **1983**, *18*, 12.

43. Blanksby, S.J., Dua, S., Bowie, J.H., Schroder, D., Schwarz, H. *J. Phys. Chem.* **2000**, *104*, 11248.

44. Luo, Y.-R. *Handbook of bond dissociation energies in organic compounds*, Boca Raton: CRC Press, **2002**.

45. Bise, R.T., Hoops, A.A., Choi, H., Neumark, D.M. *J. Chem. Phys.* **2000**, *113*, 4179.

46. Armstrong, J., Degoricija, L., Hildebrand, A., Koehne, J., Fleming, P.E. *Chem. Phys. Lett.* **2000**, *332*, 591.

47. Fleming, P.E. *Chem. Phys. Lett.* **2000**, *321*, 129.

48. Lu, W., Tosi, P., Bassi, D. *J. Chem. Phys.* **2000**, *113*, 4132.

49. Clifford, E.P., Wenthold, P.G., Lineberger, W.C., Petersson, G.A., Broadus, K.M., Kass, S.R., Kato, S., DePuy, C.H., Bierbaum, V.M., Ellison, G.B. *J. Chem. Phys. A* **1998**, *102*, 7100.

50. Krogh-Jespersen, K. *Tetrahedron Lett.* **1980**, *21*, 4553.

51. Paulett, G.S., Ettinger, R. *J. Chem. Phys.* **1963**, *39*, 825.

52. Gordon, M.S., Kass, S.R. *J. Phys. Chem.* **1995**, *99*, 6548.

53. Goldberg, N., Fiedler, A., Schwarz, H. *Helv. Chim. Acta* **1994**, *77*, 2354.

54. Cacace, F., De Petris, G., Grandinetti, F., Occhiucci, G. *J. Phys. Chem.* **1993**, *97*, 4239.

55. Bastide, J., Maier, J.P. *Chem. Phys.* **1976**, *12*, 177.

56. Wong, M.W., Wentrup, C. *J. Am. Chem. Soc.* **1993**, *115*, 7743.

57. Kawauchi, S., Tachibana, A., Mori, M., Shibusa, Y., Yamabe, T. *J. Mol. Struct. (THEOCHEM)* **1994**, *310*, 255.

58. Stafast, H., Bock, H. *Chem. Ber.* **1974**, *107*, 1882.

59. Cambi, R., Von Niessen, W., Schirmer, *J. Chem. Phys.* **1984**, *86*, 389.

60. Guimon, C., Khayar, S., Gracian, F., Begtrup, M., Pfister-Guillouzo, G. *Chem. Phys.* **1989**, *138*, 157.

61. Foster, M.S., Williamson, A.D., Beauchamp, J.L. *Int. J. Mass Spectrom. Ion Phys.* **1974**, *15*, 429.

62. Rossini, F.D., Montgomery, R.L. *J. Chem. Thermodyn.* **1978**, *10*, 465.

63. Hiroaka, K., Takao, K., Nakagawa, F., Iino, T., Ishida, M., Fujita, K., Hiizumi, K., Yamabe, S. *Int. J. Mass Spectrom.* **2003**, *227*, 391.

64. McGibbon, G.A., Kingsmill, C.A., Terlouw, J.K. *Chem. Phys. Lett.* **1994**, *222*, 129.

65. Cheng, M.-J., Hu, C.-H. *Chem. Phys. Lett.* **2000**, *322*, 83.

66. Burgers, P.C., Drewello, T., Schwarz, H., Terlouw, J.K. *Int. J. Mass Spectrom. Ion Processes* **1989**, *95*, 157.
67. Van Garderen, H.F., Ruttink, P.J.A., Burgers, P.C., McGibbon, G.A., Terlouw, J.K. *Int. J. Mass Spectrom. Ion Processes* **1992**, *121*, 159.
68. Nguyen, M.T. *Int. J. Mass Spectrom. Ion Processes* **1994**, *136*, 45.
69. Cacace, F., Grandinetti, F., Pepi, F. *J. Chem. Soc. Chem. Commun.* **1994**, 2173.
70. Pyykko, P., Runeberg, N. *J. Mol. Struct. (THEOCHEM)* **1991**, *234*, 269.
71. Woods, R.C., Dixon, T.A., Saykally, R.J., Szanto, P.G. *Phys. Rev. Lett.* **1975**, *35*, 1269.
72. Guedeman, C.S., Woods, R.C. *Phys. Rev. Lett.* **1982**, *48*, 1344.
73. Bohme, D.K., Goodings, J.M., Ng, C.-W. *Int. J. Mass Spectrom. Ion Phys.* **1977**, *24*, 25.
74. Burgers, P.C., Holmes, J.L. *Chem. Phys. Lett.* **1983**, *97*, 236.
75. Illies, A.J., Jarrold, M.F., Bowers, M.T. *J. Am. Chem. Soc.* **1983**, *105*, 2562.
76. Freeman, C.G., Knight, J.S., Love, J.G., McEwan, M.J. *Int. J. Mass Spectrom. Ion Processes* **1987**, *80*, 225.
77. Burgers, P.C., Holmes, J.L., Mommers, A.A. *J. Am. Chem. Soc.* **1985**, *107*, 1099.
78. Semo, N., Osheroff, M.A., Koski, W.S. *J. Phys. Chem.* **1983**, *87*, 2302.
79. Wagner-Redeker, W., Kemper, P.R., Jarrold, M.F., Bowers, M.T. *J. Chem. Phys.* **1985**, *83*, 1121.
80. Bohme, D.K. *Int. J. Mass Spectrom. Ion Processes* **1992**, *115*, 95.
81. Cunje, A., Rodriquez, C.F., Bohme, D.K., Hopkinson, A.C. *J. Phys. Chem. A* **1998**, *102*, 478.
82. Chalk, A.J., Radom, L. *J. Am. Chem. Soc.* **1997**, *119*, 7573.
83. Ma, N.L., Smith, B.J., Radom, L. *Chem. Phys. Lett.* **1992**, *197*, 573.
84. Matthews, C.S., Warneck, P. *J. Chem. Phys.* **1969**, *51*, 854.
85. Guyon, P.M., Chupka, W.A., Berkowitz, J. *J. Chem. Phys.* **1976**, *64*, 1419.
86. Traeger, J.C. *Int. J. Mass Spectrom. Ion Processes* **1985**, *66*, 271.
87. Bouma, W.J., MacLeod, J.K., Radom, L. *Int. J. Mass Spectrom. Ion Phys.* **1980**, *33*, 87.
88. Wesdemiotis, C., McLafferty, F.P. *Tetrahedron Lett.* **1981**, *22*, 3479.
89. Burgers, P.C., Mommers, A.A., Holmes, J.L. *J. Am. Chem. Soc.* **1983**, *105*, 5976.
90. Stahl, D., Maquin, F. *Chem. Phys. Lett.* **1984**, *106*, 531.
91. Bouma, W.J., Radom, L. *J. Am. Chem. Soc.* **1983**, *105*, 5484.
92. Bouma, W.J., Burgers, P.C., Holmes, J.L., Radom, L. *J. Am. Chem. Soc.* **1986**, *108*, 1767.
93. Ventura, O.N., Kieninger, M., Coitino, E.L. *J. Comp. Chem.* **1996**, *17*, 1309.
94. Ma, N.L., Smith, B.J., Radom, L. *Chem. Phys. Lett.* **1992**, *193*, 386.
95. Ma, N.L., Smith, B.J., Collins, M.A., Pople, J.A., Radom, L. *J. Phys. Chem.* **1989**, *93*, 7759.
96. Hiraoka, K., Kebarle, P. *J. Chem. Phys.* **1975**, *63*, 1688.
97. Dill, J.D., Fisher, C.L., McLafferty, F.P. *J. Am. Chem. Soc.* **1979**, *101*, 6531.
98. Bursey, M.M., Hass, J.R., Harvan, D.J., Parker, C.E. *J. Am. Chem. Soc.* **1979**, *101*, 5485.
99. Burgers, P.C., Holmes, J.L. *Org. Mass Spectrom.* **1984**, *19*, 452.
100. Holmes, J.L., Terlouw, J.K. *Org. Mass Spectrom.* **1986**, *21*, 776.

101. Aschi, M., Harvey, J.N., Schalley, C.A., Schroder, D., Schwarz, H. *J. Chem. Soc. Chem. Commun.* **1998**, 531.
102. Zappey, H., Ingemann, S., Nibbering, N.M.M. *J. Am. Soc. Mass Spectrom.* **1992**, *3*, 515.
103. Traeger, J.C., Holmes, J.L. *J. Am. Chem. Soc.* **1993**, *97*, 3453.
104. Sablier, M., Fujii, T. *Chem. Rev.* **2002**, *102*, 2855.
105. Johnson, R.D., Hudgens, J.W. *J. Chem. Phys.* **1996**, *100*, 19874.
106. Lossing, F.P. *J. Am. Chem. Soc.* **1977**, *99*, 7526.
107. Refaey, K.M.A., Chupka, W.A. *J. Chem. Phys.* **1968**, *48*, 5205.
108. Traeger, J.C., Holmes, J.L. *J. Phys. Chem.* **1993**, *97*, 3453.
109. Dewar, M.J.S., Rzepa, H.S. *J. Am. Chem. Soc.* **1977**, *99*, 7432.
110. Schleyer, P.V.R., Jemmis, E.D., Pople, J.A. *J. Chem. Soc. Chem. Commun.* **1978**, 190.
111. Bouma, W.J., Nobes, R.H., Radom, L. *Org. Mass Spectrom.* **1982**, *17*, 315.
112. Ferguson, E.E., Roncin, J., Bonazzola, L. *Int. J. Mass Spectrom. Ion Processes* **1987**, *79*, 215.
113. Ruscic, B., Berkowitz, J. *J. Chem. Phys.* **1991**, *95*, 4033.
114. Hiraoka, K., Kebarle, P. *J. Chem. Phys.* **1975**, *63*, 1688.
115. Bouma, W.J., MacLeod, J.K., Radom, L. *J. Am. Chem. Soc.* **1982**, *104*, 2930.
116. Holmes, J.L., Lossing, F.P., Terlouw, J.K., Burgers, P.C. *J. Am. Chem. Soc.* **1982**, *104*, 2931.
117. Gauld, J.W., Audier, H.E., Fossey, J., Radom, L. *J. Am. Chem. Soc.* **1996**, *118*, 6299.
118. Gauld, J.W., Radom, L. *J. Am. Chem. Soc.* **1997**, *119*, 9831.
119. Mourgues, P., Audier, H.E., Leblanc, D., Hammerum, S. *Org. Mass Spectrom.* **1993**, *28*, 1098.
120. Audier, H.E., Leblanc, D., Mourgues, P., McMahon, T.B., Hammerum, S. *J. Chem. Soc. Chem. Commun.* **1994**, 2329.
121. Fridgen, T.D., Parnis, J.M. *Int. J. Mass Spectrom.* **1999**, *190/191*, 181.
122. Tu, Y.-P., Holmes, J.L. *J. Am. Chem. Soc.* **2000**, *122*, 3695.
123. Huntress Jr, W.T. *Int. J. Mass Spectrom. Ion Phys.* **1973**, *12*, 1.
124. Rabrenovic, M., Beynon, J.H. *Int. J. Mass Spectrom. Ion Processes* **1984**, *56*, 85.
125. McAdoo, D.J., Zhao, G., Ahmed, M.S., Hudson, C.E., Giam, C.S. *Org. Mass Spectrom.* **1994**, *29*, 428.
126. Tajima, S., von der Greef, J., Nibbering, N.M.M. *Org. Mass Spectrom.* **1978**, *13*, 551.
127. Cao, J., Sun, W., Holmes, J.L. *Int. J. Mass Spectrom.* **2002**, *217*, 179.
128. Bennett, S.L., Field, F.H. *J. Am. Chem. Soc.* **1972**, *94*, 5188.
129. Thaddeus, P., Guelin, M., Linke, R.A. *Astrophys. J.* **1981**, *246*, L41.
130. Frost, M.J., Sharkey, P., Smith, I.W.M. *Faraday Discuss. Chem. Commun.* **1991**, *91*, 305.
131. Bursey, M.M., Harvan, D.J., Parker, C.E., Pedersen, L.G., Hass, J.R. *J. Am. Chem. Soc.* **1979**, *101*, 5489.
132. Burgers, P.C., Holmes, J.L., Mommers, A.A., Szulejko, J.E. *J. Am. Chem. Soc.* **1984**, *106*, 521.
133. Ruttink, P.J.A., Burgers, P.C., Terlouw, J.K. *Int. J. Mass Spectrom.* **1999**, *185/186/187*, 291.

134. Uggerud, E., Koch, W., Schwarz, H. *Int. J. Mass Spectrom. Ion Processes* **1986**, *73*, 187.
135. McCarthy, M.C., Thaddeus, P. *Chem. Soc. Rev.* **2001**, *30*, 177.
136. Burgers, P.C., McGibbon, G., Terlouw, J.K. *Chem. Phys. Lett.* **1994**, *224*, 539.
137. Cheung, Y.-S., Li, W.-K. *Chem. Phys. Lett.* **1994**, *223*, 383.
138. Cheung, Y.-S., Li, W.-K. *Chem. Phys. Lett.* **1995**, *333*, 135.
139. Holmes, J.L., Mommers, A.A., De Koster, C., Heerma, W., Terlouw, J.K. *Chem. Phys. Lett.* **1985**, *115*, 437.
140. Schalley, C.A., Harvey, J.N., Schroder, D., Schwarz, H. *J. Phys. Chem. A* **1998**, *102*, 1021.
141. Wang, X., Holmes, J.L. *J. Phys. Chem. A* **2006**, *110*, 8409.
142. Schalley, C.A., Schroder, D., Schwarz, H. *Int. J. Mass Spectrom. Ion Processes* **1996**, *153*, 173.
143. Van Doren, J.M., Barlow, S.E., DePuy, C.H., Bierbaum, V.M., Dotan, I., Ferguson, E.E. *J. Phys. Chem.* **1986**, *90*, 2772.
144. Kirchner, N.J., Van Doren, J.M., Bowers, M.T. *Int. J. Mass Spectrom. Ion Processes* **1989**, *92*, 37.
145. Glosik, J., Jordan, A., Skalsky, V., Lindinger, W. *Int. J. Mass Spectrom. Ion Processes* **1993**, *129*, 109.
146. Coitino, E.L., Lledos, A., Serra, R., Bertran, J., Ventura, O.N. *J. Am. Chem. Soc.* **1993**, *115*, 9121.
147. Ohkubo, K., Fujita, T., Sato, H. *J. Mol. Struct.* **1977**, *36*, 101.
148. Blanksby, S.J., Ramond, T.M., Davico, G.E., Nimlos, M.R., Kato, S., Bierbaum, V.M., Lineberger, W.C., Ellison, G.B., Okumura, M. *J. Am. Chem. Soc.* **2001**, *123*, 9585.
149. Schalley, C.A., Dieterle, M., Schroder, D., Schwarz, H., Uggerud, E. *Int. J. Mass Spectrom. Ion Processes* **1997**, *163*, 101.
150. Aschi, M., Attina, M., Cacace, F., Cipollini, R., Pepi, F. *Inorg. Chim. Acta* **1998**, *275/276*, 192.
151. Hunter, E.P., Lias, S.G. *J. Phys. Chem. Ref. Data* **1998**, *27*, 413.
152. Cao, J., Tu, Y.-P., Sun, W., Holmes, J.L. *Int. J. Mass Spectrom.* **2003**, *222*, 41.
153. Terlouw, J.K., Lebrilla, C.B., Schwarz, H. *Angew. Chem. Int. Ed. Engl.* **1987**, *26*, 354.
154. Terlouw, J.K., Sulzle, D., Schwarz, H. *Angew. Chem. Int. Ed. Engl.* **1990**, *29*, 404.
155. Gerbaux, P., Turecek, F. *J. Phys. Chem. A* **2002**, *106*, 5938.
156. Hrusak, J., McGibbon, G.A., Schwarz, H., Terlouw, J.K. *Int. J. Mass Spectrom. Ion Processes* **1997**, *160*, 117.
157. Brown, P., Djerassi, C. *J. Am. Chem. Soc.* **1966**, *88*, 2469.
158. Egsgaard, H., Carlsen, L. *J. Chem. Soc. Faraday Trans. 1* **1989**, *85*, 3403.
159. Holmes, J.L., Lossing, F.P. *Can. J. Chem.* **1982**, *60*, 2365.
160. Larson, J.W., McMahon, T.B. *J. Am. Chem. Soc.* **1982**, *104*, 6255.
161. Guelin, M., Cernicharo, J., Paubert, G., Turner, B.E. *Astron. Astrophys.* **1990**, *230*, L9.
162. Fleming, P.E., Lee, E.P.F., Wright, T.G. *Chem. Phys. Lett.* **2000**, *332*, 199.
163. Esseffar, M., Luna, A., Mo, O., Yanez, M. *J. Phys. Chem.* **1994**, *98*, 8679.
164. Pascoli, G., Lavendy, H. *J. Phys. Chem. A* **1999**, *103*, 3518.

165. Gingerich, K.A. *Thermochim. Acta* **1971**, *2*, 233.
166. Weisstein, E.W., Serabyn, E. *Icarus* **1996**, *123*, 23.
167. King, M.A., Klapstein, D., Kroto, H.W., Maier, J.P., Nixon, J.F. *J. Mol. Struct.* **1982**, *80*, 23.
168. Biczysko, M., Tarroni, R. *Phys. Chem. Chem. Phys.* **2002**, *4*, 708.
169. Ishikama, H., Field, R.W., Farantos, S.C., Joyeux, M., Koput, J., Beck, C., Schinke, R. *Ann. Rev. Phys. Chem.* **1999**, *50*, 443.
170. Lohr, L.L., Schlegel, H.B., Morokuma, K. *J. Phys. Chem.* **1984**, *88*, 1981.
171. Esteves, P.M., Laali, K.K. *Organometallics* **2004**, *23*, 3701.
172. Adams, N.G., McIntosh, B.J., Smith, D. *Astron. Astrophys.* **1990**, *232*, 443.
173. Nguyen, M.T., Creve, S., Vanquickenborne, L.G. *J. Chem. Phys.* **1996**, *105*, 1922.
174. Chow, J.R., Beaudet, R.A., Goldwhite, H. *J. Phys. Chem.* **1989**, *93*, 421.
175. Benson, S.W. *Thermochemical Kinetics*, Wiley-Interscience: New York, **1976**.
176. Schweighofer, A., Chou, P.K., Thoen, K.K., Nanayakkara, V.K., Keck, H., Kuchen, W., Kenttamaa, H.I. *J. Am. Chem. Soc.* **1996**, *118*, 11893.
177. Weger, E., Levsen, K., Ruppert, I., Burgers, P.C., Terlouw, J.K. *Org. Mass Spectrom.* **1983**, *18*, 327.
178. Keck, H., Kuchen, W., Tommes, P., Terlouw, J.K., Wong, T. *Angew. Chem. Int. Ed. Engl.* **1992**, *31*, 86.
179. Gauld, J.W., Radom, L. *J. Phys. Chem.* **1994**, *98*, 777.
180. Yates, B.F., Bouma, W.J., Radom, L. *J. Am. Chem. Soc.* **1987**, *109*, 2250.
181. Henry, D.J., Parkinson, C.J., Radom, L. *J. Phys. Chem. A* **2002**, *106*, 7927.
182. *CRC Handbook of Chemistry and Physics*, 83 ed., CRC Press LLC: Boca Raton, **2003**.
183. Prinslow, D.A., Armentrout, P.B. *J. Chem. Phys.* **1991**, *94*, 3563.
184. Coppens, P., Drowart, J. *Chem. Phys. Lett.* **1995**, *243*, 108.
185. Guedeman, C.S., Haese, N.N., Piltch, N.D., Woods, R.C. *Astrophys. J.* **1981**, *246*, L47.
186. Ruscic, B., Berkowitz, J. *J. Chem. Phys.* **1993**, *98*, 2568.
187. Butler, J.J., Baer, T. *J. Am. Chem. Soc.* **1982**, *104*, 5016.
188. Pope, S.A., Hillier, I.H., Guest, M.F. *J. Am. Chem. Soc.* **1985**, *107*, 3789.
189. Wong, M.W., Nobes, R.H., Radom, L. *J. Mol. Struct. (THEOCHEM)* **1988**, *163*, 151.
190. Curtiss, L.A., Nobes, R.H., Pople, J.A., Radom, L. *J. Chem. Phys.* **1992**, *97*, 6766.
191. Sumathi, R., Peyerimhoff, S.D., Sengupta, D. *J. Phys. Chem. A* **1999**, *103*, 772.
192. Nobes, R.H., Radom, L. *Chem. Phys. Lett.* **1992**, *189*, 554.
193. Dill, J.D., McLafferty, F.W. *J. Am. Chem. Soc.* **1978**, *100*, 2907.
194. Dill, J.D., Mclafferty, F.W. *J. Am. Chem. Soc.* **1979**, *101*, 6526.
195. Roy, M., McMahon, T.B. *Org. Mass Spectrom.* **1982**, *17*, 392.
196. Harrison, A.G. *J. Am. Chem. Soc.* **1978**, *100*, 4911.
197. Aschi, M., Grandinetti, F. *J. Chem. Phys.* **1999**, *111*, 6759.
198. Pope, S.A., Hillier, I.H., Guest, M.F. *Chem. Phys. Lett.* **1984**, *104*, 191.
199. Nobes, R.H., Bouma, W.J., Radom, L. *J. Am. Chem. Soc.* **1984**, *106*, 2774.
200. Ruscic, B., Berkowitz, J. *J. Chem. Phys.* **1992**, *97*, 1818.
201. Berkowitz, J., Ellison, G.B., Gutman, D. *J. Phys. Chem.* **1994**, *98*, 2744.

202. Chou, P.K., Smith, R.L., Chyall, L.J., Kenttamaa, H.I. *J. Am. Chem. Soc.* **1995**, *117*, 4374.
203. Cheung, Y.-S., Hsu, C.-W., Ng, C.-W. *J. Chem. Phys.* **1998**, *109*, 1781.
204. Cheung, Y.-S., Hsu, C.-W., Huang, J.-C., Ng, C.Y., Li, W.-K., Chiu, S.-W. *Int. J. Mass Spectrom. Ion Processes* **1996**, *159*, 13.
205. Martinez-Nunez, E., Vazquez, S.A. *J. Mol. Struct. (THEOCHEM)* **2000**, *505*, 109.
206. Chiang, S.-Y., Ma, C.-I., Shr, D.J. *J. Chem. Phys.* **1999**, *110*, 9056.
207. Wong, M.W., Wentrup, C., Flammang, R. *J. Phys. Chem.* **1995**, *99*, 16849.
208. Nagesha, K., Bapat, B., Marathe, V.R., Krishnakumar, E. *Chem. Phys. Lett.* **1994**, *230*, 283.
209. Vivekananda, S., Srinivas, R., Manoharan, M., Jemmis, E.D. *J. Phys. Chem. A* **1999**, *103*, 5123.
210. Nguyen, M.T., Nguyen, T.L., Le, H.T. *J. Phys. Chem. A* **1999**, *103*, 5758.
211. Ma, Z.-X., Liao, C.L., Ng, C.-Y., Cheung, Y.-S., Li, W.-K., Baer, T. *J. Chem. Phys.* **1994**, *100*, 4870.
212. Cheung, Y.-S., Li, W.-K., Ng, C.-Y. *J. Mol. Struct. (THEOCHEM)* **1995**, *339*, 25.
213. Butler, J.J., Baer, T., Evans Jr, S.A. *J. Am. Chem. Soc.* **1983**, *105*, 3451.
214. Gerbaux, P., Van Haverbeke, Y., Flammang, R., Wong, M.W., Wentrup, C. *J. Chem. Phys. A* **1997**, *101*, 6970.
215. Sulzle, D., Egsgaard, H., Carlsen, L., Schwarz, H. *J. Am. Chem. Soc.* **1995**, *112*, 3750.
216. Ricca, A. *J. Phys. Chem. A* **1999**, *103*, 1876.
217. Marshall, P., Misra, A., Schwartz, M. *J. Chem. Phys.* **1999**, *110*, 2069.
218. Bauschlicher Jr., C.W., Ricca, A. *J. Phys. Chem. A* **2000**, *104*, 4581.
219. Hrusak, J., Herman, Z., Sandig, N., Koch, W. *Int. J. Mass Spectrom.* **2000**, *201*, 269.
220. Rodriquez, C.F., Bohme, D.K., Hopkinson, A.C. *J. Phys. Chem.* **1996**, *100*, 2942.
221. Chiang, S.-Y., Bahou, M., Sankaran, K., Lee, Y.-P., Lu, H.-F., Su, M.-D. *J. Chem. Phys.* **2003**, *118*, 62.
222. Dixon, D.A., de Jong, W.A., Peterson, K.A., Francisco, J.S. *J. Phys. Chem. A* **2002**, *106*, 4725.
223. Chiang, S.-Y., Fang, Y.-S., Sankaran, K., Lee, Y.-P. *J. Chem. Phys.* **2004**, *120*, 3270.
224. Poutsma, J.C., Paulino, J.A., Squires, R.R. *J. Phys. Chem. A* **1997**, *101*, 5327.
225. Schwartz, M., Marshall, P. *J. Phys. Chem. A* **1999**, *103*, 7900.
226. Sendt, K., Bacskay, G.B. *J. Chem. Phys.* **2000**, *112*, 2227.
227. Born, M., Ingemann, S., Nibbering, N.M.M. *J. Am. Chem. Soc.* **1994**, *116*, 7210.
228. Schwartz, M., Peebles, L.R., Berry, R.J., Marshall, P. *J. Chem. Phys.* **2003**, *118*, 557.
229. Moc, J. *J. Chem. Phys.* **1999**, *247*, 365.
230. Cheong, B.-S., Cho, H.-G. *J. Phys. Chem. A* **1997**, *101*, 7901.
231. Ma, Z.-X., Liao, C.-L., Ng, C.Y. *J. Chem. Phys.* **1993**, *99*, 6470.
232. Ma, N.L., Lau, K.-C., Chien, S.-H., Li, W.-K. *Chem. Phys. Lett.* **1999**, *311*, 275.
233. Torres, I., Martinez, R., Sanchez Rayo, M.N., Castano, F. *J. Phys. B: At. Mol. Opt. Phys.* **2000**, *33*, 3615.
234. Paddison, S.J., Tschuikow-Roux, E. *J. Phys. Chem. A* **1998**, *102*, 6191.

235. Buckley, T.J., Johnson III, R.D., Huie, R.E., Zhang, Z., Kuo, S.C., Klemm, R.B. *J. Phys. Chem.* **1995**, *99*, 4879.
236. Henrici, H., Lin, M.C., Bauer, S.H. *J. Chem. Phys.* **1970**, *52*, 5834.
237. Dyke, J.M., Jonathan, N., Morris, A., Winter, M.J. *J. Chem. Soc. Faraday Trans. 2* **1981**, *77*, 667.
238. Asher, R.L., Appelman, E.H., Ruscic, B. *J. Chem. Phys.* **1996**, *105*, 9781.
239. Nicovich, J.M., Kreutter, K.D., Wine, P.H. *J. Chem. Phys.* **1990**, *92*, 3539.
240. Chien, S.-H., Lau, K.-C., Li, W.-K., Ng, C.Y. *J. Phys. Chem. A* **2000**, *103*, 7918.
241. Sherwood, P., Seddon, E.A., Guest, M.F., Parkington, M.J., Ryan, T.A., Seddon, K.R. *J. Chem. Soc. Dalton Trans.* **1995**, 2359.
242. Arnold, D.W., Bradforth, S.E., Kim, E.H., Neumark, D.M. *J. Chem. Phys.* **1995**, *102*, 3493.
243. Dibble, T.S., Francisco, J.S. *J. Phys. Chem.* **1994**, *98*, 11694.
244. Langford, M.L., Harris, F.M. *Int. J. Mass Spectrom. Ion Processes* **1990**, *96*, 111.
245. Dixon, D.A., Feller, D. *J. Phys. Chem. A* **1998**, *102*, 8209.
246. Dixon, D.A., Peterson, K.A., Francisco, J.S. *J. Phys. Chem. A* **2000**, *104*, 3227.
247. Flowers, B.A., Francisco, J.S. *J. Chem. Phys.* **1999**, *111*, 3464.
248. Parkington, M.J., Ryan, T.A., Seddon, K.R. *J. Chem. Soc. Dalton Trans.* **1997**, 251.
249. Drewello, T., Weiske, T., Schwarz, H. *Angew. Chem. Int. Ed. Engl.* **1985**, *24*, 869.
250. Bews, J.R., Glidewell, C. *J. Mol. Struct.* **1981**, *71*, 287.
251. Hop, C.E.C.A., Holmes, J.L., Lossing, F.P., Terlouw, J.K. *Int. J. Mass Spectrom. Ion Processes* **1988**, *83*, 285.
252. Li, Q., Ran, Q., Chen, C., Yu, S., Ma, X., Sheng, L., Zhang, Y., Li, W.-K. *Int. J. Mass Spectrom. Ion Processes* **1996**, *153*, 29.
253. Sadilek, M., Turecek, F. *J. Phys. Chem.* **1996**, *100*, 224.
254. Apeloig, Y., Karni, M., Ciommer, B., Frenking, G., Schwarz, H. *Int. J. Mass Spectrom. Ion Processes* **1983/1984**, *55*, 319.
255. Petersen, A.C., Hammerum, S. *Int. J. Mass Spectrom.* **2001**, *210/211*, 403.
256. Fridgen, T.D., Zhang, X.K., Parnis, J.M., March, R.E. *J. Phys. Chem. A* **2000**, *104*, 3487.
257. Lewars, E. *J. Mol. Struct. (THEOCHEM)* **1998**, *425*, 207.
258. Shin, D.N., Jung, K.-H., Ha, T.-K. *J. Mol. Struct. (THEOCHEM)* **1998**, *422*, 229.
259. Blint, R.J., McMahon, T.B., Beauchamp, J.L. *J. Am. Chem. Soc.* **1974**, *96*, 1269.
260. Halim, H., Ciommer, B., Schwarz, H. *Angew. Chem. Int. Ed. Engl.* **1982**, *21*, 528.
261. Lias, S.G., Karpas, Z., Liebman, J.F. *J. Am. Chem. Soc.* **1985**, *107*, 6089.
262. Hop, C.E.C.A., Bordas-Nagy, J., Holmes, J.L. *Org. Mass Spectrom.* **1988**, *23*, 155.
263. Yates, B.F., Bouma, W.J., Radom, L. *J. Am. Chem. Soc.* **1986**, *108*, 6545.
264. Duflot, D., Robbe, J.-M., Flament, J.-P. *J. Chem. Phys.* **1995**, *103*, 10571.
265. Bogan, D.J., Hand, C.W. *J. Phys. Chem.* **1971**, *75*, 1532.
266. Ruscic, B., Berkowitz, J. *J. Chem. Phys.* **1994**, *100*, 4498.
267. Hop, C.E.C.A., van den Berg, K.J., Holmes, J.L., Terlouw, J.K. *J. Am. Chem. Soc.* **1989**, *111*, 72.
268. Luna, A., Mebel, A.M., Morokuma, K. *J. Chem. Phys.* **1996**, *105*, 3187.
269. Belson, D.J., Strachan, A.N. *Chem. Soc. Rev.* **1982**, *11*, 41.

270. Hrusak, J., Holthausen, M.C., Goldberg, N., Iraqi, M., Koch, W., Schwarz, H. *Israel J. Chem.* **1993**, *33*, 277.
271. Ijjaali, F., Alcami, M., Mo, O., Yanez, M. *Mol. Phys.* **2001**, *99*, 1129.
272. Hop, C.E.C.A., Holmes, J.L., Ruttink, P.J.A., Schaftenaar, G., Terlouw, J.K. *Chem. Phys. Lett.* **1989**, *156*, 251.
273. McGibbon, G.A., Terlouw, J.K., Burgers, P.C. *Rapid Commun. Mass Spectrom.* **1993**, *7*, 138.
274. Kiplinger, J.P., Maynard, A.T., Bursey, M.M. *Org. Mass Spectrom.* **1987**, *22*, 534.
275. Bouchoux, G., Espagne, A. *Chem. Phys. Lett.* **2001**, *348*, 329.
276. Hop, C.E.C.A., Chen, H., Ruttink, P.J.A., Holmes, J.L. *Org. Mass Spectrom.* **1991**, *26*, 679.
277. McGibbon, G.A., Burgers, P.C., Terlouw, J.K. *Int. J. Mass Spectrom. Ion Processes* **1994**, *136*, 191.
278. Polasek, M., Sadilek, M., Turecek, F. *Int. J. Mass Spectrom.* **2000**, *195/196*, 101.
279. Polasek, M., Turecek, F. *J. Am. Chem. Soc.* **2000**, *122*, 525.
280. Ruttink, P.J.A., Burgers, P.C., Terlouw, J.K. *Int. J. Mass Spectrom. Ion Processes* **1995**, *145*, 35.
281. Schaftenaar, G., Postma, R., Ruttink, P.J.A., Burgers, P.C., McGibbon, G.A., Terlouw, J.K. *Int. J. Mass Spectrom. Ion Processes* **1990**, *100*, 521.
282. Batt, L., Milne, R.T. *Int. J. Chem. Kinet.* **1973**, *5*, 1067.
283. Tortajada, J., Leon, E., Morizur, J.-P., Luna, A., Mo, O., Yanez, M. *J. Phys. Chem.* **1995**, *99*, 13890.
284. Lin, H.-Y., Douglas, D.P., Uggerud, E., Vulpius, T. *J. Am. Chem. Soc.* **1994**, *116*, 2996.
285. Nguyen, M.T., Raspoet, G., Vanquickenborne, L.G. *J. Chem. Soc. Perkin Trans.* **1995**, *2*, 1791.
286. Burgers, P.C., Lisfshitz, C., Ruttink, P.J.A., Schaftenaar, G., Terlouw, J.K. *Org. Mass Spectrom.* **1989**, *24*, 579.
287. O'Bannon, P.E., Sulzle, D., Dailey, W.P., Schwarz, H. *J. Am. Chem. Soc.* **1992**, *114*, 344.
288. Polasek, M., Turecek, F. *J. Phys. Chem. A* **2001**, *105*, 1371.
289. Nibbering, N.M.M., de Boer, T.J., Hofman, H.J. *Rec. Trav. Chim.* **1965**, *84*, 481.
290. Sirois, M., Holmes, J.L., Hop, C.E.C.A. *Org. Mass Spectrom.* **1990**, *25*, 167.
291. Van den Berg, K.J., Lebrilla, C.B., Terlouw, J.K., Schwarz, H. *Chimia* **1987**, *71*, 122.
292. Egsgaard, H., Carlsen, L., Florencio, H., Drewello, T., Schwarz, H. *Ber. Bunsengs Phys. Chem.* **1989**, *93*, 76.
293. Beijersbergen, J.H.M., van der Zande, W.J., Kistemaker, P.G., Los, J., Drewello, T., Nibbering, N.M.M. *J. Phys. Chem.* **1992**, *96*, 9288.
294. De Bruijn, D.P., Los, J. *Rev. Sci. Instrum.* **1982**, *53*, 1020.
295. Kornig, S., Beijersbergen, J.H.M., van der Zande, W.J., Los, J. *Int. J. Mass Spectrom. Ion Processes* **1989**, *93*, 49.
296. Baer, T., Hass, J.R. *J. Phys. Chem.* **1986**, *90*, 451.
297. Egsgaard, H., Carlsen, L. *J. Chem. Soc. Faraday Trans.* **1994**, *90*, 941.
298. DePetris, G. *Org. Mass Spectrom.* **1990**, *25*, 83.
299. DePetris, G., DiMarzio, A., Grandinetti, F. *J. Phys. Chem.* **1991**, *95*, 9782.

300. Wang, X., Holmes, J.L. **2004**, unpublished data.
301. Polasek, M., Turecek, F. *J. Am. Soc. Mass Spectrom.* **2000**, *11*, 380.
302. De Petris, G., Cacace, F., Cipollini, R., Cartoni, A., Rosi, M., Troiani, A. *Angew. Chem. Int. Ed. Engl.* **2005**, *44*, 462.
303. Poppinger, D., Radom, L. *J. Am. Chem. Soc.* **1978**, *100*, 3674.
304. Nguyen, M.T., Allaf, A.W., Flammang, R., Van Haverbeke, Y. *J. Mol. Struct. (THEOCHEM)* **1997**, *418*, 209.
305. Turner, B.E. *Astrophys. J. Suppl. Ser.* **1989**, *70*, 539.
306. Kambouris, P., Plisnier, M., Flammang, R., Terlouw, J.K., Wentrup, C. *Tetrahedron Lett.* **1991**, *32*, 1487.
307. Gerbaux, P., Van Haverbeke, Y., Flammang, J.-P., Wong, M.W., Wentrup, C. *J. Phys. Chem. A* **1997**, *101*, 6970.
308. Wierzejewska, M., Moc, J. *J. Phys. Chem. A* **2003**, *107*, 11209.
309. Allaf, A.W., Suffolk, R.J. *J. Chem. Res.* **1994**, 186.
310. Allaf, A.W., Johnston, R.L., Suffolk, R.J. *Chem. Phys. Lett.* **1995**, *233*, 33.
311. McGibbon, G., Kingsmill, C.A., Terlouw, J.K., Burgers, P.C. *Int. J. Mass Spectrom. Ion Processes* **1992**, *121*.
312. Eberlin, M.N., Majundar, T.K., Cooks, R.G. *J. Am. Chem. Soc.* **1992**, *114*, 2884.
313. Flammang, R., Landu, D., Laurent, S., Barbieux-Flammang, M., Kappe, C.O., Wong, M.W., Wentrup, C. *J. Am. Chem. Soc.* **1994**, *116*, 2005.
314. Flammang, R., Gerbaux, P., Morkved, E.H., Wong, M.W., Wentrup, C. *J. Phys. Chem.* **1996**, *100*, 17452.
315. Cruz, E.M., Lopez, X., Ugalde, J.M., Cossio, F.P. *J. Phys. Chem.* **1995**, *99*, 12170.
316. Lopez, X., Cossio, F.P., Ugalde, J.M., Barrientos, C., Largo, A. *J. Phys. Chem.* **1994**, *98*, 2294.
317. Heydorn, L.N., Burgers, P.C., Ruttink, P.J.A., Terlouw, J.K. *Int. J. Mass Spectrom.* **2003**, *228*, 759.
318. Heydorn, L.N., Wong, C.Y., Srinivas, R., Terlouw, J.K. *Int. J. Mass Spectrom.* **2003**, *225*, 11.
319. Heydorn, L.N., Burgers, P.C., Ruttink, P.J.A., Terlouw, J.K. *Chem. Phys. Lett.* **2003**, *368*, 584.
320. Feng, R., Cooper, G., Brion, C.E. *Chem. Phys.* **2000**, *252*, 359.
321. Carnovale, F., Hitchcock, A.P., Cook, J.P.D., Brion, C.E. *Chem. Phys.* **1982**, *66*, 249.
322. Wang, P., Vidal, C.R. *J. Chem. Phys.* **2003**, *118*, 5383.
323. Adams, N.G., Smith, D., Grief, D. *Int. J. Mass Spectrom. Ion Phys.* **1978**, *26*, 405.
324. Fock, W., McAllister, T. *Astrophys. J.* **1982**, *257*, L99.
325. McAllister, T., Taylor, P.R., Scarlett, M. *Org. Mass Spectrom.* **1986**, *21*, 157.
326. Liu, J., Van Devener, B., Anderson, S.L. *J. Chem. Phys.* **2002**, *117*, 8292.
327. Scarlett, M., Taylor, P.R. *Chem. Phys.* **1986**, *101*, 17.
328. Saebo, S., Sanz, M.M., Foster, S.C. *Theor. Chem. Acc.* **1997**, *97*, 271.
329. Nakanaga, T., Amano, T. *Mol. Phys.* **1987**, *61*, 313.
330. Ohshima, Y., Endo, Y. *Chem. Phys. Lett.* **1996**, *256*, 635.
331. Block, E., Penn, R.E., Olsen, R.J., Sherwin, P.F. *J. Am. Chem. Soc.* **1976**, *98*, 1264.
332. Lahem, D., Flammang, R., Nguyen, M.T. *Chem. Phys. Lett.* **1997**, *270*, 93.

333. Le, H.T., Nguyen, T.L., Lahem, D., Flammang, R., Nguyen, M.T. *Phys. Chem.* **1999**, *1*, 755.

334. Nguyen, M.T., Weringa, W.D., Ha, T.-Y. *J. Phys. Chem.* **1989**, *93*, 7956.

335. Ruttink, P.J.A., Burgers, P.C., Trikoupis, M.A., Terlouw, J.K. *Chem. Phys. Lett.* **2001**, *342*, 447.

336. Ventura, O.N., Kieninger, M., Denis, P.A., Cachau, R.E. *Chem. Phys. Lett.* **2002**, *355*, 207.

337. Block, E., Corey, E.R., Penn, R.E., Renken, T.L., Sherwin, P.F., Bock, H., Hirabayashi, T., Mohmand, S., Solouki, B. *J. Am. Chem. Soc.* **1982**, *104*, 3119.

338. Turecek, F., Drinkwater, D.E., McLafferty, F.P. *J. Am. Chem. Soc.* **1989**, *111*, 7696.

339. Gozzo, F.C., Eberlin, M.N. *J. Mass Spectrom.* **1995**, *30*, 1553.

340. McGibbon, G.A., Burgers, P.C., Terlouw, J.K. *Chem. Phys. Lett.* **1994**, *218*, 499.

341. Bouchoux, G., Le, H.T., Nguyen, M.T. *J. Chem. Phys. A* **2001**, *105*, 11128.

342. Ruttink, P.J.A., Burgers, P.C., Terlouw, J.K. *Chem. Phys. Lett.* **1994**, *229*, 495.

343. Mineva, T., Russo, N., Sicilia, E., Toscano, M. *Theor. Chem. Acc.* **1999**, *101*, 388.

344. Zha, Q., Nishimura, T., Meisels, G.G. *Int. J. Mass Spectrom. Ion Processes* **1988**, *83*, 1.

345. Bouchoux, G., Salpin, J.-Y. *Rapid Commun. Mass Spectrom.* **1999**, *13*, 932.

346. Graedel, T.E. *Rev. Geophys. Space Phys.* **1977**, *15*, 421.

347. Singleton, D.L., Irwin, R.S., Cvetanovic, R.J. *Can. J. Chem.* **1983**, *61*, 968.

348. Hatakeyama, S., Akimoto, H. *J. Phys. Chem.* **1983**, *87*, 2387.

349. Balla, R.J., Heicklen, J. *J. Photochem.* **1985**, *29*, 297.

350. Block, E. *Angew. Chem. Int. Ed. Engl.* **1992**, *31*, 1135.

351. Wang, L., Zhang, J. *J. Mol. Struct. (THEOCHEM)* **2002**, *581*, 129.

352. Turecek, F. *J. Phys. Chem.* **1994**, *98*, 3701.

353. Hung, W.-C., Shen, M.-Y., Lee, Y.-P., Wang, N.-S., Cheng, B.-M. *J. Chem. Phys.* **1996**, *105*, 7402.

354. Turecek, F., Brabec, L., Vondrak, T., Hanus, V., Hajicek, J., Havlas, Z. *Coll. Czech. Chem. Commun.* **1988**, *53*, 2140.

355. Benson, S.W., Cruickshank, F.R., Golden, D.M., Haugen, G.R., O'Neil, H.E., Rodgers, A.S., Shaw, R., Walsh, R. *Chem. Rev.* **1969**, *69*, 279.

356. Benson, S.W. *Chem. Rev.* **1978**, *78*, 23.

357. Sulzle, D., Drewello, T., Van Baar, B.L.M., Schwarz, H. *J. Am. Chem. Soc.* **1988**, *110*, 8330.

358. Glukhovtsev, M.N., Bach, R.D. *J. Phys. Chem. A* **1997**, *101*, 3574.

359. Zhao, Y., Francisco, J.S. *Chem. Phys. Lett.* **1990**, *173*, 551.

360. Hrusak, J., Schwarz, H. *J. Phys. Chem.* **1993**, *97*, 4659.

361. Wiedmann, F.A., Wesdemiotis, C. *J. Am. Chem. Soc.* **1994**, *116*, 2481.

362. Zachariah, M.R., Westmoreland, P.R., Burgess Jr., D.R., Tsang, W., Melius, C.F. *J. Phys. Chem.* **1996**, *100*, 8737.

363. Keck, H., Kuchen, W., Renneberg, H., Terlouw, J.K., Visser, H.C. *Angew. Chem. Int. Ed. Engl.* **1991**, *30*, 318.

364. Vivekananda, S., Srinivas, R. *Int. J. Mass Spectrom. Ion Processes* **1997**, *171*, 79.

3.6 IONS CONTAINING TWO CARBON ATOMS

3.6.1 $[C_2,H]^+$ – m/z 25

The $\Delta_fH(C_2H^{1+}) = 1621$ kJ/mol derives from $\Delta_fH(C_2H^{1\bullet}) = 594$ kJ/mol and $IE_a = 10.64$ eV.[1] It is noteworthy that a recent theoretical calculation (G3MP2B3) gives $\Delta_fH(C_2H^{1\bullet}) = 563.49$ kJ/mol.[2] The ethynyl radical $(C_2H^{1\bullet})$ has been investigated thoroughly because it has been identified as playing a significant role in interstellar chemistry,[3,4] planetary atmospheres[5,6] and combustion chemistry.[7,8]

3.6.2 $[C_2H_2]^{+\bullet}$ – m/z 26

Isomers

There are two isomers: ionized acetylene $(HCCH^{1+\bullet})$ and the vinylidene cation, $CH_2C^{1+\bullet}$.

Identification

Recent high level computations by Jursic[9] indicate that the barrier separating the isomers is likely very small, approximately 2 kJ/mol, and so the experimental isolation of the thermochemically much less stable vinylidene is a very difficult task. This is in agreement with the conclusions of early collision-induced charge reversal mass spectrometry experiments on the CH_2C^{1-} anion[10,11] that showed that the corresponding (ground state) cation could be no more than a transient species. Evidence from the reactions of a beam of fast $CH_2CCl_2^{1+\bullet}$ ions with K and Cs indicate that the vinylidene cation may have a lifetime of at least 8.5 μs.[12] It is possible that an excited state product may be involved.

Thermochemistry

The ionization energy of acetylene (11.40 eV[1]) together with $\Delta_fH(HCCH) = 226.7$ kJ/mol[1] gives $\Delta_fH(HCCH^{1+\bullet}) = 1326.7$ kJ/mol. Computations place $CH_2C^{1+\bullet}$ at an energy of 1503 kJ/mol.[10]

3.6.3 $[C_2H_3]^+$ – m/z 27

Isomers

There is only one isomer, the vinyl cation CH_2CH^{1+}.

Identification

This ion is readily generated from an ionized vinyl halide by loss of the halogen atom. The most noteworthy feature of the ion's structure is that it is best viewed as a symmetrical proton-bridged acetylene,[13] unlike the radical $CH_2CH^{]\bullet}$ that has a conventional form. Note that the ethyl cation is also a proton-bridged species of similar shape.[14] It can easily be distinguished from the isobaric ions $HCN^{]+\bullet}$ or $HNC^{]+\bullet}$ by virtue of the presence of m/z 24, 25 and 26 in the ion's CID mass spectrum.

Thermochemistry

Although the vertical ionization energy (IE_V) of the vinyl radical has long been known (ca 8.95 eV),[15] the adiabatic IE and heat of formation of the vinyl radical have both been elusive quantities, especially the latter which has wandered to and fro from about 250 to 325 kJ/mol[16] before settling at the presently accepted 299 ± 5 kJ/mol.[1] The latter value agrees well with a computed (G3MP2B3) value of 292.63 kJ/mol.[2] The IE_a from Ref. [1] is given as 8.25 eV leading to $\Delta_f H(C_2H_3^{]+}) = 1095 \pm 5$ kJ/mol. However the calculations of Glukhovtsev[13] and Bach indicate that a higher value of 8.61 eV may be more appropriate giving $\Delta_f H(C_2H_3^{]+}) = 1130$ kJ/mol. The AE of $C_2H_3^{]+}$ from C_2H_3I, a simple bond cleavage, gives a value of $\Delta_f H(C_2H_3^{]+}) = 1108 \pm 14$ kJ/mol, a result that does not settle the question (AE $= 11.29 \pm 0.05$ eV, $\Delta_f H(C_2H_3I) = 126 \pm 6$ kJ/mol, $\Delta_f H(I^\bullet) = 107$ kJ/mol).[1] It is worth noting that NIST lists $\Delta_f H(C_2H_3^{]+}) = 1110 \pm 10$ kJ/mol.[1] The large difference between the IE_a and IE_V values is in keeping with the ground state's proton-bridged acetylene structure.

3.6.4 $[C_2H_4]^{+\bullet} - m/z$ 28

Isomers

Only ionized ethene has been investigated by experiments. The isomeric methyl carbene ion, $CH_3CH^{]+\bullet}$, has not been identified by experiment and calculations by Van der Hart[17] show that the ion does not occupy a potential well. At best it lies on a flat portion of the $C_2H_4^{]+\bullet}$ hypersurface.

Identification

If it were necessary to distinguish $CH_2CH_2^{]+\bullet}$ from other ions of the same mass, the sequence m/z 24, 25, 26 and 27 in the CID mass spectrum would suffice.

Thermochemistry

The heat of formation of the ethene ion is given by the ionization energy (10.51 eV[1]) and $\Delta_f H(C_2H_4) = 52.5$ kJ/mol,[1] which leads to $\Delta_f H(C_2H_4^{]+\bullet}) = 1067$ kJ/mol.

3.6.5 $[C_2H_5]^+$ – m/z 29

Isomers

This cation, like vinyl, has only one stable form and is a hydrogen-bridged structure that can be envisaged as a symmetrically protonated ethene. The classical $CH_3CH_2^{1+}$ structure is not a minimum on the ion's potential surface, but only a plateau on each side of the well and lying some 28 kJ/mol above the minimum.[14]

Identification

The ion gives an intense and structure-characteristic peak in its MI mass spectrum for the loss of H_2, with a small kinetic energy release ($T_{0.5} = 8$ meV).[18] The transition state (TS) for this reaction lies a computed 25 kJ/mol above the thermochemical minimum for the production of the vinyl cation, at an energy of 198 kJ/mol ($\Delta_f H(C_2H_5^{1+}) = 902$ kJ/mol and $\Delta_f H(C_2H_3^{1+}) = 1100$ kJ/mol).[1] The reaction proceeds by a late TS in which the departing H_2 molecule is already some 2.3 Å from the vinyl cation; the late TS also accounts for the lack of translational kinetic energy released from the reverse energy barrier.[18] It is worth noting that neutralization (see neutralization–reionization mass spectrometry) of the cation produces few stable $C_2H_5^{1\bullet}$ species because the neutral structure corresponding to the cation is the TS for a 1,2 H-shift in the radical.

Thermochemistry

From the proton affinity of ethene, $\Delta_f H(C_2H_5^{1+}) = 902$ kJ/mol (PA $= 680.5$ kJ/mol, $\Delta_f H(C_2H_4) = 52.5$ kJ/mol and $\Delta_f H(H^+) = 1530$ kJ/mol).[1] The NIST also gives a value for the $IE_a(C_2H_5^{1\bullet}) = 8.11 \pm 0.01$ eV from the PI/IE of the ethyl radicals generated by the (exothermic) reaction $F^\bullet + C_2H_6 \longrightarrow HF + C_2H_5^{1\bullet}$; this together with $\Delta_f H(C_2H_5^{1\bullet}) = 119 \pm 2$ kJ/mol gives the same value for $\Delta_f H(C_2H_5^{1+})$.

3.6.6 $[C_2H_6]^{+\bullet}$ – m/z 30

Isomers

Ionized ethane has no known isomers, except that ions of different symmetry appear as minima on the potential energy surface (computed using density functional theory)[19] but having very low barriers to their interconversion, so low as to preclude their experimental isolation in the gas phase.

Identification

The MI mass spectrum of $C_2H_6^{1+\bullet}$ has only one signal for the loss of H_2 and it is a flat-topped peak with $T_{0.5} = 200$ meV.[20] This reaction's pathway has been characterized by means of computational chemistry and the elimination of H_2 takes place via a nonsymmetric TS.[21] Note that the reaction is formally a 1,2-H_2 elimination; metastable $CH_3CD_3^{1+\bullet}$ ions only lose HD.[20] Ionized ethane can easily be distinguished from ions of the same nominal mass (e.g., $H_2CO^{1+\bullet}$, $[C,H_4,N]^+$) by the series of CID fragments extending from m/z 24 ($C_2^{1+\bullet}$) to m/z 30.

Thermochemistry

The heat of formation of ionized ethane is derived from $\Delta_fH(C_2H_6) = -84.2 \pm 1$ kJ/mol[1] and the experimental adiabatic IE of 11.52 ± 0.04 eV,[1] $\Delta_fH(C_2H_6^{1+\bullet}) = 1028 \pm 5$ kJ/mol. (Note, however, that the calculations referred to above[19] have placed IE_a at a lower value (11.29 eV), a result that may arise from inadequacies in the method employed; see the original paper for a full discussion of this particular difficulty.)

3.6.7 $[C_2H_7]^+$ – m/z 31

Isomers

This nonclassical ion has a long history in gas-phase ion chemistry, it being first implicated[22,23] as an unexpected intermediate in the reaction of CH_3^{1+} with CH_4 to yield $C_2H_5^{1+} + H_2$.

Identification

A recent study[24] that reviewed much of the considerable literature for experiments and calculations considered five species of which three forms of $C_2H_7^{1+}$ were found to be minima on the potential energy surface.

All are only weakly bound with respect to loss of H_2 by about 20, 37 and 50 kJ/mol.

Hiraoka and Kebarle[25] conducted high pressure mass spectrometry experiments in which they studied the equilibrium

$$C_2H_5^{1+} + H_2 \leftrightarrow C_2H_7^{1+}$$

They found that the temperature dependence of the equilibrium clearly showed that two $C_2H_7^{1+}$ structures must be participating. At lower temperatures the ion designated as structure (**2**) is formed while at higher temperatures a second species with a greater binding energy, structure (**1**), was proposed. The respective Δ_fH and Δ_fS values for (**1**) and (**2**) were 49 and 17 kJ/mol and -105 and -82 J/mol K, respectively. High level ab initio calculations[26] at the MP4(SDTQ)/6-311G*//MP2(full)6-31G** level supported the above two structure assignments and gave binding energies and entropies of 44, 26 kJ/mol and -105 and -112 J/mol K, in fair agreement with experiment. Earlier quantum theory calculations[27] had produced similar overall results, with corresponding energies of 60 and 30 kJ/mol. However, the computed energy barriers for interconversion between the isomers prohibited the experimental production of the two ions by an equilibrium experiment. The recent study[24] concluded that ion (**1**) has to surmount a small energy barrier of about 4 kJ/mol above the thermochemical dissociation limit (50 kJ/mol) in order to dissociate to $C_2H_5^{1+} + H_2$.

Identification of the individual isomers by a dissociation experiment (and employing isotopic labels) is not possible because the three structures can interconvert at a low internal energy, resulting in the loss of positional identity of the label atoms.

Thermochemistry

The latest results indicate that the proton affinity of C_2H_6 to yield ion (**1**) (the ground state) is 597 kJ/mol,[1] leading to $\Delta_fH(1) = 849$ kJ/mol $(\Delta_fH(C_2H_6) = -84$ kJ/mol,[1] $\Delta_fH(H^+) = 1530$ kJ/mol[1]).

3.6.8 $[C_2,N]^+$ – m/z 38

Isomers

There are two isomers for this ion, CCN^{1+} and CNC^{1+}. Interest in them arises from their likely presence in dense interstellar clouds[28] where they are believed to be formed by the reaction

$$C^{+\bullet} + HCN \rightarrow C_2N^{1+} + H^\bullet$$

Identification

The 50 eV electron-impact ionization of cyanogen (C_2N_2) produces a mixture of the above ions.[29] A selected ion flow tube (SIFT) study showed that the isomer CCN^{1+} is the more reactive species, being the only isomer to react with H_2, CO_2, Xe and O_2. It also reacts with H_2O, CH_4, etc. The more stable isomer (CNC^{1+}) does not react with CO_2 and CH_4 and only very slowly with H_2O. It is produced by the reaction

$$C^{+\bullet} + C_2N_2 \rightarrow CNC^+ + CN^\bullet$$

A simple mass spectrometric method for their identification has not been described.

Thermochemistry

The thermochemistry of the two isomers is not well established. Measurements of appearance energies by Harland and McIntosh[30] agreed quite well with the values of some earlier calculations.[28] $\Delta_fH(CCN^{1+}) = 1720 \pm 20$ kJ/mol and $\Delta_fH(CNC^{1+}) = 1620 \pm 20$ kJ/mol appear to be reasonable values. The computations by Knight et al.[31] give an energy difference between the isomers of 88 kJ/mol (MP4/6-311++G(df, pd)//6-31G), within the above experimental limits. The same computations gave an energy barrier to their interconversion of 195 kJ/mol above the more stable isomer, CNC^{1+}, much lower than that estimated by Haese and Woods.[28] The Δ_fH for the C_2N radical is given as 556 kJ/mol, but the structure is not certain because the same value is reported for both CNC^1 and CCN.[32] The ionization energies are therefore also uncertain.

3.6.9 $[C_2,H,N]^{+\bullet} - m/z$ 39

Isomers

Three isomers having this formula were generated by mass spectrometry, the ions $HCNC^{1+\bullet}$, $HCCN^{1+\bullet}$ and $HNCC^{1+\bullet}$ and two other minima corresponding to the cyclic structures cy–C(H)CN–$^{1+\bullet}$ and cy–N(H)CC–$^{1+\bullet}$ appear on the potential surface calculated at the B3LYP and CCSD(T) levels of theory.[33]

Identification

The ion $HCNC^{1+\bullet}$ was produced by the collision-induced charge reversal (with O_2) of the corresponding anion (produced by the reaction of $O^{-\bullet}$ with CH_3NC) and it showed intense, structure-characteristic signals at m/z 26 (CN^{1+}) and 27 ($HCN^{1+\bullet}$) and also with weak peaks at m/z 24 and 25 which can only arise from a rearrangement among highly energized ions.

The second isomer, $HCCN^{1+\bullet}$, was produced similarly from CH_3CN. It showed the same signals as the previous ion, but with much more intense peaks at m/z 24 and 25. The third ion, $HNCC^{1+\bullet}$, is generated in the normal mass spectrum of 5-chlorouracil[34] (cy–N(H)C(O)N(H)C(H)C(Cl)C(O)–, losses of HCl, CO, and HNCO) and also from ionized 5-amino-4-cyano isoxazole (cy–NC(H)C(CN)C(NH$_2$)C(O)–, loss of HNCO and then HNC).[34] The ion is characterized by the unique presence of a very weak m/z 15 (NH^{1+}). All the species had H-atom loss giving rise to the most abundant fragment ion.

With the exception of this m/z 15 peak, distinguishing between these isomers depends almost wholly on relative peak intensities and so assessing their purity is difficult. The computations show the energy manifold for all five stable forms and the energy barriers between them, the highest of these lies only 75 kJ/mol above the global minimum, HCNC^{1+}. The lowest energy dissociation, probably H$^\bullet$ loss, was not calculated and so it is likely that the ions can interconvert close to this dissociation limit, thus giving rise to the remarkably similar CID mass spectra. Note that this would easily be confirmed if the H$^\bullet$ loss is metastable and generates the same shaped peak for all isomers. That the above is likely correct was supported by the observation that the collision-induced CR mass spectrum of the HC^{13}CN anion indicated the complete loss of positional identity of the labeled atom.

Thermochemistry

The only values from an experiment are $\Delta_fH(\text{HCCN}^{1+\bullet}) = 1531$ kJ/mol and $\Delta_fH(\text{HCCN}) = 366$ kJ/mol.[30] The relative energies of the ions were computed to be as follows: with the global minimum HCNC$^{1+\bullet}$ at zero, HCCN$^{1+\bullet}$ lies at 0.7 kJ/mol and HNCC$^{1+\bullet}$ at 17 kJ/mol. The transition state energies for their interconversion and the two stable cyclic forms were also computed.[33]

3.6.10   [C$_2$,H$_2$,N]$^+$ – m/z 40

Isomers

Four isomers have been proposed for this ion, CH$_2$CN^{1+}, CH$_2$NC^{1+}, the cyclic N-bridged acetylene cy–NC(H)C(H)–$^{1+}$ and the linear HCCNH^{1+} species. Interest in these ions again comes from their involvement in interstellar chemistry.[35]

Identification and Thermochemistry

Because of conflicting results between theory and experiment it is useful to combine these sections to aid the discussion. According to the most recent computational studies,[36,37] the four stable ions should be observable by experiment. The ion HCCNH^{1+} has only been found in a matrix isolation experiment where its IR spectrum was recorded[37] but the others have some mass spectral data. The computed and experimental heats of formation are given in the table below.

Heat of Formation of the $[C_2,H_2,N]^+$ Isomers

Ion	Theory $\Delta_f H^{36}$ (kJ/mol)	Relative Energies[37] (kJ/mol)	Experiment[38,a] (kJ/mol)
cy–NC(H)C(H)–	1200	0	1170
CH_2CN	1250	48.5	1201 $(1246)^b$
HCCNH		61.2	—
CH_2NC	1265	67.5	1186

[a]The experimental data are from the AE measurements of Ref. [38] and using updated values for $\Delta_f H(CH_3CN) = 74$ kJ/mol,[1] $\Delta_f H(CH_3NC) = 164$ kJ/mol[1] and $\Delta_f H(1\text{-H-}1,2,3\text{-triazole}) = 278$ kJ/mol.[36]

[b]From a measured $IE(CH_2CN^{|\bullet}) = 10.30$ eV and $\Delta_f H(CH_2CN^{|\bullet}) = 258$ kJ/mol.[39] This is the same IE as reported earlier[40] where it was argued to be a vertical value rather than the adiabatic. Note however that the geometry difference between the ion and the radical is minimal[36] and so this is not a viable explanation.

The metastable ion and CID mass spectra of the $[C_2,H_2,N]^+$ ions derived from CH_3CN, CH_3NC, the 1-H-1,2,3-triazole and ICH_2CN have been described.[38] Note that the $[C_2,H_2,N]^+$ ion from isoxazole is also indistinguishable from that derived from C_2H_3N.[41] The metastable ion peaks for the $H^\bullet$ loss from the first three $[C_2,H_3,N]^{+\bullet}$ ions have quite large kinetic energy release values (146 ± 4 meV), usually indicative of a reverse energy barrier for the fragmentations and moreover, that a single TS may be involved. In contrast, the iodine atom loss from ICH_2CN has a kinetic energy release of only 5 meV, typical of a threshold process and a simple bond cleavage. It is attractive therefore to propose that all of these dissociations generate the global minimum cyclic isomer without a significant reverse energy barrier. However, the agreement between the calculated and measured $\Delta_f H$ values is surprisingly unsatisfactory and it is difficult to pinpoint the reason for it.

The CID mass spectra of the above ions generated in the ion source or from metastable precursor ions are essentially the same, with only minor intensity differences. Thus experimental distinction of the four ions remains problematic. To date, computations have not evaluated the heights of the isomerization barriers between the four. Before digressing from this topic, it is worth noting that the $\Delta_f H$ value for the radical $CH_2NC^{|\bullet}$ is far from settled, data from experiments occupy a range of about 100 kJ/mol with computed values lying uncertainly in the middle.[42]

3.6.11 $[C_2,H_3,N]^+ - m/z\ 41$

Isomers

Nine isomers of composition $[C_2,H_3,N]^+$ have been identified or postulated from experiment and theory. Acetonitrile (1) is the most common of the corresponding neutral molecules, but as will be illustrated below, it has been one of the most elusive small ions to characterize. Ionized ketenimine (2) can also be thought of as a distonic ion resulting from a 1,3-H shift in (1). Both have corresponding isocyano analogs ((3) and (4)).

CH$_3$CN   CH$_2$CNH   CH$_3$NC   CH$_2$NCH

1   2   3   4   5

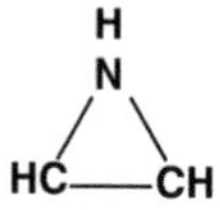

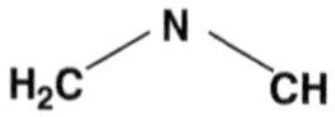

6 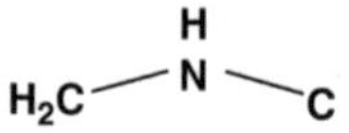   7 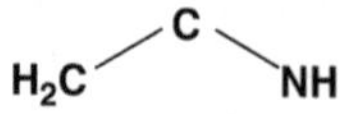   8   CHCHNH   9

Choe[43] has recently explored the $[C_2,H_3,N]^+$ potential energy surface at the B3-LYP/6-311++G(3df,3dp) level of theory and calculated the relative energies and interconversion barriers for isomers (1)–(8). Isomers (6), (7) and (8) are calculated to lie in shallow energy wells and so are unlikely to be experimentally observed. Note that isomers (6) and (8) are bent versions of the global minima (4) and (2) respectively, which have C_{2v} symmetry.

Identification

The experimental characterization of the $[C_2,H_3,N]^+$ ions has been the subject of considerable debate over the years. It was previously assumed that ionization of acetonitrile and isocyanomethane gave their corresponding ions, (1) and (3). Ions derived from the dissociative ionization of larger nitriles and isonitriles were assigned to isomers (2) and (4). These assignments were based on CID mass spectra and specifically on the ratio of two fragment ion intensities m/z 15:14, $CH_3^{1+}:CH_2^{1+\bullet}$.[44–47] On average (since the measured ratios differed from instrument to instrument), the m/z 15:14 ratio was doubled when acetonitrile was ionized vs when m/z 41 was obtained from other nitrile precursors, the connection being that (1) would give more CH_3^{1+} compared with (2). However, the ratio is always *small*, of the order of 0.04–0.1. RRKM modeling of the interconversion of (1) and (2) showed a rate constant $>10^9$ s^{-1}, just a few kilojoules per mole above the barrier to their interconversion (which lies between 70 and 82 kJ/mol above (1)).[43,47] This drew into question the purity of the ions generated when acetonitrile is ionized. The two isomers were also studied by their ion–molecule reaction with CO_2 and shown to be distinct. (1) can proton transfer to CO_2 but (2) cannot, because the proton affinity of CO_2 (540.5 kJ/mol)[1] lies between that of the carbon[1] (524 kJ/mol) and nitrogen[48] (763 kJ/mol) atoms in CH_2CNH. Using this approach, de Petris et al.[47] were able to demonstrate that the dissociative ionization of butyronitrile leads to pure (2), whereas ionization of acetonitrile leads to a mixture of (1) and (2). Chemical ionization of acetonitrile with Xe (which has a slightly lower IE than CH_3CN) yielded relatively more (1), but was still a mixture. Thus it was concluded that the CID mass spectrum of ionized acetonitrile was the product of a mixture of ions (1) and (2).

The isomerization of (3) and (4) has not been similarly studied by ion–molecule reactions, but based on the results for (1) and (2), it may be reasonable to assume that the CID mass spectra reported in the literature for (3) and (4) are also of questionable utility. The calculated barrier between (3) and (4) lies approximately 148 kJ/mol above (3), so it may be possible to generate relatively more (3) from ionizing isocyanomethane than was the case for acetonitrile. Indeed, the CID mass spectrum of ionized isocyanomethane exhibits an m/z 15:14 ratio that is greater than 1.0, indicating that much more of the ion population retains the CH_3NC connectivity.[46]

Ion (9) was assumed to be produced by the dissociative ionization of imidazole (cy–N(H)C(H)NC(H)C(H)–) and pyrazole (cy–N(H)NC(H)C(H)C(H)–).[46] The CID mass spectra of these ions are distinguishable from that of (2) only by a very small relative intensity difference for m/z 28. On the basis of Choe's calculations, (5), which is the ring closed version of (9), lies in a deep energy well of about 170 kJ/mol and therefore should be observable. The similarities in the above CID mass spectra suggest that the dissociative ionization of imidazole and pyrazole may indeed produce (2) rather than (9).

Thermochemistry

The $\Delta_f H(1) = 1251$ kJ/mol is based on $\Delta_f H(CH_3CN) = 74.04 \pm 0.37$ kJ/mol and IE $= 12.20 \pm 0.01$ eV.[1] Similarly, $\Delta_f H(3) = 1248$ kJ/mol ($\Delta_f H(CH_3NC) = 163.5 \pm 7.2$ kJ/mol and a reported photoelectron IE of 11.24 eV).[1] However, this value is in considerable disagreement with the B3-LYP/6-311++G (3df,3dp) relative energies for (1) and (3) reported by Choe,[43] (3) being 50 kJ/mol higher in energy than (1) and is likely erroneous. The $\Delta_f H(CH_3NC)$ was originally determined from a heat of isomerization, but was reevaluated by Pedley et al.[49] The only other source for the discrepancy is in the experimentally measured IE, which must be too low. An electron-impact IE measured by Harland and McIntosh[50] of 11.53 eV gives $\Delta_f H(3) = 1276$ kJ/mol, in better (but still not good) agreement with calculations.

Relative Energies of the $[C_2,H_3,N]^{+\bullet}$ Isomers[43]

Ion	Relative Energy (0 K) (kJ/mol)
(1)	0
(2)	−207
(3)	50
(4)	−165
(5)	−32
(6)	14
(7)	92
(8)	55
(9)	

3.6.12 $[C_2,H_4,N]^+$ – *m/z* 42

Isomers

Twelve isomers of composition $[C_2,H_4,N]^+$ have been identified or postulated from experiment and theory. Protonated acetonitrile (**1**) and isocyanomethane (**2**) have received the most experimental attention.

CH_3CNH **1**

CH_3NCH **2**

CH_2NCH_2 **3**

CH_2CNH_2 **4**

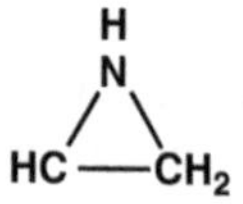

5

NH_3CCH **6**

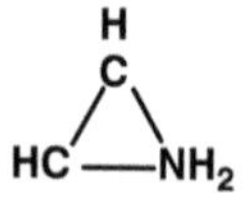

7

CH_2CHNH **8**

CH_3NHC **9**

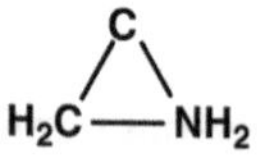

10

NH_2CHCH **11**

CH_3CHN **12**

Identification

(**1**) and (**2**) can easily be distinguished based on their ion–molecule reactivity with bases, since CH_3CN and CH_3NC have significantly different proton affinity values (see below). Knight et al.[51] demonstrated that the reaction between CH_3^{1+} and HCN produces 85% (**1**) and 15% (**2**) based on the proton-transfer reaction of the product $[C_2,H_4,N]^+$ ions with neutral molecules such as methylformate. The ions can also be distinguished by their CID mass spectra, in which (**2**) exhibits a much greater yield of CH_3^{1+} (ratio m/z 38:15 = 3.6) due to the formation of the stable HCN coproduct ((**1**) would have to form CH_3^{1+} + HNC, a much less stable neutral, resulting in a ratio m/z 38:15 = 125).[52] (**2**) also exhibits a greater propensity to undergo charge stripping.[52]

Thermochemistry

Recent calculations by Corral et al.[53] explored the $[C_2,H_4,N]^+$ potential energy surface at the B3-LYP/6-311++G(3df,2p) and G2(MP2) levels of theory and evaluated the relative energies (see table) and interconversion barriers for isomers (**1**)–(**10**). Most of the isomers appear to exist in fairly deep potential energy wells and so many should be observable. Isomers (**11**) and (**12**) have been examined only at the HF/6-31G** level of theory by Nguyen and Ha.[54] The experimental $\Delta_f H(\mathbf{1}) = 825$ kJ/mol is based on $\Delta_f H(CH_3CN) = 74.04 \pm 0.37$ kJ/mol, PA = 779 kJ/mol[1] and $\Delta_f H(H^+) = 1530$ kJ/mol.[1] Similarly, $\Delta_f H(\mathbf{2}) = 854$ kJ/mol ($\Delta_f H(CH_3NC) = 163.5 \pm 7.2$ kJ/mol, PA = 839.1 kJ/mol[1] and $\Delta_f H(H^+) = 1530$ kJ/mol[1]).

Relative Energies of the $[C_2,H_4,N]^{+\bullet}$ Isomers

| | Relative Energies[53] (kJ/mol) | |
Ion	B3-LYP/6-311+G(3df,2p)	G2(MP2)
(**1**)	0	0
(**2**)	51	43
(**3**)	57	64
(**4**)	53	67
(**5**)	180	174
(**6**)	172	161
(**7**)	195	188
(**8**)	263	272
(**9**)	287	275
(**10**)	400	388
(**11**)	270 (HF/6-31G**)[54]	
(**12**)	309 (HF/6-31G**)[54]	

Note that at both levels of theory, the relative energies of isomers (**1**) and (**2**) are significantly larger than that obtained from the experimental heat of formation values of the ions.

3.6.13 $[C_2,H_5,N]^+$ – m/z 43

Isomers

Eight isomers of composition $[C_2,H_5,N]^+$ have been identified or postulated from experiment and theory.

$$CH_2CHNH_2 \quad CH_3CNH_2 \quad CH_3CHNH \quad CH_2NCH_3 \quad NH_3CHCH$$

$$\text{1} \quad\quad\quad \text{2} \quad\quad\quad \text{3} \quad\quad\quad \text{4} \quad\quad\quad \text{5}$$

$$NH_3 \cdots CCH \quad\quad H_2C \overset{\displaystyle \overset{H}{N}}{\underset{\textstyle}{——}} CH_2 \quad\quad CH_2NHCH_2$$

$$\text{6} \quad\quad\quad\quad\quad \text{7} \quad\quad\quad\quad \text{8}$$

Identification

Metastable ions (**1**), (**3**), (**4**) and (**7**) were found to be distinguishable from the magnitude of the kinetic energy released in the $H^\bullet$ loss reaction.[55] Ion (**1**) was formed from ionized cyclobutylamine and propylamines and the $H^\bullet$ loss reaction has a $T_{0.5} = 390 \pm 20$ meV. Ion (**3**) was formed with (**1**) from the dissociation of ionized diethylamine. The contribution of (**3**) was separated by deuterium labeling studies and this ion was found to have a kinetic energy release of 200 ± 10 meV. Ion (**4**) (resulting from the dissociation of ionized trimethylamine and N-methylpiperidine (cy–$N(CH_3)(CH_2)_5$–) has a $T_{0.5} = 140 \pm 10$ meV. Ions (**7**) and (**8**) were found to be indistinguishable, with H-loss $T_{0.5}$ values of 260 ± 15 meV and they are made from aziridine, dimethylamine, pyrrolidine, and substituted piperidines. On the basis of MP4 level calculations, (**1**) was predicted to isomerize to (**2**) before H-loss.[56]

Thermochemistry

An unpublished PI/AE measurement of the ion from cyclobutylamine yields $\Delta_f H(\mathbf{1}) = 826 \pm 8$ kJ/mol.[56] Combining the heat of formation and ionization energy of CH_3CHNH leads to $\Delta_f H(\mathbf{3}) = 950$ kJ/mol ($\Delta_f H = 24 \pm 8$ kJ/mol (obtained from the ΔH_r for the deprotonation of $CH_3CHNH_2^{\vert+})^{57}$ and IE $= 9.6$ eV[58]). The thermochemistry of ion (**4**) is based on a neutral $\Delta_f H$ of 44 ± 8 kJ/mol[57] and an IE of 9.3 eV,[58] which, when combined, gives $\Delta_f H(\mathbf{4}) = 950$ kJ/mol. These experimental values are in reasonable, but not excellent, agreement with G2(MP2) and CBS-Q values calculated by Henriksen and Hammerum[59] and the relative energies predicted at the G2M level of theory by Cui and Morokuma.[60] The ion–radical complex (**6**) was calculated by the latter group to lie in a well only 30 kJ/mol deep.

Energies of the $[C_2,H_5,N]^{+\bullet}$ Isomers

Ion	$\Delta_f H_{G2(MP2)}$ (298 K)[59] (kJ/mol)	Relative Energies (0 K)[60] (G2M, kJ/mol)
(1)	849	0
(2)		78
(3)	969	117
(4)	967	
(5)		151
(6)		297

3.6.14 $[C_2,H_6,N]^+ - m/z$ 44

Isomers

Seven isomers of composition $[C_2,H_6,N]^+$ have been identified or postulated from experiment and theory.

$$CH_3CHNH_2 \quad CH_2CH_2NH_2 \quad CH_3NHCH_2 \quad CH_2CHNH_3 \quad \underset{H_2C—CH_2}{\overset{H_2N}{\triangle}}$$

$$\quad 1 \qquad\qquad 2 \qquad\qquad 3 \qquad\qquad 4 \qquad\qquad 5$$

$$CH_3CH_2NH \quad CH_3NCH_3$$

$$\quad 6 \qquad\qquad 7$$

Identification

Several attempts have been made to characterize these isomers generated from a wide variety of amine precursors.[61-63] Ions (1) and (3) have similar MI mass spectra showing losses of C_2H_2 and $H^{\bullet}$, indicating that they dissociate over the same potential energy surface.[61] However, they are distinguishable by their CID mass spectra, with (1) exhibiting a more pronounced ion at m/z 41 and being approximately 30% of base peak as opposed to only 22% of base peak in the case of (3). These were the only two distinguishable isomers made by EI from no less than 19 amino precursors.[62]

Bursey and coworkers[64] were able to generate isomer (6) by the collisional charge reversal of the negative ion. The CID mass spectrum of the resulting cation has a prominent peak at m/z 29 (36% base peak) due to the loss of $NH^{1\bullet}$ to form the ethyl cation. This fragment ion signal in the CID mass spectrum of (1) and (3) is of small relative intensity (5% and 10% base peak, respectively). Their attempt to generate (7) in the same manner resulted in a cation with a CID mass spectrum essentially the same as that of (3).

Van de Sande and coworkers[63] generated ion (5) by field ionization of aziridine (cy–C(H)$_2$C(H)$_2$N(H)–) and by 14 eV EI of 2-phenoxyethylamine.

The CID mass spectrum of (5) has a characteristic peak at m/z 30, absent in those of (1) and (3).

Thermochemistry

To date the best experimental thermochemistry for this set of isomers comes from Traeger and Harvey[65] who measured the PI/AE for (1) from five primary amines, resulting in $\Delta_f H(1) = 665 \pm 1.4$ kJ/mol. This value agrees well with a number of high level *ab initio* predictions, including a G3 value of 666.6 kJ/mol.[65] This value is some 8 kJ/mol higher than that reported by Lossing et al.[66] (657 kJ/mol) and obtained from monoenergetic EI/AE of (1) from four neutral precursors.

The NIST WebBook lists $\Delta_f H(3) = 695$ kJ/mol based on Lossing et al.'s EI/AE measurements of the ion from three precursors.[66] The best computed value has been obtained at the HF/6-31G*//HF/4-31G* level of theory, which predicts (3) to lie 40 kJ/mol higher than (1), consistent with the above experimental values.[67] Combining the measured IE and calculated (G3MP2B3) heat of formation of the radical leads to the somewhat higher value $\Delta_f H(3) = 726$ kJ/mol (IE = 5.9 eV[1] and $\Delta_f H(CH_3NHCH_2^{1\bullet}) = 156.58$ kJ/mol[68]).

The best estimate for the relative energy for (2) comes from QCISD(T)/6-311G(d,p)//QCISD/6-31G(d) calculations by Hudson and McAdoo that predict it to lie 40 kJ/mol above (1) (but separated from it by a 165 kJ/mol barrier).[69] The PI/IE of the $(CH_3)_2N$ radical formed from trimethylamine has been measured by Forde et al.[70] They measured two values, 5.17 and 9.01 eV, the former of which is listed in the NIST WebBook. G3 calculations on the states of the ion showed that the 5.17 eV most closely matches the formation of ion (3). The IE of 9.01 eV most closely matches the formation of the ^1A state of (7). These G3 calculations predicted (7) to lie 322 kJ/mol above (3), a value consistent with an IE for $(CH_3)_2N^{1\bullet}$ of 9.01 eV, but not 5.17 eV. Thus, combining 9.01 eV with $\Delta_f H((CH_3)_2N^{1\bullet}) = 142$ kJ/mol results in a $\Delta_f H(7) = 1011$ kJ/mol.

Solka and Russell reported low energy-resolution EI/AE values for isomers (1)–(7) from a variety of precursors.[71] Given that only isomers (1) and (3) are easily distinguishable, these values likely correspond to barrier heights for isomer interconversion and so will not be reported here.

Relative Energies of the $[C_2,H_6,N]^+$ Isomers

Ion	Relative Energy (kJ/mol)
(1)	0
(2)	40[69]
(3)	40[67]
(4)	79[67]
(5)	100[67]
(6)	
(7)	322[70]

3.6.15 $[C_2,H_7,N]^{+\bullet}$ – m/z 45

Isomers

Five $[C_2,H_7,N]^{+\bullet}$ isomers have been identified.

CH$_3$CH$_2$NH$_2$	CH$_3$CHNH$_3$	CH$_2$CH$_2$NH$_3$	CH$_3$NHCH$_3$	CH$_2$NH$_2$CH$_3$
1	**2**	**3**	**4**	**5**

Identification

All five isomers have distinct MI mass spectra:[72]

1$\longrightarrow$CH$_2$NH$_2$$^{1+}$ + CH$_3$$^{1\bullet}$	$T_{0.5}$ = 0.6 meV
$\longrightarrow$[C$_2$,H$_6$,N]$^+$ + H$^\bullet$	
2$\longrightarrow$CH$_2$NH$_2$$^{1+}$ + CH$_3$$^{1\bullet}$	$T_{0.5}$ = 29 meV
3$\longrightarrow$CH$_2$NH$_2$$^{1+}$ + CH$_3$$^{1\bullet}$	$T_{0.5}$ = 21 meV
$\longrightarrow$NH$_4$$^{1+}$ + C$_2$H$_3$$^{1\bullet}$	
4$\longrightarrow$[C$_2$,H$_6$,N]$^+$ + H$^\bullet$	
5$\longrightarrow$CH$_2$NH$_2$$^{1+}$ + CH$_3$$^{1\bullet}$	$T_{0.5}$ = 255 meV

Ions (**1**) and (**4**) are made by ionizing the corresponding neutral molecule. Holmes et al.[73] made (**3**) by CH$_2$NH loss from 1,3-diaminopropane. A characteristic feature of the CID mass spectrum of (**3**) is a prominent charge-stripping peak at m/z 22.5. Hammerum and coworkers[74] found that the dissociative ionization of most *n*-alkylamines produced structure (**3**), while many diamines made (**2**). Wesdemiotis et al.[75] were able to easily distinguish (**1**) and (**3**) by their NR mass spectra, with (**1**) exhibiting m/z 30 as base peak and a significant signal at m/z 18, whereas (**3**) has as base peak in its NR mass spectrum m/z 27 and also a peak at m/z 17, not present in that of (**1**). Hammerum et al.[72] produced (**5**) from the dissociative ionization of *N*-methylhexylamine. The large kinetic energy release in the methyl loss reaction was taken as evidence for an activation barrier for the methyl radical addition reaction to CH$_2$NH$_2$$^{1+}$.

Thermochemistry

Combining the ionization energy and heat of formation of ethylamine leads to $\Delta_f H(\mathbf{1}) = 802$ kJ/mol (PE/IE of approximately 8.8 eV[1] and $\Delta_f H = -47.4$ kJ/mol[49]). Similarly, $\Delta_f H(\mathbf{4}) = 776$ kJ/mol can be derived from $\Delta_f H((CH_3)_2NH) = -18.6$ kJ/mol[49] and IE = 8.24 eV.[1] G2(MP2) and CBS-Q calculations[72] of the relative energies for the five isomers give similar results and are consistent with the above relative $\Delta_f H$ values for (**1**) and (**4**). Similar relative energies were obtained earlier by Yates and Radom[76] at the MP3 level of theory.

Relative Energies of the $[C_2,H_7,N]^{+\bullet}$ Isomers

Ion	Relative Energies (0 K) (G2(MP2), kJ/mol)
(1)	0
(2)	−20
(3)	−26
(4)	−24
(5)	7

3.6.16 $[C_2,H_8,N]^+$ – m/z 58

Isomers

The two isomers are protonated ethylamine ($CH_3CH_2NH_3^{1+}$) and protonated dimethylamine (($CH_3)_2NH_2^{1+}$).

Identification

$CH_3CH_2NH_3^{1+}$ and $(CH_3)_2NH_2^{1+}$ can readily be produced by ion source protonation of the corresponding amines. They are easily distinguished by their MI mass spectra. The ion $(CH_3)_2NH_2^{1+}$ competitively loses H_2 (49%), CH_4 (36%) and $CH_3^{1\bullet}$ (15%) to yield $CH_3NHCH_2^{1+}$, $CH_2NH_2^{1+}$ and $CH_3NH_2^{1+\bullet}$ (or $CH_2NH_3^{1+\bullet}$).[77] The reactions proceed over an energy barrier sufficiently high to allow them to be energetically competitive in the MI time frame.[77] The kinetic energy release values for these are $T_{0.5} = 1960$, 165 and 35 meV, respectively, reflecting the exothermicity of the descents from the TS(s) to the products (see Thermochemistry below).

In marked contrast, the $CH_3CH_2NH_3^{1+}$ ion has only one MI reaction,[77] the loss of C_2H_4, yielding NH_4^{1+} ions with a moderate kinetic energy release ($T_{0.5} = 20$[77] or 31[78] meV). This reaction too involves a reverse energy barrier (see below), but the small KER is indicative of the TS involving an ion–molecule complex.[78] Deuterium labels in the ethyl group retain their positional identity (to ≥95%) (i.e., they are not scrambled). This distribution of label in the products is in keeping with only a simple methyl to N H-transfer, leading to the NH_4^{1+} ion. (Compare this ion with its oxy analog $CH_3CH_2OH_2^{1+}$, where H/D scrambling precedes H_3O^+ production.)

Thermochemistry

From the proton affinity of dimethylamine, $\Delta_f H((CH_3)_2NH_2^{1+}) = 582$ kJ/mol (PA$((CH_3)_2NH) = 929.5$ kJ/mol and $\Delta_f H((CH_3)_2NH) = -18.6$ kJ/mol).[1] The product energies are 695, 674.5 and 989 kJ/mol for the losses of H_2, CH_4 and $CH_3^{1\bullet}$, respectively ($\Delta_f H(CH_3CHNH_2^{1+}) = 695$ kJ/mol, ($\Delta_f H(CH_2NH_2^{1+}) = 749$ kJ/mol,[58] $\Delta_f H(CH_4) = -74.5$ kJ/mol,[1] $\Delta_f H(CH_3NH_2^{1+}) = 836$ kJ/mol[58] and $\Delta_f H(CH_3^{1\bullet}) = 147\,kJ/mol^{58}$). The energy barrier referred to above thus must have a height greater than $(989–582) = 407\,kJ/mol$, but lie well below that to the next most energy demanding fragmentation, to $CH_3^{1+} + CH_3NH_2$ at $(1070–582) = 488\,kJ/mol$, $(\Delta_f H(CH_3NH_2) = -23\,kJ/mol$ and $\Delta_f H(CH_3^{1+}) = 1093\,kJ/mol$).[58]

From the proton affinity of $CH_3CH_2NH_2$, $\Delta_f H(CH_3CH_2NH_3^{1+}) = 571\,kJ/mol$ $(PA(CH_3CH_2NH_2) = 912\,kJ/mol^1$ and $\Delta_f H(CH_3CH_2NH_2) = -47\,kJ/mol)^{58}$. The loss of C_2H_4 $(\Delta_f H = 52.5\,kJ/mol)^1$ produces NH_4^{1+} $(\Delta_f H = 630\,kJ/mol^{58})$ and so the product energy is at $685.5\,kJ/mol$. However the computed potential energy surface for this ion[78] shows that the TS for the production of the ethane–ammonium ion complex lies some $110\,kJ/mol$ above the product energies and so the reaction involves a "hidden" reverse energy barrier.[77]

3.6.17 $[C_2,N_2]^+$ – m/z 52

Isomers

Very little information is available on the two isomeric species $CNCN^{1+\bullet}$ (ionized isocyanogen) and $NCCN^{1+\bullet}$ (ionized cyanogen). The $[C_2,N_2]$ neutrals have however been extensively studied by experiments and computations.[79,80]

Identification[81]

Both ions were obtained by ionization of the corresponding neutral. Their reported MI mass spectra are composed of the same fragment peaks with slightly different relative intensities and kinetic energy releases.

MI Mass Spectrum of the $[C_2,N_2]^+$ Isomers

Ion	$12\ (C^{+\bullet})$	$24\ (C_2^{1+\bullet})$	$26\ (CN^{1+})$	$38\ (C_2N^{1+})$
		m/z		
CNCN	10	100 (31)	31 (121)	21 (127)
NCCN	16	100 (57)	74 (118)	54 (170)

Values in parentheses are kinetic energy releases ($T_{0.5}$) in milli-electronvolts.

Quite surprisingly, the most prominent fragment ion is $C_2^{1+\bullet}$, which implies that both ions rearrange before this dissociation. In order to rationalize how the dissociations to $C_2^{1+\bullet} + N_2$ compete with production of $CN^{1+} + CN$, the latter having an appearance energy significantly higher than the former (ca 3 eV), the authors proposed that the two processes take place from different electronic states of the molecular ions.[81]

It is quite evident from these data that these ions need to be investigated further before any conclusive assignment can be made on the connectivity of the ions as well as the mechanism by which they dissociate.

Thermochemistry

From the data provided by NIST,[1] $\Delta_fH(NCCN^{1+\bullet}) = 1599$ kJ/mol ($\Delta_fH(NCCN) = 308.9 \pm 1.8$ kJ/mol and IE $= 13.37 \pm 0.1$ eV). NIST[1] also lists IE(CNCN) $= 12.873$ eV and combined with an estimated $\Delta_fH(CNCN) = 413$ kJ/mol (obtained from the calculated relative energies of the isomers, ca 104 kJ/mol above NCCN (B3LYP/6-311G(d) and CCSD(T)/6-311G(d)),[79] leads to $\Delta_fH(CNCN^{1+\bullet}) = 1655$ kJ/mol.

3.6.18 $[C_2,H,N_2]^+ - m/z$ 53

Isomers

Interest in this family of cations comes from investigations of the protonation of $[C_2,N_2]$ neutrals that have been observed in interstellar clouds and in the atmosphere of Titan.[82] Petrie[83] has calculated the nine structures shown below. The most commonly studied is protonated cyanogen (NCCNH^{1+}, (1)).

NCCNH	CNCNH	HCNCN	CNNCH	CN(H)CN
1	2	3	4	5

CN(H)NC	CN···H···NC	CN···H···CN	NC···H···CN
6	7	8	9

Identification

No experimental characterization for these isomers has been made to date. Theory[84] predicts the barriers to interconversion of (1)–(4) are on the order of 140 kJ/mol and so these ions should be observable.

Thermochemistry

The experimental $\Delta_fH(1) = 1183$ kJ/mol is based on the proton affinity of the neutral cyanogen ($\Delta_fH = 309$ kJ/mol,[1] PA $= 656$ kJ/mol[85] and $\Delta_fH(H^+) = 1530$ kJ/mol[1]). Note that the first PA value 674 ± 4 kJ/mol[84,86,87] was initially questioned by G2 calculations,[83] which implied that it was closer to 655 kJ/mol. Petrie[83] points to the reevaluation of the PA values for the reference bases (which have all been lowered) used in the experimental estimates as the cause for this discrepancy. Indeed, when the PA value was subsequently remeasured with the new PA values for the reference bases, the value given above (656 kJ/mol) was obtained.[85]

Computed (G2) and Experimental Heat of Formation Values of the $[C_2,H,N_2]^+$ Isomers

Ion	$\Delta_f H_{expt}$ (298 K) (kJ/mol)	$\Delta_f H_{G2}$ (0 K)[83] (kJ/mol)
(1)	1164,[1,86,87] 1183[85]	1192
(2)		1256
(3)		1254
(4)		1427
(5)		1501
(6)		1672
(7)		1640
(8)		1773
(9)		1885

3.6.19 $[C_2,H_2,N_2]^{+\bullet}$ – m/z 54

Isomers

Seven $[C_2,H_2,N_2]^{+\bullet}$ ions have been explored by calculations and experiment.

HNC(H)CN   HCN-HCN   H$_2$CNCN   HNC(H)NC   H$_2$NCCN

1   2   3   4   5

H$_2$ ⋯ NCCN

6

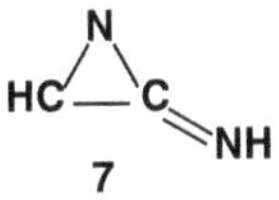

7

Identification

No experimental characterization of these isomers has been made to date. Nenner and coworkers[88] investigated the decomposition of the s-tetrazine ion (cy−N=NC(H)N=NC(H)−][+•]) into $N_2 + [C_2,H_2,N_2]^{+\bullet}$ by PEPICO spectroscopy, and relied on thermochemical estimates (see below) to determine the structure of the product ion. However, no definitive conclusion could be reached due to the poor overall state of the thermochemistry of this family of ions.

Thermochemistry

The relative energies of isomers (1)–(6) have been reported by Nenner et al.[88] They were obtained by combining the calculated relative energies of the corresponding neutral molecules[89] (obtained at the HF/DZP level of theory) with estimated IE values.

Ion	Relative Energies (kJ/mol)
(1)	0
(2)	81
(3)	27
(4)	-37 to $+41^{a}$
(5)	-37 to $+41^{a}$
(6)	$378, 571^{b}$

[a]On the basis of an estimated neutral, IE is in the 9.9–10.6 eV range.
[b]Depending on whether the IE of C_2N_2 (13.374 eV) or H_2 (15.42 eV) is chosen for the calculation.

3.6.20 $[C_2,H_3,N_2]^+$ – m/z 55

Isomers

Four $[C_2,H_3,N_2]^+$ ions have been explored by theory.

$$CH_2C(H)N_2 \quad CH_2N(H)CN \quad NH_2C(H)CN \quad CH_2CH\text{-}\text{-}N_2$$
$$1 \qquad\qquad 2 \qquad\qquad\quad 3 \qquad\qquad\quad 4$$

Identification

No experimental characterization of these isomers has been made to date.

Thermochemistry

The $\Delta_f H$ values for (1)–(3) have been obtained using a core electron replacement energy coupled with the neutral molecule $\Delta_f H$ by Jolly and Gin.[90] Using this approach, all three are predicted to have a $\Delta_f H$ of 1000 kJ/mol. The reliability of this approach to ion thermochemistry has never been established and so these values should be treated with some caution.

Glaser calculated the dissociation energy of (1) into $CH_2CH^{1+} + N_2$ to be 94 kJ/mol at the MP4(SDQ)/6-31 + G* level of theory.[91] Combined with the $\Delta_f H$ of the vinyl cation (1110 ± 10 kJ/mol),[1] this dissociation energy leads to $\Delta_f H(1) = 1006$ kJ/mol, in excellent agreement with the Jolly and Gin estimate. Glaser calculated the energy of (4) to be 92 kJ/mol higher than (1) at the same level of theory, but it is unlikely that this ion would be observable as the dissociation energy was calculated to be only 3 kJ/mol.[92]

3.6.21 $[C_2,H_4,N_2]^{+\bullet}$ – m/z 56

Isomers

Three $[C_2,H_4,N_2]^{+\bullet}$ ions can be found in the literature, although only two have been explored in any detail (1) and (2). Ion (1) is ionized aziridinimine (cy–C(H)$_2$C(NH)N(H)–$^{]+•}$) whereas (**2**) is the ring-opened distonic ion CH$_2$N(H)CNH$^{]+•}$. Ion (**3**) is ionized 2,3-diazabutadiene, CH$_2$=N–N=CH$_2$$^{]+•}$. The ionized form of the most stable neutral [C$_2$,H$_4$,N$_2$] isomer, aminoacetonitrile (NH$_2$CH$_2$CN), has never been studied.

Identification

None of the [C$_2$,H$_4$,N$_2$]$^{+•}$ ions has been the subject of experimental investigation.

Thermochemistry

The relative energies of isomers (**1**) and (**2**) have been calculated by Nguyen et al.[93] at the QCISD(T)/6-311G(d,p)//MP2/6-31G(d,p) level. The distonic ion (**2**) lies 128 kJ/mol lower in energy than (**1**). A barrier of 39 kJ/mol (relative to (**1**)) lies between them. Although an IE for neutral (**3**) is reported in the NIST database (8.95 eV),[1] no neutral $\Delta_f H$ value exists.

3.6.22 [C$_2$,H$_5$,N$_2$]$^+$ – m/z 57

Isomers

The two [C$_2$,H$_5$,N$_2$]$^+$ ions that have been investigated are the ethanediazonium ion (CH$_3$CH$_2$N$_2$$^{]+}$, (**1**)) and its nonclassical isomer (**2**), which consists of the bridged ethyl cation interacting with N$_2$.

Identification

Neither of the [C$_2$,H$_5$,N$_2$]$^+$ ions has been the subject of experimental investigation.

Thermochemistry

The relative energies of isomers (**1**) and (**2**) have been calculated by Glaser et al.[94] at the MP4(SDTQ)/6-311G**//MP2/6-31G* level of theory. The nonclassical ion (**2**) lies 57 kJ/mol higher in energy than (**1**) and only 4 kJ/mol below the dissociation threshold to C$_2$H$_5$$^{]+}$ + N$_2$.

3.6.23 [C$_2$,H$_6$,N$_2$]$^{+•}$ – m/z 58

Isomers

Only one [C$_2$,H$_6$,N$_2$]$^{+•}$ ion has been studied to any extent, ionized dimethyldiazene (or azomethane) CH$_3$NNCH$_3$$^{]+•}$.

Identification

Foster and Beauchamp[95] studied the ion–molecule reactions of ionized dimethyldiazene by ICR mass spectrometry, but the reactions of other isomeric ions have not been explored.

Thermochemistry

Foster, Williamson and Beauchamp measured the IE of dimethyldiazene to be 8.45 ± 0.05 eV by photoionization spectroscopy[96] and coupled with $\Delta_f H(CH_3NNCH_3) = 188$ kJ/mol[95] yields a $\Delta_f H(CH_3NNCH_3^{1+\bullet}) = 1003$ kJ/mol.

3.6.24 $[C_2,H_7,N_2]^+$ – m/z 59

Isomers

Four isomeric $[C_2,H_7,N_2]^+$ ions have been calculated: the protonated amidinium ion $(CH_3C(NH_2)_2^{1+}$, (**1**)) and the proton-bound pairs $[CH_3CN \cdots H \cdots NH_3]^+$ (**2**), $[H_2C=C=NH \cdots H \cdots NH_3]^+$ (**3**) and $[HC\equiv CNH_2 \cdots H \cdots NH_3]^+$ (**4**).

Identification

No experimental investigation of these isomers has been made.

Thermochemistry

Nixdorf and Grutzmacher[97] have calculated the energies of (**2**)–(**4**) relative to (**1**) to be $+23$, $+171$ and $+245$ kJ/mol, respectively, at the MP4(SDTQ)/D95** level of theory.

3.6.25 $[C_2,H_8,N_2]^{+\bullet}$ – m/z 60

Isomers

Twelve isomeric $[C_2,H_8,N_2]^{+\bullet}$ ions have been explored.

$(CH_3)_2NNH_2$	$NH_2CH_2CH_2NH_2$	$CH_3NHNHCH_3$	$CH_3CH_2NHNH_2$
1	**2**	**3**	**4**

$NH_2CHCH_2NH_3$	$NH_3CH_2CH_2NH$	$H_2N \cdots \overset{H}{} \cdots NH_2$ over $HC-CH_2$	$CH_2NH_2 \cdots NH_2CH_2$
5	**6**	**7**	**8**

$NH_2CH_2 \cdots NH_2CH_2$	$CH_2NH_2 \cdots CH_2NH_2$	$NH_3 \cdots NH_2CH=CH_2$	$NH_2CH=CH_2 \cdots NH_3$
9	**10**	**11**	**12**

Identification

There has been little in the way of experimental characterization of these isomers. Bouchoux et al.[98] examined the metastable decomposition of (**2**) (by

loss of ammonia) and concluded that it first isomerized into the distonic ion
(5) before decomposition. This was based on the calculated potential energy
surface and RRKM calculations of the resulting competing processes. Boulanger et al.[99] have stated that the metastable ions (1) decompose by loss of
$CH_3^{1\bullet}$ and $H^\bullet$.

Thermochemistry

The relative energies of many of the isomers have been calculated by
Bouchoux et al.[98] at the QCISD(T)/6-31G*//MP2/6-31G* level of theory
and are listed below. The IE for neutral (1) has been measured by charge
transfer equilibrium mass spectrometry to be 7.29 ± 0.05 eV.[1] This is probably the only way to reliably determine the adiabatic IE for hydrazines due
to the well-documented geometry change that occurs upon ionization.[99]
Combined with a neutral $\Delta_f H$ of 83 kJ/mol gives $\Delta_f H((CH_3)_2NNH_2^{1+\bullet}) =$
786 kJ/mol.[58] G3 level calculations predict a slightly higher value of
804 kJ/mol (but give a comparable IE).[99] On the basis of a neutral $\Delta_f H$
of -17 kJ/mol and an IE from photoelectron spectroscopy of 8.6 eV,[1]
$\Delta_f H(2) = 813$ kJ/mol.

Ion	Relative Energy[98] (kJ/mol)
(1)	
(2)	0
(3)	
(4)	
(5)	-38
(6)	-19
(7)	-7
(8)	22
(9)	84^a
(10)	33^a
(11)	-16^a
(12)	-72^a

^{a}MP2/6-31G* values.

3.6.26 $[C_2,H_9,N_2]^+ - m/z$ 61

Isomers

Four isomeric $[C_2,H_9,N_2]^+$ ions have been explored.

$$H_2N \overset{\cdots H \cdots}{\underset{H_2C - CH_2}{\diagdown}} NH_2$$

1

$$NH_2CH_2CH_2NH_3$$

2

$$\underset{NH_3}{\overset{NH_2}{\underset{H_2C - CH_2}{\diagdown}}}$$

3

$$((CH_3)_2NNH_2)H$$

4

Identification

Bouchoux et al.[100] have recorded the MI mass spectrum of (**1**) and observed only loss of ammonia to yield protonated aziridine (cy–C(H)$_2$C(H)$_2$N(H)$_2$–$^{1+}$, confirmed by CID mass spectrometry of the product ions). Their calculations showed that NH$_3$ loss occurs by the isomerization of (**1**) to (**2**) and (**3**) before decomposition. The TS leading to the formation of the CH$_3$CHNH$_2$$^{1+}$ product ion lies 208 kJ/mol above (**1**), which is 53 kJ/mol above the dissociation limit to protonated aziridine plus NH$_3$.

Thermochemistry

The relative energies of (**1**)–(**3**) were obtained by Bouchoux et al.[100] at the MP2/6-31G* level of theory to be 0, 47 and 110 kJ/mol, respectively. They calculated $\Delta_fH(1) = 566$ kJ/mol G2(MP2,SVP), a value close to that derived from the neutral Δ_fH and PA values (-17 kJ/mol and 951.6 kJ/mol, respectively, and $\Delta_fH(H^+) = 1530$ kJ/mol).[1]

On the basis of the reported heat of formation and proton affinity of dimethylhydrazine ((CH$_3$)$_2$NNH$_2$), $\Delta_fH((CH_3)_2NNH_3^{1+}) = 686$ kJ/mol ($\Delta_fH((CH_3)_2NNH_2) = 83$ kJ/mol[58] and PA = 927.1 kJ/mol,[1] respectively, and $\Delta_fH(H^+) = 1530$ kJ/mol[1]). Using Boulanger et al.[99] G3 value for $\Delta_fH((CH_3)_2NNH_2) = 94$ kJ/mol leads to a slightly higher $\Delta_fH((CH_3)_2NNH_3^{1+}) = 697$ kJ/mol.

3.6.27 [C$_2$,O]$^{+\bullet}$ – m/z 40

Isomers

Ionized carbonyl carbene (CCO$^{1+\bullet}$) has been investigated by mass spectrometry. Its isomer, COC$^{1+\bullet}$, is predicted by theory to be stable. Neutral CCO, ketenylidene, has been detected in dense interstellar clouds.[101–103]

Identification

CCO$^{1+\bullet}$ ions can be generated by the following methods: electron bombardment of carbon suboxide (C$_3$O$_2$) in a high-pressure ion source,[104] photodissociation of Fe(CO)$_5$$^{2+}$,[105] oxidation of C$_2$$^{1+}$ by O$_2$, the bimolecular reaction of CO$^+$ with CO[106] and the dissociation of ionized carbon suboxide.[107] The CID mass spectrum[107] of the ions generated by the latter dissociation is dominated by the m/z 28 (CO$^{1+\bullet}$) fragment peak. The remaining significant peaks at m/z 24 (C$_2$$^{1+\bullet}$) and 12 (C$^{+\bullet}$) are consistent with the CCO$^{1+\bullet}$ connectivity.

Thermochemistry

The latest experimental value for the heat of formation of $CCO^{1+\bullet}$, 1418 ± 19 kJ/mol (obtained from the measurement of the energy threshold for the formation of $C_2O^{1+\bullet}$ in the reaction $CO^{1+} + CO$),[108] is in good agreement with $\leq 1414 \pm 5$ kJ/mol (appearance energy of $C_2O^{1+\bullet}$ from $C_3O_2)^{107}$ and 1406 kJ/mol obtained by combining the calculated ionization energy (10.58 eV,[109] MP4SDQ/6-311G**//HF/6-311G** with scaled ZPVE) and heat of formation of C_2O (385 ± 19 kJ/mol[110]).

Note that although the $\Delta_f H(C_2O)$ listed above is in agreement with previous experimental[111] and theoretical values[112] and also consistent with the measured[110,113] and calculated (at a variety of levels of theory)[114] $C-CO$ bond dissociation energy in the ketenylidene molecule, NIST gives $\Delta_f H(C_2O) = 286.60$ kJ/mol.[1]

The ground state of $CCO^{1+\bullet}$ is calculated to be 148.2 kJ/mol lower in energy than the ground state of the isomer $COC^{1+\bullet}$ (QCISD(T)/6-311G(2df)),[115] in agreement with previous calculations[112] (MP4SDQ/6-311G**//HF/6-311G**).

3.6.28 $[C_2,H,O]^+ - m/z$ 41

The ketenyl cation ($HCCO^{1+}$) can be generated by dissociative ionization of $HC \equiv COCH_3$. From the measured appearance energy, $\Delta_f H(HCCO^{1+}) = 1129 \pm 15$ kJ/mol (AE = 12.44 ± 0.05 eV,[116] $\Delta_f H(CH_3^{1\bullet}) = 145.7$ kJ/mol and $\Delta_f H(HC \equiv COCH_3) = 75 \pm 10$ kJ/mol). The latter heat of formation is not recorded in the literature nor is it available from Benson's additivity.[117,118] However, the present authors propose two new additivity terms based on recent data, namely $C_t(O) = 123 \pm 5$ kJ/mol and $O(C)(C_t) = -119 \pm 5$ kJ/mol, which lead to the reevaluation of $\Delta_f H(HC \equiv COCH_3)$. Consequently, the $\Delta_f H(HCCO^{1+})$ value given above (1129 ± 15 kJ/mol) is different from the originally reported 1096 kJ/mol.[119] This new value is in adequate agreement with the 1140 kJ/mol obtained from recent G3 calculations.[120] The NIST WebBook[1] lists $\Delta_f H(HCCO^{1+}) = 1120$ kJ/mol (but no reference is given) and 1140 kJ/mol is obtained from an estimated value for the proton affinity of CCO (PA = 774.7 kJ/mol (MP2/6-31G*, MP4SDQ/6-31G*, MP4SDQ/6-31G**, and G1 levels of theory),[109] $\Delta_f H(C_2O) = 385 \pm 19$ kJ/mol[110] and $\Delta_f H(H^+) = 1530$ kJ/mol[1]).

It is noteworthy that the ketenyl radical ($HCCO^{1\bullet}$) has been identified as an intermediate in the oxidation of acetylene[121–123] and is therefore an important species in combustion reactions. The ground state for $[C_2,H,O]^\bullet$ is the doublet ketenyl radical.[124] Four low-lying isomeric forms, namely the trans- and cis-quartet states of $HCCO^{1\bullet}$, the hydroxy ethenyl radical ($CCOH^{1\bullet}$), and the cyclic oxiryl radical (cy$-$(H)CCO$-^{1\bullet}$), have all been investigated by theory.[124]

3.6.29 $[C_2,H_2,O]^{+\bullet}$ – m/z 42

Isomers

Ionized ketene ($CH_2CO^{]+\bullet}$), hydroxyacetylene (or ethynol) ($HC{\equiv}COH^{]+\bullet}$), oxirene (cy–$(H)CC(H)O–^{]+\bullet}$) and ketocarbene ($HC–C(H)O^{]+\bullet}$) have been investigated by experiment. Two more isomers, the $C=OCH_2^{]+\bullet}$ and $C=C(H)OH^{]+\bullet}$ radical cations, have been found by computations to be stable.

Identification

The m/z 42 fragment ions of $[C_2,H_2,O]$ composition are found in the mass spectra of a variety of molecular species.[125,126] Direct ionization of ketene generates the corresponding radical cation ($CH_2CO^{]+\bullet}$) whereas dissociative ionization of propiolic acid ($HC{\equiv}CC(O)OH$, loss of CO) yields the hydroxyacetylene radical cation ($HC{\equiv}COH^{]+\bullet}$).[127] The two ions' CID mass spectra are structure-indicative and distinct, permitting their facile differentiation. Both ions do however dissociate metastably by loss of CO but the kinetic energy releases are different, indicating that the isomers do not interconvert before this dissociation ($T_{0.5}(CH_2CO^{]+\bullet})=2.8$ meV[128] and $T_{0.5}(HC{\equiv}COH^{]+\bullet})=26$ meV[127]).

Partial CID Mass Spectra of the $[C_2,H_2,O]^{+\bullet}$ Isomers

Ion	12	13	14	17	21[a]	24	25	26	28	29
CH_2CO^{127}	13	41	100	—	8.2	13	24	16	69	56
$HC{\equiv}COH^{127}$	9.4	26	34	0.94	0.94	13	21	2.8	25	100
cy–$(H)CC(H)O–^{129}$	4.4	17	100	—	1.7	2.8	5.6	4.4	14	33
$HC–C(H)O^{130}$	4.7	27	54	—	—	7.3	33	70	35	100

[a]Corresponds to $[C_2,H_2,O]^{2+}$.

Dissociative ionization of molecules such as oxazole (cy–$(H)CC(H)OC(H)N–$, loss of HCN), isoxazole (cy–$(H)CC(H)ONN(H)–$, loss of N_2) and vinylene carbonate (cy–$OC(H)C(H)OC(O)–$, loss of CO_2) also yields a $[C_2,H_2,O]^{+\bullet}$ species.[129] The MI mass spectrum of this ion is identical to that of ionized ketene, but on the basis of the measured heat of formation of the ion generated (see Thermochemistry below), the production of $CH_2CO^{]+\bullet}$ was ruled out.[129] Its CID mass spectrum is also very similar to that of $CH_2CO^{]+\bullet}$ but the different fragment peak intensity ratios m/z 29:28, 29:41 and 26:25 and the lower intensity of the doubly charged ion suggest that ionized oxirene (cy–$(H)CC(H)O–^{]+\bullet}$) is present in the ion flux in a significant amount.[129] The lower intensity of the doubly charged ion in the CID mass spectrum of ionized oxirene is consistent with computations (at various of levels of theory), predicting it to be much less stable than CH_2CO^{2+}.[131] The m/z 42 ion flux generated upon collision-induced dissociation of vinylene carbonate molecular ions produced the highest yield of cy–(H)CC(H)O–$^{]+\bullet}$.[129]

Note that the CID mass spectrum of the radical cation produced by charge reversal (i.e., the removal of two electrons) of the $[C_2H_2O]$ radical anion, generated following decarboxylation of the vinylene carbonate radical anion (cy–OC(H)C(H)OC(O)–, loss of CO_2) is different from that of $CH_2CO^{]+\bullet}$ or $HC{\equiv}COH^{]+\bullet}$ and cannot be reproduced by a linear combination of the CID mass spectra of the latter two isomers. In order to account for the observed spectrum, the authors proposed that the ion generated upon charge reversal is yet another isomer, $HC–C(H)O^{]+\bullet}$, the ketocarbene radical cation (based on the structure of vinylene carbonate).[130] The intense m/z 29 and 26 fragment peaks are consistent with the proposed connectivity. However, in spite of these observations, calculations predict that this ion is not stable, readily isomerizing to lower energy species.[125] The prominent m/z 14 fragment peak ($CH_2^{]+\bullet}$) was therefore attributed to the *exothermic* isomerization of $HC–C(H)O^{]+\bullet}$ to energy-rich $CH_2CO^{]+\bullet}$, which then readily decomposes to $CH_2^{]+\bullet} + CO$. The stable m/z 42 cation generated after vertical removal of two electrons from the anion was proposed to be ionized oxirene obtained from the isomerization of $HCC(H)O^{]+\bullet}$ ions containing only a small amount of excess energy.[130]

Thermochemistry

Because of its interesting functionality, ketene, as well as its substituted analogs, has received considerable attention by both experimentalists and theoreticians.[132,133] Consensus on the value of its heat of formation has only been reached recently. The current recommended value is $\Delta_f H(CH_2CO) = -52$ kJ/mol (from both recent experiments[134,135] and computations[136,137]). This is in decent agreement with -47.7 ± 2.5 kJ/mol, the much earlier (1971) value obtained by measuring the heat of solution of ketene in aqueous sodium hydroxide.[138]

Ionized ketene is the global minimum on the potential energy surface[125] and by combining the heat of formation and ionization energy of the molecule (-52 kJ/mol and 9.613 ± 0.004 eV),[137] $\Delta_f H(CH_2CO^{]+\bullet}) = 876$ kJ/mol is obtained. This value is in agreement with both experimental[135] and computed values[137] (G3).

Calculations predict that $HC{\equiv}COH^{]+\bullet}$ lies 189 kJ/mol above $CH_2CO^{]+\bullet}$ with a computed barrier to its isomerization to ionized ketene of 330 kJ/mol (MP3/6-31G**+ZPE).[125] Note that this barrier is above the energy required for dissociation of ketene ions into $CH_2^{]+\bullet}$ and CO, and therefore insures that the ions do not interconvert before dissociation. The appearance energy of $HC\equiv COH^{1+\bullet}$ from $HC\equiv CC(O)OH$ gives $\Delta_f H(HC\equiv COH^{1+\bullet}) = 1072$ kJ/mol ($AE = 11.4 \pm 0.2$ eV,[127] $\Delta_f H(HC\equiv CC(O)OH) = -138$ kJ/mol (by additivity[117,118] and reevaluated by Holmes et al.[129] using new terms, $C_t(CO) = 117$ kJ/mol and $CO(C_t)(O) = -125$ kJ/mol. $\Delta_f H(CO) = -110.53 \pm 0.17$ kJ/mol[1])), which is only slightly higher than that predicted based on its relative energy with respect to $CH_2CO^{1+\bullet}$.

Ionized oxirene is the third lowest energy isomer and is predicted to lie 257 kJ/mol above $CH_2CO^{1+\bullet}$ (MP3/6-31G** + ZPE).[125] Vinylene carbonate loses CO_2 to produce ionized oxirene. From the measurement of the appearance energy of this ion $\Delta_f H(cy-(H)CC(H)O-^{1+\bullet}) = 1119$ kJ/mol ($AE = 11.86 \pm 0.05$ eV,[129] $\Delta_f H(cy-OC(H)C(H)OC(O)-) = -418.61^1$ and -393.51 ± 0.13 kJ/mol[1]), a result slightly lower than that based on the calculated relative energies. No transition structure for the direct isomerization of cy–(H)CC(H)O–$^{1+\bullet}$ to $CH_2CO^{1+\bullet}$ was found although a pathway involving the ketocarbene $HC-C(H)O^{1+\bullet}$ led to an estimated barrier of 161 kJ/mol (MP3/6-31G** + ZPE).[125] As mentioned above, the ketocarbene radical cation was found not to be a stable isomer but rather a transition structure. It is predicted to lie 418 kJ/mol[125] above $CH_2CO^{1+\bullet}$ (MP3/6-31G** + ZPE).[125]

The two other isomers, $C=OCH_2^{1+\bullet}$ and $C=C(H)OH^{1+\bullet}$, are computed to lie 259 and 376 kJ/mol, respectively, above $CH_2CO^{1+\bullet}$ (MP3/6-31G** + ZPE).[125] Only a small barrier of 35 kJ/mol prevents $C=C(H)OH^{1+\bullet}$ from isomerizing to $HC\equiv COH^{1+\bullet}$ whereas a larger barrier of 93 kJ/mol separates $C=OCH_2^{1+\bullet}$ from $CH_2CO^{1+\bullet}$.[125]

3.6.30 $[C_2,H_3,O]^+$ – m/z 43

Isomers

Computational chemistry finds no less than nine minima on the potential energy surface.

Of these, the acetyl (CH_3CO^{1+}, (1)), 1-hydroxyvinyl (CH_2COH^{1+}, (2)), oxiranyl (cy–$C(H)_2C(H)O-^{1+}$, (3)) and the oxygen-methylated carbon monoxide ($CH_3\cdots OC^{]+}$, (**9**)) cations have all been investigated and identified by experiment.

Identification

$[C_2,H_3,O]^+$ fragment ions are very commonly observed in the mass spectrum of a wide variety of oxygen-containing precursor molecules. Differentiation is difficult because all the isomers show common dissociation products. In addition, the dissociative ionization of many precursor molecules generates mixtures of isomeric species. Nonetheless, differentiation is possible but relies on the combination of different experimental techniques (mass spectrometric and thermochemical) and theoretical calculations.

Precursor molecules of general structure CH_3COR can be expected to give the acetyl cation by a simple bond cleavage. Indeed, biacetyl ($CH_3C(O)$-$C(O)CH_3$) produces an m/z 43 fragment ion upon electron-impact, which is expected to be $CH_3CO^{]+}$ (**1**). The metastable peak associated with this dissociation is a narrow Gaussian signal ($T_{0.5} = 0.9$ meV),[139] suggesting that this process occurs at the chemical threshold in keeping with the production of $CH_3CO^{]+}$ by a simple bond cleavage. Confirmation of the structure assignment is based on the measured heat of formation of the ion (appearance energy), which is consistent with the accepted value for $\Delta_f H(CH_3CO^{]+})$.[139] Acetyl cations dissociate metastably according to $CH_3CO^{]+} \longrightarrow CH_3^{]+} + CO$. This peak is of Gaussian profile with a $T_{0.5} = 1.5$ meV superimposed on a weak broader component.[139] This feature, the narrow Gaussian peak accompanying CO loss, is characteristic of $CH_3CO^{]+}$ ions and also (see below) is very characteristic of halogen-substituted acetyl cations. Using this as a structure-identifying tool, several precursor molecules were shown to generate the acetyl cation.[139] However, as can be seen in the following table,[140,141] these molecules and a great variety of others produce m/z 43 fragment ions having very similar overall CID characteristics. From these mass spectra, the best structure-identifying tool was found to be the m/z 29:28 fragment peak intensity ratio.[140] A ratio of ~0.67 is characteristic of source generated $CH_3CO^{]+}$ cations. However, reducing the ionizing electron energy or investigating metastably generated m/z 43 ions lowers the above ratio to ~0.45, likely showing that source ions, even those produced from acetone, are not only $CH_3CO^{]+}$ cations.[141] The charge stripping mass spectrum of this ion exhibits three significant doubly charged peaks at m/z 21.5 (43^{2+}), 21 (42^{2+}) and 20.5 (41^{2+}). The latter was however shown to be easily contaminated with isobaric $[C_3,H_5]^+$ ions, which produce an intense doubly charged peak. The peak abundance ratio m/z 21.5:21 characteristic of $CH_3CO^{]+}$ ions (obtained by metastable dissociation of ionized $CH_3C(O)C(O)CH_3$) is ~0.24.[139]

Partial CID Mass Spectra of $[C_2,H_3,O]^+$ Ions

Precursor Molecule	m/z									
	12	13	14	24	25	26	27	28	29	29:28
$CH_3C(O)CH_3$[140,141]	0.9	4.1	7.1	1.0	2.5	2.2	0.4	7.8	5.3	0.68
$CH_3C(O)C(O)CH_3$[139]	2	8	13	2	5	4	0.4	10	7	0.7
4 kV[a]	1	4	8	1	2	2		9	5	0.56
$86^{+\bullet} \longrightarrow 43^{+}$[b]	1	4	8	≤1	2	2		11	5	0.45
$CH_3C(O)OCH_3$[141]										0.67
EI at 13 eV										0.56
$74^{+\bullet} \longrightarrow 43^{+}$[b]										0.55
CH_3CH_2OD[139]										
m/z 43	1	6	10	1	3	3		8	5	0.63
m/z 44	≤1	2	4	1	3	3		4	10[c]	2.5
$ClCH_2CH(OH)CH_2Cl$[139]	1	4	13	≤1	3	4		4	16	4.0
$79^{+} \longrightarrow 43^{+}$[d]	1	3	10	≤1	2	3		4	20	5.0

[a]Ion acceleration voltage 4 kV. All others from 8 kV ions.
[b]Refers to the m/z 43 ion produced metastably.
[c]Corresponds to (HCO + DCO)/CO.
[d]m/z 79 ($ClCH_2CHOH^{1+}$) is obtained by loss of $CH_2Cl^{1\bullet}$ from the precursor molecule. $ClCH_2CHOH^{1+}$ further dissociates metastably by loss of HCl to give m/z 43 ion ($T_{0.5} = 1.5$ meV).[139]

Protonation of ketene using strong acids such as CH_5^{1+} produces the 1-hydroxyvinyl cation, CH_2COH^{1+} (2).[142] Consecutive losses of H[•] and H_2 from hydroxy deuterated ethanol (CH_3CH_2OD) also yields CH_2COD^{1+} ions.[139,143] This ion shows an m/z (29 + 30):28 peak abundance ratio of 2.5 (see table above) and the metastable peak accompanying the loss of CO is a broad Gaussian signal with a $T_{0.5} = 52$ meV.[139] Note that ionized CH_3CH_2OD also loses H[•] and HD to produce a mixture of the acetyl and oxiranyl cations.[139] The charge stripping mass spectrum of CH_2COH^{1+} shows a significantly more intense signal at m/z 21.5 (43^{2+}) and 21 (42^{2+}) than does the acetyl cation, in keeping with calculations that show it to be more stable than the doubly charged ion of acetyl (MP2/6-31G*//4-31G).[144] The m/z 21.5:21 ratio is 25.[139]

1,3-Dichloropropan-2-ol shows an m/z 43 fragment peak in its mass spectrum. The m/z 29:28 peak abundance ratio for this fragment ion is 4.0 (see table above), indicating that acetyl cations are not the major component of the ion flux, although their formation cannot be completely ruled out. From the two possible structures, CH_2CHO^{1+} and cy–$C(H)_2C(H)O-^{1+}$, the former can be eliminated based on computations that show that it rearranges without activation to the acetyl cation.[145] Therefore, the m/z 43 ion formed from 1,3-dichloropropan-2-ol was assigned the oxiranyl connectivity cy–$C(H)_2C(H)O-^{1+}$ (3).[139] The peak accompanying the metastable dissociation of this ion is non-Gaussian, with a $T_{0.5} = 20$ meV. Deconvolution of the peak yields two components: a narrow Gaussian peak, $T_{0.5} \approx 1.5$ meV, attributable to acetyl cations and a wider Gaussian, $T_{0.5} \approx 35$ meV (and therefore characteristic of cy–C(H)$_2$C(H)O–$^{]+}$).[139] The m/z 43 fragment ion from other precursor molecules such as propanal, divinylether, oxirane and methyloxirane showed the same characteristics.[139] Note that the mass spectrum of 1,3-dichloropropan-2d-ol contains a small m/z 44 fragment peak corresponding to a [C$_2$,H$_2$,D,O]$^+$ species. The metastable peak accompanying its fragmentation to CH$_2$D$^+$ + CO is essentially the same ($T_{0.5} = 55 \pm 5$ meV)[139] as that of CH$_2$COH$^{]+}$ and therefore strongly suggests that source generated m/z 43 ions from 1,3-dichloropropan-2-ol are compatible with a mixture of cy–C(H)$_2$C(H)O–$^{]+}$, CH$_3$CO$^{]+}$ and CH$_2$COH$^{]+}$ ions. In addition, 1,3-dichloropropan-2-ol dissociates by loss of CH$_2$Cl$^{]\bullet}$, producing ClCH$_2$CHOH$^{]+}$ ions that further dissociate metastably by loss of HCl to give m/z 43 ions.[139] On the basis of thermochemical considerations (described below), this fragment ion is believed to be cy–C(H)$_2$C(H)O–$^{]+}$ produced at its thermochemical threshold. The charge stripping mass spectrum of the oxiranyl cation (obtained by metastable dissociation of ClCH$_2$CHOH$^{]+}$) produces the less intense m/z 21.5 (43^{2+}) signal and the m/z 21.5:21 ratio is 0.04.[139]

The complex [CH$_3$$\cdots$OC]$^+$ (9) was generated from the dissociative ionization of dimethyl squarate according to the following reaction taking place in the ion source of the mass spectrometer:

The CID mass spectrum of the deuterated analog [CD$_3$$\cdots$OC]$^+$ (generated from dimethyl squarate-d$_6$) was slightly different from those of its isomers in that the signals corresponding to the fragment ions C$_2$$^{]+\bullet}$ and C$_2$D$^{]+}$ were very weak, indicating that the ion does not contain a C–C bond, and also contained an m/z 32 signal (CD$_2$O$^{]+\bullet}$), which was absent in the spectra of the other isomers.[146]

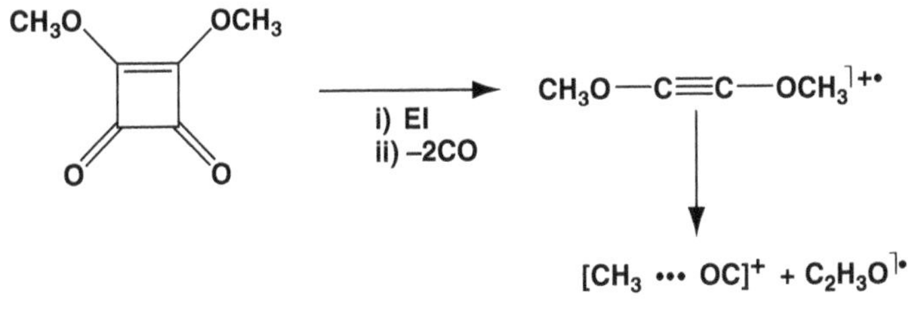

Ion–molecule reactions have also been used to differentiate between acetyl, 1-hydroxyvinyl and oxiranyl cations.[147,148] For example,[148] when the isomers are allowed to react with 1,3-dioxolane (cy–OC(H)$_2$C(H)$_2$OC(H)$_2$–) the mass spectrum of the products obtained are clearly isomer specific: CH$_3$CO$^{]+}$ produces predominantly the oxirane addition product as opposed to CH$_2$COH$^{]+}$ and cy–C(H)$_2$C(H)O–$^{]+}$, which generate species resulting from proton transfer, the yields of which are isomer dependent.

Thermochemistry

The acetyl cation is the global minimum on the $[C_2,H_3,O]^+$ potential energy surface.[145] The most recent experimental determination gives $\Delta_f H(CH_3CO^{]+}) = 659.4 \pm 1.1$ kJ/mol[149] (threshold photoelectron photoion coincidence), in good agreement with earlier measurements (657 ± 10 kJ/mol, $AE = 10.0 \pm 0.1$ eV,[139] $\Delta_f H(CH_3CO^{]•}) = -12 \pm 3$ kJ/mol[1] and $\Delta_f H(CH_3C(O)C(O)CH_3) = -326.8$ k/mol[1]) and 654.7 ± 1.1 kJ/mol[150] (photoionization mass spectrometry) and with *ab initio* calculations (658 kJ/mol[151] (G1) and 655.0 kJ/mol[152] (G2)). This value is also in keeping with the measured proton affinity of ketene, which yields a $\Delta_f H(CH_3CO^{]+}) = 653$ kJ/mol ($PA = 825.3$ kJ/mol,[1] $\Delta_f H(CH_2CO) = -52$ kJ/mol[134–137] and $\Delta_f H(H^+) = 1530$ kJ/mol[1]).

The heat of formation of $CH_2COH^{]+}$ has not been measured by experiment because no known precursor molecule exclusively generates this ion, but calculations predict it to lie 181 kJ/mol above acetyl (MP3/6-31G**).[145] It resides in a deep potential energy well and is protected against isomerization to acetyl by an additional barrier of 287 kJ/mol (this energy, ca 1127 kJ/mol, is well above the latter's dissociation limit to $CH_3^{]+} + CO$, $\Sigma\Delta_f H_{products} = 984$ kJ/mol).[1] The ion $CH_2CHO^{]+}$ is proposed as the intermediate species involved in the isomerization process.

An appearance energy measurement for the metastable dissociation of $ClCH_2CHOH^{]+} \longrightarrow$ cy–$C(H)_2C(H)O^{]+} + HCl$ gave $\Delta_f H($cy–$C(H)_2C(H)O^{]+}) = 904$ kJ/mol ($AE = 12.9 \pm 0.1$ eV),[139] in very good agreement with computations (MP3/6-31G**),[145] which place the oxiranyl cation 244 kJ/mol above acetyl. The oxiranyl cation can interconvert to acetyl over a barrier of 85 kJ/mol (i.e., at ca 989 kJ/mol, very close to dissociation to $CH_3^{]+} + CO$, ($\Delta_f H_{products} = 984$ kJ/mol)).[1] The ion $CH_2CHO^{]+}$ is again proposed to be an intermediate involved in the isomerization. For the other isomers, only the computed relative energies are available[145] and a good representation of the overall potential energy surface, including isomerization barriers, is given by Turecek and McLafferty.[141]

Computed Relative Energies of the $[C_2,H_3,O]^+$ Isomers (MP3/6-31G**)[145]

Ion	Relative Energy (kJ/mol)	Ion	Relative Energy (kJ/mol)
(1)	0		
(2)	181	(6)	330
(3)	244	(7)	425[a]
(4)	357	(8)	429[a]
(5)	358	(9)	216

[a]HF/4-31G.

3.6.31 $[C_2,H_4,O]^{+\bullet}-m/z\ 44$

Isomers

Computations at modest levels of theory find 11 minima on the potential energy surface.

CH₂CHOH   CH₃CHO   (oxirane, 3)
1   2   3

CH₂OCH₂   CH₂CH₂O   CH₃COH
4   5   6

CH₃OCH   [H₂CC ⋯ OH₂]   [HCCH ⋯ OH₂]
7   8   9

                 OH
CHCH₂OH   CH₂-CH
10   11

Ionized acetaldehyde ($CH_3CHO^{]+\bullet}$, (**2**)), its enol vinyl alcohol ($CH_2CHOH^{]+\bullet}$, (**1**)) and oxirane (cy–C(H)₂C(H)₂O–$^{]+\bullet}$, (**3**)) as well as methylhydroxycarbene ($CH_3COH^{]+\bullet}$, (**6**)), $CH_2OCH_2^{]+\bullet}$, (**4**) and methoxy-carbene ($CH_3OCH^{]+\bullet}$, (**7**)) have been characterized by mass spectrometry.

Identification

Ionization of acetaldehyde and oxirane produces the corresponding radical cations ((**2**) and (**3**), respectively). Their MI[153] (see Figure 1.17 and the discussion of these metastable peak shapes) and CID mass spectra are structure-characteristic.[154,155] The loss of D$^\bullet$ from ionized CH₃CDO displays an unusual isotope effect in that the KER for the D$^\bullet$ loss (104 meV) is appreciably larger than for H$^\bullet$ loss from ionized CH₃CHO (43 meV).[153] They are also distinct from those of ionized vinyl alcohol ($CH_2CHOH^{]+\bullet}$, (**1**)), which can easily be generated by dissociative ionization of a variety of molecular species such as aliphatic aldehydes (C4 to C8), cyclic alcohols (C4 to C6), vinyl ethers, epoxy alkanes (C4 to C8), halo ethanols (X = Cl, Br and I) and glycerol.[153–155]

The loss of CO_2 from metastable ionized pyruvic acid (CH₃C(O)C(O)OH) yields the methylhydroxycarbene radical cation ($CH_3COH^{]+\bullet}$, (**6**)),[156] whereas the methoxycarbene radical cation ($CH_3OCH^{]+\bullet}$, (**7**)) is obtained from the dissociative ionization of methyl glyoxylate (HC(O)C(O)OCH₃, loss of CO_2).[155]

Finally, dissociative ionization of 1,3-dioxolane (cy–C(H)₂C(H)₂OC(H)₂O–, loss of H₂CO) yields the isomer[157] $CH_2OCH_2^{]+\bullet}$ (**4**). It is interesting to note that the $CH_2CH_2O^{]+\bullet}$ (**5**) isomer, which could in principle be produced in the decarboxylation of ionized ethylene carbonate (cy–C(H)₂C(H)₂OC(O)O–, loss of CO_2), was proposed from ion cyclotron resonance experiments, to at least partially isomerize to a species with equivalent methylene groups, namely the $CH_2OCH_2^{]+\bullet}$ ion.[158] There is some controversy surrounding this because computations[158] later showed that the $CH_2CH_2O^{]+\bullet}$ ion collapses without activation to ionized acetaldehyde. It is however noteworthy that although the ICR experiments were unable to detect the presence of $CH_3CHO^{]+\bullet}$, the authors did not completely exclude the possibility of isomerization to other more stable $[C_2,H_4,O]^{+\bullet}$ isomeric forms.[158]

Partial CID Mass Spectra of the $[C_2,H_4,O]^{+\bullet}$ Isomers[155]

Ion	12	13	14	15	16	24	25	26	27	28	29	30	31	40	41	42	43
(1)	1.1	3.6	9.1	18	2.3	1.4	4.9	11	5.5	10	100	0.21	—	2.5	11	32	*
(2)	2.4	7.3	14	36	—	2.4	11	36	40	4.5	*	6.2	0.42	4.8	18	100	*
(3)																	
(4)	4.2	14	67	97	27	1.7	4.7	10	7.7	36	*	22	9.8	6.0	17	100	*
(5)	5.1	14	74	73	21	0.79	2.0	5.1	5.5	37	*	29	11	4.0	13	100	*
(7)	1.9	13	23	100	22	—	—	0.74	1.5	93	*	28	1.5	3.0	19	63	*

*Indicates a metastable process.

Thermochemistry

Ionized vinyl alcohol ($CH_2CHOH^{]+\bullet}$, (1)) is the lowest energy isomer. Its absolute heat of formation has been recently obtained by computation, $\Delta_f H(CH_2CHOH^{]+\bullet}) = 765$ kJ/mol (G2(MP2)).[159] This value is in excellent agreement with recent experiments that give 768 ± 5 kJ/mol (photoionization threshold energy determination of $CH_2CHOH^{]+\bullet}$ generated from cyclobutanol)[160] and 770.6 ± 1.3 kJ/mol (photoionization threshold energy determination of $CH_2CHOH^{]+\bullet}$ generated from six different precursor molecules)[161] but slightly higher than earlier appearance energy measurements.[162]

Calculated Relative Energies of $[C_2,H_4,O]^{+\bullet}$ (RHF/6-31G*//RHF/STO-3G)[163]

Ion	Relative Energies (kJ/mol)	Ion	Relative Energies (kJ/mol)
(1)	0		
(2)	52.3	(7)	136.7
(3)	182.6	(8)	212.8
(4)	124.9	(9)	233.2
(5)	237.3	(10)	311.5
(6)	72.3	(11)	219.5

The second most stable isomer, the acetaldehyde radical cation (**2**), is predicted to lie 55 kJ/mol above (**1**) (G2(MP2(full)/6-31G* level of theory),[164] in agreement with the earlier calculations[163] (see table above). Its absolute heat of formation, $\Delta_fH(CH_3CHO^{1+\bullet}) = 821$ kJ/mol, obtained by combining the heat of formation and ionization energy of the molecule ($\Delta_fH(CH_3CHO) = -166.1$ kJ/mol[49] and IE = 10.229 ± 0.0007 eV[1]) is consistent with the predicted relative energies. Note that NIST lists $\Delta_fH(CH_3CHO) = -170.7 \pm 1.5$ kJ/mol but the former value (-166.1 ± 0.5 kJ/mol),[49] together with related data for homologous aldehydes, is to be preferred, being consistent with the accepted additivity term for methylene[118] (C-(H$_2$)(C$_2$)).

Ionized oxirane (**3**) is higher in energy than both of the previous isomers. Its heat of formation, $\Delta_fH(cy-C(H)_2C(H)_2O-^{1+\bullet}) = 921$ kJ/mol, calculated by combining the heat of formation and ionization energy of the molecule, ($\Delta_fH(cy-C(H)_2C(H)_2O-) = -52.63 \pm 0.63$ kJ/mol[1] and IE = 10.15 eV[165] (in agreement with the calculated adiabatic IE, 10.2 eV (MP2/6-311G**)),[166] places it 153 kJ/mol above (**1**). This is markedly lower than the 182.6 kJ/mol predicted by theory (RHF/6-31G*//RHF/STO-3G).[163]

The symmetrical ring opened form of ionized oxirane, CH$_2$OCH$_2$$^{1+\bullet}$ (**4**), is lower in energy than the closed form with a measured $\Delta_fH(CH_2\,OCH_2^{1+\bullet}) = 856$ kJ/mol (appearance energy measured for the production of the [C$_2$,H$_4$,O]$^{+\bullet}$ ion from 1,3-dioxolane; AE = 10.87 eV,[162] $\Delta_fH(cy-OC(H)_2OC(H)_2C(H)_2-) = -301.7 \pm 2.2$ kJ/mol[1] and $\Delta_fH(H_2CO) = -108.6 \pm 0.46$ kJ/mol).[1] Therefore, the experimentally derived relative energy of the closed and open forms of oxirane is 65 kJ/mol, in agreement with the first series of computations (see table above)[163] but considerably lower than that predicted by the most recent theoretical calculations (85 kJ/mol at MP4/6-311G**[166] and 99.7 kJ/mol at CCSD(T)[167]). This wide range of results therefore deserves further investigation.

The methylhydroxycarbene radical cation (**6**) has been calculated to be 68.1 kJ/mol higher in energy than CH$_2$CHOH$^{1+\bullet}$ (G2)[164] with a computed absolute heat of formation of 839 or 835 kJ/mol, depending on the level of theory used (G2 and G2(MP2), respectively).[168] The $\Delta_fH(CH_3COH^{1+\bullet})$, evaluated from the appearance energy of the ion generated from pyruvic acid, lies in the 824–849 kJ/mol range due to the uncertainty with regard to $\Delta_fH(CH_3C(O)C(O)OH)$ (AE = 10.14 ± 0.05 eV,[56] $\Delta_fH(CO_2) = -393.51 \pm 0.13$ kJ/mol[1] and for $\Delta_fH(CH_3C(O)C(O)OH)$ refer to the section on [C$_3$,H$_4$,O$_3$]$^{+\bullet}$). The values are generally in agreement with the calculated relative energies (see table above, RHF/6-31G*//RHF/STO-3G).[163]

The measured appearance energy of the m/z 44 fragment ion from methyl glyoxylate, to which the methoxycarbene (**7**) connectivity was given, is $\Delta_fH(CH_3OCH^{1+\bullet}) = 930 \pm 5$ kJ/mol (AE = 10.32 ± 0.05 eV,[155] $\Delta_fH(HC(O)C(O)OCH_3) = -459$ kJ/mol[117,118] and $\Delta_fH(CO_2) = -393.51 \pm 0.13$ kJ/mol[1]). This value is somewhat higher than the computed values 912 and 909 kJ/mol

(G2 and G2(MP2), respectively).[168] Note that the computed values agree with the relative energies given in the table above. For the other isomers, only the computed relative energies presented in the above table are available.

3.6.32 $[C_2,H_5,O]^+$ – m/z 45

Isomers

Four ions have been characterized by experiment: the 1-hydroxyethyl cation (CH_3CHOH^{1+}, (**1**), protonated acetaldehyde), the methoxymethyl cation ($CH_3OCH_2^{1+}$, (**2**)), oxygen-protonated oxirane (cy–$C(H)_2C(H)_2O(H)$–$^{1+}$, (**3**)) and the vinyloxonium cation ($CH_2 = CHOH_2^{1+}$, (**4**)). Computations find seven other stable isomeric species.

CH_3CH_2O	CH_3COH_2	$CHCH_2OH_2$	$CH_3O(H)CH$
5	6	7	8

$[CH_3CO \bullet\bullet\bullet H_2]$	$[CH_4 \bullet\bullet\bullet HCO]$	$[CH_5 \bullet\bullet\bullet CO]$
9	10	11

Identification

The 1-hydroxyethyl cation (CH_3CHOH^{1+}, (**1**)) is generally obtained by the dissociative ionization of primary and secondary alcohols and some ethers and esters[169–179] as well as from ion–molecule reactions of acetaldehyde.[180] Fragmentation of ionized alkyl methyl ethers produces the methoxymethyl cation, $CH_3OCH_2^{1+}$ (**2**).[170,171,175,176,178,179,181] The CID mass spectra of these two isomers are very different. It is interesting to note that metastable methoxymethyl cations dissociate mainly by methane loss to generate an m/z 29 fragment ion peak ($[H,C,O]^+$) and that this peak is composite. This composite peak was initially interpreted as resulting from the generation of two reaction products, namely HCO^{1+} (major component) and HOC^{1+} (minor component).[182] Further experimental and computational results however show that only HCO^{1+} ions are generated but via two different transition states: the minor component arises from the dissociation of the complex $[CH_5 \cdots CO]^+$ (**11**) whereas the major component is from the complex $[CH_4 \cdots HCO]^+$ (**10**).[183]

  Protonated oxirane (cy–$C(H)_2C(H)_2O(H)$–$^{1+}$, (**3**)) is most directly generated by high-pressure protonation of the molecule.[184] Its CID mass spectrum resembles that of isomer (**1**) but shows a much greater m/z 31:30 fragment peak abundance ratio, consistent with its connectivity. Dissociative ionization of n-butanol and 1,3-propanediol generates a fourth isomer to which the vinyloxonium connectivity ($CH_2 = CHOH_2^{1+}$, (**4**)) was given.[185] The distinguishing feature in this ion's CID mass spectrum is the presence of an intense

m/z 19 fragment peak. The charge stripping region of the CID mass spectra (m/z 20.5–22.5) are also clearly structure-indicative.

CID Mass Spectra of the [C$_2$,H$_5$,O]$^+$ Isomers[185]

							m/z^a						
Ion	13	14	15	19	25	26	27	28	29	30	31	41	42	43
(1)	1	2	4	5	2	7	13	2	16	1	0.5	3	16	34
(2)	3	10	50	—	—	1	1	9	180	10	3	2	8	11
(3)	1	2	4	5	2	6	15	2	18	2	3	2	11	31
(4)	1	2	4	17	2	7	18	2	17	1	0.5	2	12	30

^{a}The peak abundances given are relative to the total ion abundance = 100, excluding m/z 29 for ion (2), m/z 19 for ions (1) and (3) and omitting the peak at m/z 44.

Charge-Stripping Region of the CID Mass Spectra of the [C$_2$,H$_5$,O]$^+$ Isomers[185]

			m/z		
Ion	20.5	21	21.5	22	22.5
(1)	1	13	83	3	—
(2)	—	2	—	98	—
(3)	1	10	76	3	10
(4)	0.5	3	11	6	80

Thermochemistry

The global minimum on the potential energy surface corresponds to the 1-hydroxyethyl cation (1).[186,187] Several appearance energy measurements give $\Delta_fH(CH_3CHOH^+) = 582$ kJ/mol,[188] whereas calculations yield values in the 587–609 kJ/mol range, depending on the level of theory used (G2,[187] G2(MP2), G2(MP2, SVP), CBS-Q, and CBS-q[189]). A value of 592 kJ/mol is obtained by combining the measured IE and the computed Δ_fH of the corresponding neutral (IE = 6.7 eV[1] and $\Delta_fH(CH_3CHOH^\bullet) = -54.03$ kJ/mol (G3MP2B3)[68]).

Calculations predict that the methoxymethyl cation (2) lies 72 kJ/mol (G2)[187] above (1) (not in disagreement with 85 kJ/mol from an earlier RHF study),[186] in good agreement with $\Delta_fH(CH_3OCH_2^+) = 657$ kJ/mol obtained from appearance energy measurements.[188] Calculations also yield absolute heat of formation values for this ion in the 662–694 kJ/mol range, depending on the level of theory used (G2,[187] G2(MP2), G2(MP2, SVP), CBS-Q and CBS-q[189]). A value of 667 kJ/mol is obtained by combining the measured IE and the computed $\Delta_f H$ of the corresponding neutral (IE $= 6.9$ eV[1] and $\Delta_f H(CH_3OCH_2]^{\bullet}) = 0.96$ kJ/mol (G3MP2B3)[68]). Combining the proton affinity and heat of formation of oxirane leads to $\Delta_f H(\text{cy–}C(H)_2C(H)_2O(H)\text{–}]^{1+}) = 703$ kJ/mol (PA(cy–C(H)$_2$C(H)$_2$O–) $= 774.2$ kJ/mol, $\Delta_f H(\text{cy–}C(H)_2C(H)_2O\text{–}) = -52.63 \pm 0.63$ kJ/mol and $\Delta_f H(H^+) = 1530$ kJ/mol),[1] in agreement with the calculated relative energy of this isomer (see table below). There are no absolute heats of formation values for the other isomers.

Calculated Relative Energies of the $[C_2,H_5,O]^+$ Isomers

Ion	Relative Energy[a] (kJ/mol)	Ion	Relative Energy[a] (kJ/mol)
(1)	0	(6)	322
(2)	72, 85[b]	(7)	321[c]
(3)	116, 123[b]	(8)	389
(4)	95, 101[b]	(9)	50
(5)[d]	393		

[a]G2 (0 K) values or otherwise stated.[187]
[b]MP3/6-31G**//RHF/4-31G.[186]
[c]RHF/4-31G//STO-3G.[186]
[d]Note that for the ethoxy cation, no stable minimum was found for the singlet state and therefore the relative energy listed corresponds to the *triplet state* of the cation; compare with the methoxy cation.

Calculated Barrier to Rearrangement[186]

Ion	Barrier (kJ/mol)	Ion	Barrier (kJ/mol)
3 ⟶ 2	262[a]	7 ⟶ 4	128[b]
4 ⟶ 1	220[a]	8 ⟶ 2	200[b]
6 ⟶ 4	116[b]	8 ⟶ 3	236[b]
6 ⟶ 1	196[b]		

[a]MP3/6-31G**//RHF/4-31G.
[b]RHF/4-31G//STO-3G.

3.6.33 $[C_2,H_6,O]^+$ – m/z 46

Isomers

Ionized ethanol ($CH_3CH_2OH]^{1+\bullet}$, (1)) and dimethyl ether ($CH_3OCH_3]^{1+\bullet}$, (2)) as well as the ionized ethylene/water ion–neutral complex ($[C_2H_4^{+\bullet}\cdots H_2O]$,

(3)) and the distonic ions $CH_3CHOH_2^{1+\bullet}$ (4) and $CH_3O(H)CH_2^{1+\bullet}$ (5) have all been characterized by experiments. Computations also suggest that the proton-bound species $[CH_2CH\cdots H\cdots OH_2]^{+\bullet}$ (6) is stable but is likely to be indistinguishable from isomer (1).

It is interesting to note that neutral ethanol is found in appreciable amounts in the region where stars are formed and it is believed to be produced from the dissociative recombination of the protonated form according to $CH_3CH_2OH_2^{1+} + e \longrightarrow CH_3CH_2OH + H^{\bullet}$.[190–194]

Identification

Ionization of molecular ethanol and dimethyl ether leads to the formation of ions (1) and (2). Isomers (3), (4) and (5) are generated in the dissociative ionization of 1,3-propanediol (loss of H_2CO),[195] lactic acid ($CH_3CH(OH)$-C(O)OH, loss of CO_2)[196] and methoxyacetic acid (loss of CO_2),[196] respectively. Note that the latter dissociation produces an m/z 46 peak containing four isobaric ions, namely $[C,H_2,O_2]^{+\bullet}$, $[^{13}C,O_2,H]^{+\bullet}$, $[C_2,H_6,O]^{+\bullet}$ and $[^{13}C,C,H_5,O]^{+\bullet}$ but increased mass resolution allows the latter two isomers to be resolved from the former pair. It is worth noting that isomer (6) was initially proposed to be the ion generated in the dissociative ionization of 2-methoxyethanol.[196] Further investigations however showed that this m/z 46 ion was isomer (1) and, moreover, when the precursor is collisionally induced to dissociate, the m/z 46 fragment ion is a mixture of isomers (1) and (5).[197] All five isomers have distinct and characteristic CID mass spectra.

Partial CID MS of the $[C_2,H_6,O]^{+\bullet}$ Isomers[196]

							m/z							
Ion	12	13	14	15	16	17	18	19	26	27	28	29	30	31
(1)	0.37	0.93	3.0	5.7	0.56	0.19	0.13	0.26	5.6	15	3.7	19	5.6	100
(2)	1.3	1.6	6.4	40	1.9	0.80	—	—	—	—	6.0	100	14	8.0
(3)	0.12	0.43	1.1	0.62	0.12	0.16	1.5	2.9	15	17	100	1.9	1.9	1.9
(4)	0.07	0.22	1.6	3.3	0.22	—	0.56	6.7	18	33	100	12	—	6.7
(5)ª	0.77	1.5	3.8	5.8	1.5	—	—	—	—	—	3.8	19	3.8	100

ªCorrected for the contribution from the isobaric $[^{13}C,C,H_5,O]^{+\bullet}$ ions.

It is worth noting that the metastable ion mass spectrum of the ion–neutral complex $[C_2H_4^{+\bullet}\cdots H_2O]$ ions shows a single, intense, narrow Gaussian peak at m/z 28 ($C_2H_4^{1+\bullet}$). The kinetic energy release was measured to be only about 0.2 meV[195] and is in keeping with the partners of the complex being electrostatically bound (i.e., not bound through a conventional covalent bond). The water molecule is free to roam around the ethene ion and there is no defined reaction coordinate for the dissociation, i.e., very little coupling between the ion and the neutral in the TS, resulting in the tiny KER.

Thermochemistry

Combining the heat of formation and ionization energy for both molecular species (ethanol and dimethyl ether) leads to $\Delta_f H(CH_3CH_2OH^{1+\bullet}) = 776$ kJ/mol and $\Delta_f H(CH_3OCH_3^{1+\bullet}) = 783$ kJ/mol ($\Delta_f H(CH_3CH_2OH) = -235.3 \pm 0.5$ kJ/mol and IE($\Delta_f H(CH_3CH_2OH) = 10.48 \pm 0.07$ eV; $\Delta_f H(CH_3OCH_3) = -184.1 \pm 0.5$ kJ/mol and IE($CH_3OCH_3) = 10.025 \pm 0.025$ eV).[1] The relative energies of all six isomers have been calculated. Note that the computations at the SD CI /6-31G** level of theory predict an energy difference between isomers (1) and (2), in agreement with that based on their respective absolute heat of formation.

Calculated Relative Energies of the [C$_2$H$_6$,O]$^{+\bullet}$ Isomers

Ion	Relative Energies (kJ/mol)		$\Delta_f H$(exp)(kJ/mol)
	MP3/6-31G**[198]	SD CI/6-31G**[197]	
(1)	43	31	776[a]
(2)	61	42	783[a]
(3)	0	0	740
(4)	17	11	
(5)	59	49	
(6)	—	38	
$C_2H_4^{1+\bullet} + H_2O$	83	79 (87)[b]	825[1]

[a]See text.
[b]Calculated at the UCCSD(T)6-311++G(d,p)//UB3LYP/6-31++G(d,p) level of theory.[199]

The measured appearance energy for the generation of isomer (3) from 2-methoxyethanol gives $\Delta_f H([C_2H_4^{+\bullet} \cdots H_2O]) = 740$ kJ/mol (AE = 10.45 ± 0.05,[200] $\Delta_f H(CH_3OCH_2CH_2OH) = -376.9 \pm 8.1$ kJ/mol[1] and $\Delta_f H(H_2CO) = -108.6 \pm 0.46$ kJ/mol[1]). Using this absolute heat of formation to compute the energies of isomers (1) and (2) gives support to the calculations obtained at the SD CI/6-31G** level of theory.[197]

Recent theoretical calculations[201] give absolute heats of formation for the α-distonic ions (4) and (5) in the range $\Delta_f H(4) = 747-751$ kJ/mol and $\Delta_f H(5) = 791-795$ kJ/mol, depending on the level of theory used. In the same study, the relative energy between the distonic ions and their ionized conventional counterparts was determined to $\Delta\Delta_fH(1\text{–}4) = -16$ to -22 kJ/mol and $\Delta\Delta_fH(2\text{–}5) = 9$ to 16 kJ/mol, not in significant disagreement with the earlier SD CI/6-31G** computations (see table above).

3.6.34 $[C_2,H_7,O]^+$ – m/z 47

Isomers

There are four postulated isomers, among which are the protonated forms of ethanol ($CH_3CH_2OH_2^{\rceil+}$, (1)) and dimethyl ether ($CH_3O(H)CH_3^{\rceil+}$, (2)) and the nonclassical ions $[C_2H_4\cdots H\cdots H_2O]^+$ (3) and $[CH_3\cdots HOCH_3]^+$ (4). The first three isomers have been generated and investigated by experiment.

Identification

Protonation (chemical ionization or self-protonation) of ethanol and dimethyl ether molecules in the ion source of a mass spectrometer leads to the generation of ions with mass spectral characteristics consistent with the expected connectivity. Dissociative ionization of 1-methoxy-2-propanol (CH_3OCH_2-$CH(OH)CH_3$, loss of $CH_3CO^{\rceil\bullet}$) and butane-2,3-diol ($CH_3CH(OH)CH(OH)$ CH_3, loss of $CH_3CO^{\rceil\bullet}$) also generates $CH_3O(H)CH_3^{\rceil+}$ and $CH_3CH_2OH_2^{\rceil+}$, respectively.[202] The CID mass spectra of $CH_3CH_2OH_2^{\rceil+}$ and $CH_3O(H)CH_3^{\rceil+}$ are dominated by fragment peaks at m/z 19 (H_3O^+) and m/z 29 ($C_2H_5^{\rceil+}$) with minor peaks in the m/z 26–28 region, consistent with the presence of a C–C bond, and m/z 29 ($HCO^{\rceil+}$) and 31 ($^+CH_2OH$), respectively.[77] The metastable isomers also show distinctive dissociations with (1) dissociating exclusively to $H_3O^+ + C_2H_4$ (KER = 2.5 meV) and (2) to $^+CH_2OH + CH_4$ (dished peak, KER = 800 meV), indicating that the ions do not interconvert before dissociation.[77]

It is noteworthy that in order to rationalize the observed labeling results, protonation of ethanol was proposed to yield, in addition to $CH_3CH_2OH_2^{\rceil+}$, the complex $[C_2H_4\cdots H_3O]^+$. This complex makes up the major portion of the m/z 47 ion beam generated when $C_2H_5^{\rceil+}$ ions (from C_2H_5Br) are allowed to react with H_2O in the ion source of the mass spectrometer, the ion cogenerated being $CH_3CH_2OH_2^{\rceil+}$. The MI mass spectrum of this ion is indistinguishable from that of $CH_3CH_2OH_2^{\rceil+}$ ions obtained from the dissociative ionization of butane-2,3-diol,[202] suggesting that these two ions freely interconvert before dissociation. However, the m/z 19:29 peak abundance ratio in the CID mass spectra of the m/z 47 ions depends on the origin of the ion, the ratio being larger for those ions produced by reaction of $C_2H_5^{\rceil+}$ and H_2O. This differentiates $CH_3CH_2OH_2^{\rceil+}$ from $[C_2H_4\cdots H_3O]^{+.202}$

The reactivity of isomers (1) and (2) toward several molecules of different proton affinities has been studied and shows that the ions can also be differentiated by either the reaction products or the rate with which reaction takes place.[203] It is noteworthy that the dissociations of the hypervalent dimethyloxonium radical $((CH_3)_2OH^{1\bullet})$ generated by NRMS of protonated dimethyl ether have been investigated by computations.[204]

Thermochemistry

From the proton affinity of ethanol and dimethyl ether, $\Delta_fH(CH_3CH_2OH_2^{1+}) = 518$ kJ/mol and $\Delta_fH(CH_3O(H)CH_3^{1+}) = 554$ kJ/mol ($PA(CH_3CH_2OH) = 776.4$ kJ/mol, $PA(CH_3OCH_3) = 792$ kJ/mol, $\Delta_fH(CH_3CH_2OH) = -235.3 \pm 0.5$, $\Delta_fH(CH_3OCH_3) = -184.1 \pm 0.5$ kJ/mol and $\Delta_fH(H^+) = 1530$ kJ/mol).[1] Note that the $\Delta_fH(CH_3O(H)CH_3^{1+})$ value obtained above is higher than both a computed value (539 kJ/mol, G2)[205] and a value deduced from the appearance energy of the ion generated from 1-methoxy-propan-2-ol (530 kJ/mol, using $AE = 9.54 \pm 0.05$ eV,[202] $\Delta_fH(CH_3OCH_2CH(OH)CH_3) = -402$ kJ/mol[118] and $\Delta_fH(CH_3CO^{1\bullet}) = -12$ kJ/mol[1]). The heat of formation of protonated ethanol has also been obtained by an appearance energy measurement, which leads to $\Delta_fH(CH_3CH_2OH_2^{1+}) = 520$ kJ/mol (using $AE = 10.00 \pm 0.05$ eV, $\Delta_fH(CH_3CH(OH)CH(OH)CH_3) = -481$ kJ/mol[49] and $\Delta_fH(CH_3CO^{1\bullet}) = -12$ kJ/mol[1]), in agreement with the value derived from the PA.

Computations[205] (G2) give $\Delta_fH([C_2H_4\cdots H\cdots H_2O]^+) = 563.2$ kJ/mol. This ion is separated from $CH_3CH_2OH_2^{1+}$ by a barrier to isomerization of only 27 kJ/mol, therefore below the dissociation limit to $H_3O^+ + C_2H_4$ and consequently allowing isomers (3) and (1) to interconvert before this dissociation. Isomer (4) is much higher in energy, calculated to lie 258 kJ/mol above $CH_3O(H)CH_3^{1+}$ and separated from it by a barrier of about 30 kJ/mol;[205] it is proposed as an intermediate species involved in the dissociation of $CH_3O(H)CH_3^{1+}$ to $CH_3^{1+} + CH_3OH$.[205]

3.6.35 $[C_2,O_2]^{+\bullet}$ – m/z 56

Isomers

The only possible bound isomer is the ionized oxycumulene, $OCCO^{1+\bullet}$. The ionized van der Waals complex, $(CO)_2^{1+\bullet}$, which is an important atmospheric species,[206] has also been investigated. It is worth noting that although attempts to generate the OCCO molecule date back as far as 1913,[207] it remains unisolated. Although its formation as a transient species in neutralization experiments is likely,[208] our present knowledge suggests that ethyenedione (OCCO) is intrinsically unstable.[209]

Identification

Ionized ethylenedione ($OCCO^{1+\bullet}$) can be generated by various means. For example, clustering of $CO^{1+\bullet}$ in the source of the mass spectrometer[210] and dissociative ionization of molecules containing an $-(O)CC(O)-$ moiety such as oxalic acid ($(C(O)OH)_2$), butane-2,3-dione ($(C(O)CH_3)_2$), oxalyl chloride ($(C(O)Cl)_2$) and squaric acid (cy–$(HO)CC(O)C(O)C(OH)-$) all produce ions with very similar mass spectral features.[107,211] Pandolfo et al.[212] suggested that a different $[C_2,O_2]^{+\bullet}$ species was produced on dissociation of $C_3O_2^{1+\bullet}$, but further investigation[208] clearly indicated that the ion produced was indeed the same as that generated from the other precursors. The ion, namely $OCCO^{1+\bullet}$, dissociates unimolecularly to $CO^{1+\bullet} + CO$ (KER $= 0.3$ meV).[107] Its CID mass spectrum is dominated by m/z 28 but also contains a prominent peak at m/z 40 ($C_2O^{1+\bullet}$) and minor peaks at m/z 44 ($CO_2^{1+\bullet}$), 24 ($C_2^{1+\bullet}$), 16 ($O^{+\bullet}$) and 12 ($C^{+\bullet}$).[107,211] The m/z 44 ion must result from a rearrangement of the OCCO connectivity.

It is noteworthy that the m/z 56 peak corresponding to $C_2O_2^{1+\bullet}$ often can be contaminated by the isobaric $[C_4,H_8]^{+\bullet}$ ion and therefore increased mass resolution is required to eliminate such a contribution from an isobaric ion. The use of higher mass resolution is necessarily achieved at the expense of sensitivity, which in this case has made the interpretation of the $C_2O_2^{1+\bullet}$ NRMS results more difficult.[107,213]

Thermochemistry

Appearance energy measurement gives $\Delta_f H(OCCO^{1+\bullet}) = 939 \pm 10$ kJ/mol (for $(C(O)OH)_2 \longrightarrow C_2O_2^{1+\bullet} + H_2O + O^{\bullet}$: AE $= 17.4 \pm 0.1$ eV,[107] $\Delta_f H(C(O)OH)_2 = -723 \pm 3$ kJ/mol,[58] $\Delta_f H(H_2O) = -241.826 \pm 0.040$ kJ/mol[1] and $\Delta_f H(O^{\bullet}) = 249.18 \pm 0.10$ kJ/mol;[1] ($C(O)Cl)_2 \longrightarrow C_2O_2^{1+\bullet} + 2Cl^{\bullet}$: AE $= 15.7 \pm 0.1$ eV,[107] $\Delta_f H(C(O)Cl)_2 = -335.8 \pm 6.3$ kJ/mol[1] and $\Delta_f H(Cl^{\bullet}) = 121.301 \pm 0.008$[1]) leading to a binding energy, relative to $CO+CO^{1+\bullet}$ ($\Sigma\Delta_f H_{products} = 1131$ kJ/mol),[1] of 192 kJ/mol. The ground state of the ion has been calculated to have a bent geometry and to lie between 9.09 and 9.32 eV, depending on the level of theory used,[209,214] above the ground (triplet) state of the neutral. The binding energy of the ion has also been computed to be in the range of 176–213 kJ/mol, again depending on the level of theory,[209,214] but generally in agreement with the experimental value given above.

It is worth noting that for the neutral molecule the ground triplet state is calculated to lie in the range of 305–337 kJ/mol, depending on the level of theory used,[209,214,215] below the dissociation products $CO(^3\Pi) + CO(^1\Sigma^+)$. The computed absolute heat of formation is $\Delta_f H(C_2O_2(^3\Sigma_g^-)) = 35$ kJ/mol (G2) and 23 kJ/mol (CCSD(T)).[214]

PEPICO studies indicate that the ionized van der Waals complex $(CO)_2^{1+\bullet}$ is bound by more than 124 kJ/mol.[216]

3.6.36 $[C_2,H,O_2]^+ - m/z$ 57

Because of the intrinsic instability of the C_2O_2 molecule (see discussion above), no proton affinity has been measured and there is consequently no heat of formation value for the $[C_2,H,O_2]^+$ ion. There are also no published mass spectral data pertaining to this ion. However, the corresponding neutral species, $[C_2,H,O_2]^\bullet$, has been investigated by theory as a possible intermediate species in the reaction between the ethynyl radical $(C_2H^{1\bullet})$ and O_2.[217,218]

3.6.37 $[C_2,H_2,O_2]^{+\bullet} - m/z$ 58

Isomers

Theoretical calculations predict no less than nine possible different atom connectivities.

Four ions have been identified by mass spectrometry: $HC(O)-C(O)H^{1+\bullet}$ (ionized glyoxal, (**1**)), the distonic ion $CH_2OC=O^{1+\bullet}$ (**6**), $HOC(H)C=C=O^{1+\bullet}$ (ionized hydroxyketene, (**7**)) and $HOC\equiv COH^{1+\bullet}$ (ionized dihydroxyacetylene, (**8**)).

Identification

Direct ionization of glyoxal produces ion (**1**) whereas the other isomers are obtained from dissociative ionization reactions: ethylene carbonate (cy–C(H)$_2$OC(O)OC(H)$_2$–; loss of H_2CO) yields ion (**6**),[219] glycolic acid[219] (HOCH$_2$C(O)OH; loss of H_2O) and 2-hydroxycyclobutanone[220] (cy–C(H)$_2$C(O)C(H)(OH)C(H)$_2$–; loss of C_2H_4) generate ion (**7**) and squaric acid (cy–(HO)CC(O)C(O)C(OH)–, loss of 2CO) produces ion (**8**).[220] Metastable ions (**1**), (**6**) and (**7**) predominantly dissociate by loss of CO to yield an m/z 30 fragment ion identified as (pure) ionized formaldehyde for isomer

(1),[221] predominantly $H_2CO^{1+\bullet}$ for isomer (6)[221] but a mixture of $H_2CO^{1+\bullet}$ and $HCOH^{1+\bullet}$ for isomer (7).[222] The metastable ion peak shapes and corresponding kinetic energy releases are significantly different: for isomer (1) the peak is Gaussian and very narrow ($T_{0.5} = 1.9$ meV), for isomer (6) it is flat-topped ($T_{0.5} = 270$ meV) and for isomer (7) it is Gaussian ($T_{0.5} = 16$ meV).[221]

In contrast, metastable ions (8) (produced by ionization of $HOC\equiv COH$) predominantly produce a fragment ion at m/z 29, which has been identified as COH^{1+} and not the more stable HCO^{1+} isomer.[220] The metastable ions' dissociation characteristic indicates that these isomers do not interconvert before dissociation. As expected, all four isomers have distinct and structure-characteristic CID mass spectra. It is noteworthy that $CH_2OC=O^{1+\bullet}$ and $C(H)C=C=O^{1+\bullet}$ exhibit intense doubly charged peaks (m/z 29).[221] It is also noteworthy that the mechanism by which glyoxal and glycolic acid ions generate m/z 30 fragment ions has been investigated by computations (CBS-QB3).[222]

Partial CID Mass Spectra of the $[C_2,H_2,O_2]^{+\bullet}$ Isomers[221]

	m/z									
Ion	12	13	14	28	29	30	40	41	44	56
(1)	0.20	0.36	0.34	20	52	100	—	—	—	12
(6)	1.6	2.0	8.8	48	64[a]	100	—	—	30	20
(7)	1.2	2.0	0.60	32	78[a]	100	6.0	26	—	14
(8)[220]	1.3	0.88	0.15	24	100	—	5.9	8.8	—	16

[a]Contains contribution from the doubly charged ion.

Thermochemistry

Combining the heat of formation and ionization energy listed in NIST for glyoxal, gives $\Delta_f H(HC(O)-C(O)H^{1+\bullet}) = 772$ kJ/mol ($\Delta_f H(HC(O)-C(O)H) = -212.0 \pm 0.79$ kJ/mol and IE = 10.2 eV))[1]. However, recent computations give a slightly lower IE value of 9.97 eV (B3LYP/cc-pVTZ),[223] which in turn leads to $\Delta_f H(HC(O)-C(O)H^{1+\bullet}) = 750$ kJ/mol. By using a computed value (G2)[224] for $\Delta_f H(HOC(H)C=C=CH_2^{1+\bullet})$ and assuming that the effect of OH substitution on the heat of formation is the same as in ionized ketenes and allenes, Suh et al. estimated $\Delta_f H(HOC(H)C=C=O^{1+\bullet}) = 657$ kJ/mol.[225] No absolute heat of formation value is available for the other isomeric species.

Relative Energies of the $[C_2,H_2,O_2]^{+\bullet}$ Isomers[223]

Ion	Relative Energies (kJ/mol)	
	CCSD(T)cc-pVTZ	B3LYP/cc-pVTZ
(1)	0	0
(2)	279	292
(3)	297	319
(4)	79	93
(5)	248	264
(6)	35	41
(7)	−92	−93
(8)	80	62
(9)	385	365

3.6.38 $[C_2,H_3,O_2]^+$ – m/z 59

Isomers

Theoretical calculations have found 11 stable isomeric forms, of which isomers (1)–(7) have clearly distinguishable mass spectral characteristics: cy–C(H)$_2$OC(OH)–$^{]+}$ (1), HOCH$_2$CO$^{]+}$ (2), CH$_3$OCO$^{]+}$ (4), HOCHC(O)H$^{]+}$(6) and CH$_2$OC(O)H$^{]+}$ (7).

Identification[226]

Dissociative ionization of $ICH_2C(O)OH$ (loss of $I^\bullet$) generates the hydroxyoxiranyl ion (1) (the ring opened form, $CH_2C(O)OH^{1+}$, was found by theory not to occupy a potential energy well).[226,227] The ion's CID mass spectral characteristics are in keeping with the assigned structure. Dihydroxyacetone ($HOCH_2$-$C(O)CH_2OH$, loss of $^\bullet CH_2OH$) produces upon dissociative ionization $[C_2,H_3,O_2]^+$ ions with a different CID mass spectrum. These ions could possibly have the connectivity of isomer (2) (from a direct bond cleavage) or isomer (8) (from a rearranged molecular ion). The presence in the CID mass spectrum of minor fragment ions at m/z 41 and 42 suggests that an isomer with a C–C–O backbone is present in the ion flux. This observation is consistent with the production of isomer (2) but does not exclude the cogeneration of isomer (8).

Ionized methylcarboxylates yield a single $[C_2,H_3,O_2]^+$ ion with mass spectral characteristics consistent with the methoxycarbonyl cation (4). In order to rationalize the presence of m/z 45 ($C(O)OH^{1+}$) in the CID mass spectrum of this ion, isomerization to the proton-bound species (5) has been proposed. Also, note that isomers (4) and (5) would be expected to have very similar mass spectral characteristics and therefore the identification of the latter ion in these experiments was not possible.

Partial CID Mass Spectra of the $[C_2,H_3,O_2]^+$ Isomers

Ion	m/z									
	12	13	14	15	16	17	18	28	29	30
(1)[a]	50	120	325	37	12	46	11	—	1000	345
(2)	—	1	2	—	—	—	—	26	69	26
(4)	6	12	39	(1000)	7	—	—	145	210	59
(6)[b]	7	21	46	—	—	—	—	120	1000	320
(7)	1	3	8	—	—	—	—	19	125	39

Ion	m/z							
	31	40	41	42	44	45	56	58
(1)	(500)	70	180	340	160	800	37	95
(2)	(1000)	—	4	5	—	—	—	9
(4)	—	—	—	—	410	63	11	46
(6)	600	25	115	57	—	—	75	440
(7)	(1000)	—	—	1	7	—	—	1

Values in parentheses contain contributions from metastable ion dissociation.

[a]Note that the mass spectrum also contains a broad unresolved signal in the m/z 24–26 region of intensity (ca 50).

[b]Note that the mass spectrum also contains a broad unresolved signal in the m/z 24–26 region of intensity (ca 15).

Protonation of glyoxal produces an ion with mass spectral characteristics consistent with the formation of isomer (**6**) while isomer (**7**) is obtained by the loss of $CH_3^{1\bullet}$ from ionized ethyl formate ($HC(O)OCH_2CH_3$).

Thermochemistry

Calculations find the methoxycarbonyl cation CH_3OCO^{1+} as the global minimum.[226,227] Its absolute heat of formation has recently been evaluated (G2) to be 540 kJ/mol,[227] significantly higher than 504 kJ/mol predicted by Blanchette et al.[226] The latter value was derived from $\Delta_fH(cy–C(H)_2OC(OH)–^{1+}) = 590 \pm$ 4 kJ/mol and Blanchette et al.'s computed relative energy (-86 kJ/mol at the CEPA2/6-31G*//6-31G* level above CH_3OCO^{1+}, not in disagreement with the G2 value of -76 kJ/mol).[227]

$\Delta_fH(cy–C(H)_2OC(OH)–^{1+})$ was obtained by two different routes, namely an appearance energy measurement and from a thermochemical analogy. In the experiment, the appearance energy of the ion in the reaction $ICH_2C(O)OH$ $\longrightarrow$ $cy–C(H)_2OC(OH)–^{1+} + I^\bullet$ was measured to be 10.86 ± 0.05 eV and combined with $\Delta_fH(ICH_2C(O)OH) = -351 \pm 4$ kJ/mol (additivity)[118] and $\Delta_fH(I^\bullet)$ yielded the value given above (590 ± 4 kJ/mol). To evaluate $\Delta_fH(cy–C(H)_2OC(OH)–^{1+})$ by thermochemical analogy, the effect on the heat of formation resulting from HO– substitution at a charge-bearing site is used: comparing $\Delta_fH(CH_2CH_2^{1+\bullet})$ with $\Delta_fH(CH_2CHOH^{1+\bullet})$ an enthalpy drop of -302 kJ/mol, results in $\Delta_fH(cy–C(H)_2OC(OH)–^{1+}) = 590$ kJ/mol. Note that this value should be raised further to 602 kJ/mol (and consequently $\Delta_fH(CH_3OCO^{1+}) = 516$ kJ/mol) when most recent thermochemical ancillary data are used ($\Delta_fH(CH_2CH_2^{1+\bullet}) = 1067$ kJ/mol,[1] $\Delta_fH(CH_2CHOH^{1+\bullet}) = 765$ kJ/mol[159] and $\Delta_fH(cy–C(H)_2OC(H)–) = 904$ kJ/mol[139,145]).

Several appearance energy measurements (originally from Blanchette et al.[226] and revised by Ruttink et al.[227]) give $\Delta_fH(CH_3OCO^{1+})$ values in the 519–556 kJ/mol range with an average value of 537 kJ/mol in very good agreement with the computed G2 value. Note that some of the precursor molecules used to generate the CH_3OCO^{1+} ions did show lower energy dissociations implying that a competitive shift could possibly result in upper limits for $\Delta_fH(CH_3OCO^{1+})$.[226] Because of the agreement with the G2 value, Ruttink et al.[227] have suggested this not to be the case. It is clear from this that the absolute heat of formation values for ions (**1**) and (**4**) are somewhat uncertain. Computations at the CBS-QB3 level of theory on the proton affinity of hydroxyketene yield $\Delta_fH(HOCH_2CO^{1+}) = 551$ kJ/mol,[228] in good agreement with the recommended value that is presented in the following table. For the other investigated isomers, no reliable heat of formation value could be obtained by experiment.

Relative Energies and Recommended $\Delta_f H$ Values of the $[C_2,H_3,O_2]^+$ Isomers

Ion	Relative Energies (kJ/mol) CEPA2/6-31G*//6-31G*$_{fs}$[226]	G2[227]	Recommended $\Delta_f H$[227] (kJ/mol)
(1)	86	76	615
(2)	15	6.7	548
(3)			
(4)	0	0	540
(5)	114	117	657
(6)	47	62	602
(7)	39	61	602
(8)	2.9		
(9)			
(10)		46	586
(11)		201	741

3.6.39 $[C_2,H_4,O_2]^{+\bullet}$ – m/z 60

Isomers

There are seven different species that have been investigated by experiment: $CH_3C(O)OH^{]+\bullet}$ ((1), ionized acetic acid), $CH_2=C(OH)_2^{]+\bullet}$ ((2), enol of ionized acetic acid), $HC(O)OCH_3^{]+\bullet}$ ((3), methyl formate radical cation), $HOCH_2C(O)H^{]+\bullet}$ ((4), ionized hydroxyacetaldehyde), $HOCH=CHOH^{]+\bullet}$ ((5), ethene-1,2-diol radical cation), $CH_3OCOH^{]+\bullet}$ ((6), ionized methoxy-hydroxy carbene) and the ion–molecule complex $[CH_2CO^{+\bullet}\cdots H_2O]$ (7).

Identification

Direct ionization of acetic acid, methyl formate and hydroxyacetaldehyde yields the radical cations (1), (3) and (4), respectively. Dissociative ionization of straight chain aliphatic acids (via a McLafferty rearrangement) generates the enol form of ionized acetic acid, isomer (2).[229] Dissociative ionization of glyceraldehyde ($HOCH_2CH(OH)C(O)H$) and 2-hydroxybutanal yields $HOCH=CHOH^{]+\bullet}$ ((5), by the loss of H_2CO and C_2H_4, respectively) whereas ionized 1,3-dihydroxyacetone dissociates (loss of H_2CO) to give the ionized ketene–water complex (7).[230] Finally, the carbene ion (6) is produced by loss of H_2CO from $(CH_3O)_2CO$.[156]

All the isomers can easily be identified by their distinct and structure-indicative CID mass spectra. The ions can also be distinguished using energy-resolved mass spectrometry.[231] It is noteworthy that, as expected for an ion neutral complex, metastable $[CH_2CO^{+\bullet}\cdots H_2O]$ dissociate predominantly by loss of H_2O and the kinetic energy released is negligible ($T_{0.5} \approx 0.2$ meV).[230]

CID Mass Spectra[a] of the $[C_2,H_4,O_2]^{+\bullet}$ Isomers

Ion						m/z						
	15	16	29	30	31	32	40	41	42	43	45	46
(1)	n/a	n/a	3.7	—	0.8	—	1.0	4.5	7.7	100	53	0.3
(2)	n/a	n/a	3.7	1.3	10	—	1.9	10	100	36	47	8.7
(3)	n/a	n/a	27	9.5	54	100	—	2	0.8	0	3.5	1.0
(4)	n/a	n/a	12	2.7	30	100	0.2	0.4	1.2	0.5	0.3	—
(5)	n/a	n/a	34	11	28	3.3	1.2	8.4	100	7.1	5.1	0.3
(6)[156]	7	5	10	—	3	—	—	—	—	7	100	n/a
(7)	n/a	n/a	3.7	1.1	3.6	3.0	0.5	4.0	100	2.2	2.0	0.4

[a]Spectra are from Terlouw et al.[230] or otherwise stated.

Thermochemistry

Combining the heat of formation and ionization energy of acetic acid yields $\Delta_f H(CH_3C(O)OH^{1+\bullet}) = 595$ kJ/mol ($\Delta_f H(CH_3C(O)OH) = -432$ kJ/mol[232] and IE $= 10.65 \pm 0.02$ eV[1]). Similarly, $\Delta_f H(HC(O)OCH_3^{1+\bullet}) = 690$ kJ/mol and $\Delta_f H(HOCH_2C(O)H^{1+\bullet}) = 659$ kJ/mol ($\Delta_f H(HC(O)OCH_3) = -355.5 \pm 0.8$ kJ/mol (in agreement with the G2 computed value -349 kJ/mol),[233] IE $= 10.835$ eV,[1] $\Delta_f H(HOCH_2C(O)H) = -324.8$ kJ/mol (G2)[152] and IE $= 10.20 \pm 0.10$ eV[234]). Note that NIST also lists $\Delta_f H(HC(O)OCH_3) = -362$ kJ/mol,[1] which is most likely too low.

Appearance energy measurements for the generation of the ionized enol form of acetic acid generated from several precursor molecules give, $\Delta_f H(CH_2 = C(OH)_2^{1+\bullet}) = 502 \pm 4$ kJ/mol[235] in agreement with a computed value of 510 kJ/mol (CBS).[236] The $\Delta_f H([CH_2CO^{+\bullet}\cdots H_2O]) = 579 \pm 5$ kJ/mol comes from the measurement of the appearance energy of the ion generated from 1,3-dihydroxyacetone (AE $= 10.12 \pm 0.05$ eV,[237] $\Delta_f H(HOCH_2C(O)CH_2OH) = -506$ kJ/mol (additivity)[118] and $\Delta_f H(H_2CO) = -108.6 \pm 0.46$ kJ/mol).[1] Finally, the heat of formation of the ionized carbene has been computed (G2) to be $\Delta_f H(CH_3OCOH^{1+\bullet}) = 640$ kJ/mol,[168] which is in agreement with the experimental value 650 ± 5 kJ/mol obtained from the measurement of the AE of the ion generated from dimethoxyacetone (AE $= 11.56 \pm 0.05$ eV,[156] $\Delta_f H(CH_3OC(O)OCH_3) = -574 \pm 5$ kJ/mol (additivity,[118] using a revised value for $CO(O)_2 = -129$ kJ/mol) and $\Delta_f H(H_2CO) = -108.6 \pm 0.46$ kJ/mol[1]).

3.6.40 $[C_2,H_5,O_2]^+$ – m/z 61

Isomers

In principle the protonation of acetic acid can produce two different $[C_2,H_5,O_2]^+$ ions, namely $CH_3C(OH)_2^{]+}$ ((**1**), carbonyl protonated) and $CH_3C(O)COH_2^{]+}$ ((**2**), hydroxy protonated) and the same is true for glycol-aldehyde, which in turn leads to $HC(OH)CH_2OH^{]+}$ ((**3**), carbonyl protonated) and $HC(O)CH_2OH_2^{]+}$ ((**4**), hydroxy protonated), whereas protonation of methyl formate yields $HC(OH)OCH_3^{]+}$ (**5**). Other isomeric forms have been investigated by theory: $[CH_3CO\cdots H_2O]^+$ (**6**), $CH_3O(H)C(O)H^{]+}$ (**7**), $CH_3O(H)COH^{]+}$ (**8**), $[CH_3O(H)\cdots H\cdots CO]^+$ (**9**), $CH_3OCOH_2^{]+}$ (**10**), $[CH_3OC\cdots OH_2]^+$ (**11**), $HOCH_2CHOH^{]+}$ (**12**), $CH_3OOCH_2^{]+}$ (**13**), $CH_3CHOOH^{]+}$ (**14**) and $[H_2CO\cdots H\cdots OCH_2]^+$ (**15**).

Identification

Carbonyl protonated acetic acid ($CH_3C(OH)_2^{]+}$, (**1**)) is generated either by chemical ionization[238] or by dissociative ionization of aliphatic carb-oxylic acids[239] and/or esters[240] by a double hydrogen transfer. The hydroxyl-protonated isomer ($CH_3C(O)COH_2^{]+}$, (**2**)) is obtained from the dissociative ionization of 2,4-dihydroxy-2-methylpentane (consecutive losses of $CH_3^{]\bullet}$ and C_3H_6).[240] Carbonyl-protonated glycolaldehyde ($HC(OH)CH_2OH^{]+}$, (**3**)), is obtained by chemical ionization.[152]

All three metastable ions lose H_2O but with different peak shapes and kinetic energy releases. For isomer (**1**), the peak is composite with the broad component having a $T_{(most\ probable)} = 456$ meV and the narrow component a $T_{(most\ probable)} = 20$ meV.[241] Isomers (**2**) and (**3**) show Gaussian peaks having measured kinetic energy releases ($T_{0.5}$) of less than 1.6^{240} and 41 meV,[152] respectively. The product ion obtained by loss of H_2O is $CH_3CO^{]+}$ for isomer (**2**)[240] and a mixture consisting mostly of $CH_3CO^{]+}$ ($\approx95\%$) with minor amounts of the isomers cy–$C(H)_2C(H)O-^{]+}$ and $CH_2COH^{]+}$ for isomer (**3**).[152] On the basis of the remarkably small kinetic energy release associated with the dissociation of isomer (**2**), it has been proposed that the ion generated is better represented as the ion–neutral complex (**6**).[152]

Protonated methyl formate ($HC(OH)OCH_3^{]+}$, (**5**)) is obtained by gas-phase protonation of the molecule or by dissociative ionization of the methyl esters of 3-methoxypropionic, isobutyric and hydroxyacetic acids, and meth-oxymethyl formate.[242] This ion is distinct from the others as its only meta-stable dissociation is the loss of CO, yielding $CH_3^+OH_2$.[242] Isomers (**1**), (**2**) and (**5**) have distinct CID mass spectra.

Partial CID Mass Spectra of Some $[C_2,H_5,O_2]^+$ Isomers

Ion	m/z										
	14	15	16	17	18	19	25	26	27	28	29
(1)[240]	5.0	8.1	0.7	0.6	0.7	0.2	1.0	1.6	0.7	3.0	15
(2)[240]	0.4	0.6	—	0.05	0.6	0.3	0.1	0.2	0.1	0.3	0.7
(5)[243]	6.8	34	1.7	1.7	3.4	0.7	—	—	—	—	100

Ion	m/z									
	30	31	33	40	41	42	43	44	45	46
(1)[240]	1.6	1.9	—	1.3	5.0	22	100	4.3	32	6.0
(2)[240]	0.05	0.2	0.2	0.1	0.9	3.5	100	0.1	0.2	0.03
(5)[243]	26[a]	41	60	—	—	—	—	—	31	12

[a]Approximate intensity since m/z 30 is not resolved from the m/z 29 peak.

Isomers (4) and (6) are proposed as intermediate species in the dissociation of protonated glycolaldehyde (3) to $CH_3CO^{1+} + H_2O$.[152] As for the dissociation of metastable protonated methyl formate (5), MNDO calculations[242] indicate that depending on the conformation of (5), it must first isomerize to ions (7) or (8), from which the complex (9) can be accessed, before dissociation to $CH_3^+OH_2 + CO$. In addition, to account for the results of ^{18}O labeling in the CID mass spectrum, another channel involving intermediates (10) and (11) is proposed.

Thermochemistry

The heat of formation of ions (1) and (5) can be deduced from the proton affinity of the molecules: $\Delta_f H(1) = 311$ kJ/mol and $\Delta_f H(5) = 391$ kJ/mol, the former value in agreement with a computed value of 313 kJ/mol (CBS)[236] and the latter value supported by the results of an appearance energy measurement[242] (PA($CH_3C(O)OH$) = 783.7 kJ/mol, $\Delta_f H(CH_3C(O)OH)$ = -432 kJ/mol,[232] PA($HC(O)OCH_3$) = 783.5 kJ/mol, $\Delta_f H(HC(O)OCH_3)$ = -355.5 ± 0.8 kJ/mol (in agreement with the G2 computed value -349 kJ/mol[233])) and $\Delta_f H(H^+)$ = 1530 kJ/mol[1]). Low level calculations predict that isomer (2) lies 15.9 or 28.4 kJ/mol below the dissociation products $CH_3CO^{1+} + H_2O$ (MNDO and MNDO/3, respectively).[240] G2 calculations give $\Delta_f H(3) = 426$ kJ/mol and the relative energies of isomers (3), (4) and (6) have been computed to be 0, 12 and -84 kJ/mol (MP2/6-31G*).[152] At the same level of theory, the isomerization barriers **3** $\longrightarrow$ **4** and **4** $\longrightarrow$ **6** are estimated to be 17 and 196 kJ/mol, respectively.

3.6.41 $[C_2,H_6,O_2]^{+\bullet}$ – m/z 62

Isomers

Several species have been identified by experiments, namely the molecular ions $HOCH_2CH_2OH^{]+\bullet}$ (**1**), $CH_3OOCH_3^{]+\bullet}$ (**2**) and $CH_3CH_2OOH^{]+\bullet}$ (**3**) and the unconventional isomers $[CH_3O\cdots H\cdots OCH_2]^{+\bullet}$ (**4**), $[CH_3O(H)\cdots H\cdots OCH]^{+\bullet}$ (**5**) and $[CH_2=CHOH\cdots H_2O]^{+\bullet}$ (**6**). Other species have been shown by theory to be stable and these are $(CH_3)_2OO^{]+\bullet}$ (**7**), $[CH_2O(H)\cdots H\cdots OCH_2]^{+\bullet}$ (**8**), $OCH_2CH_2OH_2^{]+\bullet}$ (**9**), $[CH_3OH_2\cdots C(H)O]^{+\bullet}$ (**10**), $CH_2O(H)-CH_2OH^{]+\bullet}$ (**11**), $[CH_3OH\cdots OCH_2]^{+\bullet}$ (**12**) and $[OCH_2\cdots(H)OCH_3]^{+\bullet}$ (**13**). Some of the latter ions have been proposed as intermediate species in the dissociation of the former.

Identification

Direct ionization of the corresponding molecular species generates molecular ions (**1**)–(**3**). These ions show distinct mass spectral characteristics consistent with their respective connectivities.

CID Mass Spectra of Some $[C_2,H_6,O_2]^{+\bullet}$ Isomers[244]

Ion	12	13	14	15	28	29	30	31	32
					m/z				
(1)	≤1	≤1	1	3	5	31	15	200	52
(2)	2	5	20	200	—	550	(270)	(920)	80
(3)	≤1	≤1	6	20	110	1000	30	5	2
(4)	≤1	≤1	2	15	9	112	53	668	166

Ion	33	42	43	44	45	46	47	61
				m/z				
(1)	(1000)	2	7	9	—	—	—	15
(2)	20	—	—	—	6	4	45	1000
(3)	7	9	100	20	(750)	—	110	20
(4)	(1000)	—	1	5	1	1	3	275

Values in parentheses contain contributions from metastable ion dissociation.

Ionized methyl glycolate $(HOCH_2C(O)OCH_3^{]+\bullet})$ loses CO to generate the proton-bound species $[CH_3O\cdots H\cdots OCH_2]^{+\bullet}$ (4).[244] Because the CID mass spectrum of isomer (4) is not significantly different from that of (1), a double-collision experiment was required to show that the m/z 32 ion produced from ionized ethelyne glycol (1) was $CH_2OH_2^{]+\bullet}$ whereas a mixture of $CH_3OH^{]+\bullet}$ and $CH_2OH_2^{]+\bullet}$ ions was found for (4).[244] The latter observation is consistent with the proposed structure (4). It is noteworthy that the dissociation mechanism for ionized ethylene glycol (1) has been investigated by a combination of experiment and theory.[245–248]

The enol of ionized methyl glycolate, generated from the dissociative ionization of $(CH_3)_2CHCH(OH)C(O)OCH_3$ (loss of $CH_3CH=CH_2$) and $CH_3OC(O)CH(OH)CH(OH)C(O)OCH_3$ (loss of $HC(O)C(O)OCH_3$), also loses CO (after a rearrangement of the ion) to yield a different proton-bound species $[CH_3O(H)\cdots H\cdots OCH]^{+\bullet}$ (5).[225] This structure assignment was based on thermochemical arguments as well as isotopic labeling studies.[225]

Under electron impact, 1,4-butanediol fragments to yield an ion consistent only with a complex of ionized vinyl alcohol and water (6).[249] This was deduced from its dissociation characteristics; both the MI and CID mass spectra of the ion are dominated by a single peak at m/z 44. From the appearance energy of the MI peak the m/z 44 ion was shown to have the $CH_2CHOH^{]+\bullet}$ structure.[249] The MI peak was reported to be very narrow ($T_{0.5}$ not given), in keeping with an ion–molecule complex involving H_2O. The $-(OD)_2$ diol gave an m/z 64 ion that lost solely D_2O, again consistent with the proposed connectivity. Computations at the 6-31G**/4-31G/BSSE level of theory[250] explored the potential surface of the various configurations of the complex (e.g., the H_2O binding to different sites in the $CH_2CHOH^{]+\bullet}$ ion) and found the minimum to be a structure of the form $[CH_2=CHO\cdots H\cdots OH_2]^{+\bullet}$.

Thermochemistry

The heats of formation for the ionized molecules are obtained by combining their respective molecular heats of formation and ionization energies. Therefore, $\Delta_f H(1) = 592$ kJ/mol $(\Delta_f H(HOCH_2CH_2OH) = -388 \pm 2$ kJ/mol and IE = 10.16 eV),[1] $\Delta_f H(2) = 752$ kJ/mol $(\Delta_f H(CH_3OOCH_3) = -126$ kJ/mol and IE = 9.1 eV)[1] and $\Delta_f H(3) = 756$ kJ/mol $(\Delta_f H(CH_3CH_2OOH) = -175 \pm 4$ kJ/mol (refer to Section 2.6) and IE = 9.65 eV (experiment and computation)).[251]

The appearance energy measurement for the production of $[CH_3O\cdots H\cdots OCH_2]^{+\bullet}$ from methyl glycolate leads to an upper limit of $\Delta_f H(4) \leq 592$ kJ/mol (AE = 10.9 $\pm$ 0.2 eV,[244] $\Delta_f H(HOCH_2C(O)OCH_3) = -570$ kJ/mol (thermochemical analogy, starting with $\Delta_f H(HOCH_2C(O)H) = -324.8$ kJ/mol (G2)[152])). A significant kinetic shift is assumed to operate in this dissociation as seen by the large kinetic energy release (140 meV) and the severe tailing observed when measuring the AE.[244] Computations at a modest level of theory (SDCI/Pople) predict isomers (4) and (5) to be 3 and 66 kJ/mol lower in energy than the dissociation products $HCO^{\uparrow \bullet} + CH_3{}^+OH_2$ ($\Sigma \Delta_f H_{products} = 619$ kJ/mol),[1] respectively, leading to estimated values for $\Delta_f H(4) = 616$ kJ/mol and $\Delta_f H(5) = 553$ kJ/mol.[252]

The appearance energy measurement for the m/z 62 ion generated from butane-1,4-diol yields $\Delta_f H(6) \leq 465$ kJ/mol (AE = 9.78 eV;[249] $\Delta_f H$(HO $(CH_2)_4OH) = -426 \pm 5.7$ kJ/mol[1] and $\Delta_f H(C_2H_4) = 52.5$ kJ/mol[1]). The upper limit arises because this AE is very close to the ionization energy of butane-1,4-diol (9.75 eV).[249] The calculated binding energy (88 kJ/mol) leads to $\Delta_f H([CH_2 = CHOH \cdots H_2O]^{+\bullet}) = 438$ kJ/mol, with $\Delta_f H(CH_2CH OH^{\uparrow +\bullet}) = 768 \pm 5$ kJ/mol[160] and $\Delta_f H(H_2O) = -241.826 \pm 0.040$ kJ/mol.[1]

As for the ionized ether oxide (7), computations give $\Delta_f H(7) = 876$ kJ/mol (B3LYP/6-311++G(d,p)).[253] Note that the same computations give $\Delta_f H(2) = 772$ kJ/mol, significantly higher than the value based on the heat of formation and ionization energy of the molecule. These two ions are separated by a barrier to interconversion of about 230 kJ/mol above the conventional ion (2).[253]

The relative energies of isomers (1), (4), (5), (8), (9), (10), (11), (12) and (13) with respect to neutral ethane-1,2-diol have also been calculated at HF/6-31G*, B3LYP/6-31G* and B3LYP/6-311+G** and are given below.[248] The calculated ionization energies are therefore 8.9, 9.1 and 9.4 eV obtained at HF/6-31G*, B3LYP/6-31G* and B3LYP/6-311 + G**, respectively, all significantly lower than that obtained by experiment.[249]

Relative Energies of $[C_2, H_6, O_2]^{+\bullet}$ Isomers[248]

Ion	HF/6-31G* (kJ/mol)	B3LYP/6-31G* (kJ/mol)	B3LYP/6-311 + G** (kJ/mol)
(1)	855.76	878.70	911.38
(4)	805.60	905.50	949.06
(5)	791.06	881.99	907.80
(8)	824.29	911.64	936.84
(9)	813.93	927.74	967.99
(10)	857.99	895.03	922.62
(11)	845.60	929.39	956.16
(12)	—	914.00	954.22
(13)	811.57	924.96	974.26

The energies are relative to neutral $HO(CH_2)_2OH$.

3.6.42 $[C_2,H_7,O_2]^+ - m/z\, 63$

Isomers

A number of isomers have been invested either by theory and/or by experiment: $HOCH_2CH_2OH_2^{1+}$ (**1**), $CH_3CH(OH)OH_2^{1+}$ (**2**), $CH_3OCH_2OH_2^{1+}$ (**3**), $CH_3O(H)CH_2OH^{1+}$ (**4**), $CH_3CH_2OOH_2^{1+}$ (**5**), $CH_3CH_2O(H)OH^{1+}$ (**6**), $CH_3OO(H)CH_3^{1+}$ (**7**), $[CH_3CH=O\cdots H\cdots OH_2]^+$ (**8**), $[CH_2=O\cdots H\cdots O(H)CH_3]^+$ (**9**), $[CH_3O(H)\cdots CH_2=COH]^+$ (**10**), $[CH_2=O\cdots CH_3OH_2]^+$ (**11**), $[CH_2=OCH_3\cdots OH_2]^+$ (**12**) and $CH_3O(OH)CH_3^{1+}$ (**13**). Attempts to generate and characterize isomers (**1**), (**2**), (**3**), (**5**), (**6**), (**7**), (**8**), (**9**) and (**13**) have been made.

Identification[254]

Protonation of glycol yields an ion whose CID mass spectrum is distinct from those of the other isomers except for the CID mass spectrum of the ion generated from the ion–molecule reaction of CH_3CHO and H_2O. On the basis of the precursor molecule (glycol), protonation is expected to produce isomer (**1**). The reaction of CH_3CHO and H_2O as well as dissociative ionization of $CH_3CH(OH)CH(CH_3)CH(OH)CH_3$ (chemical ionization followed by loss of C_4H_8 from the protonated species) produces an ion with a CID mass spectrum closely similar to $HOCH_2CH_2OH_2^{1+}$ with the small differences that isomer (**1**) shows a significant fragment peak at m/z 31 (CH_2OH^{1+}) and a peak at m/z 33 ($CH_3^+OH_2$) which is absent in this isomer. The dominant metastable dissociation of this ion is water loss. Interestingly, this dissociation is accompanied by a very small kinetic energy release ($T_{0.5} = 1$ meV) and the product ion was identified as CH_3CHOH^{1+}. Combined with the observation that a mixture of CH_3CHO and water in a chemical ionization source generates an ion with an identical CID mass spectrum (under such conditions a proton-bound pair is expected to be produced),[255] this ion was assigned the structure of the proton-bound species (**8**). Note, however, that it is not possible, based on the CID mass spectrum alone, to eliminate the possibility that isomer (**2**) is instead generated or that a mixture of the two structures is produced.

The product ion obtained following chemical ionization (using CH_5^{1+}) of dimethyl peroxide is consistent with the connectivity of isomer (**7**), the important structure-characteristic fragment ion peaks being m/z 15 (loss of CH_3OOH), 31 (loss of methanol), 47 (loss of CH_4) and 48 (loss of $CH_3^{1\bullet}$). Note that in order to rationalize the dominant peaks at m/z 31 (CH_2OH^{1+}) and 33 ($CH_3^+OH_2$), isomerization of (**7**) to (**9**) before these dissociations has been proposed.[256] The presence of a fragment peak at m/z 45 (H_2O loss) in the

CID mass spectrum of (7) is however surprising. When generated independently by the reaction of CH_3OH and H_2CO in a chemical ionization plasma, the dominant unimolecular dissociation of the proton-bound dimer $[CH_2=O\cdots H\cdots O(H)CH_3]^+$ (9) is H_2O loss. In order to account for this dissociation, isomerization to the complexes (11) and (12) before dissociation was proposed.[256]

Protonation of ethyl hydroperoxide can in principle lead to the formation of isomers (5) and (6). As was the case for methyl hydroperoxide, theoretical calculations (3-21G) predict that the α-protonated species, $CH_3CH_2O(H)OH^{1+}$ (6), is lower in energy and separated from (5) by a significant isomerization barrier (240 kJ/mol at MNDO), suggesting that if both species were formed, they would not interconvert before dissociation.[256,257] The CID mass spectrum of the ion is dominated by m/z 29 (loss of H_2O_2). Isomer (6) contains an intact H_2O_2 moiety, suggesting that it composes a significant portion of the ion beam. However, small fragment peaks at m/z 45 ($[C_2,H_5,O]^+$) and 43 (CH_3CO^{1+}) could indicate that isomer (5) is cogenerated.

Chemical ionization, followed by CO loss from $[HOCH_2C(O)OCH_3]H^+$, yields an ion whose CID mass spectrum is dominated by fragment peaks at m/z 31 (CH_2OH^{1+}) and 33 ($CH_3{}^+OH_2$). The kinetic energy released in the MI dissociation yielding m/z 33 (H_2CO loss) is very small ($T_{0.5} \approx 1.5$ meV), suggesting that the ion could have the proton-bound structure (9). However, the presence of the small peak at m/z 45 (H_2O loss), which is also present in the MI mass spectrum, indicates that a second species, possibly isomer (3), is cogenerated. The ions produced by the dissociative ionization of $HOCH_2CH(CH_3)CH_2OCH_3$ (loss of $C_3H_5{}^{1\bullet}$) and $HOCH_2-CH(OH)CH_2OCH_3$ (loss of $[C_2,H_3,O]^\bullet$) show very similar CID mass spectra with the exceptions that the ester yields more m/z 45 and, in the case of the diol, the production of m/z 45 surpasses that of m/z 33. These observations could indicate that all three precursors yield a mixture containing the same two isomers, namely (9) and (3), but in different proportions. That isomer (9) is generated in all three cases is consistent with the observed kinetic energy release of the m/z 63 $\longrightarrow$ m/z 33 + H_2CO reaction which is precursor-independent. This is however not the case for the m/z 63 $\longrightarrow$ m/z 45 + H_2O reaction and therefore could suggest that other isomeric forms are present.

Methylation of CH_3OOH under chemical ionization conditions (using CH_3F as methylating agent) generates yet another ion whose CID mass spectral characteristics are consistent with a C–O–C connectivity.[257] Isomer (13) was assumed to constitute at least a fraction of the ion beam generated. Note that no other isomeric form containing the C–O–C connectivity was considered.

CID Mass Spectra of Some $[C_2,H_7,O_2]^+$ Isomers[254]

Proposed Structures	m/z									
	13	14	15	16	17	18	19	25	26	27
(1)	1	3	8	—	—	1	9	2	8	19
(8) or (2)	—	2	5	—	—	2	21	1	5	12
(7)	—	7	210	—	—	—	—	—	—	6
(6) or (5)	1	4	9	1	1	2	16	8	38	125
(9) or (3)	—	—	15	—	—	—	—	—	—	—
(9) or (3)	—	2	13	—	—	—	3	—	—	—
(3) or (9)	—	2	15	—	—	3	7	—	—	—
(9)[256]		—	3						—	—
(13)[257,a]			340							

Proposed Structures	m/z									
	28	29	30	31	32	33	34	35	41	42
(1)	5	24	5	70	9	11	—	—	4	17
(8) or (2)	3	15	1	1	—	—	—	—	2	13
(7)	55	180	70	530	85	1000	—	—	—	—
(6) or (5)	55	1000	18	27	15	2	10	35	3	9
(9) or (3)	—	50	22	100	12	1000	—	—	—	—
(9) or (3)	3	36	15	87	9	1000	—	—	—	—
(3) or (9)	—	33	10	37	7	480	—	—	—	—
(9)[256]	—	50	40	150	20	1000				
(13)[257,a]		340	86		114					

Proposed Structures	m/z									
	43	44	45	46	47	48	59	60	61	62
(1)	42	31	1000	—	—	—	—	1	3	1
(8) or (2)	34	28	1000	—	1	1	3	3	6	13
(7)	—	—	50	4	65	110	—	6	35	175
(6) or (5)	53	5	43	—	6	—	—	3	1	12
(9) or (3)	—	—	12	—	—	—	—	—	—	24
(9) or (3)	—	4	330	—	3	—	—	—	7	9
(3) or (9)	7	16	1000	—	—	—	—	—	5	10
(9)[256]	—	—	24	—	—	—			—	1
(13)[257,a]			1000	829	86	57				

[a]Note that fragment ions produced by metastable decomposition of the precursor have not been taken into account.

Thermochemistry

From the proton affinity of the molecular species, $\Delta_f H(HOCH_2CH_2OH_2^{1+}) = 326$ kJ/mol and $\Delta_f H(CH_3OO(H)CH_3^{1+}) = 649$ kJ/mol ($\Delta_f H(HOCH_2 CH_2OH) = -388 \pm 2$ kJ/mol,[1] PA $= 815.9$ kJ/mol,[1] $\Delta_f H(CH_3OOCH_3) = -126$ kJ/mol,[1] PA $= 755 \pm 8$ kJ/mol (G2)[257] and $\Delta_f H(H^+) = 1530$ kJ/mol[1]). The latter value is in good agreement with the computed value of 656 kJ/mol (B3LYP/6-311++G**).[256] Some relative and absolute heats of formation have been computed but care must be taken since most of the computations were done at a very modest level of theory.

Computed Relative and/or Absolute Heat of Formation of the $[C_2,H_7,O_2]^+$ Isomers

Ion	3-21G[254]	Energies (kJ/mol) B3LYP/6-311++G**[256]	G2[257]
(1)	113		
(2)	47		
(3)	90		
(4)	74		
(5)	342		
(6)	326	656	
(7)	330		0
(8)	0		
(9)	58	345	
(10)		377	
(11)		416	
(12)		392	
(13)			−19

Computed Isomerization Energy Barriers of the $[C_2,H_7,O_2]^+$ Isomers

Isomers	Energy Barriers (kJ/mol) B3LYP/6-311++G**[256]	G2[257]
6 ⟶ 10	56	
10 ⟶ 9	52	
10 ⟶ 12	94	
11 ⟶ 12	24	
7 ⟶ 13		169

3.6.43 $[C_2,H_8,O_2]^{+\bullet}$ – m/z 64

Isomers

The two ion–molecule pairs $[CH_2OH_2]^{+\bullet}/CH_3OH]$ **(1)** and $[CH_3OH]^{+\bullet}/CH_3OH]$ **(2)** have been investigated by experiment. Several other isomers have been proposed as intermediates in the former ion's isomerization and dissociations. These are $[CH_3O(H)H^+\cdots OCH_3^\bullet]$ **(3)**, $[CH_3O(H)H^+\cdots(H)OCH_2^\bullet]$ **(4)**, $[(CH_3)(CH_2^\bullet)OH^+\cdots OH_2]$ **(5)** and $[CH_3OCH_3^{+\bullet}\cdots OH_2]$ **(6)**.

Identification

Tu and Holmes showed that collision-induced dissociation of the proton-bound dimer of methanol (generated under chemical ionization conditions) yields a significant fragment peak at m/z 64, which they initially proposed to be $[CH_2OH_2]^{+\bullet}/CH_3OH]$ **(1)** generated according to $[CH_3OH_2]^{+}/CH_3OH] \longrightarrow [CH_2OH_2]^{+\bullet}/CH_3OH] + H^\bullet$.[258] A labeling experiment confirmed the identity of the proton lost. This ion can also be more explicitly produced, again under CID conditions, from $[CH_3CH_2OH_2]^{+}/CH_3OH] \longrightarrow [CH_2OH_2]^{+\bullet}/CH_3OH] + CH_3^\bullet$.[258] Later calculations have however found that isomer **(1)** is a TS on the potential energy surface and that the ion is better represented as isomer **(4)**, $[CH_3O(H)H^{+\bullet}\cdots(H)OCH_2^\bullet]$.[259] This is not in disagreement with Tu and Holmes, who had indicated that there may be little barrier separating isomers **(1)** from **(4)**.[258] Using the same experimental technique, $[CH_3OH]^{+\bullet}/CH_3OH]$ **(2)** can be produced from $[(CH_3)_2OH]^{+}/CH_3OH] \longrightarrow [CH_3OH]^{+\bullet}/CH_3OH] + CH_3^\bullet$. Metastable ions of isomers **(4)** and **(2)** dissociate by the same routes, generating m/z 33 ($CH_3^+OH_2$ by loss of $[C,H_3,O]^\bullet$), 46 (loss of H_2O) and 49 (loss of $CH_3^\bullet$) fragment ions.[258] This indicates that the ions interconvert before dissociation. A distinctive feature of these two isomers is that for the methanol dimer ion **(2)**, labeling experiments have shown that all the hydrogen atoms of the ionic partner lose their positional identity before dissociation to $CH_3^+OH_2$ whereas in the case of isomer **(4)**, the same hydrogen atoms do not become equivalent before this process.[258] The mechanism by which these ions interconvert and dissociate has been investigated by labeling experiment[258] and computations (CBS-QB3).[259]

Thermochemistry

A complete potential energy surface (CBS-QB3),[259] accounting for the results of labeling experiments,[258] provides an insight into the energetics of the various isomeric species.

Calculated Energies and Isomerization Energy Barriers of the $[C_2,H_8,O_2]^{+\bullet}$ Isomers (CBS-QB3)[259]

Ion	$\Delta_f H$ (kJ/mol)	Isomers	Energy Barriers (kJ/mol)
(1)[a]	489	2 $\longrightarrow$ 3	13
(2)	540	3 $\longrightarrow$ 4	66
(3)	489	4 $\longrightarrow$ 5	109
(4)	454	5 $\longrightarrow$ 6	59
(5)	438		
(6)	482		

[a]Note that this ion is only a transition state for the degenerate isomerization of isomer (4) and not a stable ion.

3.6.44 $[C_2,H_9,O_2]^+$ – m/z 65

Isomers

Under high-pressure chemical ionization conditions, the proton-bound dimer of methanol, $[CH_3O(H)\cdots H\cdots(H)OCH_3]^+$ (1), can be generated. Computations find three other stable species: the ion–molecule complexes $[CH_3O(H)\cdots CH_3OH_2]^+$ (2), $[H_2O\cdots CH_3O(H)CH_3]^+$ (3) and $[H_2O\cdots HO(CH_3)_2]^+$ (4). Note that the complex $[H_3O\cdots O(CH_3)_2]^+$ was found to collapse to isomer (4) without activation.

Identification

As with the higher homologs,[260] metastable ions (1) show two competing dissociations: a simple bond cleavage yielding protonated methanol (loss of CH_3OH) and dehydration producing protonated dimethyl ether (loss of H_2O). These product ions are also the dominant fragment ion peaks observed in the ions' CID mass spectra: protonated methanol (m/z 33, 100%) and protonated dimethyl ether (m/z 47, 50%).[258]

Thermochemistry

Calculated Relative Energies and Isomerization Energy Barriers of the $[C_2,H_9,O_2]^+$ Isomers (MP2/6-31G*//MP2/6-31G*)[261]

Ion	Relative Energies (kJ/mol)	Isomers	Energy Barriers (kJ/mol)
$CH_3OH + CH_3OH_2$	0	1 $\longrightarrow$ 2	99
(1)	−151	2 $\longrightarrow$ 3	30
(2)	−56	3 $\longrightarrow$ 4	3
(3)	−102		
(4)	−175		

3.6.45 $[C_2,O_3]^{+\bullet} - m/z\ 72$

Isomers

The only identified cation is the $[OCO\cdots CO]^{+\bullet}$ complex.

Identification

Under chemical ionization conditions, CO_2/CO mixtures yield a $[C_2,O_3]^{+\bullet}$ ion. In order to rationalize the observed isotopic exchange reactions taking place in labeled mixtures, e.g., $C^{18}O_2/CO$ or $CO_2/^{13}CO_2$, an ion of OCOCO connectivity was proposed to be formed. The MI and CID mass spectra of the $C_2O_3^{+\bullet}$ ion show only two fragmentations corresponding to the production of $CO_2^{+\bullet}$ and $CO^{+\bullet}$.[262] These observations, combined with the results from computations, suggest that the ion is the complex $[OCO\cdots CO]^{+\bullet}$. A $[C_2,O_3]^{+\bullet}$ ion, although not isolated, is also presumed to be formed under low-pressure conditions in order to again explain the isotope exchange reactions taking place.

It is noteworthy that attempts to generate the isomeric trioxydehydroethene radical cation $(O_2C-CO^{+\bullet})$ have been made using charge reversal of the corresponding radical anion (generated by electron capture of 1,3-dioxolane-2,5-dione $(cy-C(H)_2OC(O)C(O)O-)$ followed by dissociation by loss of H_2CO). The charge reversal mass spectrum showed no peak at $m/z\ 72$ but only fragmentation products, indicating that the corresponding cation was not stable.[263]

Note also that the Van der Waals neutral complex $[CO_2\cdots CO]$, which possesses a T-shaped geometry as opposed to the linear geometry of the cation, has been extensively studied.[264–269] Several attempts ($^-NR^+$ and $^-NR^-$) to generate the bonded neutral trioxydehydroethene (O_2C-CO) have also been made.[263] On the basis of computations (CCSD-(T)/aug-cc-pVDZ//B3LYP/6-31G(d) level of theory), the triplet state of this neutral is a high energy but stable species. In these experiments, no recovery signal was observed. However, a small fragment peak at $m/z\ 56$ corresponding to $C_2O_2^{+\bullet}$ is present in the $^-NR^+$ mass spectrum. This signal is likely to come from a dissociating $[C_2,O_3]^{+\bullet}$ entity, which in turn could only be formed from the O_2C-CO transient neutral. These experiments certainly indicate that if O_2C-CO is formed, its lifetime is less than 10^{-6} s.

Thermochemistry[262]

The ground state of the complex $[OCO\cdots CO]^{+\bullet}$ is calculated to be the global minimum on the potential energy surface. It is stabilized with respect to its dissociation products $CO_2^{+\bullet}$ and CO by 139 or 115 kJ/mol depending on the level of theory (B3LYP and CCSD(T) levels, respectively). Other isomeric forms, which lie much higher in energies, have also been identified. It is noteworthy that the O-exchange reaction in the complex $[OCO\cdots CO]^{+\bullet}$ occurs over a energy barrier of 99 or 110 kJ/mol, again depending on the level of theory used (B3LYP and CCSD(T) levels, respectively).

3.6.46 $[C_2,H_3,O_3]^+$ – m/z 75

Isomers

Three isomers have been investigated by experiments, namely $HC(O)C(OH)_2^{]+}$, $[H_2CO\cdots H\cdots OCO]^+$ and $HC(OH)C(O)OH^{]+}$ (protonated glyoxylic acid).

Identification

Although the metastable ion mass spectra of ionized methyl glycolate $(HOCH_2C(O)OCH_3^{]+\bullet})$ and its enol $(HOCH=C(OH)OCH_3^{]+\bullet})$ are completely different, both isomers show the loss of $CH_3^{]\bullet}$. The $[C_2,H_3,O_3]^+$ ions produced have distinct and structure-characteristic CID mass spectra and were respectively assigned the connectivities $[H_2CO\cdots H\cdots OCO]^+$ and $HC(O)C(OH)_2^{]+}$.[225] Ionized ethyl lactate $(CH_3CH(OH)C(O)OCH_2CH_3^{]+\bullet})$ loses 43 mass units to produce an m/z 75 fragment ion. Early investigations indicated that an acetyl radical was lost accompanied by the cogeneration of protonated ethyl formate $(HC(OH)OCH_2CH_3^{]+})$.[270] However, reinvestigation has shown that in addition to $HC(OH)OCH_2CH_3^{]+}$, two other species were produced, one having a $[C_2,H_3,O_3]$ elemental composition. It was proposed, based on labeling experiments,[271] that part of the m/z 75 ion beam was composed of protonated glyoxylic acid, $HC(OH)C(O)OH^{]+}$. Metastable $HC(OH)C(O)OH^{]+}$ ions show a unique dissociation to m/z 19. Unfortunately, no CID mass spectrum for this ion is available.

Partial CID Mass Spectra for Two of the $[C_2,H_3,O_3]^+$ Isomers[225]

Ion	m/z					
	19	29	31	44	45	46
$[H_2CO\cdots H\cdots OCO]$	—	14	100	14	9.8	—
$HC(O)C(OH)_2$	69	100	—	—	40	25

Thermochemistry

The only available thermochemical data are estimates; $\Delta_f H(HC(O)C(OH)_2^{]+}) = 316$ kJ/mol[225] (based on the effect of OH substitution in $HC(O)CHOH^{]+}$ using the most recent thermochemical data: $\Delta_f H(CH_2=CHCHOH^{]+}) = 666$ kJ/mol (see section on $[C_3,H_5,O]^+$) and

$\Delta_f H$ $(CH_2=CHC(OH)_2]^{1+}) = 380$ kJ/mol (see section on $[C_3,H_5,O_2]^+$) and $\Delta_f H(HC(O)CHOH^{1+}) = 602$ kJ/mol (see section on $[C_2,H_3,O_2]^+$)) and $\Delta_f H([H_2CO\cdots H\cdots OCO]^+) = 285 \pm 4$ kJ/mol (from the Larson and McMahon relationship for estimating the heat of formation of proton-bound dimers[255] and using the proton affinities and $\Delta_f H$ values found in NIST[1]). It is interesting to note that the ion $HOC(H)C(O)OH^{1+}$ could in principle also be generated from $HOCH=C(OH)OCH_3^{1+\bullet}$ but because of its heat of formation (estimated to be 320 kJ/mol) and other thermochemical considerations, its production to any significant extent is believed not to take place.[225]

3.6.47 $[C_2,H_4,O_3]^{+\bullet} - m/z\ 76$

Isomers

Calculations have investigated eight different atom connectivities: ionized trihydroxyethene $(HOC(H)=C(OH)_2]^{1+\bullet}$, (**1**)), ionized glycolic acid $(HOCH_2 C(O)OH^{1+\bullet}$, (**2**)), the complexes $[CH_2OH\cdots OCOH]^{+\bullet}$ (**3**) and $[HOC(H)=C=O\cdots OH_2]^{+\bullet}$ (**4**), the β-distonic ion $(HO)_2COCH_2^{1+\bullet}$ (**5**), the carbene $HOC(H_2)OCOH^{1+\bullet}$ (**6**), $H_2OC(H_2)OC=O^{1+\bullet}$ (**7**) and $H_2OC(H_2)\ C(O)O^{1+\bullet}$ (**8**). Isomers (**1**)–(**4**) have all been generated and characterized.

Identification

The direct ionization of glycolic acid produces the corresponding radical cation (**2**). The metastable ionized enol of ethyl glycolate $(HOCH=C(OH)OC_2H_5^{1+\bullet})$ loses C_2H_4 to generate isomer (**1**)[272] whereas isomer (**3**) results from the decarbonylation of ionized β-hydroxypyruvic acid $(HOCH_2C(O)C(O)OH)$.[273] It is noteworthy that short-lived β-hydroxypyruvic acid radical cations (those that dissociate in the ion source) fragment to give isomer (**3**), but the longer lived ions (metastable) have time to isomerize to the enol form $(HOC(H)=C(OH)C(O)OH^{1+\bullet})$ before dissociating to isomer (**4**).[273] The first three isomers show distinctive CID characteristics and although the mass spectrum of isomer (**4**) is very similar to that of isomer (**1**), it differs from it in several ways; by the absence of the structure-characteristic peak m/z 46 $(HOCOH^{1+\bullet})$, by the presence of a fragment peak at m/z 48 (CO loss) and by the significantly reduced intensity of the m/z 19 peak.[273]

Partial CID Mass Spectra of the $[C_2,H_4,O_3]^{+\bullet}$ Isomers[273]

Ion	19	29	31	32	41	45	46	47	48	58
(**1**)	9.7	100	41.7	11	11	28	21	13	—	88
(**2**)	—	6.9	25	100	1.4	5.6	1.4	—	—	1.4
(**3**)	—	28	100	75	—	31	36	—	—	5.6
(**4**)	25	14	7.9	—	6.1	—	—	—	1.8	100

Thermochemistry

Calculations find the global minimum to be the ionized enol (isomer (**1**)),[273] with a computed (G2(MP2)) $\Delta_f H(HOC(H) = C(OH)_2]^{1+\bullet} = 300$ kJ/mol.[272] Calculations (MP2 and CCSD(T)) have also been made for the different conformers of the eight isomers.[273] For isomer (**2**) five different conformers were investigated and two of them are listed in the table below: conformer (**2a**) is believed to play a significant role in the high energy (source) reactions of ionized glycolic acid whereas conformer (**2e**) is the lowest energy species. All three conformers of isomer (**3**) are proton-bound species: (**3a**) can best be considered as a complex of $CH_2OH]^{1+}$ electrostatically bound to a $HOCO]^{1\bullet}$ radical whereas (**3c**) consists of a CO_2 molecule interacting with the $^\bullet CH_2{}^+OH_2$ distonic ion. Two conformers for isomer (**6**) were considered: (**6b**) possesses a slightly elongated C–O bond (1.587 Å) whereas in (**6a**) this bond is much longer (1.980 Å) and can be represented as $[HOC(H_2)\cdots OCOH]^{+\bullet}$. These two ions, as well as isomers (**7**) and (**8**), are considered as key intermediates in the decarboxylation of isomers (**2**) and (**3**).[273]

Relative Energies of the $[C_2,H_4,O_3]^{+\bullet}$ Isomers[273]

Ion	MP2	CCSD(T)
	Relative Energies (kJ/mol)	
(**1**)	0	0
(**2a**)	208	240
(**2e**)	125	114
(**3a**)	110	120
(**3c**)	42	61
(**4**)	91	85
(**5**)	67	74
(**6a**)	147	147
(**6b**)	174	182
(**7**)	179	196
(**8**)	200	198

3.6.48 $[C_2,H_2,O_4]^{+\bullet}$ – m/z 90

Ionization of oxalic acid generates a small molecular ion $((C(O)OH)_2]^{1+\bullet}$, 1.2%).[274,275] Metastable $(C(O)OH)_2]^{1+\bullet}$ ions dissociate by loss of $C(O)OH]^{1\bullet}$ (leading to m/z 45, which is the base peak in the normal mass spectrum) and by loss of CO_2 (leading to m/z 46).[274] It is noteworthy that $(C(O)OH)_2]^{1+\bullet}$ has been used as a precursor molecule for the generation of the ionized oxycumulene $C_2O_2]^{1+\bullet}$.[107] Combining the molecule's heat of formation and ionization energy leads to $\Delta_f H((C(O)OH)_2^{1+\bullet}) = 367$ kJ/mol $(\Delta_f H((C(O)OH)_2 = -723 \pm 3$ kJ/mol[58] and IE $= 11.30$ eV[1]).

3.6.49 $[C_2,S]^{+\bullet}$ – m/z 56

Isomers

Calculations predict that the linear and cyclic species, $CCS^{1+\bullet}$ (ionized thioxoethylidene) and cy–CCS–$^{1+\bullet}$, are both stable. The corresponding neutral C_2S species has been observed in interstellar media.[276–278]

Identification

The electron-impact mass spectrum of benzothiazole (C_7H_5NS) contains, albeit in very low abundances, signals corresponding to the polycarbon sulfide ions $C_nS^{1+\bullet}$ ($n = 2$–6).[279] The CID mass spectrum of the $[C_2,S]^{+\bullet}$ species is dominated by the m/z 44 (100%) fragment peak but also shows peaks at m/z 24 (8%) and 32 (32%).[279] This spectrum is identical to that of the ion generated from thioketene and is consistent with the C–C–S connectivity.

Thermochemistry

Calculations predict that the linear isomer is the lowest energy species,[280,281] with a binding energy of 177 kJ/mol with respect to the neutral ground state gas-phase atoms.[281] The cyclic isomer is higher in energy by 117 kJ/mol (DFT/B3LYP/6-311G*)[280] or 109 kJ/mol (B3LYP),[281] depending on the level of theory. It is noteworthy that the $CSC^{1+\bullet}$ species is unstable.[281]

3.6.50 $[C_2,H,S]^+$ – m/z 57

Isomers

Four isomeric forms of protonated C_2S have been investigated by theory: the linear and bent ions $HCCS^{1+}$ and $CCSH^{1+}$, respectively, and the cyclic ions cy–$SCC(H)$–$^{1+}$ and cy–$CCS(H)$–$^{1+}$. The carbon-protonated linear ($HCCS^{1+}$) isomer has been investigated by experiment. This is of interest in interstellar chemistry,[282] as it is believed to be produced from the reaction of $S^+ + HCCH$. In addition, its dissociative recombination with electrons is proposed to lead to C_2S.[283]

Identification[279]

The electron-impact mass spectrum of benzothiazole (C_7H_5NS) contains a fragment peak corresponding to protonated C_2S (HC_2S^{1+}). The CID mass spectrum of this ion is dominated by the fragment peak corresponding to H[●] loss, with minor peaks at m/z 33 (SH^{1+}), 44 ($CS^{1+\bullet}$) and 45 (HCS^{1+}).

Thermochemistry

Of the four isomers, $HCCS^{1+}$ is the lowest energy species. The ground state triplet is calculated to lie 521 kJ/mol below the dissociation products $H^{\bullet} + C_2S^{1+\bullet}$ (QCISD(T)/6-311G** + ZPVE).[279] The other isomers are predicted to be higher in energy by, depending on the level of theory, 138–245 kJ/mol for cy–SCC(H)$-^{1+}$, 258–296 kJ/mol for $CCSH^{1+}$ and 384–487 kJ/mol for cy–CCS(H)$-^{1+}$.[282] Although the proton affinity of C_2S is known (869.6 kJ/mol),[1] no absolute $\Delta_f H(HCCS^{1+})$ value can be obtained because no reliable $\Delta_f H(CCS)$ is available.

3.6.51 $[C_2,H_2,S]^{+\bullet}$ – m/z 58

Isomers

There are three possible isomers: the molecular ions of thioketene $(CH_2C = S^{1+\bullet})$, thiirene (cy–C(H)C(H)S$-^{1+\bullet}$) and ethynethiol (HC≡CSH$^{1+\bullet}$). Ionized thioketene has been investigated by experiment.

Identification

Under flash vacuum thermolysis (FVT), 1,2-dithioethene-3-one (cy–C(H)C(H)SSC(O)–) fragments by loss of SCO.[284] Spectroscopic (IR) analysis of the resulting product indicates that thioketene $(CH_2C = S)$ is cogenerated. Ionization of the products formed therefore leads to $CH_2C = S^{1+\bullet}$. Its CID mass spectrum is dominated by $H^{\bullet}$ loss, with other significant peaks at m/z 32 (S^+), 44 $(CS^{1+\bullet})$ and 45 (HCS^{1+}).[284] 1,2,3-thiadiazole (cy–C(H)C(H)SNN–) has also been identified as a precursor molecule for thioketene.[285] The latter is generated by thermolysis in the mass spectrometer's ion source. It is noteworthy that in order to account for labeling results that show sensitivity to the initial position of the label, an unidentified unsymmetrical species $(HC = C(H)S^{1+\bullet}$ or HC≡CSH$^{1+\bullet}$) is proposed to be present in addition to thioketene.[286]

Note also that metastable dissociation of the adduct ion $[OCS\cdots C_2H_2]^{+\bullet}$ produces a $[C_2,H_2,S]^{+\bullet}$ cation.[287] From the analysis of the kinetic energy release distribution, it is believed that a single isomer is produced. On the basis of thermochemical arguments, the generation of the HC≡CSH$^{1+\bullet}$ ion can be ruled out, but the data do not allow determining whether $CH_2C = S^{1+\bullet}$ or cy–C(H)C(H)S$-^{1+\bullet}$ is the product ion. However, later calculations (QCISD and CCSD) suggested that the ion generated was the cyclic isomer.[288]

Thermochemistry

Of the three isomers, $CH_2C = S^{1+\bullet}$ is the lowest energy species. From its computed heat of formation and ionization energy, $\Delta_f H(CH_2C = S^{1+\bullet}) = 1056$ kJ/mol ($\Delta_f H(CH_2C = S) = 193.6$ kJ/mol and IE = 8.94 eV at the G2(MP2) level of theory;[289] note that the IE is in agreement with the measured value, 8.77 eV[1]). The ions cy–C(H)C(H)S–$^{1+\bullet}$ and HC≡CSH$^{1+\bullet}$ are estimated to be higher in energy by about 33 and 136 kJ/mol, respectively (MP2/6-31+G(d)//HF/6-31+G(d) and MP2/6-31+G(d)//MP2/6-31+G(d) levels of theory).[287]

3.6.52 $[C_2,H_3,S]^+$ – m/z 59

Isomers

The latest theoretical computations find no less than 11 stable structures: CH$_2$C(H)S^{1+} (**1**), CH$_3$CS^{1+} (thioacylium, (**2**)), CH$_3$SC^{1+} (**3**), cy–C(H)$_2$C(H)S–$^{1+}$ (**4**), cy–C(H)C(H)S(H)–$^{1+}$ (**5**), cy–C(H)$_2$CS(H)–$^{1+}$ (**6**), CH$_2$CSH^{1+} (**7**), CH$_2$SCH^{1+} (**8**), [CS$\cdots$CH$_3$]$^{1+}$ (**9**), CHCS(H)$_2$$^{1+}$ (**10**), and CC(H)S(H)$_2$$^{1+}$ (**11**).

Identification

A large variety of sulfur-containing precursor molecules produce an m/z 59 fragment ion with indistinguishable CID[290,291] and charge-stripping[290] mass spectral characteristics, suggesting that a single isomer or mixture of isomers is generated. The high-energy CID mass spectra of the m/z 59 ions produced is dominated by the loss of H$^\bullet$ (m/z 58) and H$_2$ (m/z 57), with minor peaks at m/z 56 (C$_2$S$^{1+\bullet}$), 45 HCS^{1+}, 44 (CS$^{1+\bullet}$), 33 HS^{1+}, 32 (S$^{+\bullet}$), 27 (C$_2$H$_3$$^{1+}$), 26 (C$_2H_2$$^{1+\bullet}$), 25 (C$_2$H^{1+}), and 24 (C$_2$$^{1+\bullet}$). Note that a fragment ion corresponding to CH$_3$$^{1+}$ is conspicuously absent from the CID mass spectra, which therefore a priori suggests that the thioacylium ion (**2**) is not produced in any significant amount.

Under low-energy conditions,[291] again all m/z 59 ions generated from a variety of precursors showed, consistent with the above results, very similar CID mass spectra but with the important difference that the m/z 15 (CH$_3$$^{1+}$) fragment peak was a dominant feature. On the basis of this observation, the structure of the ion formed was ascribed the thioacylium (**2**) connectivity and the absence of the m/z 15 fragment peak in the high-energy CID mass spectra was rationalized by proposing that dissociation under these conditions occurred from an electronically excited state of the CH$_3$CS^{1+} cation and not the ground state.[291]

That the ion produced in the dissociative ionization of precursor molecules containing the CH$_3$–C–S moiety was in fact CH$_3$CS^{1+} had previously been implied from the results of the reactivity of the thioacylium ion produced in the ion–molecule displacement reaction CH$_3$C(O)SH + CH$_3$CO^{1+} ⟶ CH$_3$CS^{1+} + CH$_3$C(O)OH, which was the same as that for ions produced by the dissociative ionization of molecules such as CH$_3$C(S)OCH$_3$ and CH$_3$C(S)SCH$_3$.[292]

In addition, more recent reactivity experiments[147] suggest that thioacylium ions are not the only isomeric species produced. Reaction of isoprene with m/z 59 ions generated from a variety of precursor molecules are essentially identical, being dominated by a cycloadduct m/z 127 ion. In all cases, the m/z 127 ion produced was the same. However, when m-methylanisole is used as the reactant, the mass spectra of the products formed showed small, but significant precursor dependent differences. From these results, for example, the precursor molecules dithiolane (cy–SC(H)$_2$SC(H)$_2$C(H)$_2$–) and propylene sulfoxide (cy–C(H)$_2$SC(H)(CH$_3$)–) were proposed to yield predominantly the thioxiranyl cation (4).

Thermochemistry

The global minimum on the potential energy surface is the thioacylium ion (2) with a computed $\Delta_f H(CH_3CS^{1+}) = 885$ kJ/mol (G2),[293] in good agreement with the value (880 kJ/mol) obtained by combining the computed (G2(MP2))[289] proton affinity and heat of formation of thioketene (PA = 843.3 kJ/mol, $\Delta_f H(CH_2C=S) = 193.6$ kJ/mol, and $\Delta_f H(H^+) = 1530$ kJ/mol).[1] Note that the computed PA is not in great disagreement with the value given by Hunter et al. (826.2 kJ/mol).[1] The computed heat of formation of the other isomers is given below. The computational study has also investigated the transition state structures associated with the interconversion of the isomers, and for a complete listing, see Ref. [293].

Computed Heat of Formation Values of the [C$_2$,H$_3$,S]$^+$ Isomers (G2)[293]

Ion	$\Delta_f H^a$ (kJ/mol)	Ion	$\Delta_f H^a$ (kJ/mol)
(1)^b	1087	(7)	996
(3)	1230	(8)	1257^b
(4)	1007	(9)	1366
(5)	1040	(10)	1092
(6)	1259	(11)	1297

^{a}Values given are for 0 K.
^{b}The energy given is for the triplet state.

3.6.53 [C$_2$,H$_4$,S]$^{+\bullet}$ – m/z 60

Isomers

Computations find several stable isomeric species: CH$_3$CHS$^{1+\bullet}$ (ionized thioacetaldehyde, (1)), CH$_2$CHSH$^{1+\bullet}$ (ionized vinylthiol, (2)), cy–C(H)$_2$C(H)$_2$S–$^{1+\bullet}$

(ionized ethylene sulfide, (3)), $CH_2SCH_2^{1+\bullet}$ (4), $cy\text{-}C(H)C(H)_2S(H)\text{-}^{1+\bullet}$ (5) and $CH_3CSH^{1+\bullet}$ (6). However, results from experiments show that $[C_2,H_4,S]^{+\bullet}$ ions generated from a variety of precursors had very similar dissociation characteristics ($T_{0.5}$ for the loss of $H^\bullet$) and therefore suggest that the ions rearrange to a single reacting configuration before their reaction by $H^\bullet$ loss.

Identification

Pyrolysis of 2,4,6-trimethyl-1,3,5-trithiane ($cy\text{-}SC(H)(CH_3)SC(H)(CH_3)SC(H)(CH_3)-$) generates monomeric thioacetaldehyde,[294] which can then be ionized to produce the corresponding radical cation ($CH_3CHS^{1+\bullet}$). Dissociative ionization of cyclobutanethiol was used to generate $CH_2CHSH^{1+\bullet}$. These two ions, as well as those produced from several other precursors and including ionization of ethylene sulfide, dissociate metastably by loss of $H^\bullet$ and the kinetic energy release ($T_{0.5}$) associated with this dissociation is essentially the same in all cases.[295]

Precursor Molecule	$T_{0.5}$ (meV) $[C_2,H_4,S]^{+\bullet} \longrightarrow [C_2,H_3,S]^+ + H^\bullet$
$cy\text{-}SC(H)(CH_3)SC(H)(CH_3)SC(H)(CH_3)-$	133 ± 3
$cy\text{-}C(H)_2C(H)_2C(H)_2C(H)S(H)-$	127 ± 4
$CH_2CHSC_5H_{11}$	134 ± 7
$cy\text{-}C(H)_2C(H)_2S-$	152 ± 3
$-C(H)C(H)C(H)C(H)S-$	128 ± 4
$HSCH_2CH_2SH$	132 ± 6
$CH_3CH(SH)CH_3$	126 ± 6
$CH_3SC_2H_5$	126 ± 6

This observation strongly suggests that all the ions dissociate through a common reacting configuration. Deuterium labeling in the $CH_3CHS^{1+\bullet}$ and $CH_2CHSH^{1+\bullet}$ ions indicates that the former could indeed correspond to the reacting configuration.[295]

Note that reactivity experiments carried out on the $[C_2,H_4,S]^{+\bullet}$ ion generated by pentyl thiovinyl ether ($CH_2CHSC_5H_{11}$) suggest that a 1:1 mixture of $CH_3CHS^{1+\bullet}$ and $CH_2CHSH^{1+\bullet}$ was formed.[296]

Thermochemistry

The computed (G3) heats of formation values for the different isomeric ions are given below and show that $CH_3CHS^{1+\bullet}$ is the global minimum on the potential energy surface.

Ion	$\Delta_f H^{297}$ (kJ/mol)	Ion	Relative Energy[298] (kJ/mol)	Isomer	Barrier to Isomerization (kJ/mol)
(1)	936	(3)	0	3 $\longrightarrow$ 5	187
(2)	942	(6)	74	5 $\longrightarrow$ 2	18
(3)	955	$CH_3CS^{1+} + H^\bullet$	140 (146)	2 $\longrightarrow$ 6	197
(4)	987			6 $\longrightarrow$ 1	94

Value in parentheses is deduced from $\Delta_f H(CH_3CS^{1+}) + \Delta_f H(H^\bullet) - \Delta_f H(cy-C(H)_2C(H)_2S-^{1+\bullet})$; 885 kJ/mol (G2),[293] 957 kJ/mol[297] and 217.998 $\pm$ 0.006 kJ/mol,[1] respectively.

An earlier study indicated that $CH_2SCH_2^{1+\bullet}$ (4) was higher in energy than cy–$C(H)_2C(H)_2S-^{1+\bullet}$ (3) by 44 kJ/mol and separated from it by a barrier to interconversion of 180 kJ/mol (QCISD(full)/6-31G*).[299] Note that the heat of formation and ionization energy of ethylene sulfide (cy–$C(H)_2C(H)_2S-$) have been measured and together they give $\Delta_f H(cy-C(H)_2C(H)_2S-^{1+\bullet}) = 955$ kJ/mol, in excellent agreement with the calculated value shown in the table ($\Delta_f H(cy-C(H)_2C(H)_2S-) = 82 \pm 1$ kJ/mol[49] and 9.051 eV[297]).

The potential energy surface[297] for the isomerization of these species has been investigated (G3) and shows that interconversion amongst the species is possible, but some of the barriers are computed to be above the threshold for dissociation to $CH_3CS^{1+} + H^\bullet$.[297] This is in conflict with the experimental observations described above. Finally, it is worth noting that the G3 calculations could not locate a TS between ionized ethylene sulfide and its open form $CH_2CH_2S^{1+\bullet}$.[297]

3.6.54 $[C_2,H_5,S]^+ - m/z$ 61

Isomers

Four isomeric species have been investigated by experiment: $CH_3SCH_2^{1+}$ (1), CH_3CHSH^{1+} (2), cy–$C(H)_2C(H)_2S(H)-^{1+}$ (3) and $CH_3CH_2S^{1+}$ (4). Theoretical calculations find five other stable ions: $CH_2CH_2SH^{1+}$ (5), $CH_2CHSH_2^{1+}$ (6), $CH_2S(H)CH_2^{1+}$ (7), cy–$C(H)_2S(H)C(H)_2-^{1+}$ (8) and cy–$C(H)_3SC(H)_2-^{1+}$ (9).

Identification

$[C_2,H_5,S]^+$ fragment ions are very commonly found in the mass spectrum of sulfides and thiols. Isomers (1) and (2) can easily be generated by dissociative ionization, although several precursor molecules lead to mixtures.[300] For example, dissociative ionization of molecules such as $CH_3S(CH_2)_3CH_3$, CH_3SCH_2Cl, CH_3SSCH_3 yields the $CH_3SCH_2^{1+}$ ion whereas molecules such as $(CH_3)_2CHSH$, $((CH_3)_2CH)_2S$ and $(CH_3CH_2)_2S$ produce the isomer CH_3CHSH^{1+}. Chemical ionization of ethylene sulfide generates the protonated form cy–$C(H)_2C(H)_2S(H)-^{1+}$.[300]

Although the CID and MI mass spectra of a large variety of $[C_2,H_5,S]^+$ fragment ions are similar, some structure assignments have been made. A typical example of the CID mass spectra of isomers (1)–(3) is given below.

Partial CID Mass Spectra of the $[C_2,H_5,S]^{+\bullet}$ Isomers[300]

Precursor	Ion	m/z								
		25	26	27	28	32	33	34	35	44
CH_3SSCH_3	(1)	5.3	1.8	16	5.3	2.1	2.1	1.8	(34)	12
$(CH_3CH_2)_2S$	(2)	2.2	13	48	4.3	4.8	9.1	8.3	(100)	10
cy-$C(H)_2C(H)_2S(H)$	(3)	4.1	15	55	6.8	7.7	12.3	11	(100)	13

Precursor	Ion	m/z							
		45	46	47	56	57	58	59	60
CH_3SSCH_3	(1)	100	45	6.3	2.4	14	32	29	(8.4)
$(CH_3CH_2)_2S$	(2)	78	14	2.6	7.4	48	100	87	(48)
cy-$C(H)_2C(H)_2S(H)$	(3)	91	17	7.7	7.7	45	82	82	—

Values in parentheses contain contributions from metastable ion dissociation.

However, ^{13}C and deuterium labeling experiments[301,302] indicate that although the CID mass spectral data suggest a single isomeric structure, in certain cases mixtures are generated. For example, the precursor molecule $(CH_3)_2CHSH^{]+\bullet}$ loses $CH_3^{]\bullet}$ and the structure of the resulting $[C_2,H_5,S]^+$ ion was assigned, by CID, to $CH_3CHSH^{]+}$. The ^{13}C-labeling experiments show a significant preference for the loss of the CH_3 carbon and these results are best explained by evoking the presence of a mixture of ions.

As mentioned above, certain precursor molecules fragment to yield a mixture of isomeric species. This is the case of methyl ethyl sulfide (MeSEt), which was shown to generate a mixture of $CH_3SCH_2^{]+}$ and $CH_3CHSH^{]+}$ isomers.[300,302] However, a recent FT-ICR study suggests,[303] based on the reaction products obtained from the reaction of the $[C_2,H_5,S]^+$ ions generated from MeSEt, that the mixture produced not only contains the two previously mentioned isomers, but also contains a third ion assigned to $CH_3CH_2S^{]+}$. Indeed, in addition to proton transfer and hydride abstraction, which are characteristic of $CH_3CHSH^{]+}$ and $CH_3SCH_2^{]+}$, respectively, charge transfer reactions are observed when reactants possessing ionization energies below 8.8 eV are used.[303] This threshold value is consistent with the recombination energy of the triplet $CH_3CH_2S^{]+}$ cation. It is estimated that at an ionizing energy of 13 eV, the $[C_2,H_5,S]^+$ ion beam produced from MeSEt is composed of about 11% of $^3CH_3CH_2S^{\rceil+}$.[303] That $^3CH_3CH_2S^{\rceil+}$ could possess a lifetime allowing its experimental observation is supported by calculations (several levels of theory) that show that at internal energies of less than 40 kJ/mol, this ion is not expected to easily dissociate unimolecularly.[303] It is clear from these data that the mass spectral characteristics of these isomers do not allow a straightforward assignment of the connectivities of the ions.

Thermochemistry

G2 calculations find isomer (2) to be the global minimum on the potential energy surface. The data for all isomers are given below.

Ion	G2 $\Delta_f H^a$ (kJ/mol)	Ion	G2 $\Delta_f H^a$ (kJ/mol)
(1)	832 (816)[304]	(6)	871
(2)	819 (793 ± 4)[305]	(7)	1126 (1130)[302]
(3)	829 (805)[b]	(8)	1002
(4)[c]	1004 (997 ± 8)[305]	(9)	1012
(5)[c]	1096		

Values in parentheses are from experiments.

[a]The $\Delta_f H$ values are those of the most stable conformer at 298 K from Chiu et al.[306]
[b]From the PA(cy–C(H)$_2$C(H)$_2$S–) = 807.4 kJ/mol[1] and $\Delta_f H$ (cy–C(H)$_2$C(H)$_2$S–) = 82 ± 1 kJ/mol.[49]
[c]Note that the stable ion is the triplet state.[306]

The computed potential energy surface (G2)[306] for these ions is quite complicated and therefore will not be discussed here. However, it is clear from it that isomers (2) and (3) can interconvert below the dissociation limit for the production of $C_2H_2+SH_3^{\rceil+}$, consistent with the similarities of the CID mass spectra. Note that extensive computations have also been made on the fragmentation pathways[307] of $CH_3SCH_2^{\rceil+}$ and $CH_2CHSH_2^{\rceil+}$, as well as on the H_2 and CH_4 eliminations[308] from $CH_3SCH_2^{\rceil+}$ and $CH_3CHSH^{\rceil+}$.

3.6.55 $[C_2,H_6,S]^{+\bullet}$ – m/z 62

Isomers

In addition to the molecular ions $CH_3CH_2SH^{\rceil+\bullet}$ and $CH_3SCH_3^{\rceil+\bullet}$, the β-distonic ion $CH_2CH_2SH_2^{\rceil+\bullet}$ has also been characterized by experiment. Computations find two other species to be stable: the α-distonic ions $CH_2S(H)CH_3^{\rceil+\bullet}$ and $CH_3CHSH_2^{\rceil+\bullet}$.

Identification

Ionization of ethanethiol and dimethyl sulfide generates the corresponding molecular ions. These two ions have distinct CID mass spectral characteristics, which allows their differentiation.[309,310] The decarbonylation of ionized S-ethyl thioformate ($HC(O)SCH_2CH_3^{]+\bullet}$) produces an m/z 62 fragment ion whose CID mass spectrum shows small but significant differences from $CH_3SCH_3^{]+\bullet}$, namely increased m/z 46 and 34 ($SH_2^{]+\bullet}$), less intense m/z 29 ($C_2H_5^{]+}$), and the presence of a charge stripping peak at m/z 31 ($[C_2H_6,S]^{2+}$). On the basis of these observations, it was proposed that the ion flux consisted of a mixture of the conventional isomer and the β-distonic ion $CH_2CH_2SH_2^{]+\bullet}$.

Partial CID Mass Spectra of the $[C_2,H_6,S]^{+\bullet}$ Isomers

Ion	m/z											
	14	15	26	27	28	29	34	35	45	46	47	61
CH_3SCH_3 [a,309]	13	14	14	18					66	59	100	66
CH_3CH_2SH [b,310]			20	67	27	70	17	23	48	37	100	43
$CH_2CH_2SH_2$ [b,310]			27	70	37	40	30	27	43	67	100	33

[a]Helium was used as target gas.
[b]Nitrogen was used as target gas.

The CID mass spectrum of the ion generated by the transfer of ionized ethene from $CH_2CH_2OCH_2^{]+\bullet}$ to H_2S was similar to that obtained by the decarbonylation of ionized S-ethyl thioformate, therefore consistent with the production of $CH_2CH_2SH_2^{]+\bullet}$.[310] It is noteworthy that the presence of the conventional isomer in the ion beam resulting from this ion–molecule reaction was attributed to postcollisional isomerization.

Note that collisional activation studies have shown that both $CH_3CH_2SH^{]+\bullet}$[311–313] and $CH_3SCH_3^{]+\bullet}$[314] dissociate nonstatistically, the dissociation involving a C–S bond scission being favored over the fragmentation requiring C–H bond scission. This has been attributed to the more efficient translation to vibration energy transfer, via the collisional activation, for the C–S stretching modes and to the weak coupling of the C–S vibrational frequencies with that of the molecular ion.

Thermochemistry

Combining the heats of formation and ionization energies of the two molecular species leads to $\Delta_fH(CH_3CH_2SH^{]+\bullet}) = 852$ kJ/mol and $\Delta_fH(CH_3SCH_3^{]+\bullet}) = 801$ kJ/mol (CH_3CH_2SH: $\Delta_fH = -46.15$ kJ/mol and IE $= 9.31 \pm 0.3$ eV and CH_3SCH_3: $\Delta_fH = -37.5$ kJ/mol and IE $= 8.69 \pm 0.2$ eV).[1]

The α-distonic ion $CH_2S(H)CH_3^{1+\bullet}$ is computed (G3) to be higher in energy than its conventional isomer by 88 kJ/mol,[314] in agreement with earlier calculations.[315] The barrier separating the two isomers is estimated to be 184 kJ/mol above $CH_3SCH_3^{1+\bullet}$ (G3) and lies below the dissociation limit to $CH_2SH^{1+} + CH_3^{1\bullet}$, indicating that the ions can freely interconvert before this fragmentation.[314]

From the computations of Flammang et al.,[310] the α-distonic $CH_3CHSH_2^{1+\bullet}$ and β-distonic $CH_2CH_2SH_2^{1+\bullet}$ ions are computed to be higher in energy than their conventional counterpart by 64 and 55 kJ/mol, respectively (QCISD(T)/6-311++G(d,p)//UMP2/6-31G(d,p)+ZPE). $CH_3CH_2SH^{1+\bullet}$ is separated from $CH_3CHSH_2^{1+\bullet}$ and $CH_2CH_2SH_2^{1+\bullet}$ by isomerization barriers of 152 and 127 kJ/mol, respectively. Dissociation to $CH_3CHSH^{1+} + H^{\bullet}$ is predicted to go over a barrier allowing the three isomers to interconvert before this dissociation. The two distonic ions are separated by a large barrier to isomerization, 180 kJ/mol above $CH_2CH_2SH_2^{1+\bullet}$.

3.6.56 $[C_2,H_7,S]^+ - m/z\ 63$

The heats of formation of protonated dimethyl sulfide and ethanethiol are deduced from the molecules' respective proton affinities to be $\Delta_fH(CH_3CH_2SH_2^{1+}) = 693$ kJ/mol and $\Delta_fH(CH_3S(H)CH_3^{1+}) = 662$ kJ/mol (CH_3CH_2SH: PA $= 789.6$ kJ/mol, $\Delta_fH = -46.15$ kJ/mol and CH_3SCH_3: PA $= 830.9$ kJ/mol, $\Delta_fH = -37.5$ kJ/mol and $\Delta_fH(H^+) = 1530$ kJ/mol).[1] The CID mass spectrum of protonated dimethyl sulfide has been published and shows the following dissociation fragments: m/z 15 (9.7%), 27 (6.0%), 32 (3.0%), 35 (4.7%), 45 (70%), 47 (100%), 48 (50%), 58 (3.0%), 61 (47%) and 62 (40%).

3.6.57 $[C_2,S_2]^+ - m/z\ 88$

Isomers

Although preliminary computations have suggested that both linear and cyclic $[C_2,S_2]^{+\bullet}$ species are stable, only the linear heterocumulene $S=C=C=S^{1+\bullet}$ (ionized ethenedithione) ion has been generated and characterized.

Identification

The electron-impact mass spectra of several precursor molecules show a prominent m/z 88 fragment peak.[316,317] In addition, flash vacuum pyrolysis (FVP) followed by ionization of the products[317,318] is also an efficient means by which $S=C=C=S^{1+\bullet}$ ions are generated. Five specific examples are given below.

In all cases, the resulting mass spectrum was essentially indistinguishable, indicating that a single isomer was obtained in all cases. The S–C–C–S connectivity was assigned to the ion based on the results of mass spectrometry and IR and UV spectrocopies. The CID mass spectrum of $S=C=C=S^{1+\bullet}$ contains the following fragments: m/z 12 ($C^{+\bullet}$, $\leq 1\%$), 24 ($C_2^{1+\bullet}$, $\leq 1\%$), 32 ($S^{+\bullet}$, 8%), 44 ($CS^{1+\bullet}$, 100%), 56 ($C_2S^{1+\bullet}$, 39%), 64 ($S_2^{1+\bullet}$, 4%) and 76 ($CS_2^{1+\bullet}$, 6%).[316–318] A charge stripping peak ($C_2S_2^{2+}$) is also superimposed on the m/z 44 fragment peak, consistent with computations that find the doubly charged ion to be stable (G2(MP2)).[319] Note that preliminary calculations (MP2/6-31G*) indicate that other isomeric forms are stable and could account for the observed $S_2^{1+\bullet}$ (m/z 64) fragment ion peak. It is also worth noting that the NRMS contains an intense recovery signal, whose CID mass spectral characteristics are the same as the ion obtained from EI or FVP/EI and therefore confirms that the $S=C=C=S$ neutral was generated.

Thermochemistry

Modest level calculations find the linear $S=C=C=S^{1+\bullet}$ to be the global minimum on the potential energy surface (UHF/6-31G*//6-31G*).[316] This ion is computed to lie 447 kJ/mol below the fragments $CS + CS^{1+\bullet}$ (QCISD (T)/6-311 + G(2d,p) + ZPVE),[317] leading to $\Delta_f H(S=C= C=S^{1+\bullet}) = 1207$ kJ/mol ($\Sigma\Delta_f H_{products} = 1654$ kJ/mol).[1] The ion has also been investigated by ab initio calculations at the G2(MP2) and CCSD(T) levels of theory.[320] The energy of the ion relative to the dissociation products $CS + CS^{1+\bullet}$ and to 2 CS was computed to be -420 and 682 kJ/mol, respectively, whence $\Delta_f H(SCCS^{1+\bullet}) = 1238 \pm 4$ kJ/mol. This is in fair agreement with the earlier value of 1207 kJ/mol at the QCISD(T)/6-311 + G(2d,p) + ZPVE level of theory.[317]

The structure and spectroscopic properties of the neutral have been extensively studied by computations.[321,322] Both the singlet and triplet states of the neutral are predicted to be stable but there is still some uncertainty with regard to which state is lower in energy.[319,323,324]

3.6.58 $[C_2,H,S_2]^+$ – m/z 89[319]

Protonation of C_2S_2 could in principle occur at the carbon and/or sulfur atoms. Calculations however find that the C-protonated isomer $SC(H)CS^{1+}$ is lower in energy. This ion is generated by FVP of 3-phenyl or 3-methylisoxazolone derivatives (shown below) followed by chemical ionization using CH_5^{1+}.

R = CH₃ or C₆H₅

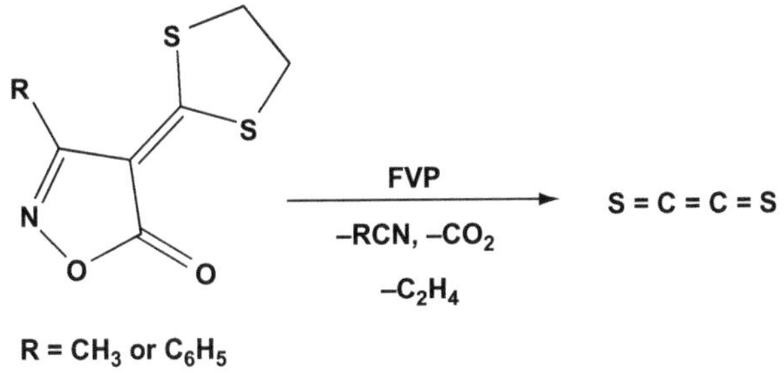

The CID mass spectrum of this ion is consistent with protonation on carbon and contains significant peaks at m/z 88 ($C_2S_2^{1+\bullet}$, 27%), 57 (loss of S, 20%), 45 (loss of CS, 100%) and 32 ($S^{+\bullet}$, 4%). The computed PA value for C_2S_2 is 842 kJ/mol (QCISD(T)/6-311 + G(2d,p + ZPVE) but since no absolute heat of formation is available for this molecule, no ionic heat of formation can be deduced.

3.6.59 $[C_2,H_2,S_2]^{+\bullet}$ – m/z 90

Ionized 1,2-dithiene (cy–CH=CHSS$^{1+\bullet}$) has been generated in solution by the reaction of HC(O)C(OH)H with sodium sulfide and sulfuric acid and identified by ESR.[325] Modest level calculations have investigated the spectroscopic properties of three different states of $CH=CHSS^{1+\bullet}$.[326] It is worth noting that for the $[C_2,H_2,S_2]$ neutrals, 33 different isomers have been investigated by computations. The results indicate a strong dependence on the basis set and therefore the relative energies of 2-dithiene ($CH=CHSS$) and dithioglyoxal ($HC(S)C(S)H$) are not well established.[327] NIST lists $IE(CH=CHSS)=$ 8.5 eV.[1]

3.6.60 $[C_2,H_3,S_2]^+ - m/z\ 91$[319]

The electron-impact mass spectrum of dithiolethione (refer to precursor (3) in the section of $[C_2,S_2]^{+\bullet}$) contains a significant peak at m/z 91. The CID mass spectrum of this ion is dominated by $CH_3^{1\bullet}$ loss (m/z 76, 100%), with minor peaks at m/z 32 ($S^{+\bullet}$, 3.8%), 44 (CS^{1+}, 13%), 45 (HCS^{1+}, 19%), 47 (CH_3S^{1+}, 11%), 59 (loss of S, 14%) and 64 ($S_2^{1+\bullet}$, 2.1%), consistent with the connectivity in $CH_3S=C=S^{1+}$. That this ion is produced, and not the C-methylated analog $S=C(CH_3)-C=S^{1+}$, is confirmed by theory (QCISD(T)/6-311 + G(2p,d) + ZPVE). At the same level of theory, $CH_3S=C=S^{1+}$ is calculated to lie 293 kJ/mol below its dissociation products $SCS + CH_3^{1+}$.

3.6.61 $[C_2,H_4,S_2]^{+\bullet} - m/z\ 92$

Isomers

There are three possible cyclic structures: $cy-SC(H)_2C(H)_2S-^{1+\bullet}$ (ionized 1,2-dithietane), $cy-C(H)_2SC(H)_2S-^{1+\bullet}$ and $cy-C(H)_2C(H)_2$ $S-S^{1+\bullet}$ (three-membered ring).

Identification

The ion–molecule reaction of thiirane ($cy-C(H)_2C(H)_2S-$) yields the dimer radical cation and one of the dimer's reactions is to undergo extrusion of ethylene accompanied by the generation of a ring-expanded radical cation (m/z 92, assigned the empirical formula $[C_2,H_4,S_2]^{+\bullet}$).[328,329] Computations (PMP2/6-31G*//6-31G* + ZPC)[330] show that when the products of the dimer's reaction are $cy-SC(H)_2C(H)_2S-^{1+\bullet}$ + ethene, all intermediates and final products are lower in energy than the reactant. However, production of $cy-C(H)_2C(H)_2S-S^{1+\bullet}$ + ethene is calculated to be endothermic, and therefore based on these data, the m/z 92 product ion was assigned to be ionized 1,2-dithietane.[330]

Thermochemistry

Ionized 1,2-dithietane ($cy-SC(H)_2C(H)_2S-^{1+\bullet}$) is calculated to be the lowest-energy structure. Its isomers, $cy-C(H)_2SC(H)_2S-^{1+\bullet}$ and $cy-C(H)_2C(H)_2$

S–S$^{]+•}$, are computed to be, respectively, 51 and 81 kJ/mol higher in energy (PMP2/6-31G*//HF/6-31G*).[330] NIST lists IE = 8.5 eV for cy–C(H)$_2$SC (H)$_2$S–(1,3-dithietane).[1]

3.6.62 [C$_2$,H$_6$,S$_2$]$^{+•}$ – m/z 94

Isomers

Three isomers have been investigated, the disulfane CH$_3$SSCH$_3$$^{]+•}$ (**1**), the distonic ion (CH$_3$)$_2$S–S$^{]+•}$ (**2**) and the molecular ion HSCH$_2$CH$_2$SH$^{]+•}$ (**3**).

Identification

Direct ionization produces the molecular ions of dimethyldisulfide (CH$_3$SSCH$_3$$^{]+•}$) and of 1,2-ethanedithiol (HSCH$_2$CH$_2$SH$^{]+•}$) whereas the distonic isomer (CH$_3$)$_2$S–S$^{]+•}$ can be generated by the ion–molecule reaction of S = C = S–S$^{]+•}$ (generated by CI of CS$_2$) or ClC≡N–S$^{]+•}$ (obtained from the dissociative ionization of cy–C(Cl)NSNC(Cl)–) with CH$_3$SCH$_3$.

Isomers (**1**) and (**2**) show the same metastable dissociation (loss of HS$^{]•}$) but with slightly different kinetic energy releases ($T_{0.5}$ = 28 and 41 meV for (**1**) and (**2**), respectively).[331] Their high and low energy CID are similar but show a structure-indicative feature: the more intense fragment peaks at m/z 62 (loss of S$^+$) and 61 (loss of HS$^{]•}$) in the spectrum of the ion obtained by the ion–molecule reactions suggest that at least some of the ions produced have the distonic ion's (**2**) connectivity.

CID Mass Spectra of the [C$_2$,H$_6$,S$_2$]$^{+•}$ Isomers[331]

	m/z											
Ion	32	44	45	46	47	48	49	61	62	64	78	79
CH$_3$SSCH$_3$	1	2	35	23	19	7	2	16	0	24	8	100
(CH$_3$)$_2$S–S	1	1	27	25	19	6	1	53	2	27	7	100

The ion–molecule reaction of the two isomers with methyl isocyanide (MeNC) was also investigated.[331] The conventional ion (**1**) does not react with MeNC. However, isomer (**2**) reacts, as is expected for distonic ions, by a sulfur transfer reaction that produces ionized dimethyl sulfide.

Ionized 1,2-ethanedithiol (**3**) formed a minor section of a study of ionized 2-mercaptoethanol.[332] The dissociation chemistry of this ion was examined by deuterium labeling and it very closely matched that of 2-mercaptoethanol, which in turn displayed the same fragmentation pathways (and mechanisms) as 1,2-ethanediol (refer to the section on $[C_2,H_6,O_2]^{+\bullet}$ and also Holmes and Terlouw).[333] The major fragment peaks in the ion's CID mass spectrum are m/z 47 (CH_2SH^{1+}, loss $CH_2SH^{1\bullet}$), 48 ($CH_2SH_2^{1+\bullet}$ with some $CH_3SH^{1+\bullet}$, loss of H_2CS), 49 ($CH_3SH_2^{1+}$, loss of $HCS^{1\bullet}$), and 60 ($CH_3CSH^{1+\bullet}$, loss of H_2S). It is noteworthy that ionized CH_3SSCH_3 has been involved in the study of isomeric $[C,H_3,S]^+$ ions.[334]

Thermochemistry

The heat of formation of the ionized molecules is deduced from their respective neutral heats of formation and ionization energies and gives $\Delta_f H(CH_3SSCH_3^{1+\bullet}) = 690$ kJ/mol and $\Delta_f H(HSCH_2CH_2SH^{1+\bullet}) = 859$ kJ/mol (CH_3SSCH_3: $\Delta_f H = -24.1 \pm 2.3$ kJ/mol and IE = 8.18 ± 0.03 eV[335] (note that the evaluated value given in the NIST, 7.4 ± 0.01 eV, is most certainly too low); $HSCH_2CH_2SH$: $\Delta_f H = -9.3 \pm 1.1$ kJ/mol and IE = 9.0 eV).[1]

The distonic ion $(CH_3)_2S-S^{1+\bullet}$ is calculated to be higher in energy than its conventional isomer (**2**) by 64 kJ/mol (UMP2/6-311G**//UMP2/6-31G*) but separated from it by a significant energy barrier to isomerization (262 kJ/mol above (**2**)).[331] The energy barrier lies very close to the dissociation limit to $[C_2,H_5,S]^+ + HS^{1\bullet}$ and therefore does not exclude the possibility of identifying both $CH_3SSCH_3^{1+\bullet}$ and $(CH_3)_2S-S^{1+\bullet}$ by experiment.

3.6.63 $[C_2,H_7,S_2]^+ - m/z$ 97

The proton affinity of dimethyldisulfide (CH_3SSCH_3) has been measured to be 815.3 kJ/mol and combined with $\Delta_f H(CH_3SSCH_3) = -24.1 \pm 2.3$ kJ/mol and $\Delta_f H(H^+) = 1530$ kJ/mol leads to $\Delta_f H(CH_3S(H)SCH_3^{1+}) = 691$ kJ/mol.[1]

3.6.64 $[C_2,S_3]^{+\bullet} - m/z$ 120

Isomers

Sulfuration of $S=C=C=S$ can in principle take place at carbon, sulfur, or at the $C=C$ double bond, leading respectively to $SC(S)CS^{1+\bullet}$, $SCCSS^{1+\bullet}$ and cy–$C(S)SC(S)-^{1+\bullet}$ (a three-membered ring), respectively. The first two isomers have been generated and characterized.

Identification[336]

Dissociative ionization of dithiolodithioles (**1**)–(**3**) generate an m/z 120 ($[C_2,S_3]^{+\bullet}$) fragment peak according to the following reactions:

The CID mass spectra of the ions generated by the three precursor molecules are very similar, all showing intense peaks at m/z 44 ($CS^{1+\bullet}$), 76 ($CS_2^{1+\bullet}$), and 88 ($C_2S_2^{1+\bullet}$). In all three cases a very small recovery peak is observed in the NRMS (1%–2%) but the overall spectra again do not allow structure elucidation.

When the $[C_2,S_3]^{+\bullet}$ ions generated from the three precursor molecules undergo ion–molecule reactions with NO, the products obtained are quite complex and consist mostly of species at m/z 106, 88, 86, 76 and 74. Note that m/z 88 and 76 are both produced by metastable $[C_2,S_3]^{+\bullet}$ ions and therefore the structure-indicative products are m/z 106, 86 and 74. The CID mass spectra of these product ions indicate that m/z 106 and 86 result from the displacement of CS and S_2 by NO, to yield respectively $[ON^+/CS_2]$ and $SCCNO^{1+}$, both consistent with the $S=C=C=S-S^{1+\bullet}$ connectivity. The latter ion, m/z 74, has been identified as $S=C=N=O^{1+}$ and is proposed to be obtained by the displacement of a CS_2 molecule in the isomeric species $S=C=CS_2^{1+\bullet}$.

Thermochemistry[336]

G2(MP2,SVP) computations show that $S=C=CS_2^{1+\bullet}$ is the lowest energy isomer with cy–$C(S)SC(S)–^{1+\bullet}$ lying slightly higher in energy (9 kJ/mol) but separated from it only by a small isomerization barrier (33 kJ/mol above cy– $C(S)SC(S)–^{1+\bullet}$). The linear species, $S=C=C=S-S^{1+\bullet}$, is calculated to be significantly higher in energy than $S=C=CS_2^{]+\bullet}$ (91 kJ/mol) but a large barrier to interconversion (72 kJ/mol above $S=C=C=S-S^{]+\bullet}$) prevents a facile 1,2-S shift from taking place. Consequently, theory predicts that both $S=C=CS_2^{]+\bullet}$ and $S=C=C=S-S^{]+\bullet}$ should be observable species. It is noteworthy that at the B3LYP level of theory,[337] the cyclic species cy–C(S)SC(S)–$^{]+\bullet}$ is calculated to be 59 kJ/mol lower in energy than $S=C=CS_2^{]+\bullet}$, in contradiction to the above results.

3.6.65 $[C_2,H_4,S_3]^{+\bullet}$ – m/z 124

The ion–molecule reaction of thiirane (cy–C(H)$_2$C(H)$_2$S–) was shown to produce ionized 1,2-dithietane (cy–SC(H)$_2$C(H)$_2$S–$^{]+\bullet}$) and ethene. When the $[C_2,H_4,S_2]^{+\bullet}$ ion further reacts with thiirane, the dimer formed again undergoes extrusion of ethene accompanied by the generation of a ring expanded radical cation (m/z 124, assigned the empirical formula $[C_2,H_4,S_3]^+$).[328,329] Computations suggest that the sulfur atom is incorporated in the ring, leading to the formation of a five-membered ring (cy–SC(H)$_2$C(H)$_2$SS–$^{]+\bullet}$) rather than the three-membered ring cy–C(H)$_2$C(H)$_2$S–S–S$^{]+\bullet}$. At the PMP2/6-31G*//HF/6-31G* level (including ZPE) of theory,[330] cy–SC(H)$_2$C(H)$_2$SS–$^{]+\bullet}$ is calculated to be lower in energy than cy–C(H)$_2$C(H)$_2$S–S–S$^{]+\bullet}$ by 97 kJ/mol.

3.6.66 $[C_2,S_4]^{+\bullet}$ – m/z 152

Isomers

Theory predicts as many as 16 covalently bonded isomeric species, the most stable is the ion–molecule complex $[CS_2\cdots CS_2]^{+\bullet}$.

Identification[338]

The m/z 152 ion obtained in the self chemical ionization of CS_2 (ion (1))[339] has mass spectral characteristics different from the ion generated by dissociative ionization of dithiolodithioles (ion (2), precursor molecules for this ion are (1) and (3) in the section on $[C_2,S_3]^{+\bullet}$). The CID mass spectrum of (1) is dominated by the peak at m/z 76 ($CS_2^{]+\bullet}$) whereas the spectrum of (2) shows significant peaks at m/z 76 ($CS_2^{]+\bullet}$, 50%) and 88 (SCCS$^{]+\bullet}$, 100%) and less intense peaks at m/z 44, 56, 64, 76 and 108. The NR mass spectra are also different, with only (2) showing a recovery signal. Ion–molecule reactions using NO$^{]\bullet}$ and CH$_3$NC also indicate that the ions are different.

All these observations indicate that chemical ionization of CS_2 generates an ion–neutral complex whereas dissociative ionization of the dithiolodithioles generates a covalently bonded species or a mixture of such species. Sixteen covalently bonded species were calculated (G2(MP2,SVP)) to be stable, and based on the connectivity in the neutral dithiolodithioles precursors and their dissociation mechanisms, three different m/z 152 species were proposed to possibly coexist in the MS experiment. These are cy–SSSC(CS)–$^{1+\bullet}$ (a four-membered ring), cy–C(S)C(S)SS–$^{1+\bullet}$ (also a four-membered ring) and the branched distonic ion SSC(S)CS$^{1+\bullet}$.

Thermochemistry[338]

The complex [CS$_2\cdots$CS$_2$]$^{+\bullet}$ is computed to lie 52 kJ/mol below cy–SSSC $=$C$=$S-$^{1+\bullet}$, the lowest energy covalently bonded species (G2(MP2,SVP)). The ions cy–C(S)C(S)SS–$^{1+\bullet}$ and SSC(S)CS$^{1+\bullet}$ are calculated to lie 49 and 57 kJ/mol, respectively, above cy–SSSC$=$C$=$S-$^{1+\bullet}$. At the same level of theory, cy–C(S)C(S)SS–$^{1+\bullet}$ is predicted to be separated from SSSC$=$C$=$S$^{1+\bullet}$ by an energy barrier of 104 kJ/mol.

3.6.67 [C$_2$,H$_4$,S$_4$]$^{+\bullet}$ – m/z 156

When the [C$_2$,H$_4$,S$_3$]$^+$ ion produced in the ion–molecule reaction of thiirane further reacts with a molecule of thiirane, the dimer formed again undergoes extrusion of ethene accompanied by the generation of a ring-expanded radical cation (m/z 156, assigned the empirical formula [C$_2$,H$_4$,S$_4$]$^+$).[328,329] Computations again favor the inclusion of sulfur in the ring, leading to the generation of a six-membered ring cy–SC(H)$_2$C(H)$_2$SSS–$^{1+\bullet}$ over the three-membered ring, cy–C(H)$_2$C(H)$_2$S(SSS)–$^{1+\bullet}$. The latter species is calculated to be 46 kJ/mol higher in energy (PMP2/6-31G*//HF/6-31G* + ZPE).[330]

3.6.68 [C$_2$,X$_n$]$^{+\bullet/+}$ where $2 \le n \le 6$ and X = F, Cl, Br and/or I

These ions do not have isomeric forms. The thermochemical data pertaining to these species are given below. The reader should take note that the thermochemistry of halogenated species is often uncertain or absent. The substitution of halogen on or α- to the formal charge site is always destabilizing. See also $\Delta_f H$ for [C$_2$, H$_n$, Cl$_{6-n}$] molecules in Section 3.6.91.

Selected Values for Ions of Empirical Formula [C$_2$,X$_n$] Where $2 \le n \le 6$ and X = F, Cl, Br and/or I

Species	$\Delta_f H$(neutral) (kJ/mol)	Ionization Energy (eV)	$\Delta_f H$(ion)^a (kJ/mol)
C$_2$F$_2$	21 ± 21	11.18	1100
C$_2$Cl$_2$	210 ± 20	10.1	1184
C$_2$Br$_2$	259	9.67	1192
C$_2$I$_2$	—	9.02	—
C$_2$F$_3$	-192 ± 8	10.2	792
C$_2$F$_4$^{b}	-659 ± 8	10.14	319
C$_2$Cl$_4$	-12 ± 2	9.33	888
C$_2$Br$_4$	—	9.1	—
C$_2$I$_4$	—	8.3	—

(continued)

(Continued)

Species	$\Delta_f H$(neutral) (kJ/mol)	Ionization Energy (eV)	$\Delta_f H$(ion)[a] (kJ/mol)
C_2F_5	−893	10.1	101
C_2F_6	−1343	13.4	−50
C_2Cl_6	−143 ± 9	11.0	918
C_2Br_6	133	—	—
C_2F_3Cl	−535 ± 30	9.81	411
$CFClCFCl$	−327	10.2	657
CF_2BrCF_2Br	−789 ± 4	11.1	282
$(CF_2Cl)_2$	−937 ± 7	12.5	269
$CFCl_2CF_2Cl$	−727 ± 4	11.99	430
C_2F_5I	−1004 ± 4	10.44	3
CF_2CCl_2	−338 ± 11	9.81	609
C_2F_3Br	—	9.67	—
C_2F_5Cl	−1114 ± 10	12.6	102
CF_3Cl_3	−725 ± 10	11.5	385
$(CFCl_2)_2$	−527 ± 10	11.3	563
$(CF_2I)_2$	−667 ± 3	10.11	308

All data are from the NIST WebBook[1] and Lias et al. compendium.[58] Note that ionization energies may not be identified as vertical or adiabatic in the above texts.

[a]Deduced from $\Delta_f H$(ion) = IE + $\Delta_f H$(neutral).

[b]The existence of the two isomeric species remains unresolved (see section below on $[C_2,F_4]^{+\bullet}$).

3.6.69 $[C_2,H,F]^{+\bullet}$ – m/z 44

The $C_2HF^{]+\bullet}$ ion has no isomers and its mass spectrometric identification by its dissociation characteristics is unlikely to be problematic. The ion is easily generated by loss of HF from ionized CH_2CF_2 and $CHFCHF$. Note that this reaction has a reverse energy barrier (see later).[340] $\Delta_f H(C_2HF^{]+\bullet}) = 1211$ kJ/mol derives from $\Delta_f H(C_2HF) = 125$ kJ/mol and its IE = 11.26 eV.[1]

3.6.70 $[C_2,H_2,F]^+$ – m/z 45

A $[C_2,H_2,F]^+$ cation has been generated by the loss of HF from metastable $CH_3CF_2^{]+}$ ions produced from ionized CH_3CHF_2 (loss of H$^\bullet$) and CH_3CF_2Cl (loss of Cl$^\bullet$).[340] Both of these reactions involve an appreciable reverse energy barrier and so any AE value will give too high a result. The elimination is believed to be a 1,2-HF loss and so the product will have the CH_2CF connectivity. The NIST WebBook provides no thermochemical data but combining the PA of $C_2HF = 686$ kJ/mol[341] with $\Delta_f H(C_2HF) = 125$ kJ/mol[1] and $\Delta_f H(H^+) = 1530$ kJ/mol[1] gives $\Delta_f H([C_2,H_2,F]^+) = 969$ kJ/mol, but the structure of this ion is not known for certain. By analogy with other halo-ions it is unlikely to have an F-bridged ethyne ion structure.

3.6.71 $[C_2,H_2,X]^+$ WHERE $X = Cl$, Br AND I – m/z 61, 105 AND 153

There are no experimental data for these ions. However, the relative energies of the isomeric forms of these ions have been evaluated by computation[342–344] and the structure $CH_2CX^{]+}$ is the global minimum. The halogen-bridged ethyne ion also is a minimum on the potential energy surface.

3.6.72 $[C_2,H_3,F]^{+\bullet}$ – m/z 46

Ionized fluoroethene ($CH_2CHF^{]+\bullet}$) is characterized by its loss of $H^\bullet$ and HF in the metastable ion time frame, the latter reaction producing a very broad peak ($T_{0.5} = 980$ meV).[340] The $[C_2,H_3,F]^{+\bullet}$ ion derived from loss of HF from ionized CH_3CHF_2 gives a slightly higher kinetic energy release ($T_{0.5} = 1020$ meV), suggesting that ionized fluoroethene is produced. It is noteworthy that the appearance energy for this HF loss reaction has been reported to be 13.4 ± 0.1 eV.[1] The energy threshold for this reaction lies significantly below at 10.9 eV ($\Sigma\Delta_f H_{products} = 1055$ kJ/mol)[1] and thus again a large fraction of the reverse energy barrier appears as product translational energy. Combining the ionization energy and heat of formation of CH_2CHF (10.36 eV and -136 kJ/mol,[1] respectively) gives $\Delta_f H(CH_2CHF^{]+\bullet}) = 864$ kJ/mol.

3.6.73 $[C_2,H_3,X]^{+\bullet}$ WHERE $X = Cl$, Br AND I – m/z 62, 106 AND 154

The ionized vinyl halides ($CH_2CHX^{]+\bullet}$) have no known isomers. Their heats of formation are $\Delta_f H(CH_2CHCl^{]+\bullet}) = 986 \pm 7$ kJ/mol (IE $= 9.99$ eV,[1] $\Delta_f H(CH_2CHCl) = 22 \pm 7$ kJ/mol; the latter value was selected on the basis of the effect of Cl substitution for H at a double bond in ethene ($\Delta\Delta_f H = -30 \pm 5$ kJ/mol) and because of the uncertainty of the numbers quoted in the NIST database; note that this general problem has been addressed earlier[345]), $\Delta_f H(CH_2CHBr^{]+\bullet}) = 1027 \pm 2$ kJ/mol (IE $= 9.82$ eV,[1] $\Delta_f H(CH_2CHBr) = 79 \pm 2$ kJ/mol;[1] note that more recent values[346] have appeared, $\Delta_f H(CH_2CHBr) = 74 \pm 3$ kJ/mol, IE $= 9.83$ eV leading to $\Delta_f H(CH_2CHBr^{]+\bullet}) = 1022 \pm 3$ kJ/mol, in satisfactory agreement with the foregoing) and $\Delta_f H(CH_2CHI^{]+\bullet}) = 1034 \pm 10$ kJ/mol (IE $= 9.32$ eV,[1] $\Delta_f H(CH_2CHI) = 135 \pm 12$ kJ/mol[345]).

3.6.74 $[C_2,H_4,F]^+$ – m/z 47

Isomers

Only one ion is stable: the 1-fluoro-ethyl cation, $CH_3CHF^{]+}$.

Identification

The MI mass spectrum of the ion produced by H-atom loss from ionized CH_3CH_2F contains a major signal for HF loss with a reported $T_{0.5} = $ ca 500 meV.[340] The $[C_2,H_4,F]^+$ ion derived from the loss of $F^{\bullet}$ from ionized CH_3CHF_2 and by $Br^{\bullet}$ loss from FCH_2CH_2Br showed the same kinetic energy release for HF loss, namely 450 ± 10 meV,[347] possibly indicating that only one structure need be considered.

Most recently Reynolds[348] has performed calculations for $[C_2,H_4,F]^+$ at the MP2/6-31G**//MP2/6-31G** level, from which it appears that only the 1-fluoro-ethyl cation is stable. This supports most previous computational chemistry investigations where the heat of formation for the cyclic ion was found to lie at least 100 kJ/mol above that of the ion $CH_3CHF^{\rceil+}$. That there is only one observable structure for this ion is supported also by the bimolecular chemistry study of Heck et al.[349] in which the ion's reactivity was independent of its precursor ion's structure, and also in the investigations of 2-phenoxyethyl fluorides.[350,351]

It has also been observed that irrespective of their precursor molecule's structure, the losses of $H^{\bullet}$ (base peak), H_2 and HF dominate the CID mass spectrum[347] of all $[C_2,H_4,F]^+$ ions, in keeping with a single structure. Moreover, the total absence of m/z 28 $(CH_2CH_2^{\rceil+\bullet})$ is in keeping with no participation of the symmetrical F-bridged ion.

Thermochemistry

The heat of formation of $CH_3CHF^{\rceil+}$ can be derived from the experimental proton affinity value for the corresponding fluoroethene. For CH_2CHF this value is 729 kJ/mol,[341] leading to $\Delta_fH(CH_3CHF^{\rceil+}) = 665$ kJ/mol. This result is in satisfactory agreement with that derived from the appearance energy measurement for loss of $H^{\bullet}$ from ionized CH_3CH_2F, namely 677 kJ/mol $(AE = 12.04 \pm 0.03$ eV,[1] $\Delta_fH(CH_3CH_2F) = -263$ kJ/mol[58] and $\Delta_fH(H^{\bullet}) = 218$ kJ/mol[1]). The intense metastable ion peak for the loss of $Br^{\bullet}$ from ionized FCH_2CH_2Br $(T_{0.5} = 33$ meV) had an appearance energy of 10.95 ± 0.2 eV[347] and combined with the ancillary thermochemical data leads to $\Delta_fH([C_2,H_4,F]^+) = 700 \pm 30$ kJ/mol for the product, certainly in keeping with it having the structure $CH_3CHF^{\rceil+}$ $(\Delta_fH(FCH_2CH_2Br) = -243 \pm 4$ kJ/mol estimated by adding 20 ± 4 kJ/mol for the replacement of H by Br) and $\Delta_fH(Br^{\bullet}) = 112$ kJ/mol).[1] This is supported by the CID mass spectrum of these metastably generated ions, which was identical with those from the other sources referred to above, except for an even more intense H-atom loss peak.[347]

3.6.75 $[C_2,H_4,X]^+$ WHERE X = Cl, Br AND I – m/z 63, 107 AND 155

Isomers

It has become clear from many studies that, in contrast with the fluoro analog, these ions have more than one stable structure. The ground state for these $[C_2,H_4,X]^+$ ions is the cyclic halonium ion (cy–C(H)$_2$C(H)$_2$Cl–$^{1+}$), with the 1-halo-ion (CH$_3$CHX^{1+}) also being a stable species.

Identification

The cyclic ions are readily produced, for example, by the loss of Y$^\bullet$ from ionized XCH$_2$CH$_2$Y,[352] whereas the 1-halo-ions can be generated, for example, from such a molecule as ClBrCHCH$_3$ by loss of Cl$^\bullet$ or Br$^\bullet$.

The isomers cannot be characterized by their MI mass spectra, both losing either HCl or HBr with tiny kinetic energy releases: for $[C_2,H_4,Cl]^+$ ions, cy–C(H)$_2$C(H)$_2$Cl–$^{1+}$ and CH$_3$CHCl^{1+}, $T_{0.5} = 1.4$ meV, and for the bromo analogs, ca 0.1 meV.[347] The CID mass spectra[347,349] are quite similar but do contain minor, structure-characteristic features such as the presence of m/z 15 (CH$_3$$^{1+}$), little CH$_2$X^{1+} ($m/z$ 39 and 83 for X = Cl and Br, respectively) for the 1-halo-ion, and a relatively significant (10 times as intense) CH$_2$X^{1+} and no m/z 15 (CH$_3$$^{1+}$) for the cyclic species. The isomers are also distinguishable by their reactivity;[349] for example, the 1-halo-ions exchange three H$^\bullet$ atoms for D$^\bullet$ when reacted with D$_2$O while the cyclic halonium ions exchange none.

Thermochemistry

The $\Delta_f H$ values for cy–C(H)$_2$C(H)$_2$X–$^{1+}$ evaluated by computation and experiment are in good agreement. Calculation gives $\Delta_f H$(cy–C(H)$_2$C(H)$_2$Cl–$^{1+}$) = 834 kJ/mol[342] compared with 831 ± 5 kJ/mol from experiment[352] (ClCH$_2$CH$_2$Cl $\longrightarrow$ cy–C(H)$_2$C(H)$_2$Cl–$^{1+}$ + Cl$^\bullet$: AE = 11.24 eV, $\Delta_f H$(ClCH$_2$CH$_2$Cl) = −137 ± 5 kJ/mol and $\Delta_f H$(Cl$^\bullet$) = 121 kJ/mol or ClCH$_2$CH$_2$Br $\longrightarrow$ cy–C(H)$_2$C(H)$_2$Cl–$^{1+}$ + Br$^\bullet$: AE = 10.68 eV, $\Delta_f H$(ClCH$_2$CH$_2$Br) = −84 ±5 kJ/mol and $\Delta_f H$(Br$^\bullet$) = 112 kJ/mol). Note that there is some uncertainty as to the best values for the heat of formation for the di-haloethanes used to generate the $[C_2,H_4,X]^+$ cations. This uncertainty therefore transposes itself to the heat of formation values deduced for the cations. The numbers given below are those selected by the authors for consistency of the halogen substitution effects. Following appearance energy measurements $\Delta_f H$(cy–C(H)$_2$C(H)$_2$Br–$^{1+}$) = 867 ± 8 kJ/mol and $\Delta_f H$(cy–C(H)$_2$C(H)$_2$I–$^{1+}$) = 890 ± 8 kJ/mol (BrCH$_2$CH$_2$Br $\longrightarrow$ cy–C(H)$_2$C(H)$_2$Br–$^{1+}$ + Br$^\bullet$: AE = 10.53 eV[352] and $\Delta_f H$(BrCH$_2$CH$_2$Br) = −42 ± 7 kJ/mol or ClCH$_2$CH$_2$Br $\longrightarrow$ cy–C(H)$_2$C(H)$_2$Br–$^{1+}$ + Cl$^\bullet$: AE = 11.15 eV;[352] ICH$_2$CH$_2$I $\longrightarrow$ cy–C(H)$_2$C(H)$_2$I–$^{1+}$ + I$^\bullet$: AE = 9.60 eV,[352] $\Delta_f H$(ICH$_2$CH$_2$I) = 70 ± 5 kJ/mol and $\Delta_f H$(I$^\bullet$) = 107 kJ/mol or BrCH$_2$CH$_2$I $\longrightarrow$ cy–C(H)$_2$C(H)$_2$I–$^{1+}$ + Br$^\bullet$: AE = 10.26 eV and $\Delta_f H$(BrCH$_2$CH$_2$I) = 13 ± 7 kJ/mol).

The heat of formation values for the 1-halo-ions deduced from appearance energy measurements[352] are of similar magnitude to the above. $\Delta_f H(CH_3CHCl^{1+}) = 826 \pm 8$ kJ/mol while $\Delta_f H(CH_3CHBr^{1+}) = 865 \pm 5$ kJ /mol $(CH_3CHClBr \longrightarrow CH_3CHCl^{1+} + Br^\bullet; \quad AE = 10.56$ eV, $\Delta_f H(CH_3CHClBr) = -81 \pm 3$ kJ/mol; $CH_3CHClBr \longrightarrow CH_3CHBr^{1+} + Cl^\bullet; \quad AE = 11.12$ eV; $CH_3CHBr_2 \longrightarrow CH_3CHBr^{1+} + Br^\bullet; \quad AE = 10.51$ eV, $\Delta_f H (CH_3CHBr_2) = -41 \pm 3$kJ/mol). It is noteworthy that the relative energies, by computation of the hydride affinities, have been given[348] and for the cyclic isomer to the 1-halo-ion are for $[C_2,H_4,Cl]^+$ 0 and -15 kJ/mol and for $[C_2,H_4,Br]^+$ 0 and $+12$ kJ/mol, respectively.

3.6.76 $[C_2,H_5,F]^{+\bullet} - m/z\ 48$

Isomers

The conventional radical cation $CH_3CH_2F^{1+\bullet}$ (ionized ethylfluoride) and the β-distonic ion $HF^+CH_2CH_2^\bullet$ have been generated and characterized.

Identification

The CID mass spectrum of $CH_3CH_2F^{1+\bullet}$ (m/z 48 generated by direct ionization of the molecule) is dominated by the very intense H-atom loss peak, but also contains a significant signal for CH_2F^{1+} (m/z 33).[353] It is noteworthy that m/z 29 is absent in the CID and normal EI mass spectra.[354] The β-distonic ion $HF^+CH_2CH_2^\bullet$ was successfully, although inefficiently, generated by the loss of H_2CO from ionized $FCH_2CH_2CH_2OH$.[355] As expected, its CID mass spectrum[355] was dominated by an extremely intense peak at m/z 28 ($C_2H_4^{1+\bullet}$) and also contained a significant m/z 20 ($HF^{1+\bullet}$) as well as m/z 14 ($CH_2^{1+\bullet}$).[354]

The MI mass spectra[355] of ionized ethyl fluoride and the β-distonic ion are similar in that both display the losses of $H^\bullet$ ($T_{0.5} = 8$ meV) and H_2 ($T_{0.5} = 310$ meV) but the latter uniquely shows an intense, broad m/z 28 peak ($T_{0.5} = 600$ meV).[354,355] Note that the conventional ion's MI mass spectrum also shows a broad but weak peak at this mass, its *origin* however is HF loss from the even electron ion $^{13}CCH_4F^{1+}$ (m/z 47, $(M–H)^{1+}$, being the base peak in the MS of ethyl fluoride and the molecular ion having a relative abundance of only ca 10%). This provides a salutary lesson in the need to carefully check for all possible origins of a signal in MI and CID mass spectra.

It is noteworthy that the α-distonic isomer of ionized ethyl fluoride $^\bullet CH_2F^+CH_3$ did not result from the loss of CO_2 from ionized $FCH_2C(O)OCH_3$,[353] as might well have been expected by analogy with the behavior of similar α-substituted esters or acids.[353,354,356] The CID mass spectrum of the $[C_2,H_5,F]^{+\bullet}$ ion so produced was very closely similar to that of ionized ethyl fluoride itself and so it was concluded that a large energy barrier precluded the required methyl shift to fluorine. The above reaction also failed for ionized $FCH_2CH_2C(O)OH$ in that it did not undergo CO_2 loss.[353] In marked contrast the reaction was a success for the corresponding bromo and chloro esters (see below).[355,356]

Thermochemistry

$\Delta_f H(CH_3CH_2F)$ is not provided in the NIST tables but a value of -263 kJ/mol comes from Lias et al.[58] The adiabatic ionization energy was measured by energy selected EI as 11.78 eV,[355] significantly lower than the reported vertical value of 12.43 eV.[1] Combining these two data yields $\Delta_f H(CH_3CH_2F^{]+\bullet}) = 873$ kJ/mol. The $\Delta_f H$ of the $[C_2,H_5,F]^{+\bullet}$ ion produced by the loss of CO_2 from ionized $FCH_2C(O)OCH_3$ ($\Delta_f H = -570$ kJ/mol[355] and $AE = 10.90$ eV, $\Delta_f H(CO_2) = -393$ kJ/mol[1]) is 875 kJ/mol, the same as that for ionized ethyl fluoride and is thus supportive of the assigned structure. The heat of formation of the β-distonic isomer ($HF^+CH_2CH_2^{\bullet}$, produced from ionized $FCH_2CH_2CH_2OH$) proved to be unmeasurable.[355] Computations at the MP2/6-31G*//3-21G* level[357] indicated that the ion $^{\bullet}CH_2CH_2F^+H$ is likely more stable than ionized ethyl fluoride by some 74 kJ/mol.

It is noteworthy that the reported AE values for the H-atom loss from CH_3CH_2F, 12.04 eV[1] and 12.14 eV,[355] lie below the listed vertical IE, a most unusual circumstance. The great stability of the ion produced, $CH_3CHF^{]+}$, accounts for the weak molecular ion in the EI mass spectrum of CH_3CH_2F.

3.6.77 $[C_2,H_5,X]^{+\bullet}$ WHERE X = Cl, Br AND I – m/z 64, 108 AND 156

Isomers

This group of ions comprises the ionized ethyl halides ($CH_3CH_2X^{]+\bullet}$) and their α- and β-distonic isomers ($CH_3X^+CH_2^{\bullet}$ and $CH_3CH^{\bullet}X^+H$ and $HX^+CH_2CH_2^{\bullet}$, respectively). It is useful to deal with all the $[C_2,H_5,X]^{+\bullet}$ ions here because the distonic forms have been generated by common preparative ion dissociations and have many similar properties.

Identification

Direct ionization of the molecular species leads to the generation of the corresponding molecular ions. Dissociative ionization of $XCH_2C(O)OCH_3$ (loss of CO_2) where $X = Cl^{354-356,358}$ and $Br^{354,358}$ yields the α-distonic ions $CH_3X^+CH_2^{\bullet}$ whereas the fragmentation of ionized $CH_3CH(X)C(O)OH$ (loss of CO_2) where $X = Cl$ and $Br^{354,358}$ produces the other α-distonic ions $CH_3CH^{\bullet}X^+H$. The β-distonic ions $HX^+CH_2CH_2^{\bullet}$ are generated from dissociative ionization of $XCH_2CH_2CH_2OH$ where $X = Cl$ and Br^{355} (loss of H_2CO). Note that the loss of CO_2 from ionized 1-halo-propanoic acids conspicuously failed to generate the desired (expected) distonic product, $HX^+CH_2CH_2^{\bullet}$.[354,358]

It is noteworthy that because the CID mass spectra are not strongly structure-characteristic, it follows therefore that the ions can at least partially interconvert before some CID processes. A description of the mass spectral characteristics for each type of ions produced is given below.

3.6.78 $CH_3CH_2X^{\rceil+\bullet}$ WHERE X = Cl, Br AND I – m/z 64, 108 AND 156

The MI mass spectra of the three ethyl halides show the following peaks with the accompanying $T_{0.5}$ values in parentheses. For $CH_3CH_2Cl^{\rceil+\bullet}$, only the loss of HCl ($T_{0.5}$ = ca 24 meV)[354,355] was observed, the very weak peak for $Cl^{\bullet}$ loss almost certainly being of collisional origin.[355] For $CD_3CH_2Cl^{\rceil+\bullet}$ ions only DCl is observed but the peak shape changes from approximately Gaussian to a clearly composite signal ($T_{0.5}$ = 69 meV).[355] The potential energy surface that leads to this reaction is not known. $CH_3CH_2Br^{\rceil+\bullet}$ ions are unusual in having a virtually peak free MI mass spectrum, there being only a weak signal for m/z 29 (loss of $Br^{\bullet}$) observable at very high gain.[355] $CH_3CH_2I^{\rceil+\bullet}$ ions also display no MI signals other than a very weak m/z 29 (loss of $I^{\bullet}$) that may be of residual collision origin.[355] The CID mass spectra are unremarkable.[355] The major peaks for the iodide and bromide are for $[M-CH_3]^+$, halogen ion and $C_2H_5^{\rceil+}$. The chloride also displays large signals for $C_2H_4^{\rceil+\bullet}$ and $C_2H_5^{\rceil+}$, the former is largely of MI origin.

3.6.79 $^{\bullet}CH_2X^+CH_3$ WHERE X = Cl AND Br – m/z 64 AND 108

A key feature of the CID mass spectra of both of these $^{\bullet}CH_2X^+CH_3$ ions is the unique, albeit weak, presence of a peak for $CH_3X^{\rceil+\bullet}$.[355] Apart from this, the spectra contain the same peaks and are distinguishable from those of the $CH_3CH_2X^{\rceil+\bullet}$ analogs only by the relative peak abundances. For example, the distonic ions show much weaker peaks in the m/z 24–29 range.

The metastable $^{\bullet}CH_2Cl^+CH_3$ ion loses HCl to give a composite signal,[355] a Gaussian peak ($T_{0.5}$ = 30 meV) atop a dished peak ($T_{0.5}$ = 450 meV). The more intense peak for loss of $Cl^{\bullet}$ has a $T_{0.5}$ = 27 meV. Labeling studies showed that the Gaussian signal involves all the $H^{\bullet}$ atoms, whereas the dished peak involved only the methyl hydrogens. It was originally proposed that two fragment ions were produced, $CH_2CH_2^{\rceil+\bullet}$ and $CH_3CH^{\rceil+\bullet}$, with the latter connected with the Gaussian peak, but the lack of a stable methylcarbene ion (qv) renders this unlikely. Again, the complexities of the potential surface for this isomer deserve investigation.

The metastable $^{\bullet}CH_2Br^+CH_3$ ion loses HBr (minor (3%), $T_{0.5}$ = 40 meV) and $Br^{\bullet}$ (major (97%), $T_{0.5}$ = 43 meV). Note that thermochemically the loss of $Br^{\bullet}$ is 10 kJ/mol favored over HBr loss, the opposite of the chloro analog where HCl loss is favored by approximately 50 kJ/mol.[355]

3.6.80 $CH_3^{\bullet}CHX^+H$ WHERE X = Cl AND Br – m/z 64 AND 108

The MI mass spectrum of $CH_3^{\bullet}CHCl^+H$ consists of a composite peak for HCl loss. However, the kinetic energy release values of the components are smaller than for the $^{\bullet}CH_2X^+CH_3$ isomer ($T_{0.5}$ = 30 and 300 meV) and so a common final TS cannot be involved. Labeling at the chloro atom with D showed that the major loss is HCl, by a factor of over 100.[355] The bromo analog, $CH_3{}^\bullet CHBr^+H$, also produced composite MI peaks with $Br^\bullet$ loss again predominant ($T_{0.5} = 24$ meV) with a minor HBr loss ($T_{0.5} = 12$ meV). This is dissimilar to the ${}^\bullet CH_2Br^+CH_3$ species and so no common final TS is obtained for these isomers too.

The CID mass spectra of these isomers[355] show more differences from the classical radical cations $CH_3CH_2X^{1+\bullet}$ than do the α-distonic ions $CH_3X^+CH_2{}^\bullet$. In particular, the former species shows a very small CH_2X^{1+} peak, a significant $HX^{1+\bullet}$ peak and, more importantly, these distonic ions show doubly charged ions of formula $[C_2,H_5,X]^{2+}$.

3.6.81 ${}^\bullet CH_2CH_2X^+H$ WHERE $X = Cl$ AND Br – m/z 64 AND 108

${}^\bullet CH_2CH_2Cl^+H$ showed no confirmable peaks in its MI mass spectrum. The MI mass spectrum of the bromo ion only showed loss of $Br^\bullet$, with a small kinetic energy release ($T_{0.5} = 7$ meV), again unlike its isomers. The CID mass spectra contain almost no m/z 15 ($CH_3{}^{1+}$) and can be differentiated from their other isomers by relative peak abundances. It should be noted that the NR mass spectra of these isomeric pairs[355] display much more significant differences with the $CH_3X^+CH_2{}^\bullet$ ions, showing no signals corresponding to C_2 fragments (i.e., m/z 24–29).

Thermochemistry

Thermochemical Data for the Isomeric Ethyl Halide Radical Cations

Ion	IE_a or AE^a (eV)	Δ_fH(molecule)[1] (kJ/mol)	Δ_fH(neutral product)[1] (kJ/mol)	Δ_fH(ion)[b] (kJ/mol)
CH_3CH_2Cl	10.98 ± 0.02	-112.3 ± 0.75		947 ± 5
CH_3CH_2Br	10.29 ± 0.1	-63.6		929 ± 5
CH_3CH_2I	9.35	-7.2 ± 0.8		895 ± 5
CH_3ClCH_2	10.7 ± 0.2	-414^c $(ClCH_2C(O)OCH_3)$	393.51 ± 0.13 (CO_2)	1011 ± 20
CH_3BrCH_2	10.8 ± 0.2	-364^c $(BrCH_2C(O)OCH_3)$	393.51 ± 0.13 (CO_2)	1071 ± 20
CH_2CH_2ClH	10.72^b	-480^c $(Cl(CH_2)_2C(O)OH)$	393.51 ± 0.13 (CO_2)	947 ± 8
CH_2CH_2BrH	10.70^b	-433^c $(Br(CH_2)_2C(O)OH)$	393.51 ± 0.13 (CO_2)	992 ± 8

[a]Values from NIST[1] or otherwise stated.
[b]Deduced from $IE = \Delta_fH(ion) - \Delta_fH(molecule)$ or $AE = \Delta_fH(ion) + \Delta_fH(neutral\ product) - \Delta_fH(molecule)$.
[c]Values estimated using additivity.[118]

It is noteworthy that computations at the MP2/6-31G*//3-21G* level[357] indicate that the β-distonic ion $^\bullet CH_2CH_2Cl^+H$ is more stable than its conventional isomer by some 15 kJ/mol. These computational results were in fair agreement with those of Clark and Simons.[359]

3.6.82 $[C_2,H_6,X]^+$ WHERE X = F, Cl, Br AND I – m/z 49, 65, 109 AND 157

Isomers

There are two known isomers for these ions, the protonated ethyl halides $(CH_3CH_2XH^{]+})$ and the dimethylhalonium ions $(CH_3X^+CH_3)$. The latter have long been known in the condensed phase.[360]

Identification

Protonation of the ethyl halides in the ion source leads to the generation of the corresponding protonated species. The isomeric dimethylhalonium ions $(CH_3X^+CH_3)$ are readily prepared in a chemical ionization ion source using the appropriate methyl halide[361] and the methyl halonium ions, produced by loss of an $X^\bullet$ atom from the dimers $(CH_3X)_2^{]+\bullet}$.[362]

The MI mass spectra of the $CH_3X^+CH_3$ ions are dominated by loss of CH_4, but the iodo ion also loses $CH_3^{]\bullet}$ and forms I^+, each to ≤5%. The NR mass spectra all have no recovery signal, showing that the corresponding neutrals are unstable.[361]

Thermochemistry

The proton affinities of the ethyl halides have all been measured and from these the heat of formation of the protonated species can be deduced.

Available Thermochemical Data for $CH_3CH_2XH^{]+}$ Ions Where X = F, Cl, Br and I

Ion	PA^{341} (kJ/mol)	$\Delta_fH(CH_3CH_2X)^1$ (kJ/mol)	$\Delta_fH(CH_3CH_2XH^{]+})^a$ (kJ/mol)
CH_3CH_2FH	683	−263	584
CH_3CH_2ClH	693	−112	725
CH_3CH_2BrH	696	−62	772
CH_3CH_2IH	725	−7	798

$^a\Delta_fH(CH_3CH_2XH^{]+}) = \Delta_fH(CH_3CH_2X) + \Delta_fH(H^+) - PA$; using $\Delta_fH(H^+) = 1530$ kJ/mol.[1]

The heats of formation of the $CH_3X^+CH_3$ ions have been reported in connection with the general topic of methyl cation affinities.[363] These values are $\Delta_f H(CH_3F^+CH_3) = 616$ kJ/mol, $\Delta_f H(CH_3Cl^+CH_3) = 752$ kJ/mol, $\Delta_f H(CH_3Br^+CH_3) = 790$ kJ/mol and $\Delta_f H(CH_3I^+CH_3) = 818$ kJ/mol.

3.6.83 $[C_2,H_2,F_2]^{+\bullet}$ – m/z 64

Isomers

Three species have been investigated by experiment: the two conformers Z- and E-CHFCHF$^{1+\bullet}$ and CH$_2$CF$_2$$^{1+\bullet}$.

Identification

The metastable ion behavior of the three isomers has been investigated in some detail[364] and was observed to be remarkably similar. The major loss was HF, accompanied by a large kinetic energy release of about 470 meV. Losses of H$^\bullet$ and F$^\bullet$ from the metastable molecular ions were of equal abundance and together made up some 40% of the MI products. Later, Cooks et al.[340] remeasured the kinetic energy release for the HF loss from ionized CH$_2$CF$_2$ and recorded a $T_{0.5} = 600$ meV. The normal EI mass spectra of the three isomers are also indistinguishable[364] and so their individual identification would be difficult. With the MI and normal mass spectra so similar it is unlikely that CID mass spectra would permit their differentiation.

Thermochemistry

From the ionization energy and heat of formation of the respective molecules, $\Delta_f H(E\text{-CHFCHF}^{1+\bullet}) = 686$ kJ/mol, $\Delta_f H(Z\text{-CHFCHF}^{1+\bullet}) = 687$ kJ/mol, and $\Delta_f H(CH_2CF_2^{1+\bullet}) = 659$ kJ/mol (E-CHFCHF: IE $= 10.15$ eV[1] and $\Delta_f H = -293$ kJ/mol;[58] Z-CHFCHF: IE $= 10.2$ eV[1] and $\Delta_f H = -297$ kJ/mol;[58] CH$_2$CF$_2$: IE $= 10.29 \pm 0.1$ eV[1] and $\Delta_f H = -334 \pm 10$ kJ/mol[1]).

The observed magnitude of the kinetic energy release for the HF losses described above strongly suggests that these processes have appearance energy values significantly above the thermochemical minimum for the production of CHCF$^{1+\bullet}$ + HF ($\Sigma\Delta_f H_{products} = 938$ kJ/mol).[1] Combining the sum of the heats of formation of the products ($\Sigma\Delta_f H_{products}$) together with the heat of formation of each of the molecules allows us to estimate the appearance energy for the loss of HF. These are respectively, 12.8, 12.8, and 13.2 eV for E-CHFCHF, Z-CHFCHF, and CH$_2$CF$_2$. The measured results are 13.7, 13.75 and 14.18 eV,[1] indeed significantly higher than the above values, showing that a large fraction of the reverse energy barrier is partitioned into translational kinetic energy. See also a detailed PIPECO study by Stadelmann and Vogt.[365]

3.6.84 $[C_2,H_2,X_2]^{+\bullet}$ WHERE X = Cl AND Br – m/z 96 AND 184

Isomers

There are three molecular ions for each dihalo-substituted ethene: ionized 1,1 dichloroethene ($CH_2CCl_2^{1+\bullet}$) and the Z- and E-conformers of 1,2-dichloro-ethene (Z-ClCHCHCl$^{1+\bullet}$ and E-ClCHCHCl$^{1+\bullet}$) and for the corresponding bromo analogs: $CH_2CBr_2^{1+\bullet}$, Z-BrCHCHBr$^{1+\bullet}$ and E-BrCHCHBr$^{1+\bullet}$.

Identification

These ions are generated by ionization of the corresponding molecules. The lowest energy dissociation channel for these ionized molecules is the loss of a halogen atom.[1] The reactivity of these molecules with ionized methyla-mine[366] studied by FT-ICR mass spectrometry followed an earlier investiga-tion of the reaction of ionized ammonia with vinyl chloride and bromide.[367] The adduct ions eliminated a halogen atom in each case to produce distonic ions. The reactivities were not sufficiently different for them to be structure-diagnostic. The same authors[368] have reported the catalyzed rearrangement of ionized 1,1-dichloroethene by methanol into the ionized 1,2-isomer.

The charge exchange reactivity of the three $[C_2,H_2,Cl_2]^{+\bullet}$ isomers have recently been investigated; the main result was the identification of long-lived (10^{-5} s or longer) excited electronic states of the isomeric molecular ions.[369]

Thermochemistry

Thermochemical Data for the $[C_2,H_2,X_2]^{+\bullet}$ Ions Where X = Cl and Br

Species	IE^a (eV)	Δ_fH(molecule)^a (kJ/mol)	Δ_fH(ion)^b (kJ/mol)
CH_2CCl_2	9.81[58]	2.6 ± 1[370]	948
E-ClCHCHCl	9.65	6 ± 2	936
Z-ClCHCHCl	9.66	4 ± 2	937
CH_2Br_2	9.78	—	—
E-BrCHCHBr	9.51[58]	106 ± 2^c	1035
Z-BrCHCHBr	9.63[58]	106 ± 2^c	1024

^{a}Values are from NIST[1] or otherwise stated.
^{b}Deduced from IE = Δ_fH(ion) − Δ_fH(molecule).
^{c}Using Δ_fH(CH$_2$CHBr) = 79 ± 2 kJ/mol[1] and the additivity term for a CH$_2$ = group (C$_d$-(H)$_2$ = 26.2 kJ/mol). Thus Δ_fH(BrCHCHBr) = (79–26.2) × 2 = 106 ± 2 kJ/mol.

3.6.85 $[C_2,H_3,F_2]^+$ – m/z 65

Isomers

The site of protonation on the $[C_2,H_2,F_2]^{+\bullet}$ isomers has been investigated by computational chemistry at a variety of levels, B3LYP/631++G(d,p),

MP2(full)/631++G(d,p) and CBS-QB3 all of which gave comparable results. Six ions are found to be local minima on the potential energy surface. These are the classical 1,1-difluoroethyl cation (CH$_3$CF$_2$$^{1+}$), the Z- and E-carbon-protonated FCH$_2$CHF^{1+} ion, the Z- and E-halogen-protonated species FCHCHF(H)$^{1+}$ and finally, the ion–dipole complex [ClCH=CH···FH]$^+$. CH$_3$CF$_2$$^{1+}$ ions have been generated and characterized.

Identification[340]

Dissociative ionization of 1,1-difluoroethane and 1-chloro-1,1-difluoroethane generates a single ion structure that dissociates metastably by loss of HF ($T_{0.5}=0.49\pm0.02$ eV). The connectivity of the ion was assumed to be that of the classical 1,1-difluoroethyl cation (CH$_3$CF$_2$$^{1+}$), the HF loss occurring by a 1,2-elimination.

Thermochemistry[371]

The PA values of the molecules were also reported and were in fair agreement with the results of the experiment. The global minimum on the potential energy surface was determined to be the classical ion CH$_3$CF$_2$$^{1+}$. From the calculated proton affinity of CH$_2$CF$_2$, $\Delta_fH(CH_3CF_2^{1+})=480$ kJ/mol (PA(CH$_2$CF$_2$)=716 kJ/mol (CBS-QB3, experimental value 734 kJ/mol),[1] $\Delta_fH(CH_2CF_2)=-334\pm10$ kJ/mol and $\Delta_fH(H^+)=1530$ kJ/mol).[1] The other isomers, Z-FCH$_2$CHF^{1+}, E-FCH$_2$CHF^{1+}, Z-FCHCHF(H)$^{1+}$, E-FCHCHF(H)$^{1+}$ and [FCH=CH···FH]$^+$ are respectively, 109, 114, 210, 224 and 190 kJ/mol higher in energy than CH$_3$CF$_2$$^{1+}$ (B3LYP/6-31++G(d,p)).

3.6.86 [C$_2$,H$_3$,Cl$_2$]$^+$ – m/z 97

Isomers

In parallel to the fluoro analog, the site of protonation on the [C$_2$,H$_2$,Cl$_2$]$^{+\bullet}$ isomers has been investigated by computational chemistry. Six ions are found to be local minima on the potential energy surface. These are the classical 1,1-dichloroethyl cation (CH$_3$CCl$_2$$^{1+}$), the bridge chloronium cation (CH$_2$(Cl)CHCl^{1+}), the carbon-protonated ClCH$_2$CHCl^{1+} ion, the cis- and trans-halogen-protonated species CH$_2$CClCl(H)$^{1+}$ and the ion–dipole complex [ClCH=CH···ClH]$^+$.

Thermochemistry[371]

The global minimum is CH$_3$CCl$_2$$^{1+}$ with a $\Delta_fH(CH_3CCl_2^{1+})=791\pm7$ kJ/mol (average computed PA=741±6 kJ/mol, $\Delta_fH(CH_2CCl_2)=2.2\pm1.4$ kJ/mol[1] and $\Delta_fH(H^+)=1530$ kJ/mol[1]). The bridged chloronium ion resulting from protonation of the ClCHCHCl isomers (E- and Z-) is also stable with $\Delta_fH(CH_2(Cl)CHCl^{1+})=852\pm8$ kJ/mol (average PA=682±5 kJ/mol,

$\Delta_f H(Z\text{-ClCHCHCl}) = 4.27$ kJ/mol[1] and $\Delta_f H(\text{H}^+) = 1530$ kJ/mol[1]), and $\Delta_f H(E\text{-CH}_2(\text{Cl})\text{CHCl}^{1+}) = 854 \pm 8$ kJ/mol (average PA $= 682 \pm 5$ kJ/mol, $\Delta_f H(E\text{-ClCHCHCl}) = 1.7$ kJ/mol[1] and $\Delta_f H(\text{H}^+) = 1530$ kJ/mol[1]). The $\text{ClCH}_2\text{CHCl}^{1+}$, $cis\text{-CH}_2\text{CClCl(H)}^{1+}$, $trans\text{-CH}_2\text{CClCl(H)}^{1+}$ and $[\text{ClCH} = \text{CH}\cdots\text{ClH}]^+$ ions are calculated to lie, respectively, 64, 128, 137 and 141 kJ/mol above $\text{CH}_3\text{CCl}_2^{1+}$ (B3LYP/6-31++G(d,p)).

3.6.87 $[C_2,H_4,X_2]^{+\bullet}$ WHERE $X = Cl$ AND Br – m/z 98 AND 186

Isomers

The isomers that have been investigated by mass spectrometry are the ionized 1,1- and 1,2-dichloroethanes (E- and Z-conformers) and the 1,2-dibromo analog.

Identification

These molecular ions are generated by ionization of the corresponding molecules. Kim et al.[372] reported the MI characteristics of $E\text{-ClCH}_2\text{CH}_2\text{Cl}^{1+\bullet}$ and $Z\text{-ClCH}_2\text{CH}_2\text{Cl}^{1+\bullet}$, both of which eliminate HCl with a significant release of kinetic energy ($T_{0.5} = $ ca 500 meV). In marked contrast, $\text{Cl}_2\text{CHCH}_3^{1+\bullet}$ loses HCl with a much smaller kinetic energy release ($T_{0.5} = $ ca 18 meV) and a Gaussian peak is observed.[372] Metastable $\text{BrCH}_2\text{CH}_2\text{Br}^{1+\bullet}$ does not eliminate HBr, but loses $\text{Br}^\bullet$ instead, with a small kinetic energy release ($T_{0.5} = 12$ meV).[373]

Thermochemistry

Thermochemical Data for the Dihaloethanes

Species	IE[1] (eV)	$\Delta_f H$(molecule)[1] (kJ/mol)	$\Delta_f H$(ion)[a] (kJ/mol)
$\text{ClCH}_2\text{CH}_2\text{Cl}$	11.07	-137 ± 5[b]	943
Cl_2CHCH_3	11.04	-127.6 ± 1.1	938
$\text{BrCH}_2\text{CH}_2\text{Br}$	10.35	-42 ± 7	957

[a]Deduced from IE $= \Delta_f H$(ion) $- \Delta_f H$(molecule).
[b]See section on $[C_2,H_4,Cl]^+$ above.

It is noteworthy that in the case of the HCl loss from the ionized dichloro compounds, the calculated appearance energy is only about 1031 kJ/mol, substantially below the minimum energy for $\text{Cl}^\bullet$ loss (1085 kJ/mol). This is consistent with the MI mass spectrum, which shows a single fragment peak. In the case of the ionized 1,2-dibromoethane, the minimum energy required for the loss of $\text{Br}^\bullet$ is 1016 kJ/mol while that for HBr loss is somewhat higher (even without a reverse energy barrier it is 1033 kJ/mol), again consistent with the MI mass spectrum.

3.6.88 $[C_2,H,F_3]^+$ – m/z 82

For ionized trifluoroethene (CHFCF$_2^{1+\bullet}$), the loss of CF$^{1\bullet}$ has been described as the most important MI process.[364] It is noteworthy that the CHF$_2^{1+}$ fragment ion dissociates by HF loss with a large kinetic energy release. Combining the ionization energy and heat of formation of the molecule (CHFCF$_2$) leads to Δ_fH(CHFCF$_2^{1+\bullet}$) = 504 kJ/mol (IE = 10.14 eV and Δ_fH(CHFCF$_2$) = -474 ± 8).[1]

3.6.89 $[C_2,H_3,Cl_3]^{+\bullet}$ – m/z 132

Isomers

Data are available for two isomers, namely the molecular ions of 1,2,2-trichloroethane (ClCH$_2$CHCl$_2^{1+\bullet}$) and 1,1,1-trichloroethane (Cl$_3$CCH$_3^{1+\bullet}$).

Identification

The metastable ion mass spectrum of ClCH$_2$CHCl$_2^{1+\bullet}$ has been described[372] and the dominant dissociation is the loss of HCl ($T_{0.5}$ = 504 meV). The labeled ion ClCH$_2$CDCl$_2^{1+\bullet}$ displayed both HCl and DCl losses but with different kinetic energy releases (450 and 567 meV, respectively). This is unlikely to be an isotope effect, but rather reflects the enthalpies of the two products formed: ClCHCHCl$^{1+\bullet}$ (Δ_fH = 936 kJ/mol, see [C$_2$,H$_2$,Cl$_2$]$^{+\bullet}$) and Cl$_2$CCH$_2^{1+\bullet}$ (Δ_fH = 948 kJ/mol, see [C$_2$,H$_2$,Cl$_2$]$^{+\bullet}$) and different energy barriers for these competing dissociations. Note, see below, that these dissociations are exothermic.

Thermochemistry

Combining the ionization energy and heat of formation values of the corresponding molecules leads to Δ_fH(ClCH$_2$CHCl$_2^{1+\bullet}$) = 913 $\pm$ 4 kJ/mol and Δ_fH(CH$_3$CCl$_3^{1+\bullet}$) = 918 $\pm$ 2 kJ/mol (Δ_fH(ClCH$_2$CHCl$_2$) = -148 ± 4 kJ/mol and IE = 11.0 eV and Δ_fH(CH$_3$CCl$_3$) = -143 ± 2 kJ/mol and IE = 11.1 eV).[1]

The loss of HCl and DCl (Δ_fH = -92 kJ/mol)[58] described above is exothermic by 68 kJ/mol for the HCl loss and 57 kJ/mol for the DCl loss.

3.6.90 $[C_2,F_4]^{+\bullet}$ – m/z 100

Isomers

There may be two isomers, namely perfluoroethene (F$_2$C=CF$_2^{1+\bullet}$) and the carbene CF$_3$CF$^{1+\bullet}$, but the existence of the two species remains unresolved.

Identification

The only MI process for ionized perfluoroethene ($F_2C=CF_2^{1+\bullet}$) is the loss of $CF^\bullet$ giving CF_3^{1+} ($T_{0.5}=5$ meV).[374] In a search for the isomer $CF_3CF^{1+\bullet}$, the ion generated by the loss of $CF_2^{1+\bullet}$ from *metastable* ionized CF_3CFCF_2 was examined and it was found to have a significantly different CID mass spectrum from that of the ion source generated $F_2C=CF_2^{1+\bullet}$. The latter was dominated by the $CF^\bullet$ loss (100%) and a small $F^\bullet$ loss (23%), whereas the latter had $F^\bullet$ loss as base peak and $CF^\bullet$ loss at 50%. The very small kinetic energy release suggests that the loss of $CF^\bullet$ is a threshold process involving the carbene isomer as the reacting configuration.

Thermochemistry

$\Delta_f H(F_2C=CF_2^{1+\bullet})=319$ kJ/mol derives from the combination of the IE $=10.14$ eV and $\Delta_f H(F_2C=CF_2)=-659\pm8$ kJ/mol.[1] The loss of $CF^{1\bullet}$ is the major fragmentation of metastable $F_2C=CF_2^{1+\bullet}$ ions (see also $[C_2,H,F_3]^{+\bullet}$ below) and the reaction has a measured AE of 13.7 eV.[1] The latter is close to the thermochemical threshold, given the uncertainties in the available data (currently,[1] $\Delta_f H(F_2C=CF_2)=-659$ kJ/mol, IE$(CF_3^{1\bullet})=8.76$ eV, $\Delta_f H(CF_3^{1\bullet})=-470$ kJ/mol and $\Delta_f H(CF^{1\bullet})=255$ kJ/mol, leading to an estimated AE of about 13.4 eV).

3.6.91 $[C_2,H_2,Cl_4]^{+\bullet}$ – m/z 166

Kim et al.[372] described the MI characteristic of ionized $Cl_2CHCHCl_2$, which loses HCl with a large kinetic energy release ($T_{0.5}=610$ meV), i.e., this is an exothermic reaction with a reaction enthalpy greater than for the trichloro ions. $\Delta_f H(CHCl_2CHCl_2^{1+\bullet})=915$ kJ/mol is obtained when combining $\Delta_f H(CHCl_2CHCl_2)=-156\pm8$ kJ/mol and IE $=$ ca 11.1 eV.[1]

One final note about the thermochemistry of these molecules is that successive Cl substitution in $[C_2,H_n,Cl_{6-n}]$ lowers $\Delta_f H$ values up to the fourth Cl atom, after which $\Delta_f H$ rises, e.g., $\Delta_f H(C_2H_2Cl_4)=-156$ kJ/mol and $\Delta_f H(C_2HCl_5)=$ ca -134 kJ/mol.

3.6.92 $[C_2,N,O]^+$ – m/z 54

Isomers

Three isomeric ions $NCCO^{1+}$ (**1**), $CNCO^{1+}$ (**2**) and $CCNO^{1+}$ (**3**) have been investigated by experiment and by computation.

Identification[375]

The starting molecules for this study were pyruvonitrile ($NCC(O)CH_3$), methylcyanoformate ($NCC(O)OCH_3$) and dichloroglyoxime (HON-$C(Cl)C(Cl)NOH$). It was concluded that loss of $CH_3^{1\bullet}$ and $CH_3O^\bullet$ from the first two ionized molecules cleanly generated ions of structure $NCCO^{1+}$ and their CID mass spectra supported this, being dominated by peaks for m/z 26 (CN^{1+}), 28 $(CO^{1+\bullet})$ and 40 $(C_2O^{1+\bullet})$.

Ion (2) was generated as a weak peak in the mass spectrum of ionized CH_3NCO (losses of $H^\bullet$ and H_2). The ion's CID mass spectrum was in keeping with the proposed structure with $CO^{1+\bullet}$ and CN^{1+} as the major peaks. Ion (3) was not cleanly produced from dichloroglyoxime in the ion source, but the CID mass spectrum of mass-selected m/z 84 $([C_2,N_2,O_2]^{+\bullet})$ and 68 $([C_2,N_2,O]^{+\bullet})$ ions therefrom produced $[C_2,N,O]^+$ ions that displayed only m/z 30 (NO^{1+}) as a major ion in their CID mass spectra together with minor peaks at m/z 24 and 26. Note that an NRMS investigation of these species showed that the neutral counterparts were also stable entities in the gas phase.

Thermochemistry

The computational chemistry results of Yu et al.[376] were conducted at a higher level of theory than that used by McGibbon et al.[375] (UMP3/6-31G*// $6\text{-}31G^* = (0.9)ZPVE$) and so the former data will be quoted here. The relative energies of *neutral* (1), (2) and (3) were 0, 77 ± 8 and 414 ± 4 kJ/mol, using the average results from G3//B3LYP, G3(MP2)//B3LYP and CASPT2//CASSCF levels of theory.

The product energies, relative to the global minimum (1), were 110 kJ/mol for $CN^{1\bullet} + CO$ $(\Sigma\Delta_f H_{products} = 325$ kJ/mol)[1] and 695 kJ/mol for $C_2 + NO$ $(\Sigma\Delta_f H_{products} = 928$ kJ/mol),[1] resulting in an estimated $\Delta_f H(NCCO) = 229 \pm 5$ kJ/mol. Consequently, from the calculated relative energies, $\Delta_f H(CNCO) = 306 \pm 13$ kJ/mol and $\Delta_f H(CCNO) = 643 \pm 9$ kJ/mol. The IE_a values were computed to be 8.84 ± 0.1, 8.59 ± 0.1 and 9.32 ± 0.1 eV, respectively, leading to ionic $\Delta_f H$ values of $\Delta_f H(NCCO^{1+}) = 1082 \pm 8$ kJ/mol, $\Delta_f H(CNCO^{1+}) = 1135 \pm 16$ kJ/mol and $\Delta_f H(CCNO^{1+}) = 1542 \pm 12$ kJ/mol. The earlier study[375] had computed relative energies of 0, 66 and 506 kJ/mol for the neutrals and an energy difference for the ions (1) and (2) of 31 kJ/mol.

3.6.93 $[C_2,H,N,O]^{+\bullet} - m/z$ 55

The simple cumulene ion $HNCCO^{1+\bullet}$ (ionized iminoethenone) is a species found in the mass spectrum of selected purines.[377] It is characterized by its CID mass spectrum, which contains peaks at m/z 40, 38, and 27, the ions $CCO^{1+\bullet}$, CCN^{1+} and $HNC^{1+\bullet}$, respectively.

The relative energy of species related to this ion was computed at the G2(MP2) level of theory.[377] For the above ion, its energy relative to the dissociation products $HNC^{1+\bullet} + CO$ was 273 kJ/mol. With $\Delta_f H(CO) = -111$ kJ/mol[1] and $\Delta_f H(HNC^{1+\bullet}) = 1349$ kJ/mol, $\Delta_f H(HNCCO^{1+\bullet}) = 965$ kJ/mol is obtained.

3.6.94 $[C_2,H_3,N,O]^{+\bullet}$ – m/z 57

Isomers

There are a large number of conventional and distonic isomeric ions to be considered here, thanks to two ab initio studies.[379,380] Experiments by Hop and Snyder attempted to prepare and compare the mass spectrometric features of some of them.[381] The 11 ions investigated are $CH_3NCO^{1+\bullet}$ (1), $CH_3OCN^{1+\bullet}$ (2), $CH_3CNO^{1+\bullet}$ (3), $HOCH_2CN^{1+\bullet}$ (4), $H_2NCCOH^{1+\bullet}$ (5), $H_2NC(H)CO^{1+\bullet}$ (6), $CH_2C(H)NO^{1+\bullet}$ (7), $CH_2N(H)CO^{1+\bullet}$ (8), $HNCH_2 CO^{1+\bullet}$ (9), $CH_3C(O)N^{1+\bullet}$ (10) and $O(H)CC(H)NH^{1+\bullet}$ (11). Note that over 20 plausible structures for this ion can be proposed.

Identification

Ion (1) is the molecular ion of methyl isocyanate and is the base peak in that molecule's mass spectrum. Its MI mass spectrum contains two signals, $H^\bullet$ loss (m/z 56, $T_{0.5} = 70$ meV) and m/z 29, whose identity was not determined and so is formally HCO^{1+} and/or $[C,H_3,N]^+$ ($T_{0.5} = 221$ meV). The CID mass spectrum contains structure-characteristic peaks at m/z 42 (NCO^{1+}) and 15 (CH_3^{1+}).

Ion (2) (ionized methyl cyanate) is argued to be produced in the mass spectrum of methyl cyanoformate ($NCC(O)OCH_3$) wherein it is a minor peak (loss of CO is only 0.4% of base peak, m/z 54). However, in principle, two rival connectivities may be proposed, namely $CH_3CNO^{1+\bullet}$ (3) or $CH_3ONC^{1+\bullet}$ (4). The MI mass spectrum of this ion contains only a peak for H-loss. The CID mass spectrum is dominated by m/z 31 (CH_2OH^{1+}).

Ion (3) ($CH_3CNO^{1+\bullet}$) and ion (10) ($CH_3C(O)N^{1+\bullet}$) are produced as a weak peak in the mass spectrum of acetohydroxamic acid ($CH_3C(O)N(H)OH$, by loss of H_2O). The MI mass spectrum of this ion is unique and contains two peaks: one at m/z 43 to which, on thermochemical grounds, the acetyl ion structure was given, and one at m/z 27, assigned to $C_2H_3^{1+}$ and not the isobaric $[H,C,N]^{+\bullet}$, also by a thermochemical rationale. The former is a dished peak ($T_{0.5} = 785$ meV) and the latter is Gaussian ($T_{0.5} = 23$ meV). This matter cannot be taken as completely settled because the metastable ion $CH_2CHNO^{1+\bullet}$ (7) also loses NO but with a tiny kinetic energy release ($T_{0.5} = 0.8$ meV). Moreover D-labeled acetohydroxamic acid ($CH_3C(O)N(D)OD^{1+\bullet}$) was observed to only lose D_2O and so ion (10) is also a possible contributor. To try to resolve the dilemma, an NRMS experiment was performed and an intense recovery signal (m/z 57) showed that at least the majority of the ions had a stable neutral counterpart.[381] Neutral (10) had been shown by calculation[379] to very readily collapse to (2), but this does not appear definitely to resolve the issue, which in our view must remain incomplete. The CID mass spectrum of the ion generated from acetohydroxamic acid is not helpful in deciding amongst the possible connectivities, except that it contains a more intense m/z 30 peak than any of its isomers.

Ion (4) (HOCH$_2$CN$^{]+\bullet}$, ionized glyconitrile) has a weak molecular ion (3.4% of base peak m/z 28). The ion has a unique MI mass spectrum with only a peak at m/z 28 ($T_{0.5} = 29$ meV). The dissociation of lowest formal energy requirement is to produce CH$_2$OH$^{]+}$ + CN$^\bullet$, and so the MI fragmentation yields HCNH$^{]+}$ + HCO$^{]\bullet}$ after two H-shifts in the molecular ion. The ion m/z 31 is only a small peak in the CID mass spectrum and this simple bond cleavage would be expected to dominate the CID reactions. No computational investigation has been performed on this ion and so the difficulty of firmly assigning a structure to this ion remains unresolved.

Ions (5) and (6) (H$_2$NCCOH$^{]+\bullet}$ and H$_2$NC(H)CO$^{]+\bullet}$) are the enol and keto forms of ionized aminoketene. The former was described by Terlouw et al.[382] and the ion was generated by the loss of two CO molecules from the ionized squaric acid derivative (cy–C(O)C(O)C(NH$_2$)C(OH)–). The CID mass spectrum is dominated by a signal at m/z 28 (100%) and minor peaks at m/z 56, 40 and 41. The MI mass spectrum is not reported. Ion (6) was generated by the loss of H$_2$O following a McLafferty-type rearrangement in ionized threonine (CH$_3$CH(OH)CH(NH$_2$)C(O)OH), a preparative method superior to one in the earlier literature that produced [C$_2$,H$_3$,N,O]$^{+\bullet}$ ions contaminated with [C$_4$,H$_9$]$^+$. The MI mass spectrum of this ion distinguishes it from the others by having two peaks, H$^\bullet$ loss (m/z 56, $T_{0.5} = 168$ meV) and m/z 29 ([H,C,O]$^+$, $T_{0.5} = 7.0$ meV) The CID mass spectrum has structure-characteristic peaks at m/z 41, 16 and 13.

Ion (7) (CH$_2$CHNO$^{]+\bullet}$) was readily produced directly from ionized nitroethene by the loss of atomic oxygen.[383] The results were confirmed[381] but only the CID mass spectrum has been reported. It is wholly dominated by m/z 27 (C$_2$H$_3$$^{]+}$), with small (5%) m/z 40 and 56 fragment peaks.

The distonic isomers (8) and (9) (CH$_2$NHCO$^{]+\bullet}$ and HNCH$_2$CO$^{]+\bullet}$) were considered as the products of CO$_2$ loss from ionized oxazolidene-2,5-dione (cy–C(O)OC(O)C(H)$_2$N(H)–). The metastable ion peak for this dissociation is composite, consisting of two broad signals of roughly equal abundance ($T_{0.5} = 128$ and 513 meV.) Consideration of the oxazolidene-2,5-dione ion's structure shows that formally two CO$_2$ loss routes are possible, one to yield the distonic ion (H$_2$CNHCO$^{]+\bullet}$, (8)) and the other giving the isomeric distonic ion (HNCH$_2$CO$^{]+\bullet}$, (9)). The ions were nicely differentiated by placing a ^{13}C label at the C5-CO group and the composite peak in the MI mass spectrum was now resolved to m/z 58 (the narrow component) and m/z 57 (the broad). The m/z 58 ion was assigned structure (9) HNCH$_2$^{13}CO$^{]+\bullet}$ and m/z 57, structure (8). The MI mass spectrum of m/z 58 contained a dished peak at m/z 29 and a smaller m/z 30 (H$_2$CNH$^{]+\bullet}$ and H^{13}CO$^{]+}$, respectively). The ion m/z 57 displays an intense, Gaussian peak at m/z 29. The CID mass spectra were also structure-supportive.

Ion (11) was prepared by the dissociative ionization of 2(3)-oxazolone (cy–C(O)OC(H)C(H)N(H)–, by loss of CO). The m/z 57 ion distinguished itself from all the others above by having a single MI process to produce an

m/z 28 ion with a tiny kinetic energy release ($T_{0.5} = 0.5$ meV), quite unlike ion (**4**) where $T_{0.5} = 29$ meV. A deuterium labeling experiment showed the m/z 28 ion to be CH_2N^{1+}, not the thermochemically disfavored $CO^{1+\bullet}$. The CID mass spectrum contains signals at m/z 38, 39 and 40 in keeping with the C–C–N connectivity.

The reactivity of ionized and neutral CH_3NCO (methyl isocyanate) has been studied by ICR techniques.[384] It reacts chiefly by charge transfer or proton acceptance. Protonation at the N atom is preferred over oxygen.

Thermochemistry

There is little thermochemical data for these ions and their neutral counterparts. An ab initio study[380] (MP2/6-31G**) of the relative stability of methyl cyanate and isocyanate showed the latter to be more stable by about 113 kJ/mol. The earlier result, 82 kJ/mol, was obtained at a lower level of theory.[379] NIST gives $\Delta_fH(CH_3NCO) = -130$ kJ/mol and an IE = 10.67 eV, making $\Delta_fH(CH_3NCO^{1+\bullet}) = 900$ kJ/mol.[1]

3.6.95 $[C_2,H_4,N,O]^+ – m/z$ 58

The proton affinity of methyl isocyanate has been determined by ICR mass spectrometry[384] to be 764.4 kJ/mol (note that the PA scale changed and the value comes from the Hunter and Lias compilation).[385] Calculations at the CEP31-G** level of theory[386] indicate that protonation at N is the preferred site and therefore when the PA is combined with $\Delta_fH(CH_3NCO) = -130$ kJ/mol and $\Delta_fH(H^+) = 1530$ kJ/mol,[1] $\Delta_fH(CH_3N^+(H)CO) = 636$ kJ/mol is obtained.

3.6.96 $[C_2,N,O_2]^+ – m/z$ 70

Isomers

Two isomeric ions have been studied: $OCNCO^{1+}$ and $OCCNO^{1+}$.

Identification

The ion $OCNCO^{1+}$ was generated by the dissociative ionization of ethoxycarbonyl isocyanate ($CH_3CH_2OC(O)NCO$) and its isomer, $OCCNO^{1+}$, was obtained likewise from ethylchlorohydroxyiminoacetate (HON = $C(Cl)C(O)OCH_2CH_3$).[387–389] The two are readily distinguished by low energy CID experiments,[389] the former producing only m/z 42 ($C_2O^{1+\bullet}$) ions and the latter giving structure-characteristic m/z 30 (NO^{1+}), 42 and 54 (NC_2O^{1+}).

Thermochemistry

Although eight isomers of this formula were considered in ab initio calculations at the G2(MP2) level of theory, the above pair were the lowest energy species.[389] The asymmetric isomer was found to lie 304 kJ/mol above $OCNCO^{1+}$. Dissociation to $NCO^{1+} + CO$ was calculated to lie 450 kJ/mol above $OCNCO^{1+}$ and only some 146 kJ/mol above the ion $OCCNO^{1+}$. With the selected $\Delta_f H(NCO^{1+}) = 1270 \pm 5$ kJ/mol (see section on $[C,N,O]^+$) and $\Delta_f H(CO) = -111$ kJ/mol,[1] $\Delta_f H(OCNCO^{1+}) = 709 \pm 5$ kJ/mol and $\Delta_f H(OCCNO^{1+}) = 1013 \pm 5$ kJ/mol are deduced. (A parallel calculation with $C_2O + NO^{1+}$ as the products produced very similar results, but with wider error limits because of the uncertainty of $\Delta_f H(C_2O)$).

3.6.97 $[C_2,H_2,N,O_2]^+ - m/z\,72$

Isomers

Two isomeric ions have been generated and investigated, namely $cy-ONOC(H)C(H)-^{1+}$ (**1**) and CH_2CONO^{1+} (**2**).

Identification

The reaction of the nitronium ion (NO_2^{1+}) with ethyne[390] produces, after stabilization, an ion assigned as $cy-ONOC(H)C(H)-^{1+}$ (note that the reaction is exothermic and without collisional stabilization the products are $CH_2CO^{1+\bullet}$ and NO). This ion results from a 1,3-dipolar cycloaddition and produces a resonance-stabilized (aromatic) ion.

The isomer CH_2CONO^{1+} is produced by the NO chemical ionization of ketene:[390]

$$CH_2CO + NO^{1+} \rightarrow CH_2CONO^{1+}$$

The MI mass spectrum of this latter ion contained a peak only at $m/z\,30$ (NO^{1+}) whereas that of ion (**2**) had peaks for $m/z\,46$ (NO_2^{1+}), 42 $(CH_2CO^{1+\bullet})$ and 30 (NO^{1+}). The kinetic energy release values were not recorded. The CID mass spectra were also in keeping with the assigned structures that for (**1**) uniquely showing $m/z\,26$ $(C_2H_2^{1+\bullet})$ and (**2**) likewise showing $m/z\,14$ $(CH_2^{1+\bullet})$.

Thermochemistry

Potential energy surfaces for this and related reactions (see below) were calculated at the MP2/6-31G* level of theory. Only relative energies were given, but some thermochemical data can be extracted from the results.

The aromatic $[C_2,H_2,N,O_2]^+$ ion (**1**) was calculated to lie 324 kJ/mol below the reactants $(\Delta_f H(C_2H_2) = 228$ kJ/mol[1] and $\Delta_f H(NO_2^{1+}) = 956$ kJ/mol (from $\Delta_f H(NO_2) = 33.1$ kJ/mol and $IE = 9.59$ eV[1])), leading to $\Delta_f H(\mathbf{1}) = 860$ kJ/mol. The ion (**2**) lies 9 kJ/mol above (**1**), and consequently, $\Delta_f H(\mathbf{2}) = 869$ kJ/mol. Ion (**2**) is separated from ion (**1**) by a barrier of 273 kJ/mol.

3.6.98 $[C_2,H_4,N,O_2]^+$ – m/z 74

Isomers

Two ions have been generated: $CH_3CHONO^{]+}$ (1) and the glycyl cation $NH_2C(H)C(O)OH^{]+}$ (2). Computations have also studied, in addition to (2), the ions $HNCH_2C(O)OH^{]+}$ (3), $[CH_2NH_2\cdots CO_2]^+$ (4) and $CH_2N(H)C(O)OH^{]+}$ (5).

Identification

The m/z 74 ion generated from the reaction of ethene with the nitronium ion $(NO_2^{]+})$ was expected to be a cyclic adduct ion, but it was shown that this ion and those produced by the nitrosation of oxirane and acetaldehyde[391,392] have indistinguishable CID mass spectra, indicating that a single structure (or a single mixture of structures) is obtained. The CID mass spectra show a very significant m/z 44 fragment peak (loss of NO) consistent with an ion containing a C–C–O–N–O moiety. On the basis of mass spectral data, reactivity patterns with $NO^{]+}$ and thermochemical data, the assigned structure was $CH_3CHONO^{]+}$.[391]

It is worth noting that the reaction of $CH_3OXNO_2^{]+}$ (X = H, NO_2) with ethene produces an m/z 74 ion distinct from that described above. Its CID mass spectrum contains a significant peak at m/z 46 $(NO_2^{]+})$, suggesting an ion with an intact NO_2 group. No structure assignment for this ion has been made.[391]

Dissociative ionization of phenyl alanine $(NH_2C(H)(C_6H_5)C(O)OH$, loss of $C_6H_5^{]•})$ produces yet another isomer, and based on the structure of the precursor, was assigned to be the glycyl cation $(NH_2C(H)C(O)OH^{]+})$. Its CID mass spectrum is different from those discussed above and identification of the fragment ions supports the proposed connectivity (i.e., m/z 57 is $NH_2C(H)CO^{]+•}$ and m/z 46 is $NH_2C(H)OH^{]+}$).[393]

Partial CID Mass Spectra of the $[C_2,H_4,N,O_2]^+$ Isomers

Ion	m/z							
	28	29	42	44	45	46	56	57
CH_3CHONO[a]		67	28	100		—		
$CH_3OXNO_2 + C_2H_4$[a]		100	20	17		37		
$NH_2C(H)C(O)OH$[b]	100	29		8.3	27	39	6.5	9.7

[a]Cacace et al.[391]
[b]Turecek and Carpenter.[393]

It is worth noting that the CID mass spectrum of the $[M–H]^+$ ions derived from glycine ($NH_2CH_2C(O)OH$) significantly differs from those above (partial CID MS: m/z 28 (36), 30 (59), 45 (9), 46 (100), 56 (20), 57 (12) and 58 (6) numbers in parentheses are relative abundances).[394] For example, m/z 30 is absent in the CID of the ion obtained from phenyl alanine assigned as $NH_2C(H)C(O)OH^{]+}$. Also noteworthy is the fact that upon charge reversal of the m/z 74 anion from glycine, yet another different CID mass spectrum is obtained (partial CID MS: m/z 28 (100), 30 (50), 42 (13), 44 (91) and 57 (2)).[394] It is most likely that in the two latter cases, the ion flux contains several isomeric species because deprotonation can, in principle, occur at three different sites (O, C and N atoms).[394]

Thermochemistry

Potential energy surfaces for the reaction of ethene with the nitronium ion and related reactions ($[C_2,H_2,N,O_2]^+$, see above) were calculated at the MP2/6-31G* level of theory.[392] Only relative energies were given, but some thermochemical data can be extracted from the results. The adduct ion $CH_3CHO–NO^{]+}$ is the global minimum and was computed to lie 309 kJ/mol below the reactants $C_2H_4 + NO_2^{]+}$ ($\Delta_fH(C_2H_4) = 52$ kJ/mol[1] and $\Delta_fH(NO_2^{]+}) = 956$ kJ/mol (from $\Delta_fH(NO_2) = 33.1$ kJ/mol and IE = 9.59 eV[1])) leading to $\Delta_fH(CH_3CHO–NO^{]+}) = 699$ kJ/mol. This ion lies 132 kJ/mol below the dissociation limit to $CH_3CHO + NO^{]+}$. The cyclic adduct ion had a heat of formation 133 kJ/mol higher (e.g. Δ_fH(cyclic ion) = 832 kJ/mol) and the barrier separating it from $CH_3CHO–NO^{]+}$ was calculated to be 150 kJ/mol.

The anti, syn conformer of (2) is calculated to be the lowest energy isomer $NH_2C(H)C(O)OH^{]+}$. It is computed to lie slightly above the dissociation products $NH_2CHOH^{]+} + CO$ and therefore must be kinetically stabilized by a high energy barrier (at 0 K it lies below the products by 13 kJ/mol (G2(MP2), 3 kJ/mol (MP2/6-311+G(2df,p), and 6 (B3LYP/6-311 +G(2df,p)).[393]

Computations at the MP2(FC) level of theory[394] find that among isomers (2)–(5), the ion–neutral complex $[CH_2NH_2\cdots CO_2]^+$ is the lowest energy species. The glycyl cation (2) lies 86 kJ/mol above it whereas isomers (3) and (5) are respectively higher in energy by 489 (the triplet state) and 86 kJ/mol.

3.6.99 $[C_2,H_5,N,O_2]^{+\bullet} – m/z$ 75

Isomers

Ionized glycine ($H_2NCH_2C(O)OH^{]+\bullet}$, (1)) and its enol isomer $H_2NCHC(OH)_2^{]+\bullet}$ (2) have received much attention. A distonic isomer

$H_3NCHC(O)OH^{]+\bullet}$ (**3**) is proposed as an intermediate ion. The extensive literature is well reviewed and summarized in the paper by Polce and Wesdemiotis.[395]

Identification

Ionization of glycine generates the corresponding molecular ion whereas the enol ion (**2**) is readily produced by a McLafferty rearrangement in ionized isoleucine ($CH_3CH(CH_2CH_3)CH(NH_2)C(O)OH$), the loss of C_4H_8 producing the desired ion.

The MI mass spectra of the isomers are characteristically different. For (**1**), the ion m/z 30 ($H_2NCH_2^{]+}$) is by far the major signal and the narrow Gaussian peak has $T_{0.5} = 3$ meV. The MI mass spectrum of (**2**) has m/z 57 (loss of H_2O) as the sole feature ($T_{0.5} = 51$ meV). This ion has the amino-ketene structure $H_2NCHCO^{]+\bullet}$ and the fragmentation is believed to involve the distonic intermediate ion $H_3NCHC(O)OH^{]+\bullet}$. All H atoms lose their positional identity before this dissociation. The CID mass spectra are in keeping with both ions retaining their initial structures; ion (**1**) shows m/z 17 ($OH^{]+}$), 45 ($C(O)OH^{]+}$), 57, 60 (loss of NH), whereas ion (**2**) has a dominant m/z 57 and relatively weak signals at m/z 19 ($H_3O^{]+}$), 29 ($HCNH^{]+}$) and 46 ($[C,O_2,H_2]^{+\bullet}$).

The computed barrier to the keto-enol tautomerization is greater than the dissociation limits of both isomers,[396] in keeping with the MI and CID mass spectra. A detailed NRMS study has also been done by Polce and Wesdemiotis.[395]

Thermochemistry

$\Delta_fH(H_2NCH_2C(O)OH^{]+\bullet}) = 464 \pm 10$ kJ/mol (from $\Delta_fH(H_2NCH_2C(O)OH)$ $= -391 \pm 5$ kJ/mol and $IE_a = 8.85 \pm 05$ eV).[1] Depke et al.,[396] at the 6-31G//STO-3G and 6-31G//3-21G level of theory, find that (**2**) is some 92 kJ/mol lower in energy than (**1**). The more recent calculations by Yu et al.,[397] at the (higher) G2(MP2) level of theory, place (**2**) 84 kJ/mol below (**1**) and so $\Delta_fH(H_2NCHC(OH)_2^{]+\bullet}) = 380 \pm 10$ kJ/mol, based on the above experimental value for glycine.

Although the latter computations gave a value of -392 kJ/mol for the Δ_fH of neutral glycine (in agreement with NIST),[1] their energy for the ion was presented as 502 kJ/mol, moving the IE significantly upward (ca 0.4 eV), a remarkably large discrepancy particularly in view of the IE_a for alanine (8.5 eV).[398] This difficulty requires resolution and the reader must decide between the experiment and calculation.

Finally, Yu et al.[397] computed $\Delta_fH(3) = 467$ kJ/mol, therefore 35 kJ/mol below (**1**). When using the experimental value $\Delta_fH(H_2NCH_2C(O)OH^{]+\bullet}) = 464 \pm 10$ kJ/mol and the relative energy, $\Delta_fH(3) = 429 \pm 10$ kJ/mol.

3.6.100 $[C_2,H_6,N,O_2]^+$ – $m/z\,76$

Isomers

Six $[C_2,H_6,N,O_2]^+$ ions have been investigated, notably protonated nitroethane (1), ethylnitrite (2), glycine (3) and isomers of (3) that have been implicated in its unimolecular dissociation (4)–(6).

$(CH_3CH_2NO_2)H$ $CH_3CH_2O(H)NO$ NH_3CH_2COOH $NH_2CH_2C(O)\text{-}OH_2$

1   2   3   4

$NH_2CH_2C(OH)_2$ CH_2NH_2 ⋯ $HC(O)OH$

5   6

Identification

Protonated ethylnitrite can be generated by CI of nitroethane and exhibits a CID mass spectrum that contains a peak ratio m/z 31:30:29 of 0:100:75.[399] Protonated ethylnitrite was generated both by CI of ethylnitrite and by NO^{1+} addition to ethanol, both sources leading to the same ion for which the CID mass spectrum has the peak ratio m/z 31:30:29 of 21:100:25.[399] Metastable protonated glycine ions lose CO to form m/z 48, accompanied by a very minor dissociation to produce m/z 30 ($CH_2NH_2^{1+}$, confirmed by isotopic labeling experiments).[400] The CID mass spectrum has the m/z 30 peak as the base peak and an intense m/z 42, an ion not present in the mass spectra of the other two isomers.[400] Ions (4)–(6) have been explored only by theory.[401,402] Ion (4) has been calculated to lie in a very shallow well close to the dissociation threshold for H_2O loss and so is unlikely to be observed directly.[401]

Thermochemistry

$\Delta_f H(2) = 603$ kJ/mol is deduced from a neutral $\Delta_f H$ of -108 kJ/mol and a PA value of 819 kJ/mol.[1] The only other $\Delta_f H$ value known is for protonated glycine, $\Delta_f H(3) = 253$ kJ/mol, obtained from a neutral molecule $\Delta_f H$ of -390.5 kJ/mol and a PA of 886.5 kJ/mol.[1]

The energies of (5) and (6) have been calculated relative to (3) at the QCISD(T)/6-31G*//B3-LYP/6-31G* level of theory to be $+113$ kJ/mol and $+33$ kJ/mol, respectively.[402] Significant barriers separate them from (3), indicating that it may be possible to independently generate them. As mentioned previously, the energy of (4) lies 141 kJ/mol above (3) and only 6 kJ/mol below the dissociation limit for H_2O loss.[401]

3.6.101 $[C_2,N_2,O]^{+\bullet}$ – m/z 68

Isomers

There are two possible ions: $NCCNO^{1+\bullet}$ and $NCNCO^{1+\bullet}$.

Identification

The photoionization mass spectrum of NCCNO yields two fragment peaks, namely m/z 30 (NO^{1+}) and 38 (NCC^{1+}), consistent with the connectivity in the neutral.[403] This ion is also proposed to be obtained by the dissociative ionization of 3,4-dicyano-1,2,5-oxadiazole (cy–ONC(CN)C(CN)N–, loss of NCCN).[404] Ionization of the thermolysis products of dicyanofuroxan (cy–ONC(CN)C(CN)N(O)–) is also believed to generate $NCCNO^{1+\bullet}$, but the presence of an m/z 40 (CO loss) fragment peak in the ion's CID mass spectrum suggests that a second isomeric species, namely $NCNCO^{1+\bullet}$, is cogenerated.[404]

Thermochemistry

For computations at the G2(MP2,SVP) level of theory refer to Flammang et al.[404]

3.6.102 $[C_2,H,N,S]^{+\bullet}$ – m/z 71[405]

Isomers

The ionized cumulene isomers $HNCCS^{1+\bullet}$ (iminoethenethione) and $HCSCN^{1+\bullet}$ (thioformyl cyanide) have been successfully generated and identified. A third isomer, $HCNCS^{1+\bullet}$, is predicted by computation to be stable.

Identification

$HNCCS^{1+\bullet}$ was produced by the losses of HNCO, CH_3CN and CO from an ionized thiazolopyrimidinedione. The CID mass spectrum was structure-indicative, showing in particular peaks for the losses of $H^\bullet$, $NH^{1\bullet}$ and HNC. $HCSCN^{1+\bullet}$ was obtained by the ionization of the molecule generated by the flash-vapor pyrolysis of allylcyanomethyl sulfide ($CH_2CHCH_2SCH_2CN$, loss of C_3H_6). It had a distinctive CID mass spectrum in keeping with the above structure, with prominent peaks for HCS^{1+} and $CS^{1+\bullet}$ as well as SCN^{1+}. The neutral analogs were also investigated by the NRMS technique.

Thermochemistry

Only relative energies of $HNCCS^{1+\bullet}$, $HCSCN^{1+\bullet}$ and $HCNCS^{1+\bullet}$ have been computed at the G2(MP2) level of theory; they are 0, 20 and 66 kJ/mol respectively.

3.6.103 $[C_2,H_2,N,S]^+$ – m/z 72[406]

Isomers

Computations find no less than six possible isomeric ions: CH_2NCS^{1+} (**1**), $HCNC(H)S^{1+}$ (**2**), $CNC(H)SH^{1+}$ (**3**), $HCNCSH^{1+}$ (**4**), $HCN(H)CS^{1+}$ (**5**) and $[HCN \cdots HCS]^+$ (**6**). Although m/z 72 ions have been generated, it is impossible, based on the mass spectral results, to unambiguously assign a structure to this ion.

Identification

In principle, isomers of this formula can be produced by the dissociative ionization of 2-mercaptoimidazole (cy–C(H)C(H)N(H)C(SH)N–). However, a combined experimental and theoretical investigation by Reddy et al.[1] showed that although no less than six isomers exist in potential wells (**1**)–(**6**), the m/z 72 ion produced in the ion source and by the metastable loss of CH_2N from the ionized imidazole had very closely similar CID mass spectra. This showed that a single ion structure, or a simple mixture of ions, was likely generated, and moreover, the CID mass spectrum displayed fragment ions fully in keeping only with structures (**1**) and (**6**). The MI mass spectrum of the m/z 72 ion showed only the loss of HCN with a very small kinetic energy release (1.3 meV), which is compatible with the dissociati of an electrostatically bound complex such as ion (**6**). This was supported by the calculations that showed that ion (**6**) indeed provided a major dissociation channel for ion (**1**).

Thermochemistry

From the computed relative energies (B3-LYP/6-311G**) of (**1**)–(**6**) (0, 75, 131, 202, 259 and 66 kJ/mol, respectively) and those of the fragmentation product pairs $HCS^{1+} + HCN$ and $HCNH^{1+} + CS$ ($\Sigma\Delta_f H_{products} = 140$ and 227 kJ/mol, respectively) it is possible to establish the absolute heats of formation for (**1**)–(**6**). Therefore, an estimated $\Delta_f H(CH_2NCS^{1+}) = 1010 \pm 5$ kJ/mol (using $\Delta_f H(HCS^{1+}) = 1019$ kJ/mol (from the PA of CS), $\Delta_f H(HCN) = 135.14$ kJ/mol, $\Delta_f H(HCNH^{1+}) = 952$ kJ/mol (from the PA of HCN) and $\Delta_f H(CS) = 280.33$ kJ/mol).[1] The calculations also indicate that the ions lie in deep potential energy wells separated by significant isomerization energy barriers.

3.6.104 $[C_2,H_3,N,S]^{+\bullet}$ – m/z 73

Isomers

The molecular ions of methyl thiocyanate and methyl isothiocyanate, $CH_3SCN^{1+\bullet}$ (**1**) and $CH_3NCS^{1+\bullet}$ (**2**), respectively, as well as the distonic ion $CH_3CNS^{1+\bullet}$ (**3**) have been investigated by experiments.

Identification

Direct ionization of the molecules leads to the production of ions $CH_3SCN^{1+\bullet}$ and $CH_3NCS^{1+\bullet}$ whereas $CH_3CNS^{1+\bullet}$ can be obtained by dissociative ionization of methyl oxathiazoles (cy–C(CH$_3$)NSC(O)O– or C(CH$_3$)NSC(CH$_3$)$_2$O)[407] and following chemical ionization of CH_3CN by CS_2.[408]

The ions are distinguished based on their respective CID mass spectra. All ions show an intense m/z 58 fragment peak corresponding to $CH_3^{1\bullet}$ loss. However, ion (**1**) is distinct from the others by its intense peaks at m/z 45 and 46, while ion (**2**) shows intense peaks at m/z 44 and 45 and ion (**3**) at m/z 32, 45, 46 and 47, in addition to a charge stripping peak at m/z 36.5.[407,408] It is noteworthy that all three isomers exhibit recovery signals in their NRMS.[407,408]

Thermochemistry

From the combined heats of formation and ionization energies of CH_3SCN and CH_3NCS, $\Delta_fH(CH_3SCN^{1+\bullet}) = 1121$ kJ/mol and $\Delta_fH(CH_3NCS^{1+\bullet}) = 1012$ kJ/mol are obtained (CH$_3$SCN: $\Delta_fH = 160$ kJ/mol[58] and IE $= 9.96 \pm 0.05$;[1] CH$_3$NCS: $\Delta_fH = 131$ kJ/mol[58] and IE $= 9.13 \pm 0.15$ eV[1]). Computations (G2(MP2,SVP)) indicate that $CH_3CNS^{1+\bullet}$ lies 312.1 kJ/mol lower in energy than the dissociation products $H^{\bullet} + CH_2CNS^{1+}$ and 459.2 kJ/mol lower than $CNS^{1\bullet} + CH_3^{1+}$.

3.6.105 $[C_2,H_4,N,S]^+$ – m/z 74

The proton affinities of CH_3NCS and CH_3SCN have been measured by ICR mass spectrometry.[386] Calculations at the CEP31-G** level of theory[386] show that for CH_3NCS protonation is at sulfur (PA $= 799.2$ kJ/mol) while for CH_3SCN the N atom is preferred (PA $= 796.7$ kJ/mol).[409] The neutral Δ_fH values[58] are 131 and 160 kJ/mol for CH_3NCS and CH_3SCN respectively, and $\Delta_fH(H^+) = 1530$ kJ/mol; thus $\Delta_fH(CH_3NCSH^{1+}) = 862$ kJ/mol and $\Delta_fH(CH_3SCNH^{1+}) = 893$ kJ/mol.

3.6.106 $[C_2,N,S_2]^+$ – m/z 102

Isomers

Two isomers have been investigated, namely $SCNCS^{1+}$ and $SNCCS^{1+}$.

Identification

An ion having the connectivity $S=C=N–C=S^{1+}$ was produced[410] by the dissociation of the ionized cyclic trimer of thiocyanuric acid (HSCN)$_3$. The CID mass spectrum of the m/z 102 ion was compatible with the proposed structure, having abundant peaks at m/z 44 (CS^{1+}), 58 ($[C,N,S]^+$) and 70 ($[C_2,N,S]^+$) and lacking signals at m/z 46 (NS^{1+}) and 56 ($C_2S^{1+\bullet}$) that would have been appropriate for the isomeric structure $S=C=C=N=S^{1+}$.

The unexpected CID peak at m/z 76 ($CS_2^{1+\bullet}$) was ascribed to a possible postcollision isomerization. An NRMS experiment confirmed the stability of the neutral analog. At the same time, Flammang et al.[412] used the same method and reported closely similar results.

The latter authors also produced the isomer $S=N-C=C=S^{1+}$ from ionized thiodiazole (cy–NSNC(C(NH$_2$)S)C(CN)–), which displays a peak at m/z 102 in its normal mass spectrum.[411] The CID mass spectrum of this peak was very closely similar to that of the $S=C=N-C=S^{1+}$ isomer except for the two minor peaks at m/z 56 (C_2S^{1+}) and 46 (NS^{1+}). This ion too produced a significant m/z 76 peak ($CS_2^{1+\bullet}$), again ascribed to a postcollision isomerization. An NRMS experiment indicated that the neutral analog was stable.

Thermochemistry

The relative energy of the above two ions was computed at the G2(MP2,SVP) level of theory. The ion $SCNCS^{1+}$ was found to have an energy 137 kJ/mol below that of $SNCCS^{1+}$.[411] There are very few data for such ions and the dissociation products that might allow a thoroughly reliable estimate of an absolute $\Delta_f H$. However, for the ion $S=C=N-C=S^{1+}$ the sum of the heats of formation of the products ($CS^{1+} + NCS^{1\bullet}$, ($\Sigma\Delta_f H_{products} = 1653$ kJ/mol from $\Delta_f H(NCS^{1\bullet}) = 280$ kJ/mol and $\Delta_f H(CS^{1+}) = 1373$ kJ/mol)[1], combined with the computed energy *difference* between the $S=C=N-C=S^{1+}$ ion and these products ($\Delta_f H = 602$ kJ/mol),[411] the value of $\Delta_f H(S=C=N-C=S^{1+})$ $= 1051$ kJ/mol is obtained. Finally then, $\Delta_f H(S=N-C=C=S^{1+})$ $= 1188$ kJ/mol can be deduced.

3.6.107 [C$_2$,H$_3$,N,S$_2$]$^+$ – m/z 105[412]

Dissociative ionization of 3-(methylthio)-1,2,5-thiadiazole (cy–NC(H)C(SCH$_3$) NS–, loss of HCN) produces an intense m/z 105 fragment peak. On the basis of the connectivity in the molecule, as well as the CID mass spectrum of this ion (m/z 46 (NS^{1+}), 58 (CNS^{1+}), 73 ($CH_3SCN^{1+\bullet}$) and an intense 90 ($N=S=C=N^{1+}$, loss of $CH_3^{1\bullet}$)), the assigned structure was $CH_3SCNS^{1+\bullet}$. This ion is also produced by sulfuration of methylthiocyante (CH_3SCN, using CS_2 under CI conditions). It is computed to lie 283.8 kJ/mol (G2(MP2,SVP) below its dissociation products $CH_3^{1\bullet} + N=S=C=N^{1+}$.

3.6.108 [C$_2$,N$_2$,S]$^{+\bullet}$ – m/z 84[412]

The base peak in the normal EI mass spectrum of 3,4-dicyano-1,2,5-thiadia-zole (cy–SNC(CN)C(CN)N–) is an m/z 84 fragment. The CID mass spectrum of this fragment ion contains the following peaks: m/z 32 ($S^{+\bullet}$), 46 (NS^{1+}), 52 (loss of S) and 58 (loss $CN^{1\bullet}$), which are indicative of the NCCNS connectivity. The NR mass spectrum contains an intense recovery signal and dissociation characteristics closely similar to those of the ion, indicating that the NCCNS molecule is stable. The ion, $NCCNS^{1+\bullet}$, is computed (G2(MP2,SVP)) to lie 351.6 kJ/mol below its dissociation products $S^{+\bullet} + NCCN$ ($\Sigma\Delta_f H_{products} = 1586$ kJ/mol),[1] yielding an estimated $\Delta_f H(NCCNS^{1+\bullet}) = 1234$ kJ/mol.

3.6.109 $[C_2,H_7,O,P]^{+\bullet} - m/z\ 78$[413]

Isomers

Two isomers have been tentatively assigned to the conventional and distonic forms $CH_3CH_2P(H)OH^{1+\bullet}$ and $(CH_3)(CH_2)P(H)OH^{1+\bullet}$.

Identification

Some radical cations of this composition were described by Keck et al.[413] The ions were produced in the EI mass spectra of $(C_2H_5)_3PO$ (1), $(n\text{-}C_3H_7)_3PO$ (2), $(C_4H_9)_3PO$ (3) and $(C_2H_5)_2P(O)SC_3H_7$ (4).

The CID mass spectra of the ions fell into two groups. The spectra of $m/z\ 78$ ions from (1) and (2) were closely similar, showing major signals at $m/z\ 47\text{--}50$, corresponding to the loss of C_2H_n fragments. On the basis of their dissociation characteristics, these $m/z\ 78$ ions were provisionally given the structure $CH_3CH_2P(H)OH^{1+\bullet}$.

The CID mass spectra of $m/z\ 78$ ions from (3) and (4) were also closely similar, but were dominated by $m/z\ 63$ (loss of $CH_3^{1\bullet}$), 45, 47, 49, 62 and 64. To these was given the distonic ion structure $CH_3(CH_2)P(H)OH^{1+\bullet}$.

3.6.110 $[C_2,H_5,O_3,P]^{+\bullet} - m/z\ 108$

Isomers

The isomeric ionized ethene phosphonate ($(cy\text{-}OC(H)_2C(H)_2O-)P(H)=O^{1+\bullet}$, K) and ionized ethene phosphite ($(cy\text{-}OC(H)_2C(H)_2O-)POH^{1+\bullet}$, E) have been investigated.[414] The ions can be regarded as "keto" (K) and "enol" (E) tautomers, respectively, and it was their tautomerism that was studied.

Identification

K ions were obtained by ionization of the stable neutral molecule while E ions were produced by selected fragmentations of related ionized molecules.[414] Structural integrity was confirmed by CID and NRMS experiments, although the mass spectral differences lie in relative peak intensities, rather than distinctive fragmentations. The chief difference lies in the presence of relatively intense peaks for the loss of $HO^\bullet$ and $C_2H_3^{1\bullet}$ from E. This difference is greatly enhanced when E ions of low internal energy, produced by the fragmentation of a metastable precursor ion, are subjected to CID. In contrast, the MI mass spectra of the two ions are indistinguishable, showing that their interconversion is possible in ions below the MI threshold for the reaction to $C_2H_4^{]+\bullet} + HOPO_2$. The TS for **K**'s rearrangement to **E** was computed (CBS-QB3) to lie only 2 kJ/mol below the last reaction's threshold.[414]

Thermochemistry

The enol ion **E** is computed to be more stable than **K** by some 143 kJ/mol (CBS-QB3).[414] With **E** requiring 205 kJ/mol to dissociate to $C_2H_4^{]+\bullet} + HOPO_2$, (calculated $\Sigma\Delta_fH_{products} = 362$ kJ/mol), the calculated $\Delta_fH(\mathbf{E}) = 157$ kJ/mol and $\Delta_fH(\mathbf{K}) = 300$ kJ/mol can be deduced. Reference values for $\Delta_fH(C_2H_4^{]+\bullet})$ and $\Delta_fH(HOPO_2)$ are 1066 kJ/mol[1] and -703 kJ/mol[415] respectively, ($\Sigma\Delta_fH_{products} = 363$ kJ/mol, in excellent agreement with the computations) and so the absolute Δ_fH for the ions **K** and **E** should lie within 5 kJ/mol of the computed values.

3.6.111 $[C_2,H_7,O_3,P]^{+\bullet} - m/z\ 110$

Isomers

The two tautomeric ions $(CH_3O)_2POH^{]+\bullet}$ (ionized dimethyl phosphite, (**1**)) and $(CH_3O)_2PH(O)^{]+\bullet}$ (ionized dimethyl phosphonate, (**2**)) have been investigated.

Identification[416]

Ion (**1**) was prepared by the loss of H_2CO from ionized trimethyl phosphate (an allowed 1,4-hydrogen shift) and ion (**2**) was generated by ionization of the molecule. The high energy CID mass spectra were closely similar except for ion (**2**) having a more abundant m/z 80 peak (loss of H_2CO, the lowest energy fragmentation) and an intensity of 162 vs 63 for (**1**) (normalized to the base peak for both ions m/z 79). The ions were much more satisfactorily differentiated by their low energy, single collision CID mass spectra. Ion (**1**) had m/z 79 as the most intense signal at translational energies from 10 eV upward, whereas ion (**2**) had m/z 80 as the intense signal at all energies. The experiments were repeated by the same authors in 1988 and there was a clear distinction between single and multiple low energy collision excitations depending on the apparatus employed.[417] Multiple collisions greatly reduced the structure-distinctive features of the spectra. These ions were thoroughly reinvestigated in 2001 by experiment and theory[418]. The rearrangement of (**2**) into (**1**) is prevented by a high barrier, 200 kJ/mol above (**1**). Ion (**2**) however can readily isomerize into a distonic isomer $CH_3OP(H)(OCH_2)(OH)^{]+\bullet}$ (**3**). A fourth isomer, an ion designated as $[P(OH)(OCH_2)^{+\bullet}\cdots HOCH_3]$ (**4**), was found to occupy a potential well. The computations revealed the existence of two ion–dipole species, not individually accessible by experiment.

Thermochemistry

At the CBS-QB3 level of theory $\Delta_f H(1) = 79$ kJ/mol, $\Delta_f H(2) = 210$ kJ/mol, $\Delta_f H(3) = 107$ kJ/mol and $\Delta_f H(4) = 203$ kJ/mol. The heats of formation of neutrals (1) and (2) were computed at the same level of theory and were established as -722 and -785 kJ/mol, respectively.

3.6.112 $[C_2,O,S]^{+\bullet} - m/z\ 72$

The $O=C=C=S^{]+\bullet}$ ion is readily produced by the dissociative ionization of 4,5,9,10-tetraoxo-2,7-dithiatricyclo$[6.2.0.0^{3,6}]$deca-1(8),3(6)-diene.[419] Its CID chemistry is unremarkable in that only ions corresponding to the original connectivity were observed (unlike the ion $OCCO^{]+\bullet}$, qv that has a $CO_2^{]+\bullet}$ peak in its CID mass spectrum). An NRMS experiment showed that the neutral analog was a stable species.

The ion $OCCS^{]+\bullet}$ has been investigated by ab initio calculations at the G2(MP2) and CCSD(T) levels of theory.[214] The energy of the ion, relative to $CO+CS^{]+\bullet}$ and to $CO+CS$, was calculated to be -229 and 873 kJ/mol, respectively. With $\Delta_f H(CO) = -110.55 \pm 0.17$ kJ/mol,[1] $\Delta_f H(CS^{]+\bullet}) = 1374$ kJ/mol[1] and $\Delta_f H(CS) = 280.33$ kJ/mol[1] (see discussion for these values in the section on C_1 ions), $\Delta_f H(OCCS^{]+\bullet}) = 1038 \pm 4$ kJ/mol.

3.6.113 $[C_2,H_2,O,S]^{+\bullet} - m/z\ 74$[420]

Isomers

Two isomeric ions have been investigated: $SCH_2CO^{]+\bullet}$ (1) and $HSCHCO^{]+\bullet}$ (2).

Identification

Observations on the loss of CH_3OH from metastable ionized methylthioglycolate $(CH_3OC(O)CH_2SH)$ and CH_3CH_2OH from the ionized ethyl analog $(CH_3CH_2O-C(O)CH_2SH)$ showed that two product ions were competitively generated. These were readily proposed as having the structures $SCH_2CO^{]+\bullet}$ and $HSCHCO^{]+\bullet}$, respectively on the basis of D-labeling experiments. The former resulted from loss of the thiol hydrogen and the latter from the loss of a methylene hydrogen. The former product ion gave rise to a broad MI peak for loss of CO whereas the latter generated a Gaussian signal. These results are in keeping with the respective products of this fragmentation being $CH_2S^{]+\bullet}$ and $HCSH^{]+\bullet}$.

Thermochemistry

At a low level of theory (B3LYP/6-31+g(d,p)//HF/6-31g(d)), $\Delta_f H$ $(SCH_2CO^{]+\bullet}) = 886$ kJ/mol and $\Delta_f H(HSCHCO^{]+\bullet}) = 832$ kJ/mol. Possibly a third stable $[C_2,H_2,O,S]^{+\bullet}$ isomer, $HSC=COH^{]+\bullet}$, is involved in this system?

3.6.114 $[C_2,H_4,O,S]^{+\bullet}$ – m/z 76

Isomers

A number of isomeric ions of this formula have been investigated by Lahem et al.: the two thioformates, the O-methyl $HC(S)OCH_3^{+\bullet}$ (1) and S-methyl species $HC(O)SCH_3^{+\bullet}$ (2), the distonic ion $HC(OH)SCH_2^{+\bullet}$ (3) and an ion–molecule complex $[CH_3SH^{+\bullet}\cdots CO]$ (4).

Identification[421]

The CID MS of ions (1) and (2) (generated by ionizing the corresponding molecules) had peaks in common, indicative of postcollision isomerization, but for (1) the base peak was m/z 45 (HCS^{+}). It is noteworthy that $HC(O)SCH_3^{+\bullet}$ has a single fragment peak in its MI mass spectrum for the loss of CO. The $[C,H_4,S]^{+\bullet}$ product ion was shown to be ionized thiomethanol and not the distonic form $CH_2SH_2^{+\bullet}$.[310]

Isomer (3) was generated by the loss of propene from ionized S-isobutylthioformate. Its CID mass spectrum is closely similar to that of ion (2), except for a peak at m/z 62 (CH_2 loss) and a significant doubly charged ion (m/z 38), both processes typical of distonic ions. The ions can also be partially distinguished by their reaction with NO, but the results are not striking.

Thermochemistry

Much more information can be derived from ab initio calculations of the potential energy surfaces for these species at the ($(U)MP2/6-31G(d,p)$ and $UQCISD(T)/6-311++G(d,p)$ levels of theory.[421] From the computed relative energies it is possible to estimate the $\Delta_f H$ values for the above four isomers. Ion (1) is computed to be 196 kJ/mol and 330 kJ/mol below its dissociation products $HCS^+ + CH_2OH^\bullet$ ($\Sigma\Delta_f H_{products} = 1019^1 + -19 \pm 1$ kJ/mol[422] $= 1001$ kJ/mol) and $CH_3OH^{+\bullet} + CS$ ($\Sigma\Delta_f H_{products} = 845 + 280.33 = 1125$ kJ/mol),[1] respectively, leading to $\Delta_f H(HC(S)OCH_3^{+\bullet}) = 800 \pm 5$ kJ/mol. Similarly, given that ion (3) lies 152 kJ/mol below its dissociation products $CH_3S^\bullet + HCO^+$ (($\Sigma\Delta_f H_{products} = 125^{423} + 825^1 = 950$ kJ/mol), $\Delta_f H(HC(OH)SCH_2^{+\bullet}) = 798$ kJ/mol. Ions (2) and (4) are computed to be respectively 4 and 55 kJ/mol more stable than ion (3) and consequently, $\Delta_f H(HC(S)OCH_3^{+\bullet}) = 794$ kJ/mol and $\Delta_f H([CH_3SH^{+\bullet}\cdots CO]) = 743$ kJ/mol can be deduced.

3.6.115 $[C_2,H_6,O,S]^{+\bullet}$ – m/z 78

Isomers

Four ion structures of this formula have been described, namely ionized 2-mercaptoethanol, $HSCH_2CH_2OH^{+\bullet}$ (1), ionized dimethylsulfoxide,

$(CH_3)_2SO^{1+\bullet}$ (**2**), its *aci*-form $CH_3S(OH)=CH_2^{1+\bullet}$ (**3**) and ionized methyl methanesulfenate, $CH_3SOCH_3^{1+\bullet}$ (**4**).

Identification

Direct ionization of the molecules produces ions (**1**), (**2**) and (**4**), whereas $CH_3S(OH)=CH_2^{1+\bullet}$ (**3**) is produced by the dissociative ionization of methyl-carboxymethyl sulfoxide ($CH_3S(O)CH_2C(O)OH$, loss of CO_2).[424]

The dissociation behavior of ionized mercaptoethanol (**1**) mirrors that of ionized ethylene glycol in that the loss of H_2O produces $CH_3CHS^{1+\bullet}$, the loss of CH_2O produces the distonic ion[73] $CH_2SH_2^{1+\bullet}$ (plus some $CH_3SH^{1+\bullet}$) and the loss of $HCO^\bullet$ generates $CH_3SH_2^{1+}$.[73,332]

The molecular ion of $(CH_3)_2SO$ (**2**) has been studied by theory and experiment,[424–427] much of the interest stemming from the problem of the identity of the fragment ion generated by the loss of $CH_3^{1\bullet}$. Computations[425] suggest that ion (**2**) is capable of rearranging to its aci-form (**3**) before $CH_3^{1\bullet}$ loss.[425]

Isomers (**2**), (**3**) and (**4**) have two common fragment ions in their MI mass spectra, loss of $CH_3^{1\bullet}$ and $^\bullet OH$, and the relative intensities of these signals serve to identify the isomers.[424] The ratios m/z 63:61 are 100:97, 33:100, and 100:58 for isomers (**2**), (**3**), and (**4**) respectively, and furthermore, the kinetic energy release values ($T_{0.5}$) are different, showing that distinct reaction pathways are involved even though the final transition states may be common. For $CH_3^{1\bullet}$ loss they are 19, 29 and 30 meV and for $HO^\bullet$ loss they are 22, 8 and 36 meV for isomers (**2**), (**3**) and (**4**), respectively. [424] In addition, ion (**4**) has a broad composite peak at m/z 48 (loss of $[C,H_2,O]$).[424]

After some controversy, the fragment ion produced at threshold from ionized dimethylsulfoxide by loss of $CH_3^{1\bullet}$ was shown to be the *O*-protonated sulfine, CH_2SOH^{1+}.[427] High energy CID of stable $(CH_3)_2SO^{1+\bullet}$ yields a mixture of $CH_2SOH^{1+\bullet}$ (60%) and $CH_3SO^{1+\bullet}$ (40%).[425] It is noteworthy that in the latter study[425] the relative energies of several $[C,H_3,O,S]^+$ isomers were also calculated (refer to the section on $[C,H_3,O,S]^+$).

Thermochemistry

Combining the IE and heat of formation of $HSCH_2CH_2OH$ yields $\Delta_fH(HSCH_2CH_2OH^{1+\bullet})=680\pm4$ kJ/mol (IE $=9.1$ eV[1] and $\Delta_fH(HSCH_2CH_2OH)$ is estimated to be -199 ± 4 kJ/mol on the basis of additivity (using $\Delta_fH(HOCH_2CH_2OH)=-388\pm2$ kJ/mol and the effect of substituting HO by SH: $\Delta_fH(C_2H_5OH)=-235.3\pm0.5$ kJ/mol and $\Delta_fH(C_2H_5SH)=-46.3\pm0.6$ kJ/mol, therefore $\Delta\Delta_fH=189$ kJ/mol).[1] Similarly, $\Delta_fH((CH_3)_2SO^{1+\bullet})=725\pm1$ kJ/mol ($\Delta_fH((CH_3)_2SO)=-151\pm1$ kJ/mol and IE $=9.08$ eV).[1]

The computational chemistry of Gozzo and Eberlin[425] at the MP2/6-31G(d,p)//6-31G9d,p)+ZPE and G2 levels of theory show that amongst isomers (**2**), (**3**) and (**4**), the global minimum is $CH_3SOCH_3^{1+\bullet}$. From their calculated relative energies, $\Delta_fH(CH_2S(OH)CH_3^{1+\bullet})=649$ kJ/mol and $\Delta_fH(CH_3SOCH_3^{1+\bullet})=620$ kJ/mol is deduced.

3.6.116 $[C_2,H_7,O,S]^+$ – m/z 79

The proton affinities of three of the above isomers, namely dimethylsulfoxide $((CH_3)_2SO)$, $CH_3S(OH)=CH_2$ and CH_3SOCH_3, have been estimated by calculations at the G2(MP2) and B3LYP levels of theory.[428] Note that there are six protonated isomers involved and that there is good agreement between the computed and experimental values for dimethylsulfoxide itself.

Ion	Computed PA[428] (kJ/mol)	Δ_fH(ion)[a] (kJ/mol)	Ion	Computed PA[428] (kJ/mol)	Δ_fH(ion) (kJ/mol)
$(CH_3)_2SOH$	886 (884)[b]	493	$CH_3S(OH_2)CH_2$	984	510
$(CH_3)_2SHO$	771	608	$CH_3SO(H)CH_3$	778	624
$CH_3S(OH)CH_2H$	1001	493	$CH_3S(H)OCH_3$	822	580

[a]Δ_fH(ion) = Δ_fH(neutral) + Δ_fH(H$^+$) – PA where Δ_fH(neutral) were deduced from the calculated relative energies[428] $(CH_3)_2SO = 0$, $CH_3S(OH)CH_2 = +115$ kJ/mol, and $CH_3SOCH_3 = +23$ kJ/mol combined with $\Delta_fH(CH_3)_2SO) = -151 \pm 1$ kJ/mol[1] and Δ_fH(H$^+$) = 1530 kJ/mol.[1]
[b]Experimental value.[409]

3.6.117 $[C_2,H_2,O,X]^+$ WHERE X = Cl, Br AND I – m/z 77, 111 AND 169

Isomers

Three halo-acetyl cations $ClCH_2CO^{1+}$, $BrCH_2CO^{1+}$ and ICH_2CO^{1+} have been generated.

Identification

Dissociative ionization of a series of molecules containing the XCH$_2$CO– group yields the three halo-acetyl cations.[429] In every case, the ion showed a significant metastable peak for the loss of CO, having, like the acetyl cation itself, a very small kinetic energy release ($T_{0.5} = 2 \pm 0.1$ meV). Moreover, the structure-characteristic m/z 29:28 ratio in the CID mass spectra was, again like CH_3CO^{1+} ions, about 0.4–0.5 (see above, $[C_2,H_3,O]^+$).

Thermochemistry

The heats of formation of the halo-acetyl cations have been measured.[429] They are appreciably higher than that of the acetyl cation itself ($\Delta_fH(CH_3CO^{1+}) = 659.4 \pm 1.1$ kJ/mol)[149] and are $\Delta_fH(ClCH_2CO^{1+}) = 708 \pm 8$ kJ/mol, Δ_fH $(BrCH_2CO^{1+}) = 750 \pm 25$ kJ/mol, and $\Delta_fH(ICH_2CO^{1+}) = 784 \pm 10$ kJ/mol. In acetyl ions halogen is always destabilizing. Among the neutral halo-acetones themselves a similar destabilizing trend obtains. $\Delta_fH(XCH_2COCH_3)$ are, for X = Cl = -225 kJ/mol; Br = -181 kJ/mol; I = -130 kJ/mol.[58]

3.6.118 $[C_2,H_4,O,F]^+$ – m/z 63

Isomers

Two isomers are proposed to have been investigated: $CH_3C(OH)F^{]+}$ and the complex $[CH_3CO\cdots FH]^+$.

Identification

An ion of this formula was observed in the mass spectrum of ionized 2-hydroxy-2-trifluoromethylpropanoic acid $(CH_3C(CF_3)(OH)C(O)OH)$. Interest in the ion lay in its reported HF loss that generated a composite metastable ion peak[430] to produce the acetyl cation $(CH_3CO^{]+})$. On the basis of computations and D-labeling experiments it was proposed that the narrow component represented the loss of HF from an ion–molecule complex of structure $[CH_3CO\cdots FH]^+$ and that the broad signal resulted from the dissociation of $CH_3C(OH)F^{]+}$. The $CH_3C(OH)F^{]+}$ ion and the ion–molecule complex originated from the intermediate ion $CH_3C(OH)CF_3^{]+}$ (generated by the loss of $C(O)OH^{]\bullet}$ from $CH_3C(CF_3)(OH)C(O)OH^{]+\bullet}$) via different competing reaction channels.

Thermochemistry

The $CH_3C(OH)F^{]+}$ ion was calculated to be higher in energy than the dissociation products $(CH_3CO^{]+} + HF)$ by 29 kJ/mol (MP2/6-311 + G(dp)).[430] With $\Delta_fH(CH_3CO^{]+}) = 659.4 \pm 1.1$ kJ/mol[149] and $\Delta_fH(HF) = -273$ kJ/mol,[1] $\Delta_fH(CH_3C(OH)F^{]+}) = 415$ kJ/mol. The complex has a computed $\Delta_fH([CH_3CO\cdots FH]^+) = 343$ kJ/mol, some 72 kJ/mol below that of the conventional ion and it dissociated without any excess activation energy being required. The TS for HF loss from the $CH_3C(OH)F^{]+}$ ion was calculated to lie 222 kJ/mol above the dissociation products, consistent with the observed magnitude of the kinetic energy release.

It is noteworthy that by using the above data, stabilization by F substitution at the charge bearing site $(CH_3CHOH^{]+}$ (598 $\pm$ 10 kJ/mol, see $[C_2,H_5,O]^+$) $\longrightarrow$ $CH_3C(OH)F^{]+}$ (415 kJ/mol)) results in the lowering of the heat of formation of 183 $\pm$ 10 kJ/mol; this seems very reasonable in comparison with, for example, $\Delta_fH(CH_2CH_2^{]+\bullet}) = 1066$ kJ/mol[1] and $\Delta_fH(CH_2CHF^{]+\bullet}) = 864$ kJ/mol (see above), which leads to a $\Delta\Delta_fH = -204$ kJ/mol.

3.6.119 $[C_2,H_5,O,X]^{+\bullet}$ WHERE X = F, Cl AND Br – m/z 64, 80 AND 124

Isomers

These ions correspond to the ionized haloethanols $FCH_2CH_2OH^{]+\bullet}$, $ClCH_2CH_2OH^{]+\bullet}$ and $BrCH_2CH_2OH^{]+\bullet}$. There is also the isomeric ionized chlorodimethyl ether $(ClCH_2OCH_3^{]+\bullet})$.

Identification

These molecular ions are generated by ionization of the corresponding molecules. Little ion chemistry is known for these species. $ClCH_2CH_2OH^{]+\bullet}$ ions specifically undergo a 1,2-HCl elimination to generate vinyl alcohol ions.[431,432]

Thermochemistry

Thermochemical Data for the $[C_2,H_5,O,X]^{+\bullet}$ Ions Where X = F, Cl and Br

Species	IE^a (eV)	$\Delta_f H$(molecule)^b (kJ/mol)	$\Delta_f H$(ion)^c (kJ/mol)
FCH_2CH_2OH	10.66	−404	625
$ClCH_2CH_2OH$	10.52	−255	760
$BrCH_2CH_2OH$	≤10.75	−212	≤825
$ClCH_2OCH_3$	10.2	−203	781

^{a}Values are from NIST[1] or otherwise stated.
^{b}Values were estimated using additivy.[118] The selected terms for C-(H$_2$)(C)(X) are −212, −63, and −20.5 kJ/mol for X = F, Cl, and Br, respectively.
^{c}Deduced from IE $= \Delta_f H$(ion) $- \Delta_f H$(molecule).

3.6.120 $[C_2,H_6,O,X]^+$ WHERE X = F, Cl AND Br – m/z 65, 81 AND 125

Isomers

These ions correspond to the protonated forms for the haloethanols and are $FCH_2CH_2OH_2^{]+}$, $ClCH_2CH_2OH_2^{]+}$ and $BrCH_2CH_2OH_2^{]+}$. There are no mass spectral data on these ions.

Thermochemistry

Species	PA^a (kJ/mol)	$\Delta_f H$(molecule)^b (kJ/mol)	$\Delta_f H$(ion)^c (kJ/mol)
$FCH_2CH_2OH_2$	715.6	−404	410
$ClCH_2CH_2OH_2$	766	−255	509
$BrCH_2CH_2OH_2$	766	−212	552

^{a}Values are from Hunter et al.[341]
^{b}Values were estimated using additivy.[118] The selected terms for C-(H$_2$)(C)(X) are −212, −63 and −20.5 kJ/mol for X = F, Cl and Br, respectively.
^{c}Deduced from $-PA = \Delta_f H$(ion) $- (\Delta_f H$(molecule) $+ \Delta_f H(H^+))$, $\Delta_f H(H^+) = 1530$ kJ/mol.[1]

Note that the presence of the halogen causes a reduction of the proton affinity of ethanol (776 kJ/mol).[341] It is also worth remarking that in general, the presence of halogen adjacent to or one atom removed from the protonation site always reduces the PA of the molecule. The effect does not, however, appear to be easy to estimate (e.g., the PA for $CH_3C(O)OH$, $FCH_2C(O)OH$, $CF_3C(O)OH$, $ClCH_2C(O)OH$ and $CCl_3C(O)OH$ are reported to be 784, 765, 712, 765 and 770 kJ/mol respectively).[341]

3.6.121 $[C_2,O,F_2]^{+\bullet}$ – m/z 78

Isomers

Only the species $CF_2CO^{]+\bullet}$ has been identified in the gas phase.[433]

Identification

$CF_2CO^{]+\bullet}$ was generated by the loss of CF_4 from ionized $CF_3OC(F)CF_2$ and it is not, as expected by analogy, produced by CF_4 loss from ionized perfluoro-acetone. Like ionized ketene, the metastable ions undergo loss of CO and with a similarly very small kinetic energy release ($T_{0.5} = 2.1$ meV, compare ionized ketene, 2.6 meV).[434]

Thermochemistry

The ionization energy for perfluoroketene is not known, but it must lie significantly below the product energies for the dissociation giving $CO + CF_2^{]+\bullet}$ ($\Sigma\Delta_f H_{products} = 786$ kJ/mol[58] $= 8.15$ eV), because the mass spectrum shows the $CF_2CO^{]+\bullet}$ ion to be quite stable with respect to the CO loss. The calculated $\Delta_f H(CF_2CO)$ is -290 ± 13 kJ/mol[435] and so the ionization energy must be less than 11 eV.

3.6.122 $[C_2,H,O,F_2]^+$ – m/z 79

Isomers

Three possible structures have been considered but calculations conclude that the following two are produced: $CF_2CHO^{]+}$ and the complex $[HF\cdots CF_2CHO]^+$.

Identification

A $[C_2,H,O,F_2]^+$ ion is produced by the loss of HF from the ion $CF_3C(H)OH^{]+}$, which in turn can be generated both by the loss of $CF_3^{]\bullet}$ from ionized 1,1,1,3,3,3-hexafluoro-2-propanol and by $CH_3^{]\bullet}$ loss from ionized $CH_3CH(OH)CF_3$.[436,437] Replacement of the hydroxyl hydrogen atom by a deuterium atom results in the specific loss of DF for the metastable $CF_3CHOD^{]+}$ ions,[437] leading to the conclusion that the product ion has the structure $CF_2CHO^{]+}$. This is an ion for which the H analog is unstable, with it rearranging to the acetyl cation. Remarkably, the MI peak corresponding to $CF_3C(H)OH^{]+} \longrightarrow [C_2,H,O,F_2]^+ + HF$ is composite, a narrow Gaussian peak ($T_{0.5} = 8$ meV) and a weaker, dished component ($T_{0.5} = 690$ meV). Also noteworthy are the reported $T_{0.5}$ values for the composite peak from the $[C_2,H_2,O,F_3]^+$ ion formed by loss of H$^\bullet$ from ionized CF_3CH_2OH, being 31 and 760 meV, respectively, similar to but not identical with those from the above propanols. Without further experimental observations and reliable thermochemical data it was difficult to draw any firm conclusions. Thus, the composite peak may reflect the generation of two fragment ion structures, e.g., $CF_2CHO^{]+}$ and $CF_2HCO^{]+}$, or even, as proposed by Tajima et al.,[437] the involvement of an isolated electronic state. The matter appears to have been resolved by the computations of Varnai et al.[438] at the MP2/6-31G*//HF/6-31G* + ZPE level of theory. They showed that the narrow component can result from the generation of a complex $[HF \cdots CF_2CHO]^+$ ion that fragments without an energy barrier and the broad component yields the isomer $CF_2HCO^{]+}$ produced exothermically from the TS. The activation energies for the two pathways were 184 and 201 kJ/mol, respectively. Relative energies can be found in the above articles but there are no absolute energies for these isomeric ions.

3.6.123 $[C_2,O,F_3]^+$ – m/z 97

Isomers

Experiments indicate that there is at least one stable isomer other than $CF_3CO^{]+}$ but the connectivity of the other isomer is not known.

Identification[374]

$CF_3CO^{]+}$ ions are produced by the loss of $CF_3^{]\bullet}$ from ionized $(CF_3)_2CO$. This metastable dissociation ($T_{0.5} \leq 0.5$ meV) is extremely collision sensitive and long before single-collision conditions have been attained, $T_{0.5}$ has increased tenfold. Like most other acetyl-type ions, $CF_3CO^{]+}$ dissociates metastably to $CF_3^{]+}$ (m/z 69) and the accompanying peak is narrow and also very collision-sensitive ($T_{0.5} = 1.9$ meV, compared with other halo-acetyl ions, which have $T_{0.5}$ values of 2.0 ± 0.2 meV). There is however a second component to this MI peak (collision insensitive), having a very large kinetic energy release ($T_{0.5} = 560$ meV).

The CID mass spectra of the centroid of this peak and the wing of the broad component are dominated by the $CF_3^{]+}$ ion (m/z 69), with only one other significant fragment peak at m/z 50 ($CF_2^{]+}$, 6%). The CID mass spectrum of ion source generated $[C_2,F_3,O]^+$ ions is quite different, with m/z 78 as the major fragment ion (100%) together with a weaker m/z 69 ion

(30%) and a trace of m/z 50. The structure of this other $[C_2,F_3,O]^+$ isomer is not known, but formally it can be $CF_2=COF^{]+}$ or $^+CF_2=CFO$, the former readily coming from the enol-type ion, $CF_2=C(F)OCF_3^{]+\bullet}$.

It is noteworthy that ionized $CF_2=C(F)OCF_3$ generates a weak m/z 97 ion whose composite MI peak for $CF_3^{]+}$ generation contains a narrow ($T_{0.5}=1.9$ meV, very collision sensitive, the same as for the hexafluoroacetone route, above) and a broad, somewhat collision sensitive component ($T_{0.5}=491$ meV). This isomer displays peaks at m/z 50 ($CF_2^{]+}$, 11%), 47 ($CFO^{]+}$, 11%) and 31 ($CF^{]+}$, 6%) in its CID mass spectrum,[374] in keeping with a connectivity CF_2-CF-O. Clearly, there remain a number of unresolved problems, but unlike the $CH_3CO^{]+}$ cation, a number of stable isomers can be generated.

Thermochemistry

Computations at the MP2/6-311G**/HF/6-31G* level of theory[439] give $\Delta_f H(CF_3CO^{]+}) = 192$ kJ/mol and later (indirectly) other computational results (MP2/6-31G(d)//MP2/6-31G(d) level) gave the same value.[440] From the bond dissociation energy, $\Delta_f H(CF_3CO^{]\bullet}) = -631 \pm 5$ kJ/mol (D(CF$_3$–CO) = 50 ± 5 kJ/mol,[441] $\Delta_f H(CF_3^{]\bullet}) = -470$ kJ/mol[1] and $\Delta_f H(CO) = -111$ kJ/mol[1]). Note that the value of Takhistov et al.[442] quoted in Luo[423] (-537 kJ/mol) would have the radical *unstable* with respect to $CF_3^{]\bullet} + CO$. Later calculations at the CBS-4 level of theory, including isodesmic considerations,[443] are in reasonable agreement with the above, giving $\Delta_f H(CF_3CO^{]\bullet}) = -605$ kJ/mol and more recently[444] -609 kJ/mol, results that lower the bond energy by some 26 kJ/mol.

Our recommended ionization energy for the CF$_3$CO radical, 8.4 ± 0.1 eV, is deduced from a compromise between theory and experiment (IE $= \Delta_f H$(ion) $- \Delta_f H$(neutral) $= 192 + 615 \pm 10$ kJ/mol $= 8.4 \pm 0.1$ eV). The reader might like to inspect the available data for the effect of a CF$_3$ group replacing H in a variety of [C,H,O] containing radicals and molecules and then decide what is a best value.

3.6.124 $[C_2,H_2,O,F_3]^+$ – m/z 99

Isomers

Two isomers have been generated and characterized, namely $CF_3C(H)OH^{]+}$ and $CF_3OCH_2^{]+}$.

Identification

The ion of structure $CF_3CHOH^{]+}$ is generated by the loss of H$^\bullet$ from trifluoro-ethanol[445] and its isomer, $CF_3OCH_2^{]+}$, is similarly produced from the ionized ether $(CF_3OCH_3^{]+\bullet})$ or by the loss of $CH_3^{]\bullet}$ and CO from ionized $CF_3C(O)OCH_2CH_3.^{440}$ They can easily be distinguished by their MI mass spectra, the former losing HF to produce a composite m/z 79 peak^{437} and the latter losing both H_2CO $(m/z$ 69, $CF_3^{]+})$ and CF_2O $(m/z$ 33, $CH_2F^{]+}).^{440}$

Thermochemistry

From the proton affinity of CF_3CHO, $\Delta_fH(CF_3CHOH^{]+}) = 84 \pm 10$ kJ/mol is obtained $(PA(CF_3CHO) = 685.5$ kJ/mol,^{1} $\Delta_fH(H^+) = 1530$ kJ/mol^1 and $\Delta_fH(CF_3CHO) = -760 \pm 10$ kJ/mol (using the $\Delta(\Delta_fH)$ from $CH_3C(O)OH$ to CH_3CHO $(-266$ kJ/mol)^1 as the correction term $(\Delta_fH(CF_3C(O)OH) = -1025 \pm 10$ kJ/mol)).^{1} The Δ_fH for the $CF_3OCH_2^{]+}$ isomer can be evaluated as 70 kJ/mol, from the potential energy diagram calculated by Sekiguchi et al.^{440} at the MP2/6-31G(d) level of theory.

It is worth noting that the stabilization of the ions $CH_3CO^{]+}$ and $CH_3CHOH^{]+}$ when CH_3 is replaced by CF_3 is about -500 kJ/mol in each case. Here the CF_3 group is attached to the charge-bearing site and the stabilization is appreciably less than when it is not so attached. Compare Δ_fH values for $CH_3OCH_2^{]+}$ and $CF_3OCH_2^{]+}$, 666 kJ/mol^1 and 70 kJ/mol,^{440} respectively, which results in a stabilization of about 600 kJ/mol, equal to that obtaining in neutrals, e.g., $\Delta_fH(CF_3C(O)OH) = -1025 \pm 10$ kJ/mol and $\Delta_fH(CH_3C(O)OH) = -432$ kJ/mol,^{1} a difference of about -590 kJ/mol.

3.6.125 $[C_2,H_3,O,F_3]^{+\bullet}$ – m/z 100

Isomers

There are two isomers, ionized CF_3CH_2OH and CF_3OCH_3.

Identification

As indicated above they are distinguishable by the markedly different properties of their $(M-H)^+$ fragment ions.

Thermochemistry

All may not be well with the thermochemistry of these two molecules. $\Delta_fH(CF_3CH_2OH) = -888$ kJ/mol^{58} is apparently the same as the computed value for $\Delta_fH(CF_3OCH_3) = -889$ kJ/mol,^{446} (compare $\Delta_fH(CH_3CH_2OH) = -235.5 \pm 0.5$ kJ/mol^1 and $\Delta_fH(CH_3OCH_3) = -184.1 \pm 0.5$ kJ/mol^1). A search of the available data^{1,49,58,422} indicates that the effect of CF_3 substitution at carbon is the same as at oxygen, as indeed it is with the CH_3 group. The $\Delta_fH(CF_3CH_2OH)$ is consistent with this, it being -650 ± 10 kJ/mol lower than that for ethanol $(\Delta_fH(CH_3CH_2OH) = -235.5 \pm 0.5$ kJ/mol).^{1} Thus, a preferred value for $\Delta_fH(CF_3OCH_3)$ would be -839 ± 10 kJ/mol

$(-184 + -650 \text{ kJ/mol})$. The IE of CF_3CH_2OH is recorded as 11.5 eV,[1] giving $\Delta_fH(CF_3CH_2OH^{1+\bullet}) = 221 \text{ kJ/mol}$.

3.6.126 $[C_2,H_4,O,F_3]^+$ – m/z 101

The proton affinities of the above two molecules have been measured and lead to $\Delta_fH(CF_3CH_2OH_2^{1+}) = -58 \pm 10$ kJ/mol and $\Delta_fH(CF_3 O(H)CH_3^{1+}) = -28 \pm 10$ kJ/mol $(PA(CF_3CH_2OH) = 700$ kJ/mol[1] and $\Delta_fH(CF_3CH_2OH_2) = -888$ kJ/mol[58] and $PA(CF_3OCH_3) = 719$ kJ/mol[1] and $\Delta_fH(CF_3OCH_3) = -839 \pm 10$ kJ/mol (see above)). Note that the presence of the halogen causes a reduction in the PA of the unsubstituted molecules.

3.6.127 $[C_2,N,O,S]^+$ – m/z 86

Isomers

Two ions have been characterized: $S=C=N-C=O^{1+}$ and $S=N-C=C=O^{1+}$.

Identification

Ionized ethoxycarbonyl isothiocyanate $(C_2H_5OC(O)NCS^{1+\bullet})$[447] and ionized rhodanine $(cy-C(H)_2S(CS)N(H)C(O)-)$[448] both generate an m/z 86 fragment ion with indistinguishable CID mass spectral characteristics. This fragment ion was assigned the $S=C=N-C=O^{1+}$ connectivity based on the ions in the CID mass spectrum: m/z 58 (NCS^{1+}), 54 $(CNCO^{1+})$, 44 (CS^{1+}), 42 (NCO^{1+}) and S^+, $CO^{1+\bullet}$ and CN^+.[447,448]

The isomer $S=N-C=C=O^{1+}$ is obtained from the dissociative ionization of the thiadiazole $cy-SNC(C(O)NH_2)C(C(O)NH_2)N-$. Its CID mass spectrum is very similar to that of $S=C=N-C=O^{1+}$, but shows two structure-identifying features: a much more intense fragment peak at m/z 44 (CS^{1+}) and a significant peak at m/z 70 corresponding to the loss of atomic oxygen.[447]

It is noteworthy that the NRMS of the ions have also been recorded and show that the neutral counterpart of $S=N-C=C=O^{1+}$ is stable.[447] There are however conflicting results concerning the stability of $S=C=N-C=O^{1\bullet}$: using Xe and N,N-dimethylaniline as neutralizing gas[447] produces a small recovery signal assigned to $S=C=N-C=O^{1+}$, but when NH_3 is used no such signal is obtained.[448]

Thermochemistry[447]

Computations at the G2(MP2,SVP) indicate that $S=C=N-C=O^{1+}$ is bent while $S=N-C=C=O^{1+}$ is linear. The former is 162 kJ/mol lower in energy than its isomer. The lowest energy dissociation for both ions is

$CO + SNC^{]+}$ and requires 410 and 316 kJ/mol for $S=C=N-C=O^{]+}$ and $S=N-C=C=O^{]+}$, respectively.

3.6.128 $[C_2,H_3,O,N,S]^{+\bullet} - m/z\ 89^{449}$

Isomers

Ionized methoxy isothiocyanate and methyl cyanate N-sulfide ($CH_3ONCS^{]+\bullet}$ (1) and $CH_3OCNS^{]+\bullet}$ (2), respectively) have been investigated. It is also possible that a third isomer, $CH_3SNCO^{]+\bullet}$ (3), has been generated.

Identification

Both isomers have been produced by the dissociative ionization of suitable precursor molecules. The triazolide (cy–NC(H)NC(H)N−C(S)N(H)(OCH$_3$)) has an m/z 89 as base peak in its normal mass spectrum and the ion was proposed to have the connectivity of ion (1). The ion has an MI mass spectrum consisting of two peaks, m/z 74 and 59, which correspond to losses of $CH_3^{]\bullet}$ and H_2CO, respectively. However, the detailed examination of the fragment ions from these dissociations showed that rearrangement must precede metastable ion dissociations.

Ion (2) was formed by the loss of CH_3OCN from ionized 3,4-dimethoxy-1,2,5-thiadiazole (cy–NC(CH$_3$O)C(CH$_3$O)NS−). The MI mass spectrum is different, showing the losses of $CH_3^{]\bullet}$ and CO, the latter indicative of a rearrangement. The connectivity is confirmed by the CID mass spectrum and the NRMS.

The most striking differences are found in their reactions with NO; ion (1) displayed peaks at m/z 30 ($NO^{]+}$), 46, 59 and 74 (loss of $CH_3^{]\bullet}$), whereas (2) showed m/z 61, 62 ($NOS^{]+}$) and 74. An attempt to make the third isomer $CH_3SNCO^{]+\bullet}$ by the loss of CH_3CN from ionized 2-methyl-5-methylthio-1,3,4-oxadiazole was likely successful, but hard to confirm.

Thermochemistry

Calculations of the potential energy surface for these ions at the G2(MP2,SVP) level of theory gave the individual ion energies relative to their dissociation products. However, the required ancillary data for the products' energies have conspicuous gaps, sufficient to prevent any assessment of the absolute $\Delta_f H$ values.

REFERENCES

1. *NIST Chemistry WebBook, NIST Standard Reference Data Base Number 69*, Gaithersburg, MD: National Institute of Standards and Technology, **2005**.
2. Janoschek, R., Rossi, M.J. *Int. J. Chem. Kinet.* **2002**, *34*, 550.
3. Tucker, K.D., Kutner, M.L., Thaddeus, P. *Astrophys. J.* **1974**, *193*, L115.

4. Jackson, W.M., Bao, Y., Urdahl, R.S. *J. Geophys. Res.* **1991**, *96*, 17569.

5. Strobel, D.F. *Planet. Space Sci.* **1982**, *30*, 839.

6. Allen, M., Yung, Y.L., Gladstone, G.R. *Icarus* **1992**, *100*, 527.

7. Shaub, W.M., Bauer, S.H. *Combust. Flame* **1978**, *32*, 35.

8. Boullart, W., Devriendt, K., Borms, R., Peeters, J. *J. Phys. Chem.* **1996**, *100*, 998.

9. Jursic, B.S. *J. Quantum Chem.* **1999**, *72*, 571.

10. Holmes, J.L., Szulejko, J.E. *Chem. Phys. Lett.* **1984**, *107*, 301.

11. Sulzle, D., Schwarz, H. *Chem. Phys. Lett.* **1989**, *156*, 397.

12. Hayakawa, S., Takahashi, M., Arakawa, K., Morishita, N. *J. Chem. Phys.* **1999**, *110*, 2745.

13. Glukhovtsev, M.N., Bach, R.D. *Chem. Phys. Lett.* **1998**, *286*, 51.

14. Carneiro, J.W.D.-M., Schleyer, P.V.R., Saunders, M., Remington, R., Schaefer III, H.F., Rauk, A., Sorenson, T.S. *J. Am. Chem. Soc.* **1994**, *116*, 3483.

15. Lossing, F.P. *Can. J. Chem.* **1971**, *49*, 357.

16. Holmes, J.L. *Int. J. Mass Spectrom. Ion Processes* **1992**, *118/119*, 381.

17. Van der Hart, W.J. *Int. J. Mass Spectrom. Ion Processes* **1995**, *151*, 27.

18. Uggerud, E. *Eur. J. Mass Spectrom.* **1997**, *3*, 403.

19. Zuilhof, A., Dinnocenzo, J.P., Reddy, A.C., Shaik, S. *J. Phys. Chem.* **1976**, *100*, 15774.

20. Lifshitz, C., Sternberg, R. *Int. J. Mass Spectrom. Ion Phys.* **1969**, *2*, 303.

21. Braten, S.V., Helgaker, T., Uggerud, E. *Org. Mass Spectrom.* **1993**, *28*, 1262.

22. Wexler, S., Jesse, N.J. *J. Am. Chem. Soc.* **1962**, *84*, 3425.

23. Field, F.H., Franklin, J.L., Munson, M.S.B. *J. Am. Chem. Soc.* **1963**, *85*, 3575.

24. Fisher, J.J., Koyanagi, G.K., McMahon, T.B. *Int. J. Mass Spectrom.* **2000**, *195/196*, 491.

25. Hiraoka, K., Kebarle, P. *J. Am. Chem. Soc.* **1976**, *98*, 6119.

26. Carneiro, J.W.D.-M., Schleyer, P.V.R., Saunders, M., Remington, R., Schaefer III, H.F., Rauk, A., Sorenson, T.S. *J. Am. Chem. Soc.* **1994**, *116*, 3483.

27. East, A.L.L., Liu, Z.F., McCague, C., Cheng, K., Tse, J.S. *J. Phys. Chem. A* **1998**, *102*, 10903.

28. Haese, N.N., Woods, R.C. *Astrophys. J.* **1981**, *246*, L51.

29. Bohme, D.K., Wlodek, S., Raksit, A.B., Schiff, H.I., Mackay, G.I., Keskinen, K.J. *Int. J. Mass Spectrom. Ion Processes* **1987**, *81*, 123.

30. Harland, P.W., McIntosh, B.J. *Int. J. Mass Spectrom. Ion Processes* **1985**, *67*, 29.

31. Knight, J.S., Petrie, S.A.H., Freeman, C.G., McEwan, M.J., McLean, A.D., DeFrees, D.J. *J. Am. Chem. Soc.* **1988**, *110*, 5286.

32. Le Teuff, Y.H. *Astron. Astrophys. Suppl. Ser.* **2000**, *146*, 157.

33. Goldberg, N., Fiedler, A., Schwarz, H. *J. Phys. Chem.* **1995**, *99*, 15327.

34. Flammang, R., Laurent, S., Barbieux-Flammang, M., Wentrup, C. *Org. Mass Spectrom.* **1993**, *28*, 1161.

35. Herbst, E. *Annu. Rev. Phys. Chem.* **1995**, *46*, 27.

36. Mayer, P.M., Taylor, M.S., Wong, M.W., Radom, L. *J. Phys. Chem. A* **1998**, *102*, 7074.

37. Frankowski, M., Sun, Z., Smith-Glickhorn, A.M. *Phys. Chem. Chem. Phys.* **2005**, *7*, 797.

38. Holmes, J.L., Mayer, P.M. *J. Phys. Chem.* **1995**, *99*, 1366.

39. Shea, D.A., Steenvoorden, J.J.M., Chen, P. *J. Phys. Chem. A* **1997**, *101*, 9728.

40. Holmes, J.L., Lossing, F.P., Mayer, P.M. *Chem. Phys. Lett.* **1993**, *212*, 134.
41. Bouchoux, G., Hoppilliard, Y. *Org. Mass Spectrom.* **1981**, *16*, 459.
42. Janoschek, R., Fabian, W.M.F. *J. Mol. Struct.* **2006**, *780/781*, 80.
43. Choe, J.C. *Int. J. Mass Spectrom.* **2005**, *235*, 15.
44. van Baar, B., Koch, W., Lebrilla, C., Terlouw, J.K., Weiske, T., Schwarz, H. *Angew. Chem. Int. Ed. Engl.* **1986**, *25*, 827.
45. Chess, E.K., Lapp, R.L., Gross, M.L. *Org. Mass Spectrom.* **1982**, *17*, 475.
46. Thuiji, J.V., Houte, J.J.V., Maquestiau, A., Flammang, R., Meyer, C.D. *Org. Mass Spectrom.* **1977**, *12*, 196.
47. de Petris, G., Fornarini, S., Crestoni, M.E., Troiani, A., Mayer, P.M. *J. Phys. Chem. A* **2005**, *109*, 4425.
48. Mayer, P.M., Glukhovtsev, M., Gauld, J.W., Radom, L. *J. Am. Chem. Soc.* **1997**, *119*, 12889.
49. Pedley, J.B., Naylor, R.D., Kirby, S.P. *Thermochemical Data of Organic Compounds*, 2nd ed., New York: Chapman and Hall, **1986**.
50. Harland, P.W., McIntosh, B.J. *Int. J. Mass Spectrom. Ion Processes* **1985**, *67*, 29.
51. Knight, J.S., Freeman, C.G., McEwan, M.J. *J. Am. Chem. Soc.* **1986**, *108*, 1404.
52. Illies, A.J., Liu, S., Bowers, M.T. *J. Am. Chem. Soc.* **1981**, *103*, 5674.
53. Corral, I., Mo, O., Yanez, M. *Int. J. Quantum Chem.* **2003**, *91*, 438.
54. Nguyen, M.T., Ha, T.-K. *J. Chem. Soc. Perkin Trans. 2* **1984**, 1401.
55. Holmes, J.L., Terlouw, J.K. *Can. J. Chem.* **1976**, *54*, 1007.
56. Bouchoux, G., Penaud-Berruyer, F., Nguyen, M.T. *J. Am. Chem. Soc.* **1993**, *115*, 9728.
57. Peerboom, R.A., Ingemann, S., Nibbering, N.M.M., Liebman, J.F. *J. Chem. Soc. Perkin Trans. 2* **1990**, 1825.
58. Lias, S.G., Bartmess, J.E., Liebmann, J.F., Holmes, J.L., Levin, R.D., Wallard, W.G. *J. Phys. Chem. Ref. Data* **1988**, *17*.
59. Henriksen, J., Hammerum, S. *Int. J. Mass Spectrom.* **1998**, *179/180*, 301.
60. Cui, Q., Morokuma, K. *J. Chem. Phys.* **1998**, *108*, 4021.
61. Bowen, R.D., Williams, D.H., Hvistendahl, G. *J. Am. Chem. Soc.* **1977**, *99*, 7509.
62. Levsen, K., McLafferty, F.W. *J. Am. Chem. Soc.* **1974**, *96*, 139.
63. Van de Sande, C.C., Ahmad, S.Z., Borchers, F., Levsen, K. *Org. Mass Spectrom.* **1978**, *13*, 666.
64. Bursey, M.M., Harvan, D.J., Parker, C.E., Darden, T.A., Hass, J.R. *Org. Mass Spectrom.* **1983**, *18*, 530.
65. Traeger, J.C., Harvey, Z.A. *J. Phys. Chem. A* **2006**, *110*, 8542.
66. Lossing, F.P., Lam, Y.-T., Maccoll, A. *Can. J. Chem.* **1981**, *59*, 2228.
67. Barone, V., Lelj, F., Grande, P., Russo, N., Toscano, M. *Chem. Phys. Lett.* **1987**, *133*, 548.
68. Janoschek, R., Rossi, M.J. *Thermochemical properties from G3MP2B3 calculations*, Wiley InterScience, **2004**.
69. Hudson, C.E., McAdoo, D.J. *Int. J. Mass Spectrom.* **2001**, *210/211*, 417.
70. Forde, N.R., Butler, L.J., Ruscic, B., Sorkhabi, O., Qi, F., Suits, A. *J. Chem. Phys.* **2000**, *113*, 3088.
71. Solka, B.H., Russell, M.E. *J. Phys. Chem.* **1974**, *78*, 1268.
72. Hammerum, S., Petersen, A.C., Solling, T.I., Vulpius, T., Zappey, H. *J. Chem. Soc. Perkin Trans. 2* **1997**, 391.

73. Holmes, J.L., Lossing, F.P., Terlouw, J.K., Burgers, P.C. *Can. J. Chem.* **1983**, *61*, 2305.
74. Hammerum, S., Kuck, D., Derrick, P.J. *Tetrahedron Lett.* **1984**, *25*, 893.
75. Wesdemiotis, C., Danis, P.O., Feng, R., Tso, J., McLafferty, F.W. *J. Am. Chem. Soc.* **1985**, *107*, 8059.
76. Yates, B.F., Radom, L. *Org. Mass Spectrom.* **1987**, *22*, 430.
77. Jarrold, M.F., Kirchner, N.J., Liu, S., Bowers, M.T. *J. Phys. Chem.* **1986**, *90*, 78.
78. Bouchoux, G., Djazi, F., Nguyen, M.T., Tortajada, J. *J. Phys. Chem.* **1996**, *100*, 3552.
79. Ding, Y.H., Huang, X.R., Li, Z.S., Sun, C.C. *J. Chem. Phys.* **1998**, *108*, 2024.
80. Ding, Y.H., Li, Z.S., Huang, X.R., Sun, C.C. *J. Chem. Phys.* **2000**, *113*, 1745.
81. Bickelhaupt, F.M., Fokkens, R.H., De Koning, L.J., Nibbering, N.M.M., Baerends, E.J., Goede, S.J., Bickelhaupt, F. *Int. J. Mass Spectrom. Ion Processes* **1991**, *103*, 157.
82. Kunde, V.G., Aikin, A.C., Hanel, R.A., Jennings, D.E., Maguire, W.C., Samuelson, R.E. *Nature (London)* **1981**, *292*, 686.
83. Petrie, S. *J. Phys. Chem. A* **1998**, *102*, 7835.
84. Petrie, S., Freeman, C.G., McEwan, M.J., Meot-Ner, M. *Int. J. Mass Spectrom. Ion Processes* **1989**, *90*, 241.
85. Milligan, D.B., Fairley, D.A., Mautner, M., McEwan, M.J. *Int. J. Mass Spectrom.* **1998**, *179/180*, 285.
86. Raksit, A.B., Bohme, D.K. *Int. J. Mass Spectrom. Ion Processes* **1984**, *57*, 211.
87. Deakyne, C.A., Fairley, D.A., Meot-Ner, M. *J. Chem. Phys.* **1987**, *86*, 2334.
88. Nenner, I., Dutiut, O., Richard-Viard, M., Morin, P., Zewail, A.H. *J. Am. Chem. Soc.* **1988**, *110*, 1093.
89. Clemmons, J.H., Jasien, P.G., Dykstra, C.E. *Mol. Phys.* **1983**, *48*, 631.
90. Jolly, W.L., Gin, C. *Int. J. Mass Spectrom. Ion Phys.* **1977**, *25*, 27.
91. Glaser, R. *J. Am. Chem. Soc.* **1987**, *109*, 4237.
92. Glaser, R. *J. Phys. Chem.* **1989**, *93*, 7993.
93. Nguyen, M.T., Keer, A.V., Vanquickenborne, L.G. *J. Chem. Soc. Perkin Trans. 2* **1987**, 299.
94. Glaser, R., Choy, G.S.-C., Hall, M.K. *J. Am. Chem. Soc.* **1991**, *113*, 1109.
95. Foster, M.S., Beauchamp, J.L. *J. Am. Chem. Soc.* **1972**, *94*, 2425.
96. Foster, M.S., Williamson, A.D., Beauchamp, J.L. *Int. J. Mass Spectrom. Ion Phys.* **1974**, *15*, 429.
97. Nixdorf, A., Grutzmacher, H.-F. *Int. J. Mass Spectrom.* **2000**, *195/196*, 533.
98. Bouchoux, G., Choret, N., Milliet, A., Rempp, M., Terlouw, J.K. *Int. J. Mass Spectrom.* **1998**, *179/180*, 337.
99. Boulanger, A.-M., Rennie, E.E., Holland, D.M.P., Shaw, D.A., Mayer, P.M. *J. Phys. Chem. A* **2006**, *110*, 8563.
100. Bouchoux, G., Choret, N., Penaud-Berruyer, F., Flammang, R. *J. Phys. Chem. A* **2001**, *105*, 9166.
101. Ohishi, M., Suzuki, H., Ishikawa, S., Yamada, C., Kanamori, H., Irvine, W.M., Brown, R.D., Godfrey, P.D., Kaifu, N. *Astrophys. J.* **1991**, *380*, L39.
102. Brown, R.D., Cragg, D.M., Godfrey, P.D., Irvine, W.M., McGonagle, D. *J. Intl. Soc. Study Origin Life* **1992**, *21*, 399.

103. Brown, R.D., Gragg, D.M., Godfrey, P.D., Irvine, W.M., McGonagle, D., Ohishi, M. *Origins of Life and Evolution of the Biosphere* **1991–1992**, *21*, 399.
104. Schildcrout, S.M., Franklin, J.L. *J. Am. Chem. Soc.* **1970**, *92*, 251.
105. Hsieh, S., Eland, J.H.D. *Int. J. Mass Spectrom. Ion Processes* **1997**, *167/168*, 415.
106. Bowers, M.T., Chau, M., Kemper, P.R. *J. Chem. Phys.* **1975**, *63*, 3656.
107. Chen, H., Holmes, J.L. *Int. J. Mass Spectrom. Ion Processes* **1994**, *133*, 111.
108. Lu, W., Tosi, P., Bassi, D. *J. Chem. Phys.* **1999**, *111*, 8852.
109. Maclagan, R.G.A.R., Sudkeaw, P. *J. Chem. Soc. Faraday Trans.* **1993**, *89*, 3325.
110. Zengin, V., Joakim Persson, B., Strong, K.M., Continetti, R.E. *J. Chem. Phys.* **1996**, *105*, 9740.
111. Becker, K.H., Bayes, K.D. *J. Chem. Phys.* **1968**, *48*, 653.
112. Walch, S.P. *J. Chem. Phys.* **1980**, *72*, 5679.
113. Choi, H., Mordaunt, D.H., Bise, R.T., Taylor, T.R., Neumark, D.M. *J. Chem. Phys.* **1998**, *108*, 4070.
114. Brown, S.T., Yamaguchi, Y., Schaefer III, H.F. *J. Phys. Chem. A* **2000**, *104*, 3603.
115. Lawrence, L.L. *J. Chem. Phys.* **1998**, *108*, 8012.
116. Lossing, F.P., Holmes, J.L. Unpublished data **1984**.
117. Benson, S.W., Cruickshank, F.R., Golden, D.M., Haugen, G.R., O'Neil, H.E., Rodgers, A.S., Shaw, R., Walsh, R. *Chem. Rev.* **1969**, *69*, 279.
118. Benson, S.W. *Thermochemical Kinetics*, New York: Wiley-Interscience, **1976**.
119. Lossing, F.P., Holmes, J.L. *J. Am. Chem. Soc.* **1984**, *106*, 6917.
120. Chiang, S.-Y., Fang, Y.-S., Sankaran, K., Lee, Y.-P. *J. Chem. Phys.* **2004**, *120*, 3270.
121. Fenimore, C.P., Jones, G.W. *J. Chem. Phys.* **1963**, *39*, 1514.
122. Schmoltner, A.M., Chiu, P.M., Lee, Y.T. *J. Chem. Phys.* **1989**, *91*, 5365.
123. Michael, J.V., Wagner, A.F. *J. Phys. Chem.* **1990**, *94*, 2453.
124. Sattelmeyer, K.W., Yamaguchi, Y., Schaefer III, H.F. *Chem. Phys. Lett.* **2004**, *383*, 266.
125. Bouma, W.J., Gill, P.M.W., Radom, L. *Org. Mass Spectrom.* **1984**, *1984*, 610.
126. Dass, C., Gross, M.L. *Org. Mass Spectrom.* **1990**, *25*, 24.
127. van Baar, B.L.M., Weiske, T., Terlouw, J.K., Schwartz, M. *Angew. Chem. Int. Ed. Engl.* **1986**, *25*, 282.
128. Hijazi, N.H., Holmes, J.L., Terlouw, J.K. *Org. Mass Spectrom.* **1979**, *14*, 119.
129. Hop, C.E.C.A., Holmes, J.L., Terlouw, J.K. *J. Am. Chem. Soc.* **1989**, *111*, 441.
130. van Baar, B.L.M., Heinrich, N., Koch, W., Postma, R., Terlouw, J.K., Schwarz, H. *Angew. Chem. Int. Ed. Engl.* **1987**, *26*, 140.
131. Koch, W., Maquin, F., Stahl, D., Schwarz, H. *J. Chem. Soc. Chem. Commun.* **1984**, 1679.
132. Patai, S. *The Chemistry of Ketenes, Allenes and Related Compounds II*, New York: John Wiley & Sons, **1980**.
133. Tidwell, T.T. *Ketenes*, New York: John Wiley & Sons, **1995**.
134. Aubry, C., Holmes, J.L., Terlouw, J.K. *J. Phys. Chem. A* **1997**, *101*, 5958.
135. Traeger, J.C. *Int. J. Mass Spectrom.* **2000**, *194*, 261.
136. Scott, A.P., Radom, L. *Int. J. Mass Spectrom. Ion Processes* **1997**, *160*, 73.
137. Chiang, S.-Y., Fang, Y.-S., Bahou, M., Sankaran, K. *J. Chinese Chem. Soc.* **2004**, *51*, 681.

138. Nuttall, R.L., Laufer, A.H., Kilday, M.V. *J. Chem. Thermodyn.* **1971**, *3*, 167.

139. Burgers, P.C., Holmes, J.L., Szulejko, J.E., Mommers, A.A., Terlouw, J.K. *Org. Mass Spectrom.* **1983**, *18*, 254.

140. Terlouw, J.K., Heerma, W., Holmes, J.L. *Org. Mass Spectrom.* **1981**, *16*, 306.

141. Turecek, F., McLafferty, F.P. *Org. Mass Spectrom.* **1983**, *18*, 608.

142. Vogt, J., Williamson, A.D., Beauchamp, J.L. *J. Am. Chem. Soc.* **1978**, *100*, 3478.

143. Terlouw, J.K., Heerma, W., Dijkstra, G. *Org. Mass Spectrom.* **1980**, *15*, 660.

144. Koch, W., Schwarz, H., Maquin, F., Stahl, D. *Int. J. Mass Spectrom. Ion Processes* **1985**, *67*, 171.

145. Nobes, R.H., Bouma, W.J., Radom, L. *J. Am. Chem. Soc.* **1983**, *105*, 309.

146. van Baar, B., Burgers, P.C., Terlouw, J.K., Schwarz, H. *J. Chem. Soc. Chem. Commun.* **1986**, 1607.

147. Eberlin, M.N., Majumdar, T.K., Cooks, R.G. *J. Am. Chem. Soc.* **1992**, *114*, 2884.

148. Eberlin, M.N., Cooks, R.G. *Org. Mass Spectrom.* **1993**, *28*, 679.

149. Fogleman, E.A., Koizumi, H., Kercher, J.P., Sztaray, B., Baer, T. *J. Phys. Chem. A* **2004**, *108*, 5288.

150. Traeger, J.C., McLoughlin, R.G., Nicholson, A.J.C. *J. Am. Chem. Soc.* **1982**, *104*, 5318.

151. Smith, B.J., Radom, L. *Int. J. Mass Spectrom. Ion Processes* **1990**, *101*, 209.

152. Bouchoux, G., Penaud-Berruyer, F., Bertrand, W. *Eur. J. Mass Spectrom.* **2001**, *7*, 351.

153. Holmes, J.L., Terlouw, J.K. *Can. J. Chem.* **1975**, *53*, 2076.

154. Van de Sande, C.C., McLafferty, F.P. *J. Am. Chem. Soc.* **1975**, *97*, 4613.

155. Buschek, J.M., Holmes, J.L., Terlouw, J.K. *J. Am. Chem. Soc.* **1987**, *109*, 7321.

156. Terlouw, J.K., Wezenberg, J., Burgers, P.C., Holmes, J.L. *J. Chem. Soc. Chem. Commun.* **1983**, 1121.

157. Bouma, W.J., MacLeod, J.K., Radom, L. *J. Chem. Soc. Chem. Commun.* **1978**, 724.

158. Baumann, B.C., MacLeod, J.K. *J. Am. Chem. Soc.* **1981**, *103*, 6223.

159. Turecek, F., Cramer, C.J. *J. Am. Chem. Soc.* **1995**, *117*, 12243.

160. Bouchoux, G., Alcaraz, C., Dutuit, O., Nguyen, M.T. *Int. J. Mass Spectrom. Ion Processes* **1994**, *137*, 93.

161. Traeger, J.C., Djordjevic, M. *Eur. J. Mass Spectrom.* **1999**, *5*, 319.

162. Holmes, J.L., Terlouw, J.K., Lossing, F.P. *J. Phys. Chem.* **1976**, *80*, 2860.

163. Bouma, W.J., MacLeod, J.K., Radom, L. *J. Am. Chem. Soc.* **1979**, *101*, 5540.

164. Bertrand, W., Bouchoux, G. *Rapid Commun. Mass Spectrom.* **1998**, *12*, 1697.

165. Holmes, J.L., Lossing, F.P. *Org. Mass Spectrom.* **1991**, *26*, 537.

166. Nobes, R.H., Bouma, W.J., MacLeod, J.K., Radom, L. *Chem. Phys. Lett.* **1987**, *135*, 78.

167. Hudson, C.E., Alexander, A.J., McAdoo, D.J. *Tetrahedron* **1998**, *54*, 5065.

168. Flammang, R., Nguyen, M.T., Bouchoux, G., Gerbaux, P. *Int. J. Mass Spectrom.* **2000**, *202*, A8.

169. Van Raalte, D., Harrison, A.G. *Can. J. Chem.* **1963**, *41*, 3118.

170. Harrison, A.G., Ivko, A., Van Raalte, D. *Can. J. Chem.* **1966**, *49*, 1625.

171. Shannon, T.W., McLafferty, F.P. *J. Am. Chem. Soc.* **1966**, *88*, 5021.

172. McLafferty, F.P., Pike, W.T. *J. Am. Chem. Soc.* **1967**, *89*, 5951.

173. Harrison, A.G., Keves, B.G. *J. Am. Chem. Soc.* **1968**, *90*, 5046.

174. Refaey, K.M.A., Chupka, W.A. *J. Chem. Phys.* **1968**, *48*, 5205.
175. Beauchamp, J.L., Dunbar, R.C. *J. Am. Chem. Soc.* **1970**, *92*, 1477.
176. McLafferty, F.P., Kornfeld, R., Haddon, W.F., Kevsen, K., Sakai, I., Bente, P.F., Tsai, S.C., Schuddemage, H.D.R. *J. Am. Chem. Soc.* **1973**, *95*, 3886.
177. Keyes, B.G., Harrison, A.G. *Org. Mass Spectrom.* **1974**, *9*, 221.
178. Solka, B.H., Russel, M.E. *J. Phys. Chem.* **1974**, *78*, 1268.
179. Lossing, F.P. *J. Am. Chem. Soc.* **1977**, *99*, 7526.
180. Pritchard, H., Harrison, A.G. *J. Chem. Phys.* **1968**, *48*, 5623.
181. Botter, R., Pechine, J.M., Rosenstock, H.M. *Int. J. Mass Spectrom. Ion Phys.* **1977**, *25*, 7.
182. Audier, H.E., Bouchoux, G., McMahon, T.B., Milliet, A., Vulpius, T. *Org. Mass Spectrom.* **1994**, *29*, 176.
183. Fridgen, T.D., Holmes, J.L. *Eur. J. Mass Spectrom.* **2004**, *10*, 747.
184. Van de Graff, B., Dymerski, P.P., McLafferty, F.P. *J. Chem. Soc. Chem. Commun.* **1975**, 978.
185. Burgers, P.C., Terlouw, J.K., Holmes, J.L. *Org. Mass Spectrom.* **1982**, *17*, 369.
186. Nobes, R.H., Rodwell, W.R., Bouma, W.J., Radom, L. *J. Am. Chem. Soc.* **1981**, *103*, 1913.
187. Curtiss, L.A., Lucas, D.J., Pople, J.A. *J. Chem. Phys.* **1995**, *102*, 3292.
188. Lossing, F.P. *J. Am. Chem. Soc.* **1977**, *99*, 7526.
189. Hammerum, S. *Int. J. Mass Spectrom. Ion Processes* **1997**, *165/166*, 63.
190. Irvine, W.M., Goldsmith, P.F., Hjalmarson, A. *Interstellar Processes*, Ed. Thronson, H.A.J., Dordrecht: Reidel, **1987**, p. 561.
191. Turner, B.E. *Astrophys. J. Suppl. Ser.* **1991**, *76*, 617.
192. Charnley, S.B., Kress, M.E., Tielens, A.G.G.M., Millar, T.J. *Astrophys. J.* **1995**, *448*, 232.
193. Herbst, E. *Astrophys. J.* **1987**, *313*, 867.
194. Herbst, E., Leung, C.M. *Astrophys. J. Suppl. Ser.* **1989**, *69*, 271.
195. Terlouw, J.K., Heerma, W., Dijkstra, G. *Org. Mass Spectrom.* **1981**, *16*, 326.
196. Burgers, P.C., Holmes, J.L., Terlouw, J.K., van Baar, B. *Org. Mass Spectrom.* **1985**, *20*, 202.
197. Postma, R., Ruttink, P.J.A., van Baar, B., Terlouw, J.K., Holmes, J.L., Burgers, P.C. *Chem. Phys. Lett.* **1986**, *123*, 409.
198. Bouma, W.J., Nobes, R.H., Radom, L. *J. Am. Chem. Soc.* **1983**, *105*, 1743.
199. Qu, Z.-W., Zhu, H., Zhang, X.-K., Zhang, Q.-Y. *Chem. Phys. Lett.* **2002**, *354*, 498.
200. Holmes, J.L., Lossing, F.P., Terlouw, J.K., Burgers, P.C. *J. Am. Chem. Soc.* **1982**, *104*, 2931.
201. Jensen, L.B., Hammerum, S. *Eur. J. Mass Spectrom.* **2004**, *10*, 775.
202. Sirois, M., George, M., Holmes, J.L. *Org. Mass Spectrom.* **1994**, *29*, 11.
203. Matthews, K.K., Adams, N.G. *Int. J. Mass Spectrom. Ion Process.* **1997**, *163*, 221.
204. Turecek, F., Reid, P.J. *Int. J. Mass Spectrom.* **2003**, *222*, 49.
205. Fairley, D.A., Scott, G.B.I., Freeman, C.G., Maclagan, R.G.A.R., McEwan, M.J. *J. Phys. Chem. A* **1997**, *101*, 2848.
206. Ferguson, E.E. *Kinetics of Ion–Molecule Reactions*, Ed. Ausloos, P., New York: Plenum, **1979**.

207. Staudinger, H., Anthes, E. *Ber. Dtsch. Chem. Ges.* **1913**, *46*, 1426.
208. Dawson, D.F., Chen, H., Holmes, J.L. *Eur. J. Mass Spectrom.* **1996**, *2*, 373.
209. Schroder, D., Heinemann, C., Schwarz, H., Harvey, J.N., Dua, S., Blanksby, S.J., Bowie, J.H. *Chem. Eur. J.* **2005**, *4*, 2550.
210. Meot-Ner, M., Field, F.H. *J. Chem. Phys.* **1966**, *61*, 3742.
211. Sulzle, D., Weiske, T., Schwarz, H. *Int. J. Mass Spectrom. Ion Processes* **1993**, *125*, 75.
212. Pandolfo, L., Paiaro, G., Catinella, S., Traldi, P. *J. Mass Spectrom.* **1996**, *31*, 209.
213. Schroder, D., Schwarz, H. *Int. J. Mass Spectrom. Ion Processes* **1995**, *146/147*, 183.
214. Maclagan, R.G.A.R. *J. Mol. Struct. (THEOCHEM)* **2005**, *713*, 107.
215. Talbi, D., Chandler, G.S. *J. Chem. Phys. A* **2000**, *104*, 5872.
216. Norwood, K., Guo, J.-H., Luo, G., Ng, C.Y. *J. Chem. Phys.* **1989**, *90*, 6026.
217. Sumathi, R., Peeters, J., Nguyen, M.T. *Chem. Phys. Lett.* **1998**, *287*, 109.
218. Li, L., Deng, P., Tian, A., Xu, M., Wong, N.-B. *J. Chem. Phys. A* **2004**, *108*, 4428.
219. McLafferty, F.P., Turecek, F. *Interpretation of Mass Spectra*, 4th ed., Mill Valley: University Science Books, **1993**.
220. Terlouw, J.K., Burgers, P.C., van Baar, B.L.M., Weiske, T., Schwarz, H. *Chimia* **1986**, *40*, 357.
221. Polce, M.J., Song, W., Cerda, B.A., Wesdemiotis, C. *Eur. J. Mass Spectrom.* **2000**, *6*, 121.
222. Wong, C.Y., Ruttink, P.J.A., Burgers, P.C., Terlouw, J.K. *Chem. Phys. Lett.* **2004**, *387*, 204.
223. Vijay, D., Sastry, G.N. *J. Mol. Struct. (THEOCHEM)* **2005**, *714*, 199.
224. McKee, M.L., Radom, L. *Org. Mass Spectrom.* **1993**, *28*, 1238.
225. Suh, D., Burgers, P.C., Terlouw, J.K. *Rapid Commun. Mass Spectrom.* **1995**, *9*, 862.
226. Blanchette, M.C., Holmes, J.L., Hop, C.E.C.A., Lossing, F.P., Postma, R., Ruttink, P.J.A., Terlouw, J.K. *J. Am. Chem. Soc.* **1986**, *108*, 7589.
227. Ruttink, P.J.A., Burgers, P.C., Fell, L.M., Terlouw, J.K. *J. Phys. Chem. A* **1999**, *103*, 1426.
228. Ervasti, H.K., Burgers, P.C., Ruttink, P.J.A. *Eur. J. Mass Spectrom.* **2004**, *10*, 791.
229. Levsen, K., Schwarz, H. *J. Chem. Soc. Perkin Trans. 2* **1976**, *11*, 1231.
230. Terlouw, J.K., de Koster, C.G., Heerma, W., Holmes, J.L., Burgers, P.C. *Org. Mass Spectrom.* **1983**, *18*, 222.
231. Jiang, Y.-X., Wood, K.V., Cooks, R.G. *Org. Mass Spectrom.* **1985**, *21*, 101.
232. Cox, J.D., Pilcher, G. *Thermochemistry of Organic and Organometallic Compounds*, New York: Academic Press, **1970**.
233. Kondo, S., Takahashi, A., Tokuhashi, K. *J. Hazard. Mater.* **2002**, *A94*, 37.
234. Ptasinska, S., Denifl, S., Scheir, P., Mark, T.D. *Int. J. Mass Spectrom.* **2005**, *243*, 171.
235. Holmes, J.L., Lossing, F.P. *J. Am. Chem. Soc.* **1980**, *102*, 3732.
236. Lee, R., Ruttink, P.J.A., Burgers, P.C., Terlouw, J.K. *Can. J. Chem.* **2005**, *83*.
237. Postma, R., Ruttink, P.J.A., Terlouw, J.K., Holmes, J.L. *J. Chem. Soc. Chem. Commun.* **1986**, 683.

238. Davidson, W.R., Yau, Y.K., Kebarle, P. *Can. J. Chem.* **1978**, *56*, 1016.
239. Halim, H., Schwarz, H., Terlouw, J.K., Levsen, K. *Org. Mass Spectrom.* **1983**, *18*, 147.
240. De Koster, C.G., Terlouw, J.K., Levsen, K., Halim, H., Schwarz, H. *Int. J. Mass Spectrom. Ion Processes* **1984**, *61*, 87.
241. Midlemiss, N.E., Harrison, A.G. *Can. J. Chem.* **1979**, *57*, 2827.
242. van Baar, B., Halim, H., Terlouw, J.K., Schwarz, H. *J. Chem. Soc. Chem. Commun.* **1986**, 728.
243. Suh, D., Kingsmill, C.A., Ruttink, P.J.A., Burgers, P.C., Terlouw, J.K. *Int. J. Mass Spectrom. Ion Processes* **1995**, *146/147*, 305.
244. Burgers, P.C., Holmes, J.L., Hop, C.E.C.A., Postma, R., Ruttink, P.J.A., Terlouw, J.K. *J. Am. Chem. Soc.* **1987**, *109*, 7315.
245. Cao, J., George, M., Holmes, J.L., Sirois, M., Terlouw, J.K., Burgers, P.C. *J. Am. Chem. Soc.* **1992**, *114*, 2017.
246. Audier, H.E., Milliet, A., Leblanc, D., Morton, T.H. *J. Am. Chem. Soc.* **1992**, *114*, 2020.
247. Ruttink, P.J.A., Burgers, P.C., Fell, L.M., Terlouw, J.K. *J. Phys. Chem. A* **1998**, *102*, 2976.
248. Li, Y., Baer, T. *J. Phys. Chem. A* **2002**, *106*, 8658.
249. Terlouw, J.K., Heerma, W., Burgers, P.C., Holmes, J.L. *Can. J. Chem.* **1984**, *62*, 289.
250. Postma, R., Ruttink, P.J.A., van Duijneveldt, F.B., Terlouw, J.K., Holmes, J.L. *Can. J. Chem.* **1985**, *63*, 2798.
251. Li, Y.-M., Sun, Q., Li, H.-Y., Ge, M.-F., Wang, D.X. *Chinese J. Chem.* **2005**, *23*, 993.
252. Ruttink, P.J.A., Burgers, P.C. *Org. Mass Spectrom.* **1993**, *28*, 1087.
253. Schalley, C.A., Harvey, J.N., Schroder, D., Schwarz, H. *J. Phys. Chem. A* **1998**, *102*, 1021.
254. van Driel, J.H., Heerma, W., Terlouw, J.K., Halim, H., Schwartz, M. *Org. Mass Spectrom.* **1985**, *20*, 665.
255. Larson, J.W., McMahon, T.B. *J. Am. Chem. Soc.* **1982**, *104*, 6255.
256. Schalley, C.A., Dieterle, M., Schroder, D., Schwarz, H., Uggerud, E. *Int. J. Mass Spectrom. Ion Processes* **1997**, *163*, 101.
257. Aschi, M., Attina, M., Cacace, F., Cipollini, R., Pepi, F. *Inorg. Chim. Acta* **1998**, *275/276*, 192.
258. Tu, Y.-P., Holmes, J.L. *J. Am. Chem. Soc.* **2000**, *122*, 3695.
259. Burgers, P.C., Ruttink, P.J.A. *Int. J. Mass Spectrom.* **2005**, *242*, 49.
260. McCormack, J.A.D., Mayer, P.M. *Int. J. Mass Spectrom.* **2001**, *207*, 183.
261. Bouchoux, G., Choret, N. *Rapid Commun. Mass Spectrom.* **1997**, *11*, 1799.
262. de Petris, G., Cartoni, A., Marzio, R., Troiani, A., Angelini, G., Ursini, O. *Chem. Eur. J.* **2004**, *10*, 6411.
263. Peppe, S., Dua, S., Bowie, J.H. *J. Phys. Chem. A* **2001**, *105*, 10139.
264. Legon, A.C., Suckley, A.P. *J. Chem. Phys.* **1989**, *91*, 4440.
265. Randall, R.W., Summersgill, J.P.L., Howard, B.J. *J. Chem. Soc. Faraday Trans.* **1990**, *186*, 1943.
266. Raducu, V., Gauthier-Roy, B., Dahoo, R., Abouaf-Marguin, L., Langlet, J., Caillet, J., Allavena, M. *J. Chem. Phys.* **1995**, *102*, 9235.

267. Xu, Y., McKellar, A.R.W., Howard, B.J. *J. Mol. Spectrosc.* **1996**, *179*, 345.
268. Muenter, J.S., Bhattacharjee, R. *J. Mol. Spectrosc.* **1998**, *190*, 290.
269. Langlet, J., Caillet, J., Allavena, M., Raducu, V., Gauthier-Roy, B., Dahoo, R., Abouaf-Marguin, L. *J. Mol. Struct.* **1999**, *484*, 145.
270. Takahashi, Y., Higuchi, T., Sekiguchi, O., Hoshino, M., Tajima, S. *Int. J. Mass Spectrom.* **1989**, *181*, 89.
271. Tajima, S., Watanabe, D., Nakajima, S., Sekiguchi, O., Nibbering, N.M.M. *Int. J. Mass Spectrom.* **2001**, *207*, 217.
272. Suh, D., Francis, J.T., Terlouw, J.K., Burgers, P.C., Bowen, R.D. *Eur. J. Mass Spectrom.* **1995**, *1*, 545.
273. Fell, L.M., Burgers, P.C., Ruttink, P.J.A., Terlouw, J.K. *Can. J. Chem.* **1998**, *76*, 335.
274. Holmes, J.L., St-Jean, T. *Org. Mass Spectrom.* **1970**, *3*, 1505.
275. Holmes, J.L. *Org. Mass Spectrom.* **1973**, *7*, 341.
276. Cernicharo, J., Guelin, M., Hein, H., Kahane, C. *Astron. Astrophys.* **1987**, *181*, L9.
277. Saito, S., Kawaguchi, K., Yamamoto, S., Ohishi, M., Suzuki, H., Kaifu, N. *Astrophys. J.* **1987**, *317*, L115.
278. Yamamoto, S., Saito, S., Kawaguchi, K., Kaifu, N., Suzuki, H., Ohishi, M. *Astrophys. J.* **1987**, *317*, L119.
279. Flammang, R., Van Haverbeke, Y., Wong, M.W., Wentrup, C. *Rapid Commun. Mass Spectrom.* **1995**, *9*, 203.
280. Pascoli, G., Lavendy, H. *Int. J. Mass Spectrom.* **1988**, *181*, 135.
281. Tang, Z., BelBruno, J.J. *Int. J. Mass Spectrom.* **2001**, *208*, 7.
282. Barrientos, C., Largo, A. *Chem. Phys. Lett.* **1991**, *184*, 168.
283. Smith, D., Adams, N.G., Giles, K., Herbst, E. *Astron. Astrophys.* **1988**, *200*, 191.
284. Jorgense, T., Pedersen, C.T., Flammang, R., Wentrup, C. *J. Chem. Soc. Perkin Trans. 2* **1997**, 173.
285. Zeller, K.-P., Meier, H., Muller, E. *Org. Mass Spectrom.* **1971**, *5*, 373.
286. Bouchoux, G., Hoppilliard, Y., Golfier, M., Rainteau, D. *Org. Mass Spectrom.* **1980**, *15*, 483.
287. Graul, S.T., Bowers, M.T. *J. Phys. Chem.* **1991**, *95*, 8328.
288. Muntean, F., Armentrout, P.B. *Zeit. fur Phys. Chem.* **2000**, *214*, 1035.
289. Ma, N.L., Wong, M.W. *Eur. J. Org. Chem.* **2000**, 1411.
290. Paradisi, C., Scorrano, F., Daolio, S., Traldi, P. *Org. Mass Spectrom.* **1984**, *19*, 198.
291. Cooks, R.G., Mabud, M.A., Horning, S.R., Jiang, X.-Y., Paradisi, C., Traldi, P. *J. Am. Chem. Soc.* **1989**, *111*, 859.
292. Caserio, M.C., Kim, J.K. *J. Am. Chem. Soc.* **1983**, *105*, 6896.
293. Chiu, S.-W., Lau, K.-C., Li, W.-K. *J. Phys. Chem. A* **2000**, *104*, 3028.
294. Kroto, H.W., Landsberg, B.M., Suffolk, R.J., Vodden, A. *Chem. Phys. Lett.* **1974**, *29*, 265.
295. Holmes, J.L., Wolkoff, P., Terlouw, J.K. *J. Chem. Soc. Chem. Commun.* **1977**, 492.
296. Tomer, K.B., Djerassi, C. *J. Am. Chem. Soc.* **1973**, *95*, 5335.
297. Chiang, S.-Y., Fang, Y.-S. *J. Electron Spectrosc. Rel. Phenom.* **2005**, *144/147*, 223.

298. Fang, Y.-S., Lin, I.-F., Lee, Y.-C., Chiang, S.-Y. *J. Chem. Phys.* **2005**, *123*, 054312.
299. BelBruno, J.J. *Chem. Phys. Lett.* **1996**, *254*, 321.
300. Van der Graaf, B., McLafferty, F.P. *J. Am. Chem. Soc.* **1977**, *99*, 6806.
301. Broer, W.J., Weringa, W.D. *Org. Mass Spectrom.* **1977**, *12*, 326.
302. Broer, W.J., Weringa, W.D., Nieuwpoort, W.C. *Org. Mass Spectrom.* **1979**, *14*, 543.
303. de Moraes, P.R.P., Linnert, H.V., Aschi, M., Riveros, J.M. *J. Am. Chem. Soc.* **2000**, *122*, 10133.
304. Nourbakhsh, S., Norwood, K., Yin, H.-M., Liao, C.-L., Ng, C.Y. *J. Chem. Phys.* **1991**, *95*, 5014.
305. Ma, Z.-X., Liao, C.-L., Yin, H.-M., Ng, C.Y., Chiu, S.-W., Ma, N.L., Li, W.-K. *Chem. Phys. Lett.* **1993**, *213*, 250.
306. Chiu, S.-W., Cheung, Y.-S., Ma, N.L., Li, W.-K., Ng, C.Y. *J. Mol. Struct.* (*THEOCHEM*) **1998**, *452*, 97.
307. Chiu, S.-W., Cheung, Y.-S., Ma, N.L., Li, W.-K., Ng, C.Y. *J. Mol. Struct.* (*THEOCHEM*) **1999**, *468*, 21.
308. Chiu, S.-W., Lau, K.-C., Li, W.-K., Ma, N.L., Cheung, Y.-S., Ng, C.Y. *J. Mol. Struct.* (*THEOCHEM*) **1999**, *490*, 109.
309. Wagner, W., Heimbach, H., Levsen, K. *Int. J. Mass Spectrom. Ion Phys.* **1980**, *36*, 125.
310. Flammang, R., Lahem, D., Nguyen, M.T. *J. Phys. Chem. A* **1997**, *101*, 9818.
311. Chen, Y.-J., Fenn, P.T., Stimson, S., Ng, C.Y. *J. Chem. Phys.* **1997**, *106*, 8274.
312. Fenn, P.T., Chen, Y.-J., Stimson, S., Ng, C.Y. *J. Phys. Chem.* **1997**, *101*, 6513.
313. Chen, Y.-J., Stimson, S., Fenn, P.T., Ng, C.Y., Li, W.-K., Ma, N.L. *J. Chem. Phys.* **1998**, *108*, 8020.
314. Chen, Y.-J., Fenn, P.T., Lau, K.-C., Ng, C.Y., Law, C.-K., Li, W.-K. *J. Phys. Chem. A* **2002**, *106*, 9729.
315. Nobes, R.H., Bouma, W.J., Radom, L. *J. Am. Chem. Soc.* **1984**, *106*, 2774.
316. Sulzle, D., Schwarz, H. *Angew. Chem.* **1988**, *100*, 1384.
317. Wong, M.W., Wentrup, C., Flammang, R. *J. Phys. Chem.* **1995**, *99*, 16849.
318. Wentrup, C., Kambouris, P., Evans, R.A., Owen, D., Macfarlane, G., Chuche, J., Pommelet, J.C., Cheick, A.B., Plisnier, M., Flammang, R. *J. Am. Chem. Soc.* **1991**, *113*, 3130.
319. Wong, M.W. *J. Mass Spectrom.* **1995**, *30*, 1144.
320. Maclagan, R.G.A.R. *J. Mol. Struct.* (*THEOCHEM*) **2005**, *713*, 107.
321. Kim, K.-H., Lee, B., Lee, S. *Chem. Phys. Lett.* **1998**, *297*, 65.
322. Wang, H.-Y., Lu, X., Huang, R.-B., Zheng, L.-S. *J. Mol. Struct.* (*THEOCHEM*) **2002**, *593*, 187.
323. Maier, G., Reisenauer, H.P., Schrot, J., Janoschek, R. *Angew. Chem. Int. Ed. Engl.* **1990**, *29*, 1464.
324. Janoschek, R. *J. Mol. Struct.* (*THEOCHEM*) **1991**, *232*, 147.
325. Russell, G.A., Tanikaga, R., Talaty, E.R. *J. Am. Chem. Soc.* **1972**, *94*, 6125.
326. Morihashi, K., Kushihara, S., Inadomi, Y., Kikuchi, O. *J. Mol. Struct.* (*THEOCHEM*) **1997**, *418*, 171.
327. Vijay, D., Sastry, G.N. *J. Mol. Struct.* (*THEOCHEM*) **2005**, *732*, 71.
328. Baykut, G., Wanczek, K.-P., Hartmann, H. *Adv. Mass Spectrom.* **1980**, *8A*, 186.

329. Baykut, G., Wanczek, K.-P., Hartmann, H. *Dynamics Mass Spectroscopy*, London: Heyden and Son, **1981**.

330. Ekern, S., Illies, A.J., McKee, M.L., Peschke, M. *J. Am. Chem. Soc.* **1993**, *115*, 12510.

331. Gerbaux, P., Salpin, J.-Y., Bouchoux, G., Flammang, R. *Int. J. Mass Spectrom.* **2000**, *195/196*, 239.

332. Sekiguchi, O., Kosaka, T., Kinoshita, T., Takjima, S. *Int. J. Mass Spectrom. Ion Processes* **1995**, *145*, 25.

333. Holmes, J.L., Terlouw, J.K. *The Encyclopedia of Mass Spectrometry*. Ed. Nibbering, N.M.M., Amsterdam: Elsevier, **2005**, Vol. 4, p. 287.

334. Dill, J.D., McLafferty, F.W. *J. Am. Chem. Soc.* **1979**, *101*, 6526.

335. Li, W.-K., Chiu, S.-W., Ma, Z.-X., Liao, C.-L., Ng, C.Y. *J. Chem. Phys.* **1993**, *99*.

336. Gerbaux, P., Flammang, R., Pedersen, C.T., Wong, M.W. *J. Phys. Chem. A* **1999**, *103*, 3666.

337. Pascoli, G., Lavendy, H. *Int. J. Mass Spectrom.* **2001**, *206*, 153.

338. Gerbaux, P., Flammang, R., Wentrup, C., Wong, M.W. *Int. J. Mass Spectrom.* **2001**, *210/211*, 31.

339. Praet, M.-T., Delwiche, J.P. *Adv. Mass Spectrom.* **1974**, *6*, 829.

340. Cooks, R.G., Kim, K.C., Beynon, J.H. *Int. J. Mass Spectrom. Ion Phys.* **1974**, *15*, 245.

341. Hunter, E.P., Lias, S.G. *J. Phys. Chem. Ref. Data* **1998**, *27*.

342. Rodriquez, C.F., Bohme, D.K., Hopkinson, A.C. *J. Org. Chem.* **1993**, *58*, 3344.

343. Okazaki, T., Lalli, K.K. *J. Org. Chem.* **2005**, *70*, 9139.

344. Campos, P.J., Rodriguez, M.A. *J. Chem. Soc. Chem. Commun.* **1995**, 143.

345. Luo, Y.-R., Holmes, J.L. *J. Phys. Chem.* **1992**, *96*, 9568.

346. Lago, A.F., Baer, T. *J. Phys. Chem. A* **2006**, *110*, 3036.

347. An, Y. In Chemistry, Ph.D. Thesis: Study of Organic Cations in the Gas Phase by Tandem Mass Spectrometry; University of Ottawa: Ottawa, **1995**, pp. 241.

348. Reynolds, C.H. *J. Am. Chem. Soc.* **1992**, *114*, 8676.

349. Heck, A.J.R., de Koning, L.J., Nibbering, N.M.M. *Org. Mass Spectrom.* **1993**, *28*, 235.

350. Ciommer, B., Schwarz, H. *Z. Naturforsch. Teil B.* **1983**, *38*, 635.

351. Nguyen, V., Cheng, X., Morton, T.H. *J. Am. Chem. Soc.* **1992**, *114*, 7127.

352. Holmes, J.L., Lossing, F.P., McFarlane, R.A. *Int. J. Mass Spectrom. Ion Processes.* **1988**, *86*, 209.

353. Ciommer, B., Frenking, G., Schwarz, H. *Int. J. Mass Spectrom. Ion Processes* **1984**, *57*, 135.

354. Blanchette, M.C. In Chemistry, M.Sc. Thesis: The Structure Assignment of Isomeric Gas Phase Organic Ions and Neutrals; University of Ottawa: Ottawa, **1987**, pp. 142.

355. Blanchette, M.C., Holmes, J.L., Lossing, F.P. *Org. Mass Spectrom.* **1987**, *22*, 701.

356. Apeloig, Y., Ciommer, B., Frenking, G., Karni, M., Mandelbaum, A., Schwarz, H., Weisz, A. *J. Am. Chem. Soc.* **1983**, *105*, 2186.

357. Yates, B.F., Bouma, W.J., Radom, L. *Tetrahedron* **1986**, *22*, 6225.

358. Holmes, J.L., Burgers, P.C., Terlouw, J.K., Schwarz, H., Ciommer, B., Halim, H. *Org. Mass Spectrom.* **1983**, *18*, 208.

359. Clark, T., Simons, M.C.R. *J. Chem. Soc. Chem. Commun.* **1986**, 96.

360. Olah, G.A. *Halonium Ions.* **New** York: Wiley-Interscience, **1975**.

361. Zappey, H.W., Drewello, T., Ingemann, S., Nibbering, N.M.M. *Int. J. Mass Spectrom. Ion Processes* **1992**, *115*, 193.

362. Nichols, L.S., McKee, M.L., Illies, A.J. *J. Am. Chem. Soc.* **1998**, *120*, 1538.

363. McMahon, T.B., Heinis, T., Nicol, G., Hovey, J.K., Kebarle, P. *J. Am. Chem. Soc.* **1988**, *110*, 7591.

364. Jennings, K.R. *Org. Mass Spectrom.* **1970**, *3*, 85.

365. Stadelmann, J.-P., Vogt, J. *Int. J. Mass Spectrom. Ion Phys.* **1980**, *35*, 83.

366. Nixdorf, A., Grutzmacher, H.-F. *Eur. Mass Spectrom.* **1999**, *5*, 93.

367. Nixdorf, A., Grutzmacher, H.-F. *J. Am. Chem. Soc.* **1997**, *119*, 6544.

368. Nixdorf, A., Grutzmacher, H.-F. *Int. J. Mass Spectrom.* **2002**, *219*, 409.

369. Kim, M., Choe, J.C., Kim, M.S. *J. Am. Soc. Mass Spectrom.* **2004**, *15*, 1266.

370. Pedley, J.B., Naylor, R.D., Kirby, S.P. *Thermochemical Data of Organic Compounds*; 2nd ed., London: Chapman and Hall, **1986**.

371. Frash, M.V., Hopkinson, A.C., Bohme, D.K. *J. Phys. Chem. A* **1999**, *103*, 7872.

372. Kim, K.C., Beynon, J.H., Cooks, R.G. *J. Chem. Phys.* **1974**, *61*, 1305.

373. Stahl, D., Schwarz, H. *Int. J. Mass Spectrom. Ion Phys.* **1980**, *34*, 387.

374. Dawson, D.F. In Chemistry, Ph.D. Thesis: Collision Induced Emission Spectroscopy of Small Polyatomic Molecules and Tandem Mass Spectrometry of Perfluorinated Organics; University of Ottawa: Ottawa, **2000**, pp. 413.

375. McGibbon, G., Kingsmill, C.A., Terlouw, J.K., Burgers, P.C. *Int. J. Mass Spectrom. Ion Process.* **1992**, *121*.

376. Yu, G., Ding, Y., Huang, X., Bai, T., Sun, C. *J. Phys. Chem. A* **2005**, *109*, 2364.

377. Flammang, R., Van Haverbeke, Y., Laurent, S., Barbieux-Flammang, M., Wong, M.W., Wentrup, C. *J. Phys. Chem.* **1994**, *98*, 5801.

378. Hansel, A., Scheiring, C., Glantschnig, M., Lindinger, W., Ferguson, E.E. *J. Chem. Phys.* **1998**, *109*, 1748.

379. Poppinger, D., Radom, L. *J. Am. Chem. Soc.* **1978**, *100*, 3674.

380. Mack, H.-G., Oberhammer, H. *Chem. Phys. Lett.* **1989**, *157*, 436.

381. Hop, C.E.C.A., Snyder, D.F. *Org. Mass Spectrom.* **1993**, *28*, 1245.

382. Terlouw, J.K., Burgers, P.C., van Baar, B.L.M., Weiske, T., Schwarz, H. *Chimia* **1986**, *40*, 357.

383. Egsgaard, H., Carlsen, L. *J. Chem. Res.* **1987**, 18.

384. Karpas, Z., Stevens, W.J., Buckley, T.J., Metz, R. *J. Am. Chem. Soc.* **1985**, *89*, 5274.

385. Hunter, E.P.L., Lias, S.G. *J. Phys. Chem. Ref. Data* **1998**, *27*, 413.

386. Karpas, Z., Stevens, W.J., Buckley, T.J., Metz, R. *J. Phys. Chem.* **1985**, *89*, 5274.

387. Sulzle, D., O'Bannon, P.E., Schwarz, H. *Chem. Ber.* **1992**, *125*, 279.

388. Yang, S.S., Chen, G., Ma, S., Cooks, R.G., Gozzo, F.C., Eberlin, M.N. *J. Mass Spectrom.* **1995**, *30*, 807.

389. Carvalho, M.C., Juliano, V.F., Kascheres, C., Eberlin, M.N. *J. Chem. Soc. Perkin Trans. 2* **1997**, 2347.

390. Bernadi, F., Cacace, F., de Petris, F., Rossi, I., Troiani, A. *Chem. Eur. J.* **2000**, *6*, 537.

391. Cacace, F., de Petris, G., Pepi, F., Rossi, I., Venturini, A. *J. Am. Chem. Soc.* **1996**, *118*, 12719.

392. Bernadi, F., Cacace, F., de Petris, G., Rossi, I., Troiani, A. *Chem. Eur. J.* **2000**, *6*, 537.

393. Turecek, F., Carpenter, F.H. *J. Chem. Soc. Perkin Trans. 2* **1999**, 2315.

394. O'Hair, R.A.J., Blanksby, S.J., Styles, M., Bowie, J.H. *Int. J. Mass Spectrom.* **1999**, *182/183*, 203.

395. Polce, M.J., Wesdemiotis, C. *J. Mass Spectrom.* **2000**, *35*, 251.

396. Depke, G., Heinrich, N., Schwarz, H. *Int. J. Mass Spectrom. Ion Processes* **1984**, *62*, 99.

397. Yu, D., Rauk, A., Armstrong, D.A. *J. Am. Chem. Soc.* **1995**, *117*, 1789.

398. Powis, I., Rennie, E.E., Hergenhahn, U., Kugeler, O., Bussy-Socrate, R. *J. Phys. Chem. A* **2003**, *107*, 25.

399. de Petris, G. *Org. Mass Spectrom.* **1990**, *25*, 557.

400. Beranova, S., Cai, J., Wesdemiotis, C. *J. Am. Chem. Soc.* **1995**, *117*, 9492.

401. Uggerud, E. *Theor. Chem. Acc.* **1997**, *97*, 313.

402. O'Hair, R.A., Boughton, P.S., Styles, M.L., Frank, B.T., Hadad, C.M. *J. Am. Soc. Mass Spectrom.* **2000**, *11*, 687.

403. Pasinszki, T., Westwood, N.P.C. *J. Phys. Chem.* **1996**, *100*, 16856.

404. Flammang, R., Barbieux-Flammang, M., Gerbaux, P., Wentrup, C., Wong, M.W. *Bull. Soc. Chim. Belge* **1997**, *106*, 545.

405. Flammang, R., Landu, D., Laurent, S., Barbieux-Flammang, M., Kappe, C.O., Wong, M.W., Wentrup, C. *J. Am. Chem. Soc.* **1994**, *116*, 2005.

406. Reddy, P.G., Srikanth, R., Bhamuprakash, R. *Rapid Commun. Mass Spectrom.* **2004**, *18*, 1939.

407. Kambouris, P., Plisnier, M., Flammang, R., Terlouw, J.K., Wentrup, C. *Tetrahedron Lett.* **1991**, *32*, 1487.

408. Gerbaux, P., Van Haverbeke, Y., Flammang, R., Wong, M.W., Wentrup, C. *J. Chem. Phys. A* **1997**, *101*, 6970.

409. Hunter, E.P., Lias, S.G. *J. Phys. Chem. Ref. Data* **1998**, *27*, 413.

410. Srinivas, R., Suma, K., Vivekananda, S. *Int. J. Mass Spectrom. Ion Processes* **1996**, *152*, L1.

411. Wong, M.W., Wentrup, C., Morkved, E.H., Flammang, R. *J. Phys. Chem.* **1996**, *100*, 10536.

412. Flammang, R., Gerbaux, P., Morkved, E.H., Wong, M.W., Wentrup, C. *J. Phys. Chem.* **1996**, *100*, 17452.

413. Keck, H., Kuchen, W., Kuhlborn, S. *Org. Mass Spectrom.* **1988**, *23*, 594.

414. Heydorn, L.N., Burgers, P.C., Ruttink, P.J.A., Terlouw, J.K. *Int. J. Mass Spectrom.* **2003**, *227*, 453.

415. Haworth, N.L., Bacsay, G.B. *J. Chem. Phys.* **2002**, *117*, 11175.

416. Kenttamaa, H.I., Cooks, R.G. *J. Am. Chem. Soc.* **1985**, *107*, 1881.

417. Brodbelt, J.S., Kenttamaa, H.I. *Org. Mass Spectrom.* **1988**, *23*, 6.

418. Heydorn, L.N., Ling, Y., de Oliviera, G., Martin, J.M.L., Lifshitz, C., Terlouw, J.K.Z. *Phys. Chem.* **2001**, *215*, 141.

419. Sulzle, D., Terlouw, J.K., Schwarz, H. *J. Am. Chem. Soc.* **1990**, *112*, 628.

420. Sekiguchi, O., Tajima, S. *J. Am. Soc. Mass Spectrom.* **1997**, *8*, 801.

421. Lahem, D., Flammang, R., Le, H.T., Nguyen, T.L., Nguyen, T.N. *J. Chem. Soc. Perkin Trans. 2* **1999**, 821.

422. Traeger, J.C., Holmes, J.L. *J. Am. Chem. Soc.* **1993**, *97*, 3453.
423. Luo, Y.-R. *Handbook of Bond Dissociation Energies in Organic Compounds*, Boca Raton: CRC Press, 2003.
424. Carlsen, L., Egsgaard, H. *J. Am. Chem. Soc.* **1988**, *110*, 6701.
425. Gozzo, F.C., Eberlin, M. *J. Mass Spectrom.* **1995**, *30*, 1553.
426. Ruttink, P.J.A., Burgers, P.C., Terlouw, J.K. *Chem. Phys. Lett.* **1994**, *229*, 495.
427. McGibbon, G.A., Burgers, P.C., Terlouw, J.K. *Chem. Phys. Lett.* **1994**, *218*, 499.
428. Turecek, F. *J. Phys. Chem. A* **1998**, *102*, 4703.
429. Holmes, J.L., Dakubu, M. *Org. Mass Spectrom.* **1989**, *24*, 461.
430. Sekiguchi, O., Watanabe, D., Nakajima, S., Tajima, S., Uggerud, E. *Int J. Mass Spectrom.* **2003**, *222*, 1.
431. Van de Sande, C.C., Mclafferty, F.W. *J. Am. Chem. Soc.* **1975**, *97*, 4613.
432. Holmes, J.L., Terlouw, J.K. *Can. J. Chem.* **1975**, *53*, 2076.
433. Dawson, D.F., Holmes, J.L. *J. Phys. Chem. A* **1999**, *103*, 5217.
434. Hop, C.E.C.A., Holmes, J.L., Terlouw, J.K. *J. Am. Chem. Soc.* **1989**, *111*, 441.
435. Zachariah, M.R., Westmoreland, P.R., Burgess, D.R. Jr., Tsang, W., Melius, C.F. *J. Phys. Chem.* **1996**, *100*, 8737.
436. Carbini, M., Conte, G., Gambaretto, S.C., Traldi, P. *Org. Mass Spectrom.* **1992**, *27*, 1248.
437. Tajima, S., Shirai, T.,Tobita, S., Nibbering, N.M.M. *Org. Mass Spectrom.* **1993**, *28*, 473.
438. Varnai, P., Nyulaszi, L., Veszpremi, T., Vekey, K. *Chem. Phys. Lett.* **1995**, *233*, 340.
439. Cacace, F., Crestoni, M.E., Fornarini, S. *J. Phys. Chem.* **1994**, *98*, 1641.
440. Sekiguchi, O., Tajima, S., Koitabashi, R. Tajima, S. *Int. J. Mass Spectrom.* **1998**, *177*, 23.
441. Maricq, M.M., Szente, J.J., Khitrov, G.A., Dibble, T.S., Francisco, J.S. *J. Phys. Chem.* **1995**, *99*, 11875.
442. Takhistov, V.V., Rodin, A.A., Pashina, T.A., Ismagilov, N.G., Orlov, V.M., Barabinov, V.G. *Zh. Org. Khim.* **1995**, *31*, 1786.
443. Viscolcz, B., Berces, T. *Phys. Chem. Chem. Phys.* **2000**, *2*, 5430.
444. Janoschek, R., Rossi, M. *J. Int. J. Chem. Kin.* **2004**, *36*, 661.
445. Yamaoka, H., Fokkens, R.H., Tajima, S., Yamataka, H., Nibbering, N.M.M. *Bull. Soc. Chim. Belges* **1997**, *106*, 399.
446. Lazarou, Y.G., Papagiannakopoulos, P. *Chem. Phys. Lett.* **1999**, *301*, 19.
447. Wong, M.W., Wentrup, C., Morkved, E.H., Flammang, R. *J. Phys. Chem.* **1996**, *100*, 10536.
448. Vivekananda, S., Srinivas, R., Terlouw, J.K. *Int. J. Mass Spectrom. Ion Processes.* **1997**, *171*, L13.
449. Flammang, R., Gerbaux, P., Barbieux-Flammang, M., Pedersen, C.T., Bech, A.T., Morkved, E.H., Wong, M.W., Wentrup, C. *J. Chem. Soc. Perkin Trans. 2* **1999**, 1683.

3.7   IONS CONTAINING THREE CARBON ATOMS

3.7.1 $[C_3,H]^+$ – m/z 37

Isomers

Three isomers of $[C_3,H]^+$ have been identified by computation: the linear $CCCH^{1+}$ ion, a bent $CC(H)C^{1+}$ ion and a truly cyclic cy-C_3H^{1+} ion. There is experimental evidence only for the linear isomer. $[C_3,H]^+$ cations are key interstellar species responsible for the formation of $C_3H_2^{1+}$, $C_3H_3^{1+}$ and neutral C_3H_2 and C_3H_3 species.[1] Two neutral $[C_3,H]$ forms, CCCH and cy-C_3H, have been identified in interstellar space.[1]

Identification

$[C_3,H]^+$ cations have been investigated by ion–molecule reaction kinetic studies. Indeed, there have been a variety of studies on the reactions of $[C_3,H]^+$ ions generated from the electron-impact dissociation of alkenes and alkynes and in no case has the telltale sign of the presence of isomeric reactants, namely nonlinear kinetics plots, been observed.[2–7] It has been deduced that the isomer present is the linear isomer, $CCCH^{1+}$, because the $CC(H)C^{1+}$ isomer was found to lie very high in relative energy. Baker et al.[8] have placed this isomer (which has a bent CCC backbone and is protonated at the central carbon, $<CCC = 121°$) 253 kJ/mol above the linear form at the MP4(STDQ)/6-31G**//MP2/6-31G** level of theory. A barrier of 336 kJ/mol, relative to $CCCH^{1+}$, separates the two ions. A truly cyclic isomer, lying 100 kJ/mol above $CCCH^{1+}$, may not exist in a potential energy well because the transition state for the interconversion to the linear isomer has virtually the same energy. Thus, it is unlikely that the $CC(H)C^{1+}$ ion or the cy-C_3H^{1+} ion could be generated by experiment.

Thermochemistry

The $\Delta_f H$ value for the linear $CCCH^{1+}$ ion has been estimated by photoionization threshold measurements and threshold photoelectron photoion coincidence (TPEPICO) measurements.[9,10] TPEPICO onsets for the formation of $CCCH^{1+}$ from allene and propyne give $\Delta_f H$ values of ≤ 1606 and ≤ 1620 kJ/mol, respectively, assuming that the reaction products in each case are $C_3H^{1+} + H_2 + H^{\bullet}$. It must be noted that the reaction is not the lowest energy dissociation in either system and thus suffers from a significant competitive shift. Wong and Radom[11] calculated the $\Delta_f H(CCCH^{1+}) = 1599$ kJ/mol at the G2 level of theory, in good agreement with the above experimental values.

3.7.2 $[C_3,H_2]^{+\bullet}$ – m/z 38

Isomers

Two isomers of $[C_3,H_2]^{+\bullet}$ have been identified by experiment, the cyclopropenylidene radical cation, cy-$C_3H_2^{1+\bullet}$ (**1**) and the linear $HCCCH^{1+\bullet}$ (**2**). Wong and Radom have calculated two additional isomers, $CH_2CC^{1+\bullet}$ (**3**) and $CC(H_2)C^{1+\bullet}$ (**4**). $C_3H_2^{1+\bullet}$ cations are key interstellar species involved in the formation of the stable $C_3H_3^{1+}$ ion and neutral C_3H_2 and C_3H_3 species.[1] They are also intermediates in combustion reactions. Of the three known neutral forms of $[C_3,H_2]$, cy-C_3H_2 is a widespread interstellar molecule.[1]

Identification

The lowest energy isomers are (**1**) and (**2**) and evidence for their formation has come from ion–molecule reaction kinetics. Smith and Adams and coworkers reacted $C_3H_2^{1+\bullet}$ ions generated by electron-impact ionization of acetylene or methylpropyne with CO and C_2H_2 in a selected ion flow tube (SIFT) instrument.[4,5] The ions reacted with CO but the kinetic plots were nonlinear, indicating the presence of two $[C_3,H_2]^{+\bullet}$ isomers. The authors attributed the greater reactivity to (**2**) and the less reactive ion as (**1**). Branching ratios between the two reaction rates indicated that the ion flux was about 20% cy-$C_3H_2^{1+\bullet}$. The two isomeric forms were confirmed by Prodnuk et al.[6,7] by the reaction with C_2H_4. Only cy-$C_3H_2^{1+\bullet}$ can react to form cy-$C_3H_3^{1+}$ as the reaction is exothermic for (**1**) whereas that for (**2**) is endothermic.

Thermochemistry

The $\Delta_f H$ of ions (**1**) and (**2**) have been derived from experiment. Smith and Adams and coworkers[4,5] determined that the reaction of C_3H^{1+} with H_2 to form $C_3H_2^{1+\bullet} + H^\bullet$ was endothermic by 4 kJ/mol by examining the temperature dependence of the reaction. This allowed them to obtain $\Delta_f H(\mathbf{1}) = 1377 \pm 21$ kJ/mol (this was originally assigned to (**2**)[4] but further ion–molecule studies showed it to be for ion (**1**)[5]). Prodnuk et al. bracketed the $\Delta_f H$ of (**1**) and (**2**) to be 1347 ± 17 and 1397 ± 17 kJ/mol, respectively.[6] There have been a number of appearance energy measurements using either photoionization or electron ionization[9,10,12,13] but they all have suffered from the formation of a mixture of the two isomers and from the reactions leading to $C_3H_2^{1+\bullet}$ not being the lowest energy processes. Consequently, all attempts gave upper limits for $\Delta_f H(C_3H_2^{1+\bullet})$. The most recent NIST compendium lists the $\Delta_f H(\mathbf{1}) = 1180 \pm 10$ kJ/mol but gives no reference for the value and is clearly in error. The experimental values given above are in favorable agreement with the best computed values available, G2 results of Wong and Radom.[11] They calculated $\Delta_f H(\mathbf{1}) = 1387$ kJ/mol, $\Delta_f H(\mathbf{2}) = 1418$ kJ/mol and $\Delta_f H(\mathbf{3}) = 1583$ kJ/mol. The energy of (**4**) relative to (**1**) was found to be 484 kJ/mol at the CISD(Q)/6-311G** level of theory.[14]

3.7.3 $[C_3, H_3]^+$ – m/z 39

Isomers

Four $[C_3,H_3]^+$ isomers have been identified by experiment: the cyclopropenium cation, cy-$C_3H_3^{1+}$ (**1**), and the linear isomers CH_2CCH^{1+} (**2**) (the propargyl cation), CH_2CHC^{1+} (**3**) and CH_3CC^{1+} (**4**).[15] Radom and coworkers[16] and Li and Riggs[17] have performed calculations on these isomers as well as on $cy-CC(H)_2C(H)-^{1+}$ (**5**), $cy-C(H)_3CC-^{1+}$ (**6**) and $CHCH_2C^{1+}$ (**7**). Ions (**5**)–(**7**) were found to be minima at the HF/4-31G level[16] of theory but saddle points at the HF/6-31G* level.[17] The $[C_3,H_3]^+$ cations are key interstellar species involved in the formation of the stable C_3H_2 and C_3H_3 species.[1] They are also intermediates in combustion reactions and are ubiquitous in the mass spectra of organic compounds containing more than two carbon atoms.

Identification

The existence of two stable isomers generated by the electron-impact ionization of a variety of precursor molecules was evident from the reactions of $C_3H_3^{1+}$ ions with a wide variety of neutral molecules. Ausloos and Lias showed that the cyclic (**1**) and propargyl (**2**) ions (generated by EI on propargyl halides, XCH_2CCH, $X = Cl$, Br) could be distinguished by their reactions with the precursor propargyl halide. Only the propargyl ion (**2**) undergoes the displacement reaction to form $C_6H_6^{1+\bullet}$ and $X^{\bullet}$.[18] In addition, the linear ion was found to be more reactive and to undergo association reactions with C_2H_2,[19,20] CH_4[4,5] and other neutral molecules.[7] By comparing the ratios of (**1**) with (**2**) made in the flowing afterglow ion source, Fetterolf and coworkers[19] found that ICH_2CCH made a larger amount of the linear ion (**2**), consistent with metastable ion mass spectrometry experiments and appearance energy measurements.

Burgers et al.[15] demonstrated the existence of ions (**1**)–(**4**) by CID mass spectrometry of a number of $[C_3,H_3]^+$ ions generated by the dissociative ionization of alkene and alkyne precursors. The cyclopropenium ion (**1**) is made in pure form only upon metastable dissociation of precursor C_4H_6 molecules. Source generated $[C_3,H_3]^+$ ions from these precursors were mixtures of (**1**) and (**2**), similar to the results from ion–molecule reaction studies discussed above. The propargyl ion (**2**) was generated both by in-source and metastable dissociation of ICH_2CCH. The less common isomers (**3**) and (**4**) were generated by the collision-induced charge reversal ($^-CR^+$) of the corresponding anions formed in the ion source by OH^- abstraction of D from CH_3CCD and cyclopropene. The four ions are distinguishable by their CID mass spectra.

Partial CID Mass Spectra for the $[C_3,H_3]^+$ Isomers

Ions	m/z							
	12	13	14	15	24	25	26	27
(1) (5.8 kV)	36	54	10	—	10	39	46	5
(2) (6 kV)	29	39	32	—	15	44	39	1
(3) (8 kV)	35	32	32	1	11	33	35	20
(4) (8 kV)	37	9	13	41	32	14	32	22

Ion (**1**) has a larger m/z 13:14 ratio than (**2**), as would be expected from its connectivity. Ion (**4**) displays a dominant m/z 15, a group not present in the other structures. Ion (**3**) tends to lose $H^\bullet$ more than ions (**1**) and (**2**).

Koppel and McLafferty showed that the doubly charged ions resulting from charge stripping of the $[C_3,H_3]^+$ ions were also diagnostic for the four ions, with each displaying a unique ratio for m/z 19.5:19 ($[C_3,H_3]^{2+}$ and $[C_3,H_2]^{2+}$).[21]

Thermochemistry

The $\Delta_f H$ of ions (**1**) and (**2**) have been derived from experiments by Lossing.[22] Using monoenergetic beams of electrons, the appearance energy method was used to determine the $\Delta_f H$ of the $[C_3,H_3]^+$ ion formed from a variety of alkyne and alkene precursors. All of the precursors gave a similar result, 1071 ± 10 kJ/mol, which was assigned to the cy-$C_3H_3^{1+}$ ion. These electron impact results agreed very well with photoionization results. The ionization energy of the propargyl radical was measured to be 8.68 eV[22] and 8.67 ± 0.02 eV[23] yielding $\Delta_f H(CH_2CCH^{1+}) = 1176$ kJ/mol, using $\Delta_f H(CH_2CCH^{1\bullet}) = 339 \pm 4$ kJ/mol,[24] which is in agreement with the computed value of 346.37 kJ/mol (G3MP2B3).[25]

In a similar appearance energy study,[26] it was found that the isomeric ionized halo-compounds $ClCH_2CCH$, $BrCH_2CCH$, CH_3CCCl, and CH_3CCBr must all generate cy-$C_3H_3^{1+}$ ions by halogen atom loss, but with significant reverse energy barriers, as shown by the very broad, dished metastable ion peaks for these dissociations. Note that these peaks were the same for each halogen, indicating that rearrangement to a single dissociating ion structure preceded fragmentation. The kinetic energy releases were so great that the only possible fragment ion was cy-$C_3H_3^{1+}$. In marked contrast, the corresponding ionized iodo analogs dissociated with narrow but different Gaussian metastable ion peaks and the appearance energy values, 10.5 ± 0.05 eV for ICH_2CCH and 10.7 ± 0.05 eV for CH_3CCI resulted in $\Delta_f H(C_3H_3^{1+})$ of 1176 ± 5 and 1186 ± 5 kJ/mol, respectively ($\Delta_f H(ICH_2CCH) = 270$ kJ/mol (estimated), $\Delta_f H(CH_3CCI) = 280$ kJ/mol (estimated) and $\Delta_f H(I^\bullet) = 106.76 \pm 0.04$ kJ/mol).[24] The former is indeed compatible with the simple bond cleavage taking place at the thermochemical threshold whereas the latter requires some excess energy for the necessary rearrangement.

The best theoretical estimates of the energies for ions (**2**)–(**4**) relative to (**1**) are 115–130 kJ/mol for (**2**),[17,27] 292 kJ/mol for (**3**)[17] and 580 kJ/mol for (**4**).[17] It is unlikely that the $\Delta_f H$ values for ions (**3**) and (**4**) could be accessed by experiment.

3.7.4 $[C_3,H_4]^{+\bullet}$ – m/z 40

Isomers

$[C_3,H_4]^{+\bullet}$ cations comprise the molecular ions of propyne ($CH_3CCH^{1+\bullet}$), allene ($CH_2CCH_2^{1+\bullet}$) and cyclopropene (cy-$C_3H_4^{1+\bullet}$).

Identification

It is known from CID mass spectrometry[28] and from the photodissociation of the ions in an ICR cell[29] that ionized propyne and allene can interconvert if the timescale is long enough but that this is not an issue for threshold generated ions. Propyne and allene ions can be distinguished by their CID mass spectra.[28,30] The most unique feature in their CID mass spectra is the ratio of m/z 14:15, for which $CH_3CCH^{]+\bullet}$ has a value of 1:1 while the allene ion has a ratio of 10:1. These two isomers also have unique charge stripping (CS) features in their CID mass spectra[28,30] and are distinguishable by charge reversal ($^+CR^-$) mass spectra.[31] On the basis of the m/z 14:15 ratio, Wagner et al.[30] were able to determine that the isomer made in the dissociation of a wide variety of ionized dienes and alkynes had the allene structure.

The identification of the cyclopropene ion is more difficult in that its CID mass spectrum is very close to that of $CH_2CCH_2^{]+\bullet}$.[28] It could be that once ionized, the cyclic structure isomerizes to the $CH_2CCH_2^{]+\bullet}$ ion. However, the authors noted that the CID mass spectrum did not change upon threshold ionization. Mommers et al.[28] claimed that the dissociative ionization of 1-chloro-1-propene generated the cyclopropene ion upon loss of HCl due to the measured appearance energy for the process. The measured AE (11.54 ± 0.05 eV) together with Δ_fH values for 1-chloro-1-propene (2 kJ/mol, additivity) and HCl (-92.31 ± 0.10 kJ/mol)[24] gave a Δ_fH for the product $[C_3,H_4]^{]+\bullet}$ ion of 1208 kJ/mol, the same as that for cy-$C_3H_4^{]+\bullet}$. Thus the identification of the cyclopropenium ion rests largely on energy measurements for the process generating it. However, NIST reports $\Delta_fH(CH_3CHCHCl) = -15$ kJ/mol,[24] a value that may be slightly too low, on consideration of the effects of halogen substitution at an unsaturated cation. The above value gives $\Delta_fH = 1190 \pm 10$ kJ/mol for the $C_3H_4^{]+\bullet}$ product, high above ionized propyne (whose structure it cannot have) but significantly below ionized cyclopropene.

Thermochemistry

The Δ_fH of the three ions can be derived from the corresponding neutral Δ_fH and the measured IE.[24,32] The values thus obtained are $\Delta_fH(CH_3CCH^{]+\bullet}) = 1185 \pm 2$ kJ/mol, $\Delta_fH(CH_2CCH_2^{]+\bullet}) = 1126 \pm 1$ kJ/mol and $\Delta_fH(cy\text{-}C_3H_4^{]+\bullet}) = 1210$ kJ/mol.

3.7.5 $[C_3,H_5]^+$ – m/z 41

Isomers

$[C_3,H_5]^+$ cations are ubiquitous in the mass spectra of organic compounds containing more than two carbon atoms. Two isomers of $[C_3,H_5]^+$ have been identified by experiment, the allyl cation $CH_2=CH=CH_2^{]+}$ and the 2-propenyl ion $CH_3C=CH_2^{]+}$. HF level calculations by Radom and coworkers predicted five structures, the above two, plus $CH_3CHCH^{]+}$, the cyclopropylium ion and corner protonated cyclopropene.[33] The most recent and highest level calculations, G2(MP2)//MP2/6-31G* results of Fairley et al.,[34] found that the cyclopropylium ion is only a transition state at MP2/6-31G*.

Identification

The allyl and 2-propenyl cations are distinguishable by their CID mass spectra and by their reactivity with neutral molecules. Bowers et al.[35] obtained CID mass spectra for a variety of $C_3H_5^+$ ions generated by electron ionization of a variety of precursor molecules and found that the ratio of fragment ions m/z 27:26 was diagnostic for the ion structure. They assigned a ratio of 1:1 to the allyl cation and 1:2 to the 2-propenyl cation. They also found that protonation of allene and propyne gave mostly the 2-propenyl cation when mild protonating agents such as H_3O^+ and $C_3H_4^+$ were used, while the use of CH_5^+ and $C_2H_5^+$ gave relatively greater amounts of the allyl cation. Burgers et al.[36] studied metastable and isotopically labeled precursor ions and found that pure allyl and 2-propenyl cations were generated only from metastable precursor ions. The metastable dissociation of 1-iodopropene resulted in the allyl cation while that of 2-iodopropene gave the 2-propenyl cation. The 2 kV CID mass spectra (due to the large mass difference between $C_3H_5^+$ and $I^\bullet$, the ions from the precursor molecule, C_3H_5I mass 168, have kinetic energy $41 \times 8000/168 = 1.95$ kV) of the two $[C_3,H_5]^+$ ions show distinctive m/z 27:26 ratios of 1:1.9 (allyl cation) and 1:0.54 (2-propenyl cation). Ion source generated $[C_3,H_5]^+$ ions generally are a mixture of these two isomers.

Fairley et al.[34] have shown that the two isomers react differently with a wide variety of neutrals. For instance, the reaction of the allyl cation with methanol occurs with an overall rate constant of 7.3×10^{-10} cm^3 s^{-1} and forms four products: $[C_3,H_7]^+$, $CH_3{}^+OH_2$, $[C_4,H_7]^+$ and the encounter complex $C_3H_5(CH_3OH)^+$. The 2-propenyl cation only forms $CH_3{}^+OH_2$ and the encounter complex with a rate constant of 1.7×10^{-9} cm^3 s^{-1}.

It is noteworthy that metastable $[C_3,H_5]^+$ ions, irrespective of their origin, fragment by the loss of H_2 to generate an unusual composite metastable ion peak, having two dished components resulting from the generation of cy-$C_3H_3^+$ and the $HCCCH_2^+$, propargyl cation.[15]

Thermochemistry

The $\Delta_f H$ of the allyl and 2-propenyl cations have been derived from experiments by Holmes and Lossing and coworkers.[22,36] Using monoenergetic beams of electrons, the appearance energy method was used to determine the $\Delta_f H$ of the $[C_3,H_5]^+$ ion formed from a variety of alkene precursors. All the results gave a $\Delta_f H$ of 941 ± 10 kJ/mol, which was assigned to the allyl cation. These results are consistent with photoionization results of Traeger.[37] The EI/AE of the $[C_3,H_5]^+$ ion from 2-bromo- and 2-iodopropene gave $\Delta_f H(\text{ion}) = 970 \pm 10$ kJ/mol, which was assigned to the 2-propenyl cation.[36]

The NIST web site lists a $\Delta_f H$ for the cyclopropylium ion of 1070 kJ/mol but no reference is given.[24] In light of the calculations of Fairley et al.,[34] it is unlikely that this ion corresponds to an equilibrium structure and that the quoted value of 1070 kJ/mol is a misprint. The best theoretical estimates of the relative energies for the 2-propenyl ion relative to the allyl cation is 33.4 kJ/mol,[34] in excellent agreement with experiment. They also calculated the barrier to interconversion of the two ions and found it to lie 109 kJ/mol above the allyl cation.

3.7.6 $[C_3, H_6]^{+\bullet} - m/z$ 42

Isomers

$[C_3,H_6]^{+\bullet}$ cations comprise the molecular ions of propene $(CH_3CH=CH_2^{1+\bullet})$ and cyclopropane $(cy\text{-}C_3H_6^{1+\bullet})$ and the ionized carbene $(CH_3)_2C^{1+\bullet}$. The trimethylene radical cation $(CH_2CH_2CH_2^{1+\bullet})$ has not been observed and indeed does not appear to be an equilibrium structure on the $[C_3,H_6]^{+\bullet}$ potential energy surface. Calculations by Skancke[38] show that cyclopropane partially ring-opens when it is ionized to form a bent trimethylene structure $CH_2CH_2CH_2^{1+\bullet}$. This means that ionized cyclopropane and ionized trimethylene are essentially the same structure and cannot be distinguished experimentally.

Identification

Ionized propene and cyclopropane are difficult to distinguish, their MI and CID mass spectra being virtually identical.[39–43] The two ions have been found to interconvert on the microsecond timescale and their identification has rested on the relative abundance of the charge stripping (CS) peaks in their CID mass spectra. Neither ion yields a strong M^{2+} ion; rather they are distinct only in the ratio of $[38^{2+}]{:}[41^{2+}]$ (m/z 19:20.5). Bowen and coworkers reported a ratio of 5:1 for ionized propene while for ionized cyclopropane the value was only 2:1.[41] Holmes et al.[42] also showed the utility of the CS region in distinguishing these two ions, but the reported ratios were different from those reported by Bowen et al., about 1.5:1 for propene and 3:1 for cyclopropane. More recently, Bouchoux and coworkers reported CS mass spectra for the two ions in which the above ratio is almost identical for the two ions, their only distinguishing feature being the intensity of m/z 21 $[42^{2+}]$ (which was more intense for propene by a factor of 2).[43] It is clear from these studies that the observations may depend critically on experimental conditions. Bouchoux et al. were also able to distinguish the two isomers based on the intensity of the recovery signal in their respective NR mass spectra, ionized propene yielding a recovery m/z 42 almost 3 times as intense as that from ionized cyclopropane.[43]

    The two ions can also be distinguished based on their reactivity with neutral molecules. Gross and McLafferty observed that the reaction of ionized cyclopropane with ammonia yielded $[C,H_4,N]^+$ and $[C,H_5,N]^+$, but ionized propene was unreactive.[44] Mourgues et al. also observed isomer-specific reactions with methanol and water.[45] The formation of $[C_2,H_5,O]^+$ in the reaction with methanol was found to be diagnostic for ionized cyclopropane.

Aubry et al. confirmed the generation of $(CH_3)_2C^{]+\bullet}$ from 2,2,4,4-tetra-methyl-1,3-cyclobutadione based on its CS mass spectrum.[46] The above molecular ion fragments into two $[C_4,H_6,O]^{+\bullet}$ ions, which then lose CO to form only a $[C_3,H_6]^{+\bullet}$ ion (from high resolution mass spectrometry). Ion source generated m/z 42 from this compound had a small but significant m/z 21 ($[42^{2+}]$) in its CID mass spectrum, but was otherwise indistinguishable from ionized propene. Decomposition of metastable dimethylketene ions produces a composite peak at m/z 42. The CID mass spectrum of the centroid component of the composite peak ($T_{0.5} = 18$ meV) displayed an intense m/z 21 ($[42^{2+}]$), unlike ionized propene and cyclopropane. These ions were assigned to $(CH_3)_2C^{]+\bullet}$. Interestingly, the CID mass spectrum of ions from one edge of the broad component ($T_{0.5} = 200$ meV) of the composite peak was identical to that of ionized propene.

Thermochemistry

$\Delta_f H(CH_3CH=CH_2^{]+\bullet}) = 959$ kJ/mol is obtained from IE($CH_3CH=CH_2$) $= 9.73 \pm 0.01$ eV[24,37] and $\Delta_f H(CH_3CH=CH_2) = 20.41$ kJ/mol.[24] Ionization of neutral cyclopropane ($\Delta_f H = 53.3 \pm 0.59$ kJ/mol,[32] IE $= 9.86 \pm 0.04$ eV[24]) yields a $\Delta_f H(cy\text{-}C_3H_6^{]+\bullet}) = 1005$ kJ/mol (which has a partially ring-opened structure).[38] Ionized dimethylcarbene $(CH_3)_2C^{]+\bullet}$ has a $\Delta_f H$ of ≤ 1062 kJ/mol (derived from the EI/AE of m/z 42 from 2,2,4,4-tetramethyl-1,3-cyclobutanedione).[46] A G2 level of theory calculation placed its $\Delta_f H$ at 1053 kJ/mol.[46]

3.7.7 $[C_3,H_7]^+$ – m/z 43

Isomers

$[C_3,H_7]^+$ cations are ubiquitous in the mass spectra of organic compounds containing more than two carbon atoms. There are nominally three isomeric forms: the primary 1-propyl cation $CH_3CH_2CH_2^{]+}$, the secondary 2-propyl cation $(CH_3)_2CH^{]+}$ and protonated cyclopropane ($CPH^{]+}$). However, there is no experimental evidence for the generation of the 1-propyl cation as a stable species. Moreover, calculations at the MP2/6-311G(d,p) level of theory indicate that $CH_3CH_2CH_2^{]+}$ is only a transition state for the interconversion of $(CH_3)_2CH^{]+}$ structures leading to C and H scrambling in these cations.[47] The same level of theory predicts $CPH^{]+}$ to be a local minimum. Density functional, MP2 and CCSD level calculations predict two forms of protonated cyclopropane, corner protonated and edge protonated, with the former being slightly more thermodynamically stable.[48,49]

Identification

The most commonly observed isomer is the 2-propyl cation. It is the most stable isomer and generally is the product from any attempt to generate the 1-propyl cation and protonated cyclopropane. McLafferty and coworkers protonated cyclopropane in the ion source of a double focusing mass spectrometer and showed that the CID mass spectrum of the resulting $[C_3,H_7]^+$ ion was indistinguishable from that of the 2-propyl cation.[50] McAdoo and McLafferty performed ion–molecule reactions on isomeric and isotopically labeled $[C_3,H_7]^+$ ions and found that the reactivity of the putative CPH^+ with methanol was identical to that of the 2-propyl cation.[51] Both of these experimental results indicate that on timescales greater than $\sim 10^{-5}$ s CPH^{1+} isomerizes to the 2-propyl cation.

In a different approach, Attina and coworkers[52] demonstrated that protonated cyclopropane can exist separately from the 2-propyl cations by observing different reactivity of the two species with benzene and other arenes. Protonated cyclopropane and propene were generated by radiolysis at 1 atm. The resulting ions alkylate the aromatic compound to form either ionized 1-C_3H_7-arene or 2-C_3H_7-arene as products, the relative abundances of which were followed by GC-MS. They found that the ratio of these two products depended on the $[C_3,H_7]^+$ isomer, with the formation of the 1-C_3H_7-arene strongly indicative of the involvement of protonated cyclopropane. Chiavarino et al.[49] have further explored the generation of protonated cyclopropane in radiolysis experiments and its interconversion with the 2-propyl cation at the CCSD(T)/cc-pVTZ//CCSD(T)/cc-pVDZ + ZPVE(MP2(full)/6 − 311G**) level of theory. The results of their experiments supported the earlier work of Attila et al.[52] By computation they found that the barrier to interconversion lies 88 kJ/mol above the 2-propyl cation and 55 kJ/mol above corner-protonated cyclopropane. So, when protonated cyclopropane is generated at relatively low pressures, thus having significant internal energy, it rearranges to the 2-propyl cation in less than 10^{-5} s. When generated under conditions of thermal equilibrium a large fraction of the ions retain their structure.

Thermochemistry

The $\Delta_f H((CH_3)_2CH^{1+}) = 801 \pm 4$ kJ/mol, calculated from the 2-propyl radical heat of formation and ionization energy (90 ± 2 kJ/mol (in agreement with the computed value of 90.39 kJ/mol (G3MP2B3))[25] and 7.37 ± 0.02 eV, respectively).[24] Threshold PEPICO[53] and MATI[54] spectroscopy determinations of the appearance energies for 2-propyl cations from 2-iodopropane give 0 K $\Delta_f H$ values of 818 ± 4 and 822 ± 4 kJ/mol, respectively. While there have been determinations of the ionization energy of the 1-propyl radical, the resulting ion has been shown to be the 2-propyl cation in every case. The $\Delta_f H$ of protonated cyclopropane is 833 kJ/mol, obtained from $\Delta_f H$(cyclopropane) $= 53.30 \pm 0.59$[24,32] and PA(cyclopropane) $= 750.3$ kJ/mol.[24] The best computed estimate of the relative energy of corner-protonated cyclopropane and the 2-propyl cation place the former 33 kJ/mol above the latter, resulting in a 0 K $\Delta_f H$ of 853 kJ/mol.[49] Edge-protonated cyclopropane was calculated to lie 7 kJ/mol higher than corner-protonated cyclopropane, but the barrier between them is virtually nonexistent, implying that edge-protonated cyclopropane will not be observable by experiment.

3.7.8 $[C_3,H_8]^{+\bullet} - m/z$ 44

There is only one stable isomer of composition $[C_3,H_8]^{+\bullet}$, ionized propane $(CH_3)_2CH_2^{]+\bullet}$. $\Delta_f H((CH_3)_2CH_2^{]+\bullet}) = 951$ kJ/mol is obtained from IE(propane) = 10.94 ± 0.05 eV[24,37] and the $\Delta_f H$(propane) = −104.7 ± 0.50.[24]

3.7.9 $[C_3,H_9]^+ - m/z$ 45

Isomers

There are two $[C_3,H_9]^+$ isomers that are close in energy, protonated propane (**1**) and the 2-propyl cation/H_2 complex (**2**). A second ion–molecule complex, between the nonclassical ethyl cation and methane (**3**), is also predicted to lie in a shallow energy well.

Identification

The location of the proton in protonated propane is difficult to establish by experiment but calculations at the MP4(SDTQ)/6-311++G**// MP2(full)/6-31G** level of theory predict that protonation occurs on the C−C bond.[55] Structures generated by protonation of either the central methylene group or a terminal methyl group lie high in energy and exist in shallow energy wells and are not likely to be observed. The second isomer consists of a complex between the 2-propyl cation and H_2; however, the structure is predicted to be bound with respect to dissociation by only 2.5 kJ/mol, making it unlikely to be observed.

Hiraoka and Kebarle[56] performed high pressure mass spectrometry experiments to make equilibrium measurements for the reaction of the 2-propyl cation with H_2. Over the temperature range of their experiments, no $[C_3,H_9]^+$ ion was observed. This allowed them to estimate the dissociation threshold for the 2-propyl cation/H_2 complex to be less than 10.5 kJ/mol. This is consistent with the calculated value of 2.5 kJ/mol.[55] The equilibrium of $C_2H_5^{]+}$ with methane yielded a $[C_3,H_9]^+$ product ion, which was assumed to be protonated propane.

Thermochemistry

$\Delta_f H(\mathbf{1}) = 800$ kJ/mol is calculated from $\Delta_f H$(propane) = −104.7 ± 0.50 kJ/mol,[24] $\Delta_f H(H^+) = 1530$ kJ/mol[24] and the experimentally measured PA(propane) = 625.7 kJ/mol.[24] The equilibrium measurements of Hiraoka and Kebarle led to a $\Delta_f H$ of 814 kJ/mol.[56] The calculations of Esteves et al.[55] predict that isomer (**2**), the complex between the 2-propyl cation and H_2, lies only 1.2 kJ/mol above protonated propane and that ions (**1**) and (**2**) are separated by a barrier of 29 kJ/mol. These calculations also show that the complex (**3**) is separated from (**1**) by a barrier of only 0.8 kJ/mol.[55]

3.7.10 $[C_3,N]^+$ – m/z 50

Isomers

There are nominally three isomeric forms, $CCCN^{1+}$ (**1**), $CCNC^{1+}$ (**2**) and the cyclic cy–$CCCN^{-1+}$ (**3**) ion. The most recent computational evaluation of the $[C_3,N]^+$ potential energy surface (CCSD(T)/6-311G(d)//B3-LYP/6-311G(d))[57] divides the cyclic isomer into two subclasses, which we will call (**3a**) and (**3b**):

| 3a | 3b |

In addition, a three-membered ring, isomer (**4**) was also calculated to be stable, but resides in a potential well less than 2 kJ/mol below a transition state leading to (**1**), so is unlikely to be observed:

4

Other three-membered ring isomers were found to be saddle points on the surface.

Identification

The $[C_3,N]^+$ family of cations is formed in the dissociative ionization of HC_nN and C_nN_m species and has been postulated to be present in interstellar space.[58] Isomers (**1**) and (**3**) have been observed in the dissociative electron impact of HCCCN.[58–60] SIFT studies of the reactivity of the $[C_3,N]^+$ ions formed from the above molecule show that 90% of the ion flux is reactive toward H_2 and CH_4, while 10% was inactive.[58] Since the carbene-like ion (**1**) was assumed to react with these neutral substrates, the unreactive portion of the flux was ascribed to ion (**3**) and not to ion (**2**). This assignment is supported by EI/AE measurements of the $[C_3,N]^+$ ions from HCCCN (see below).[59,60] Ion (**2**) was assumed to be the $[C_3,N]^+$ ion generated from an ion–molecule reaction of C_2^{1+} with NCCN.[61]

Thermochemistry

The experimental EI/AE curve for the formation of $[C_3,N]^+$ from HCCCN exhibits two thresholds, one at 17.78 ± 0.08 eV and a second (a discontinuity) at 18.64 ± 0.08 eV.[59] Harland and Maclagan estimated $\Delta_f H$

(HCCCN) = 354 kJ/mol, from the measured appearance for the formation of $CN^{1+} + HCC$ (19.96 ± 0.08 eV) and ancillary thermochemistry.[60] This neutral molecule $\Delta_f H$ value, combined with the above two AE thresholds, yields $[C_3,N]^+$ $\Delta_f H$ values of 1850 kJ/mol and 1935 kJ/mol.[60] Provided that the lower value corresponds to ion (1), the higher value should be for ion (3a) since the difference in $\Delta_f H$ value (85 kJ/mol) is closer to the calculated value between (1) and (3a) (112 kJ/mol) than to that between (1) and (2) (44 kJ/mol).[57] On the basis of the calculated relative energy, ion (2) should have a $\Delta_f H$ of 1894 kJ/mol. Theory predicts (3b) to lie 155 kJ/mol above (1) and (4) to be 151 kJ/mol higher than (1).

3.7.11 $[C_3,H,N]^{+\bullet} - m/z\ 51$

Isomers

The $[C_3,H,N]^{+\bullet}$ family of cations consists of six known isomers, ionized propiolonitrile, $HCCCN^{1+\bullet}$ (1), isocyanoacetylene, $HCCNC^{1+\bullet}$ (2), $CCCNH^{1+\bullet}$ (3) and three cyclic ions:

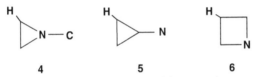

4   **5**   **6**

Only isomers (1) and (2) have been observed by experiment.

Identification

Scott et al.[62] assumed the formation of (1) in the ion–molecule reaction between C_3H^{1+} and atomic nitrogen. Apart from the ionization energies, the mass spectrometries of isomers (1) and (2) have not been described. Isomers (3)–(6) have been investigated only by computations and there is no information as to their interconversion.

Thermochemistry

The NIST reported IE for HCCCN, 11.62 ± 0.03 eV, is an estimated value.[24] Combined with the $\Delta_f H$(HCCCN) estimated by Harland and Maclagan[60] (354 kJ/mol) yields a $\Delta_f H$(1) = 1475 kJ/mol. Using the IE measured by Harland[59] (11.56 ± 0.04 eV) gives 1469 kJ/mol. Note that these values both depend on the reliability of the estimated neutral $\Delta_f H$ of HCCCN. The IE of HCCNC, 11.23 eV,[63] combined with an estimated $\Delta_f H$(HCCNC) = 444 kJ/mol, results in $\Delta_f H$(2) = 1528 kJ/mol, in fair agreement with the calculated relative energies (see text below), $\Delta\Delta_f H$(experiment) = 59 kJ/mol. In two theoretical studies of $[C_3,H,N]^{+\bullet}$ isomers at the G2 and CCSD(T)/6-311G(d,p)//B3-LYP/6-311G(d,p) levels of theory, the relative energies of isomers (1)–(6) were found to be (for ions (1)–(6), respectively, G2 values in text, CCSD(T) values in parentheses, in kJ/mol) 0 (0), +51 (+63), +21 (+45), +213, +209 (+195), and +127 (+149).[64,65]

3.7.12 $[C_3,H_2,N]^+ - m/z\,52$

Isomers

Only one $[C_3,H_2,N]^+$ isomer has been identified by experiment, protonated cyanoacetylene $HCCCNH^+$ (**1**). Several other isomers have been investigated by computation.

Identification

Petrie et al.[66] found that the $[C_3,H_2,N]^+$ ion generated by dissociative ionization of acrylonitrile ($CH_2 = CHCN$) had the same reactivity as (**1**) toward neutral reagents in SIFT experiments. The isomers have been tackled by computations from the point of view of following the ion–molecule reactions between CN and $C_2H_2^{1+\bullet}$ and between $C_3H_2^{1+}$ and N.[64,65] The latter study focused only on the triplet surface of the resulting $[C_3,H_2,N]^+$ isomers.[65] The singlet state isomers that have been modeled with G2 theory are shown below along with their relative energies.[64]

HCCCNH	H$_2$CCCN	H$_2$CCNC	HCCHCN
1 (0)	2 (152)	3 (194)	4 (249)

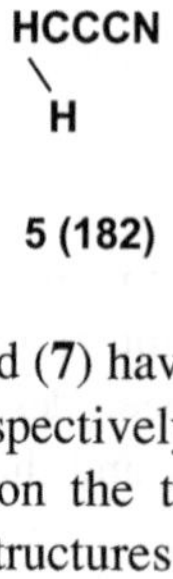

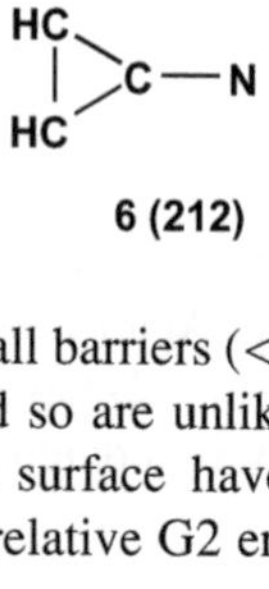

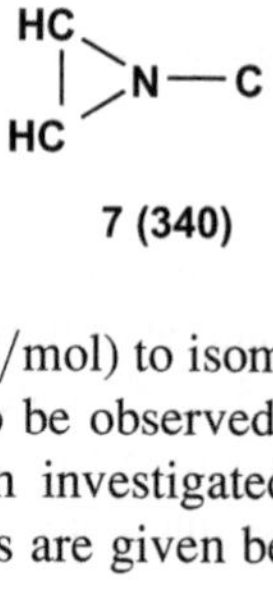

5 (182)   6 (212)   7 (340)

Ions (**5**), (**6**) and (**7**) have small barriers (<16 kJ/mol) to isomerization to (**2**), (**5**) and (**3**), respectively, and so are unlikely to be observed by experiment. Nine isomers on the triplet surface have been investigated by Barrientos et al.[65] Their structures and relative G2 energies are given below.

HCCCNH	H$_2$CCCN	H$_2$CCNC	HCCHCN
1^t (103)	2^t (1)	3^t (49)	4^t (31)

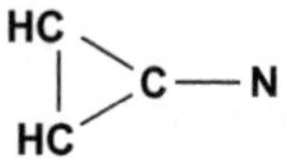

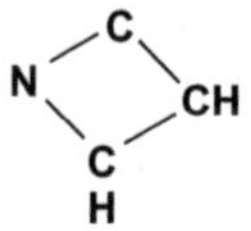

6^t (0)   8^t (60)   9^t (138)   10^t (165)   11^t (84)

Calculated transition states for the interconversion of these triplet state isomers suggest that all reside in reasonable potential energy wells.

Thermochemistry

The $\Delta_f H(\mathbf{1}) = 1132$ kJ/mol can be derived from the PA(cyano-acetylene) $= 751.2$ kJ/mol,[24] $\Delta_f H(H^+) = 1530$ kJ/mol[24] and the estimated $\Delta_f H$(cyanoacetylene) $= 354$ kJ/mol.[59] The G2 relative energies should be sufficient to obtain the $\Delta_f H$ values of all other isomers.

3.7.13 $[C_3,H_3,N]^{+\bullet} - m/z\ 53$

Only one $[C_3,H_3,N]^{+\bullet}$ cation has been the subject of investigation by experiment, ionized acrylonitrile, $CH_2 = CHCN^{]+\bullet}$. McEwan and coworkers have studied the reactivity of ionized acrylonitrile in SIFT experiments with a variety of neutral reagents.[66,67] $\Delta_f H(CH_2 = CHCN^{]+\bullet}) = 1232$ kJ/mol, using $\Delta_f H(CH_2 = CHCN) = 179.7$ kJ/mol and $IE(CH_2 = CHCN) = 10.91 \pm 0.01$ eV.[24]

3.7.14 $[C_3,H_4,N]^+ - m/z\ 54$

Isomers

The $[C_3,H_4,N]^+$ ion family contains a number of unsaturated compounds ranging from protonated acrylonitrile, alkylnitriles and alkyneamines. While a large number of isomers have been identified by calculations, relatively few have been observed by experiment. The figure below shows all 25 structures calculated by Heerma et al.[68]

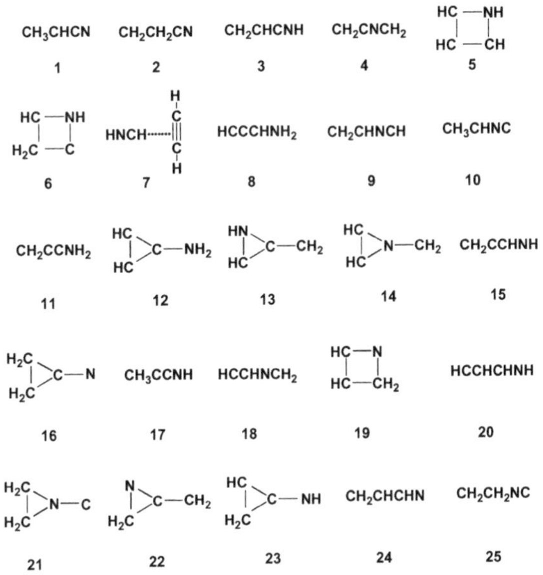

Identification

Six of these structures, namely (1), (3), (4), (8), (9) and (10), have been characterized by CID mass spectrometry and the differences in the resulting CID mass spectra are however quite small. Apart from the above CID mass spectrometry investigation, (4) has been the most widely studied isomer due to its formation when ionized acetonitrile reacts with acetonitrile in an ICR experiment or when acetonitrile is used as a CI reagent gas.[69–74] (4) is also the isomer generated from the dissociative ionization of long chain nitriles.[68] Ions (3) and (9) are formed from the protonation of acrylonitrile[66,68] and CH_2CHNC, respectively.[68] Ions (1) and (10) can both be formed from the dissociative ionization of substituted CH_3CHXCN or CH_3CHXNC molecules, respectively.[68] They are characterized by intense m/z 15 peaks in their CID mass spectra.

Isomer (8) is the only $[C_3,H_4,N]^+$ amine characterized by experiment. Its CID mass spectrum is unique in that it is the only isomer to form m/z 17 ($NH_3^{+\bullet}$), 18 (NH_4^+) and 29 ($HCNH_2^+$).[68] Isomer (7) has been proposed to be an intermediate in the formation of (3) in the ion–molecule reaction between $HCNH^+$ and C_2H_2.[75] In SIFT experiments, the $[C_3,H_4,N]^+$ products from the above ion–molecule reaction undergo both fast ion–molecule reactions with ethanol (assumed to be due to structure (3)) and much slower reactions (proposed to be (7)).[75]

Thermochemistry

The relative energies of all 25 isomers were obtained at the semiempirical MINDO/3 and MNDO levels of theory.[68] The results are not reported here because of the well-known inaccuracies of these methods for gas-phase ion energy calculations.

Isomer (3) is the only one to have a measured heat of formation, $\Delta_f H(3) = 925$ kJ/mol (based on the neutral $\Delta_f H = 179.7$ kJ/mol, $\Delta_f H(H^+) = 1530$ kJ/mol and a measured PA = 784.7 kJ/mol).[24] The IE of the neutral radical analogs of (1) and (2) has been measured by Pottie and Lossing to be 9.76 ± 0.1 eV and 9.85 ± 0.1 eV, respectively,[76] with the free radical $\Delta_f H$ values of 227 and 255 kJ/mol,[77] leading to $\Delta_f H(1) = 1169$ and $\Delta_f H(2) = 1205$ kJ/mol. For the relative energies of the other isomers we must rely on ab initio calculations. Milligan et al.[75] calculated G2(MP2) $\Delta_f H$ values for ions (3) (935 kJ/mol), (5) (1105 kJ/mol), (6) (1206 kJ/mol), and (7) (1136 kJ/mol), the first value being in reasonable agreement with the value based on protonating acrylonitrile. Hopkinson and Lien calculated the relative energies of (3), (8), (9) and (11) at the HF/6-31G* level of theory to be 0, 7, 41 and 48 kJ/mol, respectively.[78]

3.7.15 $[C_3,H_5,N]^{+\bullet}$ – m/z 55

Isomers

Of the 15 calculated isomers[79] in the $[C_3,H_5,N]^{+\bullet}$ family, five have been produced by experiment: (1), (2), (5), (8) and (9).

CH$_3$CNCH$_2$ CH$_3$NCCH$_2$ CH$_2$CHNCH$_2$ CH$_3$CHNCH   CH$_3$CHCNH

1 (0)   2 (–1)   3 (27)   4 (43)   5 (6)

CH$_2$CH$_2$NCH   CH$_2$CH$_2$CNH   CH$_3$CH$_2$CN   CH$_3$CH$_2$NC $\begin{array}{c} HC \\ \quad \\ HN \end{array}\!\!>\!C\!-\!CH_3$

6 (8)   7 (50)   8 (228)   9 (247)   10 (147)

CH$_2$CHCHNH   CH$_3$CNHCH   CH$_2$NHCCH$_2$ $\begin{array}{c} H_2C-CH \\ | \quad\quad | \\ H_2C-N \end{array}$ CH$_2$CHNHCH

11 (96)   12 (447)   13 (99)   14 (134)   15 (131)

Identification

Ion (**8**) (ionized propionitrile) is characterized by an intense (base peak) m/z 28 in its high energy CID mass spectrum (due to loss of HCN) and an m/z 27:28 ratio of 1:6, whereas (**9**) (isocyanoethane) has a ratio of nearly 1:1.[80] Ion (**9**) is the only one of the five isomers to exhibit an intense m/z 29 ($C_2H_5^{1+}$) ion in its CID mass spectrum nominally due to the weaker C_2H_5–NC bond versus the C_2H_5–CN bond in (**8**).[80] Ions (**8**) and (**9**) exhibit both charge-exchange and proton-transfer reactions with neutral pyridine, but to different extents (30:70 for (**8**) and 60:40 for (**9**)).[81] Ions (**1**) and (**2**) do not charge exchange with pyridine but rather transfer $CH_2^{1+\bullet}$.[81] Upon reaction with CH_3SSCH_3, (**8**) and (**9**) only charge exchange whereas (**1**) and (**2**) undergo different degrees of CH_3S-addition (for (**1**) CH_3S-addition is 93% of the observed reaction while for (**2**) it is only 8%, the rest being charge-exchange).[81] The high energy CID mass spectra of (**1**) and (**2**) are very similar with (**1**) exhibiting a CS peak at m/z 27.5 ($CH_3CNCH_2^{12+}$) and (**2**) displaying a larger peak at m/z 15. Their low energy CID mass spectra are dominated by H$^\bullet$ loss, but the remaining products are more distinct, with (**2**) showing a propensity for CH_3^{1+} formation while (**1**) generates m/z 42 (loss of $CH^{1\bullet}$).[81]

Salpin et al.[79] have investigated the above isomers from the point of view of rationalizing their observed CID and NR mass spectrometry. From their calculations it appears that ions (**1**)–(**4**), (**6**), (**13**) and (**15**) all reside in reasonably deep potential energy wells. No such information is available for the other isomers.

Thermochemistry

On the basis of reported neutral molecule $\Delta_f H$ values and IE values,[24] the $\Delta_f H$(**8**) and $\Delta_f H$(**9**) are 1195 and 1222 kJ/mol, respectively. The $\Delta_f H$(**5**) has been determined by Chess et al.[80] to be 1117 kJ/mol based on an EI/AE for the ion from 2-cyanobutane of 11.5 eV ($\Delta_f H$ by group additivity is 6.3 kJ/mol). Ions (**1**) and (**2**) are the lowest energy isomers (having roughly the same energy), lying some 228 kJ/mol below (**8**) at the QCISD(T)/6-31G(d,p)//MP2/6-31G(d,p) level of theory.[79] The relative energies at this level of theory for all the isomers are shown in the above figure.

3.7.16 $[C_3,H_6,N]^+$ – m/z 56

Isomers

The $[C_3,H_6,N]^+$ ion family consists of at least 42 structures that have been accessed by computational chemistry, although relatively few of these have been characterized by experiment.

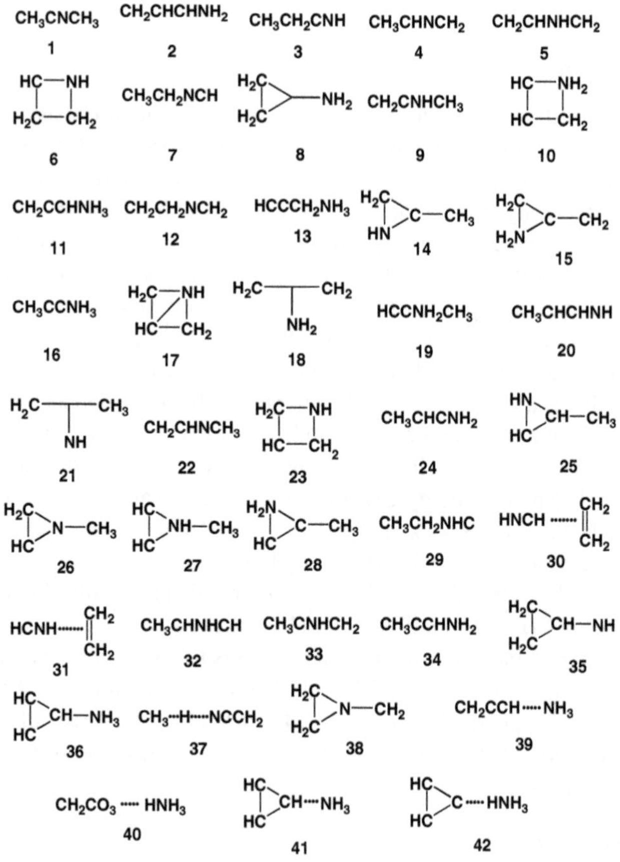

CH₃CNCH₃	CH₂CHCHNH₂	CH₃CH₂CNH	CH₃CHNCH₂	CH₂CHNHCH₂
1	**2**	**3**	**4**	**5**

HC—NH ‖ ‖ H₂C—CH₂   **6**

CH₃CH₂NCH   **7**

H₂C⟍ / —NH₂ (ring) CH₂   **8**

CH₂CNHCH₃   **9**

HC—NH₂ ‖ ‖ HC—CH₂   **10**

CH₂CCHNH₃   **11**

CH₂CH₂NCH₂   **12**

HCCCH₂NH₃   **13**

H₂C⟍ C—CH₃ / HN   **14**

H₂C⟍ C—CH₂ / H₂N   **15**

CH₃CCNH₃   **16**

H₂C—NH △ HC—CH₂   **17**

H₂C—CH₂ | NH₂   **18**

HCCNH₂CH₃   **19**

CH₃CHCHNH   **20**

H₂C—CH₃ | NH   **21**

CH₂CHNCH₃   **22**

H₂C—NH ‖ ‖ HC—CH₂   **23**

CH₃CHCNH₂   **24**

HN⟍ CH—CH₃ / HC   **25**

H₂C⟍ N—CH₃ / HC   **26**

HC⟍ NH—CH₃ / HC   **27**

H₂N⟍ C—CH₃ / HC   **28**

CH₃CH₂NHC   **29**

HNCH·····CH₂‖CH₂   **30**

HCNH·····CH₂‖CH₂   **31**

CH₃CHNHCH   **32**

CH₃CNHCH₂   **33**

CH₃CCHNH₂   **34**

H₂C⟍ CH—NH / H₂C   **35**

HC⟍ CH—NH₃ / HC   **36**

CH₃···H···NCCH₂   **37**

H₂C⟍ N—CH₂ / H₂C   **38**

CH₂CCH····NH₃   **39**

CH₂CO₃ ···· HNH₃   **40**

HC⟍ CH····NH₃ / HC   **41**

HC⟍ C····HNH₃ / HC   **42**

Identification

The ion that has perhaps been subjected to most interest is (**1**). This ion is formed by the associative ion–molecule reaction between CH_3^+ and CH_3CN or CH_3NC. This reaction was originally thought to be the interstellar source of protonated propionitrile (**3**), but (**1**) was confirmed by isotopic labeling studies and ion–molecule reaction chemistry.[82–87] Ions (3) and (7) are simply protonated propionitrile and isocyanoethane, respectively. Isomer (2) has been generated by H• loss from allylamine while (5) can be formed by ethene loss from ionized piperidine,[88,89] although it is possible that the latter process makes (6) (the ring-closed analog). Ion (2) has also been generated by H• loss from cyclopropyl amine.[90]

Many of the isomers in the above figure were calculated in order to track the interconversion and dissociation of the most stable ions. For example, (3) and (7) were found to interconvert via the two electrostatically bound complexes (30) and (31). (30) is bound by 40 kJ/mol while (31) is only bound by 25 kJ/mol and so they are unlikely to be easily accessible by experiment.[91] Indeed, the reaction between HCNH^{1+} and C_2H_4 yields an adduct whose reactivity is consistent only with (6).[75] Isomers (39)–(42) were investigated in order to explain the ion–molecule reaction between ammonia and the propargyl cation.[92] Probably the most complete examination of the isomers and their interconversion was made by Liu et al. where 42 minima and 57 transition states were found on the calculated potential energy surface at the QCISD(T)/6-311+G(d,p)//B3-LYP/6-31G(d,p) level of theory.[93]

Ions (1)–(3), (5) and (7) have been characterized by their CID mass spectra.[88] (5) (or possibly (6)) is the only ion with an intense m/z 30 fragment ion (HCNH$^+$). (1) is the only isomer that has a strong m/z 15 peak in its CID mass spectrum while (7) is characterized by a peak at m/z 16 (the only isomer having such a fragment ion). (2) displayed no low mass fragments in the m/z 12–16 region but has a unique small signal at m/z 36. The CID mass spectrum of protonated propionitrile (3) is dominated by m/z 28 and 29 but m/z 30 is conspicuously absent. These isomers have also been distinguished by their ion–molecule reactivity with isoprene (2-methyl-1,4-butadiene).[94] (1) and (2) are unreactive with isoprene while (3) shows only proton transfer. Both (5) and (7) undergo a [4+2] cyclo-addition reaction with isoprene.

Thermochemistry

There are few $\Delta_f H$ values for the $[C_3,H_6,N]^+$ ion family derived from experiment. Using the proton affinity and neutral molecule $\Delta_f H$ values, $\Delta_f H(3)$, $\Delta_f H(7)$ and $\Delta_f H(13)$ are 787, 824 and 879 kJ/mol, respectively (from NIST,[24] $\Delta_f H(CH_3CH_2CN) = 51.46$ kJ/mol, PA(CH$_3$CH$_2$CN) = 794.1, $\Delta_f H(CH_3CH_2NC) = 145 \pm 6$ kJ/mol (estimated from known thermochemical data[24] and experimental value[95]), PA(CH$_3$CH$_2$NC) = 851.3 kJ/mol and PA(HCCCH$_2$NH$_2$) = 887.4 kJ/mol). Note that $\Delta_f H(HCCCH_2NH_2) = 236$ kJ/mol was estimated by adding $\Delta_{vap} H$ (30 kJ/mol) to the liquid phase $\Delta_f H$ listed by Pedley et al. (206 kJ/mol).[32] The $\Delta_{vap} H$ was itself estimated by comparison with the corresponding values of other primary amines. The relative energies of the other isomers have been calculated at various levels of theory. The table below summarizes the most reliable data.

Relative Energies[a] for the $[C_3,H_6,N]^+$ Isomers

Ions	MP4/6-31G* //HF/3-21G[b]	QCISD(T)/6-311 + G(d,p) //B3-LYP/6-31G(d,p)[c]	CCSD(T)/6-311 + G(2df,2p) //MP2(full)/6-31 + G(d,p)[d]	G2[e]
(1)	0	0		5
(2)	8	−2	0	0
(3)	43	31	39	39
(4)	60	57		211
(5)	61	52		54
(6)	63	56	62	58
(7)	70	64		74
(8)	98	89		91
(9)	108	96		297
(10)	108	98		98
(11)	127	110	114	
(12)	129			
(13)	133	112		
(14)	145	142		
(15)	149	138		
(16)	180	159		
(17)	180	173		
(18)	185			
(19)	212	191		
(20)	221(s), 322(t)	211	220	
(21)	273(s), 316(t)			
(22)	273(s), 312(t)	262		
(23)	335	297		
(24)				Not a minimum
(25)		180		185
(26)		179		184
(27)				155
(28)			179	164
(29)		279		
(30)		190	203	
(31)		167		
(32)		321		
(33)		320		
(34)		241	242	
(35)		223	238	
(36)		150	156	
(37)		363		
(38)		129		
(39)			315	
(40)			306	
(41)			213	
(42)			232	

[a]In kJ/mol.
[b]Ref. [88].
[c]Ref. [93].
[d]Ref. [92]. Values are $\Delta_f H$ values corrected to 298 K.
[e]Ref. [89].

3.7.17 $[C_3,H_7,N]^{+\bullet}$ – $m/z\ 57$

Isomers

The $[C_3,H_7,N]^{+\bullet}$ family of ions has not received as much attention as some of the other series of isomers. That being said, there are still 15 known isomers.

Identification

Ions (1)–(3) exhibit unique metastable ion mass spectra with $H^\bullet$ and $CH_3^{\bullet}$ loss being the two dominant pathways.[96] The mass spectrum of metastable (3) shows comparable $H^\bullet$ and $CH_3^{\bullet}$ loss peaks and the latter is narrow with a small KER ($T_{0.5} = 3$ meV). Ion (1) does not have a strong $H^\bullet$ loss channel while (2) again has comparable $H^\bullet$ and $CH_3^{\bullet}$ losses. The $CH_3^{\bullet}$ loss peak for (1) and (2) is broad and composite with similar KER values, indicating that these two ions interconvert before this loss. The CID mass spectra of (1) and (2) are virtually identical, reinforcing the proposal that they interconvert on the microsecond timescale. High level calculations place the barrier to their interconversion 63 kJ/mol above (1), but the barrier to $CH_3^{\bullet}$ loss lies 225 kJ/mol above (2), allowing (1) to isomerize to (2) before dissociation.

Ionized cyclopropylamine (4) dissociates by loss of $H^\bullet$, a process that starts with the opening of the ring to form the distonic ion $^\bullet CH_2CH_2{}^+CHNH_2$ (8).[90] Bouchoux et al. have reported that attempts to find an equilibrium structure for (4) by ab initio calculations failed, indicating that (4) may ring-open immediately upon ionization.[90] $H^\bullet$ atom loss from (8) is preceded by isomerization to (9).[90]

Thermochemistry

One problem facing the thermochemistry of $[C_3,H_7,N]^{+\bullet}$ ions is a dearth of neutral molecule $\Delta_f H$ values. The IE of a variety of isomers has been measured but the $\Delta_f H$ values for the neutral molecules are largely estimates. The table below lists the known thermochemistry. Hammerum et al.[96] have calculated the $\Delta_f H$ values for ions (1)–(3) at the CBS-RAD, CBS-Q, G2(MP2) and G3 levels of theory, but as all the values were quite consistent, only the G3 values are reported below. Bouchoux et al.[90] used experimental PEPICO data to obtain $\Delta_f H(8) = 840$ kJ/mol. Using the calculated relative energies (PUMP4STDQ/6-311G**// UMP2/6-31G*) gives the $\Delta_f H$ values for (9)–(11) (see the table). The same authors also predicted the neutral $\Delta_f H$ values of (5), (9)–(11) to be 55, 30, 160 and 10 kJ/mol. The IE of (12)–(15) has been measured but there are no neutral molecule $\Delta_f H$ values on which to base ionic heats of formation.

Heat of Formation for the $[C_3,H_7,N]^{+\bullet}$ Isomers

Ions	$\Delta_f H(\text{expt})^{a,b}$ (kJ/mol)	$\Delta_f H(\text{G3})^c$ (kJ/mol)	$\Delta_f H(\text{theory})^d$ (kJ/mol)
(1)	964[b]	964	
(2)		855	
(3)		920	
(4)	926[a]		
(5)	893[b]		
(6)	898[b]		
(7)	961[b]		
(8)	840[d]		
(9)			770
(10)			922
(11)			884
(12)	n/a		
(13)	n/a		
(14)	n/a		
(15)	n/a		

[a]NIST.[24]
[b]Ref. [95]. Most are based on estimated neutral $\Delta_f H$ values.
[c]Hammerum et al.[96]
[d]Bouchoux et al.[90] Theoretical numbers based on PUMP4STDQ/6-311G**//UMP2/6-31G* relative energies.

3.7.18 $[C_3,H_8,N]^+$ – m/z 58

Isomers

The $[C_3,H_8,N]^+$ ion family consists of a variety of protonated neutral $[C_3,H_7,N]$ molecules and ion–molecule complexes. Those that have been examined by experiment and theory are shown below.

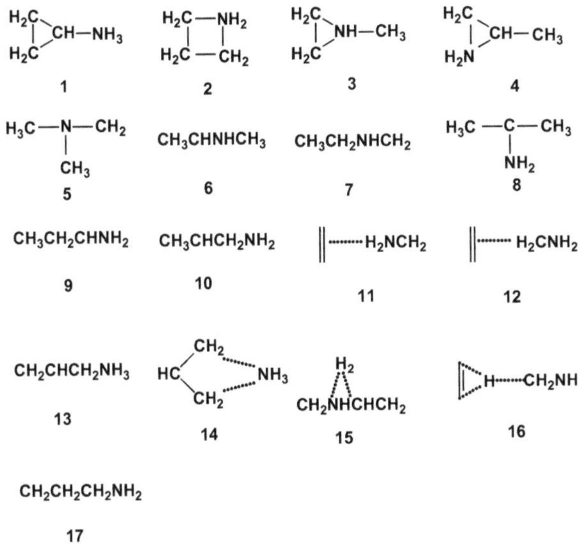

H_2C—CH—NH_3 (with H_2C cyclopropane ring) **1**

H_2C—NH_2 / H_2C—CH_2 **2**

H_2C>NH—CH_3 (H_2C) **3**

H_2C>CH—CH_3 / H_2N **4**

H_3C—N—CH_2 / CH_3 **5**

$CH_3CHNHCH_3$ **6**

$CH_3CH_2NHCH_2$ **7**

H_3C—C—CH_3 / NH_2 **8**

$CH_3CH_2CHNH_2$ **9**

$CH_3CHCH_2NH_2$ **10**

‖·······H_2NCH_2 **11**

‖·······H_2CNH_2 **12**

$CH_2CHCH_2NH_3$ **13**

HC<(CH_2)(CH_2)>NH_3 **14**

H_2 / $CH_2NHCHCH_2$ **15**

▷H·····CH_2NH **16**

$CH_2CH_2CH_2NH_2$ **17**

Identification

Ions (**1**)–(**4**) are protonated cyclic molecules whose ion chemistry has not been explored. Ions (**5**)–(**9**) have received the most attention with regard to their structural characterization. Metastable ions (**5**)–(**9**) lose ammonia, ethene and H_2 to differing extents. (**8**) and (**9**) exhibit the same MI mass spectra as does the pair (**5**) and (**6**) (indicating that (**8**)/(**9**) and (**5**)/(**6**) interconvert on the microsecond timescale) while (**7**) is unique, only losing ethene.[97] However, their respective CID mass spectra have significant differences (see table below).[98]

CID Mass Spectra for Some $[C_3,H_8,N]^+$ Isomers

						m/z								
Ions	15	18	27	28	29	30	39	41	42	43	54	55	56	57
(5)	7	1	3	6	8	(52)	1	<6	52	21	2	1	(16)	(20)
(6)	5	1	10	24	15	(109)	1	12	16	25	3	2	(63)	(29)
(7)	2	1	16	29	32	(256)	1	6	11	6	2	1	16	(17)
(8)	5	(10)	4	6	<2	(5)	15	(70)	49	17	3	1	4	(33)
(9)	3	3	10	26	<10	(16)	12	(112)	19	22	4	1	40	(25)

Levsen and McLafferty[98] found that trimethylamine dissociates to form (**5**), N-methyl amines produce (**6**), substituted ethylamines generate (**7**), propyl and butyl amines make (**8**) and larger alkyl amines tend to form (**9**). These five ions have received considerable attention from the point of view of their ion chemistry, especially the role of ion–molecule complexes in their dissociation.[99–103] Ions (11) and (12) are only weakly bound with respect to dissociation and appear on the potential energy surface as preludes to the metastable loss of ethene.[99,102,103] Since the dissociations of (7) and (9) occur over high energy barriers,[99,102,103] (11) and (12) are not formed as stable intermediates. They can be generated only by the ion–molecule reaction between $CH_2NH_2^{1+}$ and ethene (note that the authors could distinguish (11) from (12) by an ion–molecule reaction).[99]

Thermochemistry

The $\Delta_f H$ values for ions (1)–(10) are listed below and were derived from either PA values[24,95] or EI/AE measurements.[104] PA values are only useful if the $\Delta_f H$ of the corresponding neutral molecule is known. Several $\Delta_f H([C_3,H_8,N])$ values are not included in the most recent NIST collection[24] but estimated values can be found in the earlier book by Lias et al.[95] The latter values were used here when we needed to obtain data for the table. Ion (17) has been calculated to lie 275 kJ/mol above (9) (MP2/6-311 + G**//HF/6-31G*).[99] The complex (16) lies 255 kJ/mol above (7) (QCISD(T)/6-311G (d, p)//QCISD/6-31G(d)).

Heat of Formation for the $[C_3,H_8,N]^+$ Isomers

Ions	$\Delta_f H$(expt) (kJ/mol)	$\Delta_f H$(G2)[a] (kJ/mol)
(1)	702[b]	
(2)	686[b]	697
(3)	722[b]	
(4)	696[b]	690
(5)	661,[c] 664[d]	
(6)	615[c]	632
(7)	653[c]	671
(8)	590,[c] 615[b]	
(9)	636[c]	644
(10)	676[b,e]	835
(11)		762
(12)		776
(13)		679
(14)		860
(15)		804

[a]From Chalk et al.[103]

[b]Values obtained from $\Delta_f H([C_3,H_7,N])$ and PA$([C_3,H_7,N])$.[24,95]

[c]Derived from EI/AE measurements.[104]

[d]Obtained by combining the PI/IE (5.35 eV)[24] and calculated $\Delta_f H(^•CH_2N(CH_3)_2)$ (147.97 kJ/mol,[105] G3MP2B3).

[e]Note that protonation of $CH_2CHCH_2NH_2$ is reported to give (10).

3.7.19 $[C_3,H_9,N]^{+\bullet}$ – m/z 59

Isomers

The $[C_3,H_9,N]^{+\bullet}$ family of ions includes ionized trimethylamine (**1**), 1-propylamine (**2**), 2-propylamine (**3**), ethylmethyl amine (**4**) and the distonic ions $CH_3CHCH_2NH_2^{\bullet+}$ (**5**) and $CH_2CHCH_2NH_3^{\bullet+}$ (**6**).

Identification

Metastable ions (**1**)–(**4**) undergo the loss of $H^\bullet$, $CH_3^{\bullet}$ and $C_2H_5^{\bullet}$ while (**5**) (generated by C_3H_6 loss from 2-methylpentylamine) exhibits only NH_4^{+} formation. Ion (**6**) is reported to be formed by γ-cleavage from ionized 1-hexanamine, but has been suggested to isomerize to (**2**) before its loss of $C_2H_5^{\bullet}$.[106,107]

Thermochemistry

$\Delta_fH(\mathbf{1}) = 734$ kJ/mol, $\Delta_fH(\mathbf{2}) = 777$ kJ/mol and $\Delta_fH(\mathbf{3}) = 746$ kJ/mol respectively, based on neutral molecule Δ_fH values and IE measurements ($\Delta_fH(N(CH_3)_3) = -23.7 \pm 0.75$ kJ/mol and $IE(N(CH_3)_3) = 7.85 \pm 0.05$ eV; $\Delta_fH(CH_3CH_2CH_2NH_2) = -70 \pm 1$ kJ/mol and $IE(CH_3CH_2CH_2NH_2) = 8.78 \pm 0.02$ eV; $\Delta_fH((CH_3)_2CHNH_2) = -83.7 \pm 0.8$ kJ/mol and $IE((CH_3)_2CHNH_2) = 8.6 \pm 0.1$ eV).[24] Discrepancies in the IE values mean that the $\Delta_fH(\mathbf{2})$ could be as low as 750 kJ/mol.

3.7.20 $[C_3,H_{10},N]^+$ – m/z 60

Isomers

The $[C_3,H_{10},N]^+$ ion group consists of protonated trimethylamine, 1-aminopropane and 2-aminopropane. No mass spectrometric characterization has been carried out on these isomers.

Thermochemistry

Combining the neutral molecule Δ_fH (see $[C_3,H_9,N]^{+\bullet}$ ions above) and corresponding proton affinity values leads to $\Delta_fH((CH_3)_3NH^+) = 557$ kJ/mol, $\Delta_fH((CH_3CH_2CH_2NH_3^+) = 542$ kJ/mol and $\Delta_fH((CH_3)_2CHNH_3^+) = 523$ kJ/mol PA$((CH_3)_3N) = 948.9$ kJ/mol, PA$(CH_3CH_2CH_2NH_2) = 917.8$ kJ/mol, PA$((CH_3)_2CHNH_2) = 923.8$ kJ/mol and $\Delta_fH(H^+) = 1530$ kJ/mol).[24]

3.7.21 $[C_3,N_2]^{+\bullet}$ – m/z 64[108]

Isomers

Three isomeric $[C_3,N_2]^{+\bullet}$ cations have been theoretically investigated: ionized dicyanocarbene $NCCCN^{+\bullet}$ (**1**), $NCCNC^{+\bullet}$ (**2**) and $CNCNC^{+\bullet}$ (**3**).

Identification

Ion (1) has been generated by the dissociation of ionized tetracyanoethene. Its CID mass spectrum is dominated by m/z 38 due to loss of CN. The resulting m/z 38 ion has a CID mass spectrum indicative of a CCN connectivity, thus eliminating the possibility that ion (3) was formed from the ionized tetracyanoethene. Ion (2) was similarly eliminated by the results from ion–molecule reactions of the ions with acetone.

Thermochemistry

The relative energies of (1)–(3) have been calculated at the CASPT/cc-pVTZ level of theory to be 0, 78 and 148 kJ/mol, respectively.

3.7.22 $[C_3,H_2,N_2]^{+\bullet} - m/z$ 66

Ionized dicyanomethane, $CH_2(CN)_2^{1+\bullet}$, has a $\Delta_f H = 1501 \pm 20$ kJ/mol, based on $\Delta_f H(CH_2(CN)_2) = 266.3$ kJ/mol and $IE(CH_2(CN)_2)$ that lies between 12.7 and 12.88 eV.[24]

3.7.23 $[C_3,H_3,N_2]^+ - m/z$ 67

Isomers

Five $[C_3,H_3,N_2]^+$ cations have been investigated but isomers (3) and (5) are postulated to be indistinguishable.

Identification

Ion (1) is protonated dicyanomethane while ions (2)–(5) are derived from the dissociation of substituted pyrazoles.[109] Loss of NO_2 from 1-nitropyrazole leads to (2), whereas loss of $Br^\bullet$ from 3-bromopyrazole gives (4). On the basis of isotopic labeling and the propensity for C_3^{1+} and $[C_3,H]^+$ formation in the CID mass spectra, the authors postulated that (3) (formed by $Br^\bullet$ loss from 4-bromopyrazole) ring opens to (5) before dissociation.[109]

CID Mass Spectra for the $[C_3,H_3,N_2]^+$ Isomers

Ions	m/z									
	28	36	37	38	39	40	41	42	65	66
(2)	2.1	1.2	6.6	18.5	53.2	6.8	0.6	—	0.7	2
(3)/(5)	4.0	0.5	1.1	15.2	16.9	31.6	2.7	—	1.7	12.6
(4)	3.5	1.1	3.9	18.6	18.4	18.9	1.9	0.7	2.7	16.9

Thermochemistry

The $\Delta_f H(1) = 1073$ kJ/mol can be derived from the PA of dicyanomethane (723 kJ/mol), its $\Delta_f H$ (266.3 kJ/mol) and $\Delta_f H(H^+) = 1530$ kJ/mol.[24]

3.7.24 $[C_3,H_4,N_2]^{+\bullet} - m/z\ 68$

Isomers

The nine isomers investigated in the literature are shown below.

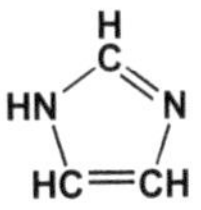

1 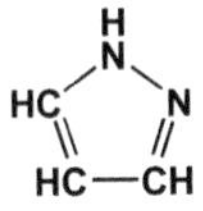   2   3   4 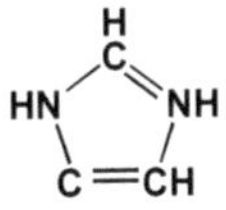   5

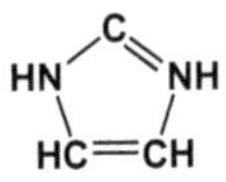

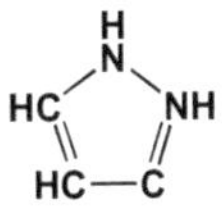

6 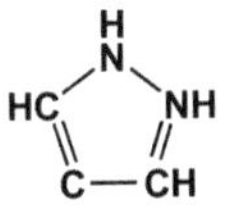   7 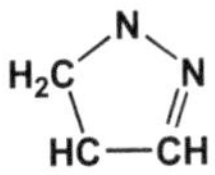   8 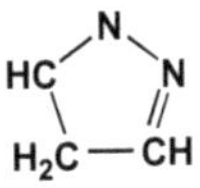   9

Identification

Ion (1) is ionized imidazole while (2) is ionized pyrazole. van Tilborg et al. derived ions (4) and (5) from the dissociation of 2- and 3-vinylimidazole, respectively, although the authors suggest that the CID mass spectrum of (5) is more consistent with an unspecified ring opened structure because of the abundance of C_3^{1+} and $[C_3,H]^+$ ions.[110] Partial CID mass spectra for ions (1), (4) and (5) are listed below. Other workers have assumed that (5) retains a ring-closed structure.[111–113] Flammang and coworkers[112,113] generated (4) (5), (6) and (7) from protonated nitro-imidazoles (4) and (5) and protonated halo- and nitro-pyrazoles (6, 7). The CID mass spectra reported for (1), (4) and (5) are similar to those obtained by van Tilborg, as shown below. (1) and (4) also exhibit different reactivities with CH_3SSCH_3, with only (4) abstracting a CH_3S radical.[114]

Partial CID Mass Spectra for Some of the $[C_3,H_4,N_2]^{+\bullet}$ Isomers[110]

Ions	38	39	40	41	42
(1)	4.6	6.4	31.7	27.9	0.5
(4)	4.6	5.9	20.9	18.4	17.0
(5)	9.4	14.7	16.7	22.7	0.7

Partial CID Mass Spectra for Some of the $[C_3,H_4,N_2]^{+\bullet}$ Isomers[113]

	m/z			
Ions	38	39	40	41
(2)	35	50	83	100
(6)	43	33	100	96
(7)	86	83	88	100

Thermochemistry

The $\Delta_f H(1) = 989 \pm 6$ kJ/mol is based on a neutral $\Delta_f H$ (139 kJ/mol) and an IE of 8.81 eV.[24] A higher neutral molecule $\Delta_f H$ (145 $\pm$ 2 kJ/mol) has been reported by Lias et al.,[95] and the above limits on $\Delta_f H(1)$ reflect the uncertainty in the neutral $\Delta_f H$. A similar discrepancy between the two compendia concerning the $\Delta_f H$ of neutral pyrazole yields a $\Delta_f H(2) = 1071 \pm 6$ kJ/mol (based on an IE of 9.25 $\pm$ 0.01 eV and $\Delta_f H$(neutral) $= 179.4 \pm 0.8$ kJ/mol).[24] The IE of (3) has been measured to be 10.2 eV.[24] Flammang and coworkers have calculated the relative energies of (1), (4) and (5) to be 0, 59 and 57 kJ/mol, respectively, at the B3-LYP/6-311++G(d,p) level of theory. The relative energies at this level of (2), (6), (7), (8) and (9) are 0, 50, 54, 114 and 136 kJ/mol, respectively. Both the imidazole derivatives and pyrazole-related ions have large 1,2-H shift barriers (over 240 kJ/mol in all cases) separating them.[112,113]

3.7.25 $[C_3,H_5,N_2]^+ - m/z$ 69

Isomers

The six isomers reported in the literature are shown below.

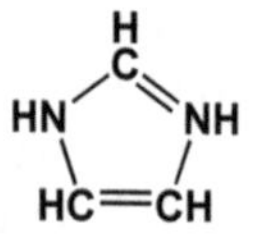

1 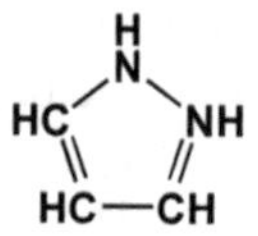   2 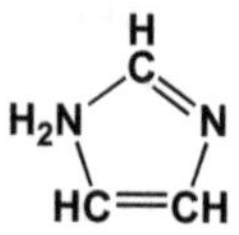   3 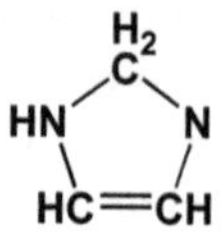   4 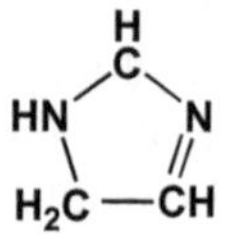   5

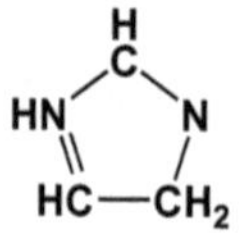

6

Ions (3)–(6) are isomers of protonated imidazole (1) while (2) is protonated pyrazole.

Thermochemistry

$\Delta_f H(\mathbf{1}) = 726$ kJ/mol, based on a neutral $\Delta_f H$ (139 kJ/mol; see [C_3,H_4,N_2] above), a PA of 942.8 kJ/mol and $\Delta_f H(H^+) = 1530$ kK/mol.[24] $\Delta_f H(\mathbf{2}) = 815$ kJ/mol based on a neutral $\Delta_f H$ (179 kJ/mol), a PA of 894 kJ/mol and $\Delta_f H(H^+) = 1530$ kJ/mol.[24] Calculations place (**1**) as the most stable protonated imidazole followed by (**6**), (**4**), (**5**) and (**3**) having relative energies of 0, +137, +150, +150 and +217 kJ/mol, respectively, at the MP2/6-311G(2d,p)//HF/6-31G(d,p).[115]

3.7.26 $[C_3,H_6,N_2]^{+\bullet}$ – m/z 70

Isomers

The only ions in this family that have been investigated are dimethylcyanamide (**1**), 3,3-dimethyldiazirine (**2**) and 2-diazopropane (**3**).

Thermochemistry

$\Delta_f H(\mathbf{1})$ is estimated to be 1007 kJ/mol based on an IE of 9.0 eV[24] and an estimated neutral $\Delta_f H$ of 139 kJ/mol.[95] $\Delta_f H(\mathbf{2})$ is an upper limit of 1118 kJ/mol, based on a vertical IE of 9.76 eV[24] and an estimated neutral $\Delta_f H$ of 176 kJ/mol.[95] The IE of neutral (**3**) is 7.88 eV.[24]

3.7.27 $[C_3,H_7,N_2]^+$ – m/z 71

Isomers

Protonation of dimethylcyanamide and 3-aminopropionitrile yields the two isomeric ions $[(CH_3)_2NCN]H^+$ and $[NH_2CH_2CH_2CN]H^+$.

Thermochemistry

$\Delta_f H([(CH_3)_2NCN]H^+) = 817$ kJ/mol using PA($(CH_3)_2NCN) = 852$ kJ/mol,[24] $\Delta_f H(H^+) = 1530$ kJ/mol[24] and an estimated neutral $\Delta_f H$ of 139 kJ/mol[95]. $\Delta_f H([NH_2CH_2CH_2CN]H^+) = 753$ kJ/mol based on a PA($NH_2CH_2CH_2CN$) of 866.4 kJ/mol[24] and a neutral $\Delta_f H$ of 90 kJ/mol[24].

3.7.28 $[C_3,H_8,N_2]^{+\bullet}$ – m/z 72

Three $[C_3,H_8,N_2]^{+\bullet}$ isomers have only estimated $\Delta_f H$ values.

Ion (**1**) has a $\Delta_f H \leq 1025$ kJ/mol (based on an estimated neutral $\Delta_f H$ of 141 kJ/mol and a vertical IE of 9.16 eV), whereas (**2**) has a $\Delta_f H \leq 1152$ kJ/mol (based on an estimated neutral $\Delta_f H$ of 243 kJ/mol and a vertical IE of 9.42 eV).[24] The adiabatic IE of (**3**) has been calculated at the CBS-4 level of theory to be 7.31 eV.[116] Until adiabatic IE values are available the first two data are upper limits only.

3.7.29 $[C_3,H_{10},N_2]^{+\bullet}$ – m/z 74

The $\Delta_f H$ of trimethylhydrazine is 849 kJ/mol (based on a neutral $\Delta_f H$ of 87 kJ/mol and IE of 7.9 ± 0.1 eV).[24,95]

3.7.30 $[C_3,H,O]^+$ – m/z 53

Isomers

There is only one known $[C_3,H,O]^+$ isomer, $HCCCO^{1+}$. $[C_3,H,O]^+$ cations have been implicated in the interstellar formation of CCCO and HCCCHO.

Thermochemistry

The value listed in the Lias et al. compendium, $\Delta_f H(HCCCO^{1+}) = 971$ kJ/mol, is based on EI/AE determinations.[95] However, the present authors have reevaluated the heats of formation of the precursor molecules and the new data are as follows: $\Delta_f H(HCCCO^{1+})$ was derived from the appearance energy for the loss of $CH_3^{1\bullet}$ from $CH_3C(O)CCH$. The value $\Delta_f H(CH_3C(O)CCH) = 73$ kJ/mol was obtained at the G3 level of theory[117] and combined with the ancillary thermochemical data (AE = 11.14 eV[118] and $\Delta_f H(CH_3^{1\bullet}) = 146$ kJ/mol[24]) gives $\Delta_f H(HCCCO^{1+}) \leq 1001$ kJ/mol. Note that the new $\Delta_f H(CH_3C(O)CCH)$ value leads to two new useful Benson additivity terms: $CO - (C_t) = -113$ kJ/mol and $C_t - (CO) = 117$ kJ/mol. This is in fair agreement with the result from the appearance energy for the loss of $CH_3^{1\bullet}$ from CH_3CHCCO, which leads to $\Delta_f H(HCCCO^{1+}) = 979 \pm 5$ kJ/mol (AE = 11.02 ± 0.05 eV,[119] $\Delta_f H(CH_3CHCCO) = 63$ kJ/mol obtained at the G3 level of theory[117] and $\Delta_f H(CH_3^{1\bullet}) = 146$ kJ/mol[24]).

3.7.31 $[C_3,H_2,O]^{+\bullet}$ – m/z 54

Isomers

Five $[C_3,H_2,O]^{+\bullet}$ isomers have been the subject of experiment and computational investigations: ionized methylene ketene $(CH_2CCO^{1+\bullet})$ (**1**), $CHCHCO^{1+\bullet}$ (**2**), cyclopropenone cy–C(H)C(H)C(O)–$^{1+\bullet}$ (**3**), $CHCCOH^{1+\bullet}$ (**4**) and propynal, $CHCCHO^{1+\bullet}$ (**5**). Only the first three ions have been generated.

Identification

Ionization of CH_2CCO and cy–C(H)C(H)C(O)– yields radical cations (**1**) and (**3**), respectively. Dissociative ionization of maleic anhydride (cy–C(H)C(H)C(O)OC(O)–, loss of CO_2) and cyclopentene-3,5-dione (loss of CH_2CO) produces isomer (**2**).[120] Ionized methylene ketene has been characterized by its CID mass spectrum, which contains peaks at m/z 14 ($CH_2^{1+\bullet}$) and 40 (loss of CH_2). The assignment of the isomer generated by the dissociative ionization of maleic anhydride and cyclopentene-3,5-dione to ion (**2**) was based principally on thermochemical grounds (see below) and that its CID mass spectrum contained m/z 13 (which by itself is insufficient to distinguish ions (**2**) and (**3**)). While ion (**3**) has been generated by photoionization, no further mass spectrometric characterization has been performed.[121]

Thermochemistry

Combining measured IE values with neutral heats of formation give $\Delta_f H(\mathbf{1}) = 975$ and $\Delta_f H(\mathbf{3}) = 1052$ kJ/mol.[121,122] The EI/AE of ion (**2**) from cyclopentene-3,5-dione (11.8 ± 0.2 eV),[120] combined with an estimated (additivity)[123] $\Delta_f H(\text{cyclopentene-3,5-dione}) = -232$ kJ/mol and the $\Delta_f H(CH_2CO) = -52$ kJ/mol,[124–127] yields a $\Delta_f H(\mathbf{2}) = 958 \pm 20$ kJ/mol.[120] The heats of formation for ions (**4**) and (**5**) have not been measured, but calculations at the CI/6-31G**//HF/3-12G level of theory place the energies of the five isomers (relative to (**1**)) at 21 kJ/mol (**2**), 114 kJ/mol (**3**), 145 kJ/mol (**4**) and 241 kJ/mol (**5**). The IE of ion (**5**) was measured by von Neissen et al., which gave a $\Delta_f H$ of 1157 kJ/mol (+182 kJ/mol relative to (**1**)), but the similarity of this value with that calculated for (**4**) suggests an erroneous structure assignment.[128]

3.7.32 $[C_3,H_3,O]^+$ – m/z 55

Isomers

Qu et al.[129] have identified no fewer than 22 isomeric $[C_3,H_3,O]^+$ cations at the B3-LYP/6-31G(d,p) level of theory. By experiment, Holmes et al. have characterized two of the three most stable isomers: CH_2CHCO^{1+} (**1**), $CHCCHOH^{1+}$ (**2**) and CH_2CCOH^{1+} (**3**).[130]

CH$_2$CHCO	CHCCHOH	CH$_2$CCOH	$\begin{array}{c}\text{HC}\\ \\ \text{HC}\end{array}\!\!>\!\!\text{C}\!-\!\text{OH}$	CH$_2$CCHO
1 (0)	2 (152)	3 (201)	4 (105)	5 (250)

CH$_2$CHOC	CHCHCHO	$\begin{array}{cc}\text{O}\!-\!\text{CH}\\ \\ \text{HC}\!-\!\text{CH}\end{array}$	CH$_2$OCCH	CCHCHOH
6 (220)	7 (250)	8 (247)	9 (322)	10 (334)

CHCH$_2$CO	$\text{HC}\!-\!\text{C}\!<\!\!\begin{array}{c}\text{CH}_2\\ \text{O}\end{array}$	$\begin{array}{c}\text{O}\\ \text{HC}\end{array}\!\!\!<\!\!\!\begin{array}{c}\text{CH}\\ \text{CH}\end{array}$	CH$_3$–C$<\!\!\begin{array}{c}\text{O}\\ \text{C}\end{array}$	CHCCOH$_2$
11 (356)	12 (402)	13 (395)	14 (393)	15 (446)

$\begin{array}{cc}\text{C}\!-\!\text{CH}_2\\ \\ \text{HC}\!-\!\text{O}\end{array}$	CH$_2$COCH	CHCHOCH	CHCHCOH	CCCHOH$_2$
16 (451)	17 (512)	18 (639)	19 (416)	20 (495)

Identification[130]

Metastable ions (1) (derived from CH$_2$=CHCOCH$_3^{1+\bullet}$, loss of CH$_3^{1\bullet}$) exhibit one dissociation channel to produce CH$_2$=CH^{1+}. Ions with structure (2) were obtained from the dissociation of ionized propargyl alcohol and its methyl homolog (CH≡CCH$_2$OH and CH≡CH(CH$_3$)OH) by simple bond cleavage and by protonation of CH≡CCHO. These [C$_3$,H$_3$,O]$^+$ ions undergo metastable loss of CO and C$_2$H$_2$. Ions (1) and (2) can also be distinguished by their CID mass spectra. Holmes et al.[130] were unable to produce pure (3) for identification.

Partial CID Mass Spectra for [C$_3$,H$_3$,O]$^+$ Isomers

						m/z						
Ions	24	25	26	27	29	30	36	37	38	52	53	54
(1)	2	9	36	100	3	—	1	4	3	5	23	37
(2)	29	55	75	100	196	29	31	114	90	27	255	235

Thermochemistry

The $\Delta_f H$ of (1) was determined by EI/AE of methyl vinyl ketone. The measured appearance energy (10.44 ± 0.05 eV)[130] and a calculated (G3) $\Delta_f H$ of methyl vinyl ketone[117] of −111 kJ/mol lead to $\Delta_f H$(1) = 749 ± 5 kJ/mol.

This is lower than 764 kJ/mol, the value obtained by combining the measured IE and calculated $\Delta_f H(CH_2CHCO^{\rceil\bullet})$ (IE = 7.0 eV[24] and $\Delta_f H(CH_2CHCO^{\rceil\bullet})$ = 88.53 kJ/mol (G3MP2B3)[105]). The $\Delta_f H$ values for ions (**2**), (**3**) and (**4**) were estimated only at a low level of theory (LCAO-SCF-MO) to be 833, 882 and 858 kJ/mol, respectively.[130] More reliable relative energies, based on the QCISD/6-31G(d,p)//B3-LYP/6-31G(d,p) calculations of Qu et al.,[129] are shown in the figure in parentheses. Qu et al. also calculated no less than 34 transition states for the interconversion of the isomers along with dissociation product energies. They concluded that ions (**1**)–(**4**) are the four lowest energy isomers that should be thermally stable (i.e., exist in a reasonable potential energy well). They also identified ions (**9**), (**14**), (**15**) and (**20**) as high energy isomers that have sufficient well depths to be potentially thermally stable. Other calculations on selected isomers have been published.[131–133]

3.7.33 $[C_3,H_4,O]^{+\bullet}$ – *m/z* 56

Isomers

Over the years 17 isomeric $[C_3,H_4,O]^{+\bullet}$ cations have been investigated by experiment and by calculations. Of these, the most stable are the methylketene ion (**1**) and the distonic $CH_2CH_2CO^{\rceil+\bullet}$ ion (**2**).

CH_3CHCO 1

CH_2CH_2CO 2

$CH_2{=}C{=}CHOH$ 3

$CH_2{=}CHCH{=}O$ 4

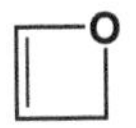

5

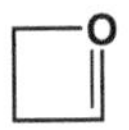

6

CHCHCHOH 7

$(CH_2CH)(HCO)$ 8

$(C_2H_2)(CH_2O)$ 9

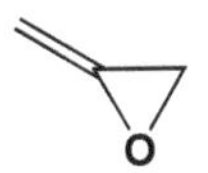

10

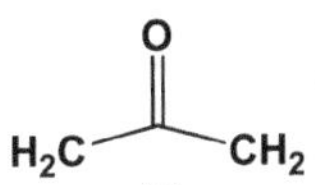

11

CH_2COCH_2 12

$(CH_2CH_2)(OC)$ 13

CH_3OCCH 14

CH_3CCOH 15

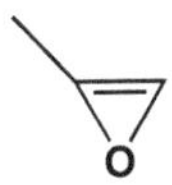

16

CH_3CCHO 17

A likely isomer, ionized cyclopropanone, is the transition state for the interconversion of distonic ions (**2**) and (**12**). The only isomers to have been characterized by experiment are (**1**), (**2**), (**4**), (**10**) and (**14**)–(**16**).

Identification

The CID mass spectrum of ionized methylketene (**1**) has an intense m/z 41 peak due to the loss of the methyl group.[134–136] The distonic ion's (**2**) CID mass spectrum is dominated by m/z 28 $(C_2H_4^{\vert+\bullet})$[135,136] whereas that of the acrolein ion (**4**) has m/z 55 (loss of $H^{\bullet}$) as base peak.[134–136] The CID mass spectrum of ionized methyleneoxirane (**10**) has two intense peaks at m/z 28 and 42 of almost equal intensity.[136] Turecek at al.[137] also obtained the CID mass spectra of ions (**14**)–(**16**). The CID mass spectrum of CH_3OCCH (**14**) contains an intense m/z 41 due to methyl loss but has an intense $H^{\bullet}$ loss peak at m/z 55 distinguishing it from ionized methylketene (**1**).[137] Ion (**15**), with its methyl group, does not dissociate by loss of $CH_3^{\vert\bullet}$, but rather loses $H^{\bullet}$ (m/z 55). Ion (**16**) is also characterized by a dominant loss of $H^{\bullet}$, but expansion of the lower mass region of its CID mass spectrum shows that it also yields m/z 28 and 43 as characteristic peaks.[137]

Thermochemistry

The heats of formation of the $[C_3,H_4,O]^{+\bullet}$ ions are listed below. The only $\Delta_f H$ values that have been experimentally determined are that of methylketene and isomer (**4**). For methylketene, EI/AE measurements place its $\Delta_f H$ at 765 ± 5 kJ/mol[138] while the most recent PI/AE by Traeger yielded a value of 783 kJ/mol.[139] Note that $\Delta_f H(3)$ is an estimate.

Heat of Formation for the $[C_3,H_4,O]^{+\bullet}$ Isomers

Ions	$\Delta_f H$(expt) (kJ/mol)	$\Delta_f H$(theory) (kJ/mol)
(**1**)	783.5 ± 0.3;[139] 765 ± 5[138]	797(G2);[140] 799(DFT)[141]
(**2**)		843.5(G2);[140] 857(DFT)[141]
(**3**)	895 (est)[118]	906.0(G2);[140] 892(DFT)[141]
(**4**)	898[95]	916.3(G2);[140] 904(DFT);[141] 900(MP4)[136]
(**5**)		920.1(G2);[140] 932(DFT)[141]
(**6**)		999.2(G2)[140]
(**7**)		936(DFT)[141]
(**8**)		1065(DFT)[141]
(**9**)		1069(DFT)[141]
(**10**)		931(MP4)[136]
(**11**)		1028(MP4)[136]
(**12**)		1060(MP4)[136]
(**13**)		895(MP4)[136]
(**14**)		1010(MP4)[137]
(**15**)		928(AM1)[137]
(**16**)		970(MP4)[137]
(**17**)		1078(MP4)[137]

3.7.34 $[C_3,H_5,O]^+$ – m/z 57

Isomers

Bouchoux et al. have identified 15 isomeric $[C_3,H_5,O]^+$ cations by computation, albeit at the modest CIPSI/4-31G//HF/4-3G level of theory (CIPSI being an implementation of MP2 theory).[142,143] Only the two lowest energy isomers have been conclusively characterized by experiment, the acylium ion $CH_3CH_2CO^{1+}$ (1) and protonated acrolein, $CH_2=CH=CHOH^{1+}$ (2).

			O—CH₂	
CH₃CH₂CO	CH₂CHCHOH	CH₂CHOCH₂	HC—CH₂	CH₃-C(O)CH₂
1 (0)	2 (57)	3 (149)	4 (157)	5 (203)

CHCHCH₂OH	CH₂CCH₂OH	CH₂CH₂CHO	CH₃CHCHO	CH₃CCHOH
6 (328)	7 (274)	8 (250)	9 (246)	10 (269)

CH₃CHCOH	H₂C,C—OH (H₂C)	CH₂C(OH)CH₂	CH₃C(O)CH₂
11 (173)	12 (203)	13 (161)	14 (303)

Identification

The ions (1) and (2) both exhibit loss of CO in their respective metastable ion mass spectra, but ion (2) has characteristic peaks at m/z 31, 36–38 and 55 in its CID mass spectrum, which are absent in that of (1).[144–148] Numerous studies have been made on the production of ions (1) and (2) from various precursors. In general, most primary alcohols and aldehydes (molecules with O on C1) generate (2) in the ion source and metastably in the first field-free region (FFR) of a mass spectrometer, whereas ketones and esters tend to form (1).[145,147] Zwinselmann and Harrison showed that internal energy content can play a role in the isomer formed in a dissociation. 3-hydroxy-1-pentene ions dissociate to form (2) in the ion source, but the fraction of (1) formed increased as the internal energy of the molecular ions was decreased (they generated ions by methane CI, argon charge-exchange, EI and via metastable ion dissociation).[146]

Thermochemistry

PI/AE measurements of the ion from several precursors lead to $\Delta_f H(1) = 591 \pm 2$ kJ/mol,[149] slightly lower than 604 kJ/mol, the value obtained by combining the measured IE and computed $\Delta_f H$ of $CH_3CH_2CO^{1\bullet}$ (IE = 6.6 eV[24] and $\Delta_f H$ of $CH_3CH_2CO^{1\bullet} = -32.83$ kJ/mol (G3MP2B3)[105]). The $\Delta_f H(2) = 666$ kJ/mol is based on the PA of acrolein (PA = 797 kJ/mol[24] and $\Delta_f H(CH_2=CHC(O)H = -67 \pm 4$ kJ/mol from revised additivity ($C_d-(H)(CO) = +21$ kJ/mol; CO–$(C_d)(H) = -115$ kJ/mol, obtained by a critical review of the available reliable thermochemical data) and $\Delta_f H(H^+) = 1530$ kJ/mol[24]. While the heats of formation of the other isomers could be estimated from the calculated relative energies (listed in parentheses below each structure, see above),[142] this should be done with caution due to the low level of theory employed in the study.

3.7.35 $[C_3,H_6,O]^{+\bullet}$ – m/z 58

Isomers

Eighteen isomeric $[C_3,H_6,O]^{+\bullet}$ cations have been investigated by experiment and by computations.

The two isomers that have received the most attention have been ionized acetone (1) and its enol form (8). It is known that neutral acetone is thermodynamically more stable than its enol form, propen-2-ol.[24] This relative stability is reversed upon ionization with (8) lying 52 kJ/mol lower in energy than (1) (see the table).

Identification

Although ionized acetone (1) and its enol (2) (obtained from the dissociative ionization of alkanones such as 2-hexanone and 4-octanone and 1-cyclobutanol)[150] exhibit similar metastable ion mass spectra (loss of $CH_3^{\bullet}$ and CH_4),[150] ionized acetone does not tautomerize to (8) before dissociation as the required 1,3-hydrogen shift lies too high in energy. Isomer (8), however, has been shown to isomerize to (1) before methyl loss,[151] but isotopic labeling and kinetic energy release experiments demonstrate that this process proceeds nonergodically, one of the very few examples that have been observed for polyatomic organic ions.[152,153] The two ions can be distinguished based on their CID mass spectra insofar as (8) exhibits an intense peak m/z 39 (loss of $H_2 + {}^{\bullet}OH$) but a significantly less intense m/z 15 (CH_3^{1+}).

Isomers with a CCCO skeleton include the conventional ions (2), (5), (9) and (10) (obtained by direct ionization of the corresponding molecules) plus the distonic isomer (12) (produced essentially pure from the dissociative ionization of α-hydroxy-γ-butyrolactone by loss of CO_2)[154] and carbene (16) (tentatively assigned as the fragment ion generated by the loss of HC(O) OCH_3 from ionized 2-methoxymethyl propionate).[155] These ions all show H$^{\bullet}$ loss in their metastable ion mass spectra.[150,155] While neutral E- and Z-propen-1-ol are distinct molecules with different heats of formation (-169 and -174 kJ/mol,[24] respectively), they have different ionization energies resulting in a single ion heat of formation. So ions (9) and (10) are indistinguishable, ionization resulting in free rotation about the formal double bond and will be referred to hereafter as (9)/(10).[148] Allyl ethers may initially generate (5), ionized propen-2-ol, but isomerize to (9)/(10) before CID.[156] Isomers (2), (9)/(10) and (12) exhibit similar metastable ion mass spectra (loss of $H^{\bullet}$) although that of the distonic ion (12) has a weak signal at m/z 30.[154] The CID mass spectra of (9)/(10) and (12) exhibit an m/z 29:31 ratio of 2:1 while that for (2) is 10:1.[154,156] Most substituted propanols dissociate to form (9)/(10).[156]

The closed ring isomers consist of (3), (4), (7) and (18). Ions (3), (4) and (7) have stable neutral counterparts while (18) is an ionized carbene that has only been studied in computations by Bouma et al.[157] Early CID results assumed that ionizing propene oxide generated (3), but the isotopic labeling results of Turecek and McLafferty[158] showed that (3) undergoes a ring-opening isomerization to ionized vinyl methyl ether (6) before dissociation. The CID mass spectra of $[C_3,H_6,O]^{+\bullet}$ ions generated from a variety of ionized cyclic ethers and terminal epoxides were found to be consistent with the formation of isomer (3) or (6).[156] The best distinguishing feature of (3) and (4) is that the metastable ion mass spectrum of (3) is dominated by $CH_3^{\ ]\bullet}$ loss while that of (4) is dominated by $H^{\bullet}$ loss.

Isomer (11) has been generated by the dissociative ionization of 1,4-dioxane. It was distinguished from (6) by its ability to charge transfer to acetonitrile.[161]

CID Mass Spectra for the $[C_3,H_6,O]^{+\bullet}$ Isomers[a]

Ions	15	26	27	28	29	30	31	39	40	41	42	43	55	56	57
(1)	15	6.0	15	4.2	13	(0.8)	0.9	9.9	4.2	13	(105)	(861)	1.0	0.2	(13)
(2)	1.2	3.8	11	16	37	(13)	4.5	5.3	2.2	2.4	(3.2)	(2.2)	6.1	4.6	(648)
(3)	6.8	6.6	17	33	19	(8.4)	12	0.5	0.5	1.4	(16)	(51)	0.4	0.3	(35)
(4)	0.6	4.5	9.2	53	23	(37)	4.7	0.9	0.4	0.3	(1.0)	(0.9)	1.4	0.7	(298)
(6)	8.8	6.3	22	21	18	(3.7)	11		0.4	0.5	2.1	(21)	(31)	1.0	1.3	(18)
(8)[b]	4.2	2.8	8.5	3.1	16	(2.7)	13		20		5.5	8.6	(21)	(109)	4.6	0.3	(35)
(9)/(10)[b,c]	1.1	4.8	9.2	7.3	25	(11)	13	11		3.4	2.3	(3.3)	93.4)	9.3	4.7	(197)
(12)[c,d]	20	44	55	45	100	46	47	47	15	10	15	7	24	20	n/a
(16)[e]	n/a	—	18	12	16	35	10	13	—	—	—	—	—	—	100

[a]All spectra are from Ref. [156] unless otherwise stated. 3.9 kV accelerating potential, He target gas, 35% beam transmission. Values are normalized to the total ion abundance ($= 100$), excluding peaks with metastable ion (in parentheses).
[b]From Ref. [154], 8 kV accelerating potential, O_2 target gas, unknown % beam transmission. Values are relative to m/z 29.
[c]See also Ref. [159].
[d]See also Ref. [160].
[e]From Ref. [155].

Relative Energies for the $[C_3,H_6,O]^{+\bullet}$ Isomers

Ions	Δ_fH(expt) (kJ/mol)	Relative Energy[157] RHF/4-31G//RHF/STO-3G[a]
(1)	717[b]	20
(2)	772[b]	93
(3)	891[b]	207
(4)	851[b]	157
(5)	809,[b] 847[c]	108
(6)		54
(7)		229
(8)	654,[b] 670 ± 20,[b] 665 ± 10[d]	0
(9)	665[b]	13
(10)	665,[b] 665 ± 10[d]	15
(11)		148
(12)	757[e]	
(13)		111
(14)		110
(15)		112
(16)	805[f]	123
(17)		174
(18)		247

[a]Energies relative to isomer (8) in kJ/mol.
[b]Values obtained by combining neutral Δ_fH and quoted IE values.[24]
[c]Estimated value.[162]
[d]Average values from EI/AE measurements of a variety of saturated ketones and aldehydes.[163]
[e]PI/AE of $12 + H_2CO$ from $CH_3OCH_2CH_2CHO$ (est. $\Delta_fH = -319.1$ kJ/mol).[160]
[f]Calculated value (G3).[164]

Thermochemistry

3.7.36 $[C_3,H_7,O]^+$ – m/z 59

Isomers

The $[C_3,H_7,O]^+$ cations, like their $[C_3,H_6,O]^{+\bullet}$ counterparts, have been one of the most extensively studied series of isomers with work dating back to the very beginnings of modern gas-phase ion chemistry. Many of the isomers are simply protonated analogs of stable neutral $[C_3,H_6,O]$ molecules while others are unique. Although 20 isomers have been postulated, relatively few have been thoroughly characterized by mass spectrometry. Most work has centered on isomers (1)–(4), with the remaining receiving only tangential treatment. Notable for their apparent instability are the oxy cations (18) and (19). While the corresponding radicals are well known, upon ionization they appear to isomerize (to (2) and (1), respectively) without an activation barrier.[165]

OH
$\big;\!\!=\!\!\bigg/$ (1) $\diagup\!\!\diagup^{OH}$ (2)   CH$_3$CHOCH$_3$ (3)   CH$_3$CH$_2$OCH$_2$ (4)   (5)

CH$_3$CHCH$_2$OH (6) $\square$—OH (7)   CH$_2$CH$_2$CH$_2$OH (8) $\triangle$—CH$_2$ / OH (9) $\diagup\!\!\diagup^{OH_2}$ (10)

$\bigg/\!\!=\!\!\bigg/^{OH_2}$ (11) $\diagup\!\!=\!\!\diagup^{OH_2}$ (12)   (13)   CH$_3$ (14)   CH$_2$CH$_2$OCH$_3$ (15)

(16)   OH / H$_3$C $^{\diagdown}$CH$_2$ (17)   CH$_3$CH$_2$CH$_2$O (18) $\bigg\rangle\!\!-\!\!O$ (19) $\|$ ---- HOCH$_2$ (20)

Identification

Ab initio calculations for these isomers[166,167] and CID mass spectra[165] have suggested that while equilibrium structures for (6), (8), (9), (15), (20) can be located, the barriers to their isomerization are so small that it is unlikely that they could be isolated (6→2, 8→6, 9→2, 15→3, 14 and 20→4). This is consistent with (8) and (15) being primary carbocations and therefore unstable toward rearrangement. Rather, these structures are intermediates in the interconversion of the other isomers. From their survey of CID mass spectra, McLafferty and Sakai concluded that most secondary and tertiary alcohols dissociate to form (1) and that ethers (other than diethyl ether) and hydroxy ethers generate (3).[165] Protonating propylene oxide and oxetane, which should yield (5) and (7) respectively, generate ion (2) instead.[165,168] Some (7) has been observed from this process when less exothermic proton-transfer reactions are used to protonate oxetane.[168]

Isomers (1)–(4) have significantly different MI and CID mass spectra. All four metastable ions lose H$_2$O and C$_2$H$_4$ while (3) has the additional channel of H$_2$CO loss.[169] Isomers (2) and (4) (generated from dissociative ionization of *n*-propanol and diethylether, respectively) exhibit a predominant loss of water while (1) (formed by H$^{\bullet}$ loss from ionized isopropanol) prefers to lose ethene.[170] These differences have been used to probe the internal energy dependence of the unimolecular fragmentations.[171,172] The CID mass spectra for the four isomers are distinct. In addition to the metastable ion components described above, (1) displays an intense m/z 43, (2) is dominated by H$^{\bullet}$ loss, (3) contains a unique m/z 33 peak and (4) has m/z 29 as its dominant CID channel.[165] While the four ions have unique CID mass spectra, their respective dissociation channels do involve their interconverting to common structures. Bowen and coworkers proposed a detailed potential energy surface for the main [C$_3$,H$_7$,O]$^+$ cations based on experimental appearance energy values and isotopic labeling experiments.[173,174]

This work was updated by Holmes et al.[175] and more recent ab initio calculations by Bouchoux et al.[167] and Hudson and McAdoo[176] have produced the most up-to-date surface for the chemistry of $[C_3,H_7,O]^+$ cations.[167]

Thermochemistry

There is generally good agreement between the thermochemical data obtained from PA values and appearance energy measurements. Where available, the experimental and calculated $\Delta_f H$ values (obtained at high level of theory) are also in good agreement. In general, the relative energies calculated by Nobes and Radom[166] agree with those obtained from the experimental $\Delta_f H$ values, the only exception being ion (**10**), for which the experimental value was estimated by Bowen and Williams.[177]

Heat of Formation and Relative Energies[a] for the $[C_3,H_7,O]^+$ Isomers

Ions	$\Delta_f H$(expt)	$\Delta_f H$(theory)[178] G2(MP2), CBS-Q	Relative Energy[166] MP3/6-31G*//HF/3-21G
(1)	499.5[b], 502[c]	496, 502	0
(2)	555.3,[b] 552 ± 8,[c] 561[d]	559, 568	57 (0)[e]
(3)	556 ± 4[c]	561, 571	51
(4)	602 ± 13[c]	611, 622	99 (48)
(5)	632[b]		143
(6)	674[f]		181[g] (116)
(7)	648.2[b]		145 (66)
(8)	720[e]		264 (208)
(9)			199 (66)
(10)	602[f]		146 (74)
(11)			145[g]
(12)			156[g]
(13)			169[g]
(14)			172
(15)			299
(16)			177
(17)			
(18)			
(19)	802[h]		
(20)			(137)

[a]In kJ/mol.
[b]Based on PA of the corresponding unprotonated molecules and their $\Delta_f H$ values.[24]
[c]EI/AE measurements.[179]
[d]Values in parentheses at MP2/6-31G*//HF/6-31G* are referenced to isomer (**2**).[167]
[e]PI/AE measurements.[180]
[f]Estimated.[177]
[g]Lowest energy conformation.
[h]Obtained by combining the measured IE and calculated $\Delta_f H$ of $(CH_3)_2C(H)O^{1\bullet}$ (IE = 9.20 ± 0.05 eV[24] and $\Delta_f H((CH_3)_2C(H)O^{1\bullet}) = -85.19$ kJ/mol (G3MP2B3)[105]).

3.7.37 $[C_3,H_8,O]^{+\bullet} - m/z$ 60

Isomers

Eleven isomers have been postulated in the study of the ion chemistry of the $[C_3,H_8,O]^{+\bullet}$ cations.

Identification

Ionized *n*-propanol (**1**), *iso*-propanol (**2**) and methyl ethyl ether (**3**) are all equilibrium structures. All three can be distinguished by their MI and CID mass spectra.[181,182] Ion (**1**) predominantly loses H$^{\bullet}$, H$_2$O (m/z 42) and C$_2$H$_5$$^{\bullet}$ (m/z 31) whereas (**2**) prefers to lose a methyl radical (m/z 45) and a molecule of methane (m/z 44). The ionized ether (**3**) is notable for the small m/z 31 in its CID mass spectrum and a significant m/z 29. A complete discussion of the *n*-propanol ion's chemistry can also be found in Ref. [183].

The distonic ion (**4**) can be generated by the dissociative ionization of 1,4-butanediol and this ion shares the strong water loss channel with (**1**), but does not display m/z 31.[182] Unlike (**1**), the NR mass spectrum of (**4**) exhibits a peak due to ionized water (m/z 18).[184] Isomer (**5**) has also been generated by the loss of formaldehyde from ionized 2-methylpropane-1,3-diol (it exhibits loss of water with a $T_{0.5}$ of only 0.6 meV),[182] but based on this $T_{0.5}$ and calculated geometries, isomers (**5**)–(**7**) can best be described as complexes between propene and water. Furthermore, it is expected that the barriers to their interconversion will be low and therefore it is unlikely that they can be distinguished based on their mass spectra.

Isomer (**8**), the complex between water and ionized cyclopropane (see section on $[C_3,H_6]^{+\bullet}$), was originally proposed as the intermediate in the water loss of (**1**),[185,186] but is unlikely to be made independently due to a small barrier for isomerization to (**4**).[43] Ion (**1**) dissociates by loss of water to form a mixture of ionized cyclopropane and ionized propene by first isomerizing to (**4**).[43,182,187] Ion (**4**) then can either dissociate to ionized cyclopropane via (**8**) or isomerize to (**6**) before forming ionized propene.[43]

The distonic ion (**9**) was prepared from the dissociative ionization of dimethoxyethane and is characterized by three metastable processes, m/z

42, 31 and 28 (the only isomer having a metastable channel forming m/z 28) and its CID mass spectrum exhibits a strong series of peaks at m/z 31–26, unique to this group of isomers.[181] The ylid ion (10) can be made from ionized 2-ethoxy ethanol and is characterized by CID processes forming m/z 59, 45 and 42.[188] It has been argued that (10) can rearrange to (1) without the need for an intervening complex (11).[189,190]

Thermochemistry

The thermochemistry of this class of ions is given below. There are discrepancies concerning the $\Delta_f H(1)$ that stem from the use of the experimental IE(1-propanol) = 10.22 eV ± 0.06,[24] which may not be the adiabatic value. Calculated values for the IE and $\Delta_f H$[191] are more in line with the fitted value from PEPICO data.[186] Experiments, usually EI/AE studies, place the $\Delta_f H$ values for (1)–(5), (9) and (10) in a very narrow range of 20 kJ/mol. The calculated value for (4) (688 kJ/mol) is significantly lower than the EI/AE value (714 kJ/mol), which points to the need for further study. It should be noted that the G2[MP2,SVP] level of theory is significantly less reliable than the higher homogs G2 and G2[MP2]. The largest discrepancy in the table is for ion (5), which is predicted to have a $\Delta_f H$ value of approximately 660 kJ/mol (based on a CBS-QB3 energy relative to (1) of −50 kJ/mol), while the experimental EI/AE value is 721 kJ/mol.[182]

Heat of Formation and Relative Energies for the $[C_3,H_8,O]^{+\bullet}$ Isomers

Ions	$\Delta_f H$ (kJ/mol)	Relative Energies (kJ/mol) G2[MP2,SVP][192] (CBS-QB3)[183]
(1)	731 ± 6,[a] 709,[b] 703,[c] 713 ± 4[d]	0
(2)	708 ± 2[a]	
(3)	721 ± 7[a]	
(4)	714 ± 5,[e] 688 ± 4[d]	−18
(5)	721 ± 5[f]	(−50)
(6)		−48
(7)		(−38)
(8)		7 (0)
(9)	712 ± 10[g]	
(10)	724 ± 10[h]	
(11)		

[a]$\Delta_f H$ values based on neutral molecule $\Delta_f H$ value and experimental IE.[24]
[b]$\Delta_f H$ values based on neutral molecule $\Delta_f H$ value and CBS-QB3 IE.
[c]$\Delta_f H$ values based on fit to k(E) vs E data with RRKM theory.[186]
[d]G2[MP2,SVP] values.[191]
[e]EI/AE measurement from 1,4-butanediol.[182]
[f]EI/AE measurement from 2-methyl-1,3-propanediol.[182]
[g]EI/AE measurement from dimethoxyethane.[189]
[h]Estimated.[189]

3.7.38 $[C_3,H_9,O]^+ - m/z$ 61

Isomers

The $[C_3,H_9,O]^+$ cations consist of the protonated analogs of $[C_3,H_8,O]$ neutral molecules n-propanol (**1**), isopropanol (**2**) and methyl ethyl ether (**3**) with at least one ion–molecule complex between the isopropyl cation, $i\text{-}C_3H_7^{1+}$, and water (**4**).

Identification

Protonated isopropanol (**2**) undergoes loss of water to form the $i\text{-}C_3H_7^{1+}$ cation,[193] a process that involves significant but incomplete hydrogen scrambling. It appears from CID mass spectrometry that when $i\text{-}C_3H_7^{1+}$ and water are brought together to form a complex (which would nominally be (**4**)), the resulting ion is indistinguishable from (**2**),[194] thus calling into question an earlier study that proposed that (**4**) is a unique species on the potential energy surface of the $[C_3,H_9,O]^+$ cations.[195] The CID mass spectra of (**1**) and (**2**) are distinguishable based on the propensity for loss of methane (m/z 45, preferred by (**2**)) and ethane (m/z 31, preferred by (**1**)).

Thermochemistry

The heat of formation values for the three isomers (**1**)–(**3**) are deduced from the PA of the corresponding neutral molecule and are $\Delta_f H(\mathbf{1}) = 488$ kJ/mol, $\Delta_f H(\mathbf{2}) = 464$ kJ/mol and $\Delta_f H(\mathbf{3}) = 505$ kJ/mol (using the following ancillary data[24]: $PA(CH_3CH_2CH_2OH) = 786.5$ kJ/mol, $\Delta_f H(CH_3CH_2CH_2OH) = -255.6$, $PA((CH_3)_2CHOH) = 793$ kJ/mol, $\Delta_f H((CH_3)_2CHOH) = -272.8$, $PA(CH_3O CH_2CH_3) = 808.6$ kJ/mol and $\Delta_f H(CH_3OCH_2CH_3) = -216.4$ kJ/mol).

3.7.39 $[C_3,O_2]^{+\bullet} - m/z$ 68[196]

Ionized carbon suboxide is generated by electron impact of the products resulting from the pyrolysis of malonic acid $(HOC(O)CH_2C(O)OH)$ in the presence of P_2O_5. Its MI mass spectrum shows a single peak at m/z 40 $(C_2O^{1+\bullet})$ with a very small kinetic energy release ($T_{0.5} = 1.8$ meV). Its CID mass spectrum contained the structure-characteristic fragment peaks m/z 52 $(C_3O^{1+\bullet})$, 36 (C_3^{1+}), 28 $(CO^{1+\bullet})$, 24 $(C_2^{1+\bullet})$ and 12 (C^+).

Combining the neutral molecule's heat of formation and ionization energy leads to $\Delta_f H(C_3O_2^{1+\bullet}) = 929$ kJ/mol $(\Delta_f H(C_3O_2) = -93.64$ kJ/mol and IE $= 10.60$ eV).[24]

3.7.40 $[C_3,H,O_2]^+ - m/z$ 69

Isomers

The $HOCCCO^{1+}$ and $OCC(H)CO^{1+}$ ions, resulting from the protonation of carbon suboxide at the central carbon and oxygen atoms, respectively, have been generated and characterized. A third species, obtained from the protonation of one of the equivalent carbon atoms, is also considered by theory.

Identification[197]

Chemical ionization of C_3O_2 neutrals, generated by flash vacuum pyrolysis of diacetyltartaric acid (cy–C(O)C(H)(C(O)CH$_3$)CH(C(O)CH$_3$)C(O)O–), generates an ion with a different CID mass spectra than the ion obtained by dissociative ionization of squaric acid (cy–C(OH)C(O)C(O)C(OH)–). The former ion, assigned as carbon protonated C_3O_2 (OCC(H)CO^{1+}), shows structure-characteristic peaks at m/z 25 (C_2H^{1+}), 28 (CO$^{1+\bullet}$) and 53 (loss of atomic oxygen), whereas the latter ion, assigned to be the oxygen protonated analog HOCCCO^{1+}, shows intense m/z 24 ($C_2^{1+\bullet}$), 29 ([H,C,O]$^{1+}$) and 52 (loss of HO$^\bullet$), consistent with the proposed connectivity.

It is noteworthy that OCC(H)CO^{1+} was previously proposed to be formed by electron impact of some aryl-substituted β-diketones[198] and also that the dissociative ionization of the hydroxymethylene Meldrum's acid (cy–C(O)C(C(H)OH)C(O)OC(CH$_3$)$_2$O–) leads to a mixture of both ions, as evident from its CID mass spectrum, which is a superpostition of the spectra for OCC(H)CO^{1+} and HOCCCO^{1+}.

Thermochemistry[197]

Calculations predict the central carbon as being the favored protonation site. The proton affinity of carbon suboxide has been measured as 791 kJ/mol, consistent with a computed value of 789 kJ/mol (MP4(SDTQ)/6-31G**//6-31G** + ZPE(6-31G**)) and combined with $\Delta_fH(C_3O_2) = -93.64$ kJ/mol,[24] leads to $\Delta_fH(OCC(H)CO^{1+}) = 645$ kJ/mol; the isomer HOCCCO^{1+} was computed to lie 132 kJ/mol higher in energy. Note that a third species, resulting from protonation at one of the two equivalent carbons (epoxide structure), is calculated to be significantly higher in energy, 348 kJ/mol above OCC(H)CO^{1+}.

3.7.41 $[C_3,H_4,O_2]^{+\bullet}$ – m/z 72

Isomers

Four $[C_3,H_4,O_2]^{+\bullet}$ isomers have been postulated (shown below).

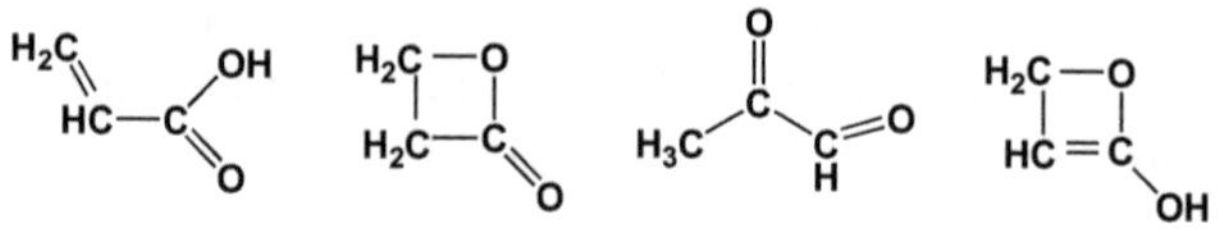

The tautomerization of the neutral analog of (**2**) (to its enol, the neutral analog of (**4**)) has been the subject of computations by Turecek et al.[199]

Thermochemistry

The Δ_fH of (**1**)–(**3**) has been derived by combining the heat of formation of the molecule with the measured IE values[24] and give $\Delta_fH(\mathbf{1}) = 692$ (or 686) kJ/mol,

$\Delta_f H(2) = 653$ and $\Delta_f H(3) = 655$ kJ/mol ($\Delta_f H(CH_2 = CHC(O)OH) = -330.7 \pm$ 4.2 or -336.9 ± 2.3 kJ/mol, IE($CH_2 = CHC(O)OH) = 10.60$ eV, $\Delta_f H$(cy–C(H)$_2$C(H)$_2$OC(O)–) $= -282.9 \pm 0.84$ kJ/mol, IE(cy–C(H)$_2$C(H)$_2$OC(O)–) $= 9.70 \pm 0.01$ eV, $\Delta_f H(CH_3C(O)C(O)H) = -271 \pm 5$ kJ/mol and IE(CH$_3$C(O)C(O)H) $= 9.60 \pm 0.06$ eV).

3.7.42 $[C_3,H_5,O_2]^+$ – m/z 73

Isomers

Fifteen $[C_3,H_5,O_2]^+$ isomers have been identified by computation.[200]

Of these, ions (1), (3)–(5), (7) and (11)–(14) have all been generated and characterized in the gas phase.[200]

Identification[200]

The metastable ion mass spectra of ions (1), (3), (4) and (7) are indistinguishable (all ions exhibit the same losses of H_2O, H_2CO, CO and CO_2) indicating that these ions can interconvert freely on the microsecond timescale. These ions were respectively generated by protonation of β-propiolactone, protonation of propenoic acid, by loss of I$^•$ from 2-iodopropanoic acid, and by loss of CH$_3$O$^•$ from HOCH$_2$CH$_2$C(O)OCH$_3$$^{1+•}$. The metastable ion mass spectrum of (5) (generated by protonation of acrylic acid) is unique in that it shows only the loss of water. Suh et al. calculated a high barrier for the isomerization of this ion to (4). The unique CID mass spectra of (4), (5) and (7) provide evidence that these three ions are stable species on the potential energy surface and can be generated independently. The CID mass spectrum of (4) shows a unique peak at m/z 56 due to $^\bullet$OH loss, while (7) does not lose $^\bullet$OH but rather forms an intense m/z 31 ($^+$CH$_2$OH). Again, (5) is unique in that it is the only isomer that forms m/z 46, proposed to be C(OH)$_2^{\cdot+}$. Ions (8) and (9) were originally investigated as possible intermediates in the dissociation of ions (1)–(7), but calculations at the MP3/6-31G*//HF/4-31G level of theory failed to find an equilibrium structure corresponding to (8) and the energy of (9) indicated that it likely is not involved in the ion chemistry of (1)–(7). There is only a small barrier separating (3) and (4) and so (3) is unlikely to be generated as an observable ion in the gas phase.

Ions (12) (produced by loss of I$^\bullet$ from ionized ICH$_2$C(O)OCH$_3$) and (13) (made by CH$_3^{\mid\bullet}$ loss from ethylacetate) exhibit unique mass spectra, indicating that the two structures do not interconvert. (12) has dished peaks in its metastable ion mass spectrum due to losses of CO and H$_2$CO, whereas (13) was found to dissociate to CH$_3$CO$^{\mid+}$ + H$_2$CO only via a complex, (15). Ion (14) was made from ethoxy radical loss from diethylcarbonate and its metastable ion mass spectrum exhibits only loss of CO. Note that isomer (11) was proposed as an intermediate in the dissociative ionization of ICH$_2$C(O)OCH$_3$, which produces ion (12).

Heat of Formation and Relative Energies[a] of the [C$_3$,H$_5$,O$_2$]$^+$ Isomers

Ions	$\Delta_f H$(expt)[b]	Relative Energy MP3/6-31G*//HF/4-31G
(1)		23
(2)		39
(3)	473	166
(4)	530	153
(5)	380[c]	0
(6)		50
(7)	459	62
(8)		
(9)	542	154
(10)		
(11)	559	169
(12)		
(13)	436	42
(14)		53
(15)		55

[a]In kJ/mol.
[b]Unless otherwise stated. Values from experiments are based on EI/AE appearance energy measurements combined with estimated $\Delta_f H$ and $\Delta_{acid} H$ values (Suh et al.[200]).
[c]This value is based on an extrapolated gas-phase basicity for the neutral acid (Bouchoux et al.[201]).

Thermochemistry

3.7.43 $[C_3,H_6,O_2]^{+\bullet} - m/z\ 74$

Isomers

Twenty-three $[C_3,H_6,O_2]^{+\bullet}$ isomers have been either characterized by experiment or investigated by theory.

Identification

The table given below lists all of the available CID mass spectra. Ionized propanoic acid (**1**) has been shown by isotopic labeling studies to isomerize with (**11**) and (**12**) before dissociation (primarily H$^{\bullet}$ loss).[202] The barriers to their interconversion lie below the thresholds for dissociation.[203] The other main dissociation channel is water loss to form m/z 56 (ratio m/z 56:73 is 0.11:1).[204] This ratio increases to 0.30:1 in ion (**3**) and 0.17:1 in the diol (**11**).[204] Metastable ions (**3**) also lose formic acid.[205]

Considerably more attention has been paid to ionized methylacetate (**2**) and its enol isomer (**6**). The tautomerization of the two structures was shown, again from isotopic labeling studies, to take place by two sequential 1,4-hydrogen shift reactions, rather than the more direct 1,3-hydrogen shift.[206] Ions (**2**), (**5**) and (**6**) dissociate by a common loss of [C,H$_3$,O]$^{\bullet}$ to form the acetyl cation, CH$_3$CO^{1+}. Rather than generate exclusively CH$_3$O$^{\bullet}$, CIDI mass spectrometry studies have shown that the radical produced from (2) is a mixture of $CH_3O^{\bullet}$ and $^{\bullet}CH_2OH$, with different groups observing different proportions of the two isomers.[207–210] To form the $^{\bullet}CH_2OH$ radical, (2) isomerizes to (5), a distonic ion. Ion (5) can then either undergo a 1,4-H shift to complete the enolization to (6) or isomerize further to complex (7), which then dissociates via an avoided crossing with an excited state having the correct electron distribution.[211–213]

Ionized hydroxy acetone (9) is also a unique structure and does not participate in the chemistry of (2), (5)–(7).[214] Ion (9) exhibits metastable losses of $^{\bullet}CH_2OH$ and $HCO^{]\bullet}$, the latter being the most abundant, which is unique to this ion.[205] The reactivity of (9) shows a tendency to transfer acetyl radicals to neutral ketones, aldehydes and alcohols, which initially suggested that it isomerized to (7).[215] However, labeling studies indicated that this rearrangement does not take place.[214] Ion (9) dissociates via (18) and (19), but not (7).[214]

Ion (14) has an NR mass spectrum consistent with a complex between vinyl alcohol and formaldehyde.[214] Ion (20), ionized glycidol, has an abundant peak due to the loss of water in its metastable ion mass spectrum.[214] Ionized dimethoxy carbene (23) was generated from a substituted oxadiazoline and has a unique CID mass spectrum containing essentially a single peak at m/z 59 corresponding to methyl radical loss (with only minor peaks in the m/z 15–44 region).[216]

The metastable ion mass spectrum of (10) has, in addition to loss of $^{\bullet}CH_2OH$ and $HCO^{]\bullet}$, a strong H atom loss signal, distinguishing it from the other isomers.[212]

CID Mass Spectra for Some $[C_3,H_6,O_2]^{+\bullet}$ Isomers[a]

Ions							m/z					
	13	14	15	26	27	28	29	30	31	37	39	40
(1)[b]				22	53	37	32	25	7			
(2)	2.7	9.8	20	6.3	17	13	62	19	100	<1	<1	4.5
(3)[c]				3	6	61	6	3	1			
(5)	1.3	5.1	13	5.1	9.6	12	88	19	100	<1	1.3	1.3
(6)	2.7	12	23	8.1	16	12	62	22	100	<1	1.4	5.4
(9)	2.2	8.1	14	12	32	31	76	29	100	2.9	4.4	2.1
(10)	1	3.1	9.4	5.2	8.3	6.3	45	15	100	<1	3.1	3.7

Ions							m/z							
	41	42	43	44	45	47	53	55	56	57	58	59	60	73
(1)[b]					60	3	7	47	100					
(2)	26	266	2089	87	85	<1	<1	1	1	1.8	22	38	9.8	1.8
(3)[c]	—	0.6	3	3	10	10			56					100
(5)	13	100	1159	292	126	<1	<1	5.8	7.1	3.2	7.7	19	3.9	9.6
(6)	35	386	2008	70	99	<1	<1	1.4	2.7	6.8	72	30	26	9.5
(9)	11	66	2344	19	646	<1	1	4.2	2.1	3.1	14	<1	<1	2.1
(10)	20	143	1676	397	148	5.9	5.9	60	64	3.7	13	11	5.9	35

[a]Data are from Ref. [212] unless otherwise stated.
[b]From Ref. [213]
[c]From Ref. [204].

Thermochemistry

Heat of Formation and Relative Energies for the $[C_3,H_6,O_2]^{+\bullet}$

		Relative Energy (theory) (kJ/mol)			
Ions	$\Delta_f H$ (expt) (kJ/mol)	QCISD(T)/6-311G(d) //QCISD/6-31G(d)[a]	MP4/6-31G* //MP2/6-31G*[b]	MP2/6-31G** //HF/6-31G*[c]	MP3/6-31G* //HF/4-31G[d]
(1)		58		0	
(2)	579,[e] 577[f]		0	−35	
(3)	626[e]				
(4)	653[e]				
(5)	527,[g] 543[h]		−39	−95	
(6)	495,[g] 477[f]		−61		
(7)	577[i]		10	−43	−8
(8)			10		
(9)	577[i]				0
(10)	560[j]				
(11)		−53			
(12)		0			
(13)	548[i]			−56	−29
(14)	536[i]			−73	−42
(15)				36	
(16)	510[i]			−93	−71
(17)				−172	
(18)					
(19)					
(20)	736[i]				
(21)	686[i]				
(22)	619[i]				52
(23)					

[a]Ref. [203].
[b]Ref. [207].
[c]Ref. [215].
[d]Ref. [214].
[e]Ref. [24].
[f]From EI/AE measurements.[217]
[g]Values based on fits to log(k) vs E curves from PEPICO experiments.[207]
[h]From EI/AE measurements.[213]
[i]Estimated values.[214]
[j]Estimated values.[212]

3.7.44 $[C_3,H_7,O_2]^+$ – m/z 75

Isomers

Two $[C_3,H_7,O_2]^+$ isomers have been characterized: namely $CH_3OCHOCH_3^{]+}$ (**1**) and the proton-bound complex of acetaldehyde and formaldehyde, $[CH_3CHO\cdots H\cdots OCH_2]^+$ (**2**).[218]

Identification

The two ions exhibit similar metastable ion mass spectra but different CID characteristics. The CID mass spectrum of (**1**) is dominated by H$^\bullet$ loss (m/z 74) and CO loss (m/z 47) whereas that for ion (**2**) is dominated by m/z 45

(H_2CO loss) as expected from its structure and relative proton affinities of acetaldehyde and formaldehyde (768.5 and 712.9 kJ/mol, respectively).[24] Ion (**2**) also exhibits H_2 loss, which is absent in the mass spectrum of (**1**).

3.7.45 $[C_3,H_8,O_2]^{+\bullet} - m/z\ 76$

Isomers

In all, 16 $[C_3,H_8,O_2]^{+\bullet}$ isomers have been either characterized by experiment or investigated by computational chemistry. Most are ion–molecule complexes that play a role in the dissociation of the covalently bound isomers.

Identification

Most of the ion–molecule complexes and distonic ions ((**9**), (**10**) and (**12**) listed above have been invoked in the unimolecular dissociation reactions of metastable ionized 1,2-propanediol (**4**), which has the unusually large number of *five* competing dissociation channels.[219,220] The CID mass spectra for isomers (**3**) and (**4**) (made by the loss of ethene from 4-methoxy-1-butanol) differ most significantly in the observed ratio of m/z 44:43 (2:1 for (**3**) and 0.8:1 for (**4**)).[220] No metastable ion mass spectrum can be obtained for (**1**) due to an extremely low intensity molecular ion.[221]

It is noteworthy that the MI mass spectrum of isomer (**4**) is most unusual, with no less than five competing dissociations:[222]

$$[CH_2CHOH \cdots HOCH_3]^{+\bullet} \rightarrow [C_2,H_5,O_2]^+ + CH_3^{\rbrack\bullet}$$
$$\rightarrow [C_3,H_6,O]^{+\bullet} + H_2O$$
$$\rightarrow CH_2CHOH^{\rbrack+\bullet} + CH_3OH$$
$$\rightarrow [C_2,H_3,O]^+ + H_2O + CH_3^{\rbrack\bullet}$$
$$\rightarrow CH_3^+OH_2 + [C_2,H_3,O]^\bullet$$

The kinetic energy release for the metastable reaction (3.3) is small (3 meV),[222] typical of an ion–molecule complex. The identity of the ionic fragment was determined from its CID characteristics to be ionized vinyl alcohol.[222] The relative abundances of the five peaks depend on the timescale of the observations, with (3.2) dominant at the shortest times. The possible participation of the hemi-acetal ion $CH_3OCH(OH)CH_3^{\rbrack+\bullet}$ was rejected[222] but its participation on the potential energy surface for this ion's fragmentations cannot be entirely ruled out. Note that the product energies for reactions (3.3)–(3.5) are almost identical, assuming that the $[C_2,H_3,O]$ species is an acetyl ion or radical. The structures of the $[C_2,H_5,O_2]^+$ and $[C_3,H_6,O]^{+\bullet}$ ions were not pursued.

Thermochemistry

Heat of Formation and Relative Energies for the $[C_3,H_8,O_2]^{+\bullet}$ Isomers

Ions	$\Delta_f H$(expt) (kJ/mol)	Relative Energy[a] (kJ/mol)
(1)	617[b]	
(2)	600[b]	
(3)	468 ± 5,[c] 482[d]	−59
(4)		0
(5)		−21
(6)		0
(7)		
(8)		
(9)		33
(10)		32
(11)		13
(12)		13
(13)		
(14)		
(15)		
(16)		−42

[a]Based on neutral molecule $\Delta_f H$ and IE values.[24]
[b]QCISD(T)/6-31G*//B3-LYP/6-31G * values.[219]
[c]Measured from the AE of the m/z 76 ion from $CH_3O(CH_2)_4OH$; AE = 9.64 ± 0.05 eV,[222] $\Delta_f H(C_5H_{12}O_2) = -410$ kJ/mol (by additivity),[123] and $\Delta_f H(C_2H_4) = 52.5$ kJ/mol.[24]
[d]From the ab initio binding energy of 84 kJ/mol[223] and $\Delta_f H(CH_2CHOH^{\rbrack+\bullet}) = 768 \pm 5$ kJ/mol[224] and $\Delta_f H(CH_3OH) = -201$ kJ/mol.[24]

3.7.46 $[C_3,H_9,O_2]^+$ – m/z 77

Isomers

Work on $[C_3,H_9,O_2]^+$ isomers stems largely from considering the site of protonation of appropriately selected bifunctional molecules.

Identification

Ion (**1**) is protonated 2-methoxy-1-ethanol and both experiment and calculations predict that the proton is stabilized by interaction with both oxygen atoms.[225] However, theory (HF/6-31G*) has indicated that ion (**5**), protonated dimethoxymethane, is not symmetric.[226] Protonated 1,3-propanediol (**2**) dissociates by loss of water only ($T_{0.5} = 21$ meV).[227] It does so by first isomerizing to (**3**) and then to (**4**).

Thermochemistry

The $\Delta_f H$ for (**1**) and (**2**) depend on the value of PA chosen for 2-methoxy-1-ethanol and 1,3-propanediol, respectively. NIST provides values of 768.8 and 876.2 kJ/mol, respectively, yielding $\Delta_f H(\mathbf{1}) = 384$ kJ/mol and $\Delta_f H(\mathbf{2}) = 246$ kJ/mol ($\Delta_f H(\text{HOCH}_2\text{CH}_2\text{OCH}_3) = -376.9 \pm 8$ kJ/mol and $\Delta_f H(\text{HOCH}_2\text{CH}_2\text{CH}_2\text{OH}) = -392 \pm 3$ kJ/mol).[24] A G2(MP2) PA value of 852 kJ/mol for 1,3-propanediol calculated by Bouchoux and Berruyer-Penaud[228] gives a higher $\Delta_f H$ value for (**2**), 270 kJ/mol. A high pressure mass spectrometer equilibrium PA of 836 kJ/mol for 2-methoxy-1-ethanol results in a lower $\Delta_f H$ value for (**1**) of 317 kJ/mol.[225] The agreement between the data is evidently poor. The relative energies of (**2**)–(**4**) have been calculated at the MP2(fc)/6-311G**//MP2(fu)/6-31G * level of theory to be 0, +76 and +95 kJ/mol, respectively.[227]

3.7.47 $[C_3,H_2,O_3]^+$ – m/z 86

The only known isomer, 1,3-dioxol-2-one (cy–OC(O)OCHCH–) has an upper limit to its heat of formation of 554 kJ/mol, based on a neutral $\Delta_f H$ of -418.61 kJ/mol and the lowest quoted vertical IE of 10.08 eV.[24]

3.7.48 $[C_3,H_3,O_3]^+$ – m/z 87

Isomers

There are two proposed structures for $[C_3,H_3,O_3]^+$, namely, HOCCHC(O)OH (**1**) and OCCH$_2$C(O)OH (**2**).

Identification

The CID mass spectra of two isomeric ions (**1**) and (**2**) have been determined by Fell et al.[229] The former ion's mass spectrum is dominated by a fragment at m/z 69 (loss of water) and a doubly charged ion at m/z 43.5 (HOCCHC(O)OH^{2+}), whereas the CID mass spectrum of the latter ion has only a very weak m/z 69 and no charge stripping peak.[229]

Thermochemistry

There are no thermochemical data for these ions except that (**2**) is estimated to be the more stable, by about 145 kJ/mol.[229]

3.7.49 $[C_3,H_4,O_3]^+$ – m/z 88

Isomers

These ions comprise ionized pyruvic acid (CH$_3$C(O)C(O)OH$^{+\bullet}$) and its enol tautomer ionized α-hydroxyacrylic acid (CH$_2$=C(OH)C(O)OH$^{+\bullet}$) and other ionized molecules such as ethylene carbonate (cy–C(H)$_2$C(H)$_2$OC(O)O–$^{+\bullet}$) and methyl glyoxylate (CH$_3$OC(O)C(O)H$^{+\bullet}$).

Identification

The CID mass spectra of pyruvic acid and its enol tautomer have been measured by Fell et al.[229] They are distinct in that the former does not exhibit two intense peaks present in the mass spectrum of the enol ion, m/z 69 and 42.

Thermochemistry

Calculations for pyruvic acid at the CBS-4 level of theory (based on B3LYP/6-31G* optimized geometries) give $\Delta_f H$(CH$_3$C(O)C(O)OH) $= -548$ kJ/mol and combined with a revised measured IE of 10.4 ± 0.1 eV[229] (the 9.9 eV value[24] is arguably too low when compared with IE values for similar species); this leads to $\Delta_f H$(CH$_3$C(O)C(O)OH$^{+\bullet}$) $= 455 \pm 10$ kJ/mol. The enol isomer, also calculated at the CBS-4 level, has a $\Delta_f H$(CH$_2$=C(OH)C(O)OH) $= -518$ kJ/mol. Again, combining this value with a measured IE $= 9.5 \pm 0.1$ eV gives $\Delta_f H$(CH$_2$=C(OH)C(O)OH$^{+\bullet}$) $= 399 \pm 10$ kJ/mol. The energy difference between the keto and enol ions, 56 ± 20 kJ/mol, is typical of such stabilizations. Note that using revised additivity terms gives $\Delta_f H(CH_3C(O)C(O)OH) = -529$ kJ/mol and $\Delta_f H(CH_2 = C(OH)C(O)OH) = -492$ kJ/mol, the former in agreement with a G3 value of -523 kJ/mol[117] (using the reevaluated terms $CO–(C)(CO) = -121$ kJ/mol, $CO–(CO)(O) = -123$ kJ/mol and $O–(H)(C_d) = -188$ kJ/mol). These results indicate that the absolute heats of formation of these molecules are uncertain.

Combining the heat of formation and ionization energy of ethylene carbonate (-503 ± 4 kJ/mol and 10.4 eV, respectively)[24] gives $\Delta_f H(cy–C(H)_2C(H)_2OC(O)O–^{]+\bullet}) = 500$ kJ/mol. NIST lists $\Delta_f H(CH_3OC(O)C(O)H^{]+\bullet}) = 937 \pm 4$ kJ/mol,[24] an impossible value. By additivity, $\Delta_f H(CH_3OC(O)C(O)H) = -451$ kJ/mol and the IE (estimated) is about 10.3 eV, whence $\Delta_f H(CH_3OC(O)C(O)H^{]+\bullet}) = 543$ kJ/mol.

3.7.50 $[C_3,H_5,O_3]^{+\bullet}$ – m/z 89

The $\Delta_f H$ of protonated ethylene carbonate is 213 kJ/mol based on a neutral $\Delta_f H$ of -503 kJ/mol and a PA of 814.2 kJ/mol.[24]

3.7.51 $[C_3,H_6,O_3]^{+\bullet}$ – m/z 90

Of the possible isomers of $[C_3,H_6,O_3]^{+\bullet}$ only ionized 1,3,5-trioxane and the methyl ester of hydroxyacetic acid have known $\Delta_f H$ values, 528 and 448 kJ/mol, respectively (from the combination of $\Delta_f H(1,3,5-$trioxane$) = -465.76 \pm 0.50$ kJ/mol and IE(1,3,5-trioxane) = 10.3 eV and $\Delta_f H(HOCH_2C(O)OCH_3) = -556.9 \pm 6.4$ kJ/mol and IE(HOCH$_2$C(O)OCH$_3$) = 10.42 ± 0.05 eV).[24] In addition, the vertical IE of dimethoxyacetone ($(CH_3O)_2CO$) has been measured to be 11 eV,[24] which when combined with an estimated neutral $\Delta_f H((CH_3O)_2CO) = -581$ kJ/mol[95] gives $\Delta_f H((CH_3O)_2CO^{]+\bullet}) \leq 480$ kJ/mol.

3.7.52 $[C_3,H_4,O_4]^{+\bullet}$ – m/z 104

Ionized malonic acid, $HOC(O)CH_2C(O)OH$, undergoes metastable dissociation by loss of CO_2 and CO_2 plus H_2O.[230] The IE of malonic acid has been measured to be 11.05 eV.[24] The $\Delta_f H$ value for malonic acid, by Benson-type additivity,[123] gives -808 ± 10 kJ/mol, leading to an ion $\Delta_f H$ of 258 ± 10 kJ/mol.

3.7.53 $[C_3,H_8,P]$ – m/z 75[231]

The PA of allyl phosphine has been determined by FTICR MS and by computational chemistry at the G2(MP2) and B3LYP/6-311 + G(3df,2p) levels of theory. The experimentally derived gas-phase basicity of allyl phosphine was measured to be 828 kJ/mol, PA = 861 kJ/mol. The most stable form of the ion was computed to be a cyclic species, cy–C(H)$_2$C(H)$_2$P(H)$_2$C(H)$_2$–$^{]+}$, having a PA of about 890 kJ/mol, but the conventional P-protonated form, which lies 28 kJ/mol higher in energy (and is separated from the cyclic ion by a large barrier, 397 kJ/mol), is very likely the species produced by protonation of allyl phosphine.

3.7.54 $[C_3,H_6,S]^{+\bullet}$ – m/z 74

Isomers

Twenty isomeric $[C_3,H_6,S]^{+\bullet}$ cations have been investigated by experiment and by computation.

SH	SH	SH	S	S
1	**2**	**3**	**4**	**5**

		CH₃CSCH₃	SH₂ / CH₂	CH₃CHSCH₂
6	**7**	**8**	**9**	**10**

CH₃CH₂CSH	CH₂CH₂SCH₂	S (ring)	CH₃CH₂SCH	S (ring)
11	**12**	**13**	**14**	**15**

(ring) SH	S—CH₂	CH₂CH₂CHSH	SCH₂CHCH₃	SCH₂CH₂CH₂
16	**17**	**18**	**19**	**20**

Identification

The mass spectrometry of (**7**), (**12**), (**13**) and (**15**) has been studied by Hop[232] and Polce and Wesdemiotis.[233] All four ions give the same MI mass spectra, indicating that they isomerize to a common species before dissociation. The CID mass spectra are distinguishable and thus they can all be generated experimentally. The distinguishing features are as follows: the distonic ion (**12**) has a CID mass spectrum very similar to trimethylenesulfide (**13**) (both have m/z 46 as the dominant channel) but has a distinctive CS peak that the mass spectrum of (**13**) lacks;[233] the CID mass spectrum of (**15**) exhibits intense peaks at m/z 41, 45 and 59 while the CID mass spectrum of (**7**) is dominated by m/z 71 and 59.[232,233] Hop found that H_2 loss from methyl ethyl thioether, $CH_3CH_2SCH_3$, produced a different ion (having a CID mass spectrum dominated by m/z 71), which he proposed was (5), while H_2CO loss from 1,4-thioxane was assumed to produce the distonic ion (12).[232] The latter observation was confirmed by Polce and Wesdemiotis, who showed that 1,4-disulfur heterocycles produce (12), while the dissociation of ionized 1,3-sulfur heterocycles gives (13) (but the results indicate that a mixture of (12) and (13) is likely formed).[233] The calculated barrier between (12) and (13) is 62 kJ/mol.[234] The ion formed in the dissociation of ionized thiane was assigned to a mixture of (2) and (3).[233]

Thermochemistry

Heat of Formation for the $[C_3,H_6,S]^{+\bullet}$ Isomers

Ions	$\Delta_fH(\text{expt})^a$ (kJ/mol)	$\Delta_fH(\text{G3})^b$ (kJ/mol)
(1)		884
(2)		871
(3)		868
(4)	≤821[c]	859
(5)	865[d]	869
(6)		902
(7)	≤956	928
(8)		964
(9)		960
(10)		918
(11)		997
(12)		981
(13)	892	904
(14)		1021
(15)	865	900
(16)		938
(17)		966
(18)		972
(19)		1028
(20)		1151

[a]Values obtained by combining neutral Δ_fH values and quoted IEs in the NIST database,[24] unless otherwise stated.
[b]G3 values based on MP2/6-31G(d) optimized geometries.[234]
[c]See also Butler and Baer.[235] The Δ_fH of neutral (4) has been reported to be −9 kJ/mol[236] but adding the vertical IE of 8.6 eV[24] leads to an upper limit to the Δ_fH of (4) of 821 kJ/mol, in clear disagreement with the G3 results. The neutral Δ_fH is likely in error as the G3 calculated value is 26.7 kJ/mol and is consistent with the acidity of neutral (4).
[d]Values found in Lias et al.[95]

3.7.55 $[C_3,H_7,S]^+ - m/z\ 75$

Isomers

Of the 11 $[C_3,H_7,S]^+$ cations shown in the figure below, ions (1)–(8) have been adequately characterized.

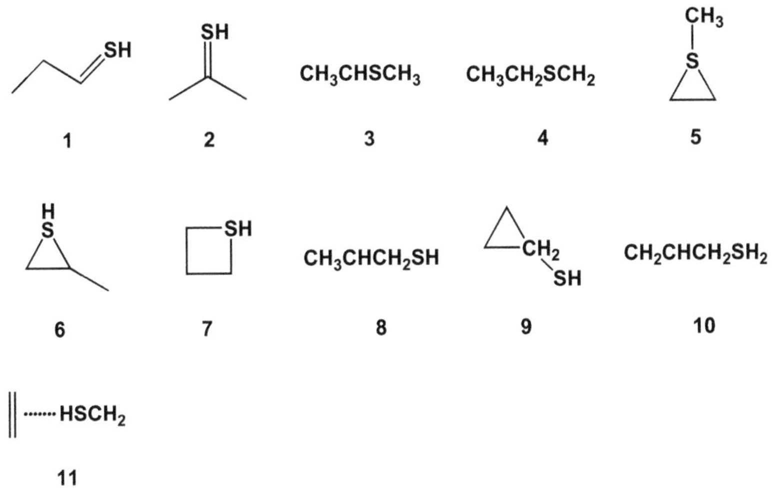

Identification

Ions (1)–(4) give the same MI mass spectrum (loss of C_2H_4 and H_2S), which indicates that they can interconvert on the microsecond time-scale.[237,238] The ring opened form of (7), $CH_2CH_2CH_2SH^{1+}$, was proposed to be the reacting configuration for the loss of C_2H_4 from (1). Isomers (3) and (4) interconvert with (1) before dissociation. Ion (2) however was shown to at least partially dissociate directly without isomerizing to (1).[239] van de Graaf and McLafferty[240] and Levsen et al.[241] studied the CID mass spectra of 24 precursor ions, leading to structures (1)–(4), (6) and (7) and found that ions (1)–(4) were the most commonly formed $[C_3,H_7,S]^+$ ions. Thioethers tend to produce (3) and (4), primary and secondary thiols generally form (1) while ionized $(CH_3)_3CSH$ directly dissociates to (2). They found very few sources of pure ions. Ions (6) and (7) could only be made in a pure form by protonating the corresponding neutral. They were unable to generate pure (5). The characteristic CID mass spectra are summarized in the table below.

CID Mass Spectra for Some of the $[C_3,H_7,S]^+$ Isomers

	m/z											
Ions	**27**	**29**	**39**	**45**	**46**	**57**	**58**	**59**	**60**	**69**	**71**	**73**
(**1**)	5.3	2	20	24	5.1	5.5	12	10	2.8	2.7	4.7	4.7
(**2**)	2.5	0.2	21	9.2	1.3	7.7	16	34	1.4	2.2	2.4	0.6
(**3**)	5.7	0.9	2.0	23	6.6	5.2	14	22	9.4	0.5	0.8	1.6
(**4**)	9.7	12	1.6	31	23	2.8	6.5	7.4	2	0.6	0.8	1.1
(**6**)	6.5	0.7	23	23	5.7	5.2	11	10	1.6	2.4	3.5	4
(**7**)	7	0.5	18	29	21	2.8	5.6	4	0.7	2.4	3.7	3.6

Thermochemistry

Although some thermochemical data were given in references,[237,238] the best source of thermochemistry for this family of ions is a G2 level study by Chalk et al.[103] Of the isomers they investigated, they found that ions (**9**) and (**11**) both have only small barriers leading to (**1**), suggesting that these ions may not be easy to generate by experiment. Ion (**8**) was also found to be separated from (**1**) and (**6**) by small barriers.

Heat of Formation of the $[C_3,H_7,S]^+$ Isomers

Ions	$\Delta_f H(\text{G2})^{103}$ **(kJ/mol)**
(**1**)	765
(**2**)	
(**3**)	731
(**4**)	770
(**5**)	
(**6**)	749
(**7**)	769
(**8**)	865
(**9**)	891
(**10**)	801
(**11**)	892

3.7.56 $[C_3,S_2]^{+\bullet}$ – *m/z* 100

An ion of structure $SCCCS]^{+\bullet}$ has been described by Wong et al.[242] following an earlier study by Sulzle et al.[243] It was prepared by the dissociative ionization of dithiolethione (cy–SC(S)C(SCH_3)C(SCH_3)S–), a species whose EI mass spectrum contained a number of other noteworthy cumulene-type ions. The connectivity of m/z 100 was confirmed by its CID mass spectrum, showing major peaks for the losses of S, CS and C_2S. A signal corresponding to the loss of S_2 was ascribed to postcollisional isomerization.

Calculations performed at the QCISD(T)/6-311+G(2d,p)+ZPVE level of theory[242] gave $\Delta_f H(SCCCS^{1+\bullet})$ relative to its dissociation products, but the absence of data for them precludes the estimate of an absolute $\Delta_f H$. $SCCCS^{1+\bullet}$ is calculated to be lower in energy than its products: 572 kJ/mol for $CS+CCS^{1+\bullet}$, 697 kJ/mol for $S+CCCS^{1+\bullet}$, 740 kJ/mol for $CCS+CS^{1+\bullet}$ and 744 kJ/mol for $CCCS+S^+$.

3.7.57 $[C_3,H,S_2]^+ - m/z\ 101^{242}$

Isomers

The two isomeric ions correspond to S- and C-protonated SCCCS. For the C-protonated ion, the proton resides on the central carbon.

Identification

Wong et al. have obtained C-protonated SCCCS by methane chemical ionization of the neutral.

Thermochemistry

The C-protonated SCCCS lies 69 kJ/mol lower in energy than the S-protonated isomer (calculated at the QCISD(T)/6-311+G(2d,p) level of theory). At the same level of theory, the proton affinity of the SCCCS molecule was calculated to be 818 kJ/mol. Without a proper heat of formation for the molecule, no value can be deduced for the protonated species.

3.7.58 $[C_3,H_3,S_2]^+ - m/z\ 103^{242}$

Isomers

Two isomeric $[C_3,H_3,S_2]^+$ ions have been explored: S- and C-methylated SCCS.

Identification

Wong et al. have obtained C-methylated SCCS from methyl cation chemical ionization of the neutral, whereas S-methylated SCCS was generated by the dissociative ionization of 1,2-dithiol-3-thione and dithietane.

Thermochemistry

The C-methylated SCCS ion was calculated to be 192 kJ/mol lower in energy than the S-methylated isomer at the QCISD(T)/6-311 + G(2d,p) level of theory.

3.7.59 $[C_3,F_n]^{+/+\bullet}$

The mass spectrometry of a number of these ions has been described in some detail in a thesis[244] and such salient features that permit the identification of isomers are reproduced here. In principle it might be supposed that the behaviors of perfluoro cations may reflect those of their $[C_3,H_n]^+$ analogs. Sometimes the parallels are striking but others are unique. In many of the examples given below the ions are prepared by the dissociative ionization of a related species containing other atoms. Two striking common dissociations are the frequent appearance of the ion CF_3^{1+} as a major fragment and also the easy loss of the CF_2 radical, both very unlike the behavior of hydrocarbon analogs. The thermochemistry for these ions is sparse but will be described at the end of this section.

3.7.60 $[C_3,F_2]^{+\bullet}$ – m/z 74^{244}

This ion was conveniently generated in the ion source by the loss of CF_4 from ionized CF_3CCCF_3 (in whose mass spectrum this ion has an abundance of 15%). Its MI mass spectrum showed only loss of $F^\bullet$ and the CID mass spectrum contains C_3F^{1+} (100%), $CF_2^{1+\bullet}$ (3%), C_2F^{1+} (3%), CF^{1+} (15%) and the doubly charged ion $C_3F_2^{12+}$ (m/z 37, 1.5%). The lowest energy isomer is likely the ion $FCCCF^{1+\bullet}$, which is isoelectronic with the molecule OCCCO. The generation of the minor $CF_2^{1+\bullet}$ peak in the CID mass spectrum may well indicate that the isomer $^\bullet CF_2CC^+$ is accessible.

3.7.61 $[C_3,F_3]^+$ – m/z 93^{244}

This ion was produced in the ion source from ionized CF_3CCCF_3 by the loss of $CF_3^{1\bullet}$, the dissociation that produces the base peak in its normal EI mass spectrum and is the only MI dissociation ($T_{0.5} = 31$ meV). By analogy with its hydrocarbon isomer it is proposed that the cyclic perfluoro cyclopropenium ion is produced. The $C_3F_3^{1+}$ ion has two peaks in its MI mass spectrum, m/z 74 (loss of $F^\bullet$, 2%) and m/z 31 (loss of C_2F_2, 100%). In the CID mass spectrum the $F^\bullet$ loss becomes base peak and there is a significant $C_3F_3^{12+}$ ion (m/z 46.5, 10%).

3.7.62 $[C_3,F_4]^{+\bullet}$ – m/z 112[244]

This ion is a significant feature (16%) of the EI mass spectrum of perfluoro-2-butyne (CF$_3$CCCF$_3$). Its MI mass spectrum contains only one signal, m/z 93 (C$_3$F$_3^{1+}$). The CID mass spectrum shows peaks for C$_3$F$_3^{1+}$ (100%), C$_3$F$_2^{1+\bullet}$ (19%), C$_2$F$_2^{1+\bullet}$ (9%) and CF^{1+} (11%), all in keeping with a cy–C(F)C(F)C(F)$_2$–$^{1+\bullet}$ structure. Minor signals for CF$_3^{1+}$ (2%) and C$_2$F^{1+} (2%) indicate that the isomer CF$_3$CCF$^{1+\bullet}$ may be accessible to these energized ions.

3.7.63 $[C_3,F_5]^+$ – m/z 131[244]

First it should be noted that ionized C$_3$F$_6$ (unlike the hydrocarbon analog) does not lose F$^\bullet$ as a significant dissociation to generate a $[C_3,F_5]^+$ species, but rather loses the CF$_2^{1\bullet}$ radical instead. Also, $[C_3,F_5]^+$ is not produced by loss of F$_2$ from $[C_3,F_7]^+$ ions. However, C$_3$F$_5^{1+}$ is a major peak in the mass spectrum of ionized n- and sec-C$_3$F$_7$I arising from the loss of IF from the fragment ion C$_3$F$_6$I^{1+}. The C$_3$F$_5^{1+}$ ions so produced (i.e., from the two C$_3$F$_7$I isomers) have identical MI and CID mass spectra, indicating a single structure or a common mixture of isomers. The former contains only the ion CF$_3^{1+}$ from the loss of C$_2$F$_2$ ($T_{0.5}$ = 30 meV). This peak is moderately collision gas sensitive, indicating either that isomeric forms such as CF$_3$CCF$_2^{1+}$ or CF$_2$CFCF$_2^{1+}$ undergo facile interconversion or that only the former ion is present. The CID mass spectrum contains peaks for C$_3$F$_3^{1+}$ (35%), C$_2$F$_3^{1+}$ (25%), C$_3$F$_5^{12+}$ (6%), C$_2$F$_2^{12+}$ (9%) and CF^{1+} (15%).

3.7.64 $[C_3,F_6]^{+\bullet}$ – m/z 150[244]

The two neutral isomers, perfluoropropene (CF$_3$CF=CF$_2$) and perfluorocyclopropane, when ionized by electron impact, exhibit identical infrared photodissociation spectra, indicating that ionized perfluorocyclopropane rapidly isomerizes to ionized perfluoropropene.[244]

The C$_3$F$_6^{1+\bullet}$ radical cation has only one peak in its MI mass spectrum for the loss of CF$_2$ ($T_{0.5}$ = 12 meV). The CID mass spectrum has CF$_3^{1+}$ as its second peak (50%), C$_2$F$_4^{1+\bullet}$ corresponding to the base peak (100%). All other signals are less than 10%. It is worth noting that the C$_2$F$_4^{1+\bullet}$ ion generated from the metastable C$_3$F$_6^{1+\bullet}$ ion has been assigned the carbene structure CF$_3$CF$^{1+\bullet}$ whereas that from the ion source has been proposed to have the CF$_2$CF$_2$ connectivity (see under C$_2$F$_2$). This may well arise from the presence in the ion source of both CF$_3$CFCF$_2^{1+\bullet}$ and ring opened cy–C(F)$_2$C(F)$_2$C(F)$_2$–$^{1+\bullet}$ ions.

3.7.65 $C_3F_7^{1+}$ – m/z 169[244]

The isomers CF$_3$CF$_2$CF$_2^{1+}$ and (CF$_3$)$_2$CF^{1+} are produced by the loss of I$^\bullet$ from n-C$_3$F$_7$I and sec-C$_3$F$_7$I precursors, respectively. They show only one MI peak, leading to the CF$_3^{1+}$ ion. The kinetic energy release associated with the generation of this fragment ion, $T_{0.5}$ = 12 meV, is the same for both ions except that for n-C$_3$F$_7^{1+}$ the peak is strongly collision gas sensitive, but barely so for the (CF$_3$)$_2$CF^{1+} ion. This is taken to indicate that the reacting configuration is the *n*-cation in each case, the *sec*-ion having to rearrange at an energy below but near to the dissociation limit for the production of $CF_3^{1+} + CF_2CF_2$. The other significant peaks in the CID mass spectrum of $CF_3CF_2CF_2^{1+}$ are only the ions $C_2F_5^{1+}$ and $C_2F_4^{1+\bullet}$ (whether the product ion's connectivity is $CF_2CF_2^{1+\bullet}$ or $CF_3CF^{1+\bullet}$ is discussed under $C_2F_4^{1+\bullet}$ isomers). The CID mass spectrum for the other isomer is similar but with $C_2F_5^{1+}$ as a less important species.

3.7.66 THERMOCHEMISTRY OF $[C_3,F_n]^{+/+\bullet}$

The data (in eV or kJ/mol) given in the table below are from Ref. [24] except where otherwise noted. Some of the IE values reported may not be the adiabatic IE. Note that the ring strain in cy-C_3F_6 (170 kJ/mol for the molecule and 228 kJ/mol for the ion) is apparently much greater than in the hydrocarbon analog (115 kJ/mol and 45 kJ/mol, respectively); this could arise from uncertainties in both data for the C_3F_6 species.

Ions	$\Delta_f H$(neutral) (kJ/mol)	IE_a (eV)	$\Delta_f H$(ion) (kJ/mol)
CF_2CCF_2	-594^{95}	10.88	456
CF_2CFCF_2	ca -700	8.44	ca 90^{244}
CF_3CFCF_2	-1152	10.6	-129
cy$-C(F)_2C(F)_2C(F)_2-$	980	11.18	99
$CF_3CF_2CF_2$	-1315 ± 20	10.06	-344 ± 20
$(CF_3)_2CF$	-1330 ± 20	10.5	-317 ± 20
C_3F_8	-1785	13.38	-494

3.7.67 $[C_3,H_2,F_3]^+ - m/z$ 98

Isomers

A series of trifluorinated alkanes was investigated by McAllister et al.[245] at the MP2/6-31G**//HF/6-31G* level of theory.

CHFCHCF$_2$ CF$_3$CHCH   CF$_3$CCH$_2$

1 (0)   2 (45)   3 (45)   4 (51)

CF$_2$CFCH$_2$

5 (14)   6 (65)   7 (45)   8 (45)

Thermochemistry

The relative energies are listed below each structure. On the basis of calculated barrier heights, it was concluded that (3) can easily isomerize to (1) or

(5) and that (2) isomerizes to (1) via the transition state (8). Structure (7) is a transition state between (2) and (3). The relative energies of (2), (3), (7) and (8) indicate that (2) and (3) are not distinguishable structures.

3.7.68 $[C_3,H_6,F]^+$ – m/z 61

Isomers

Work on the fluoro-substituted alkyl cations is largely limited to computational investigations of their relative energies. Thirteen structures have been thus identified and they are shown below.

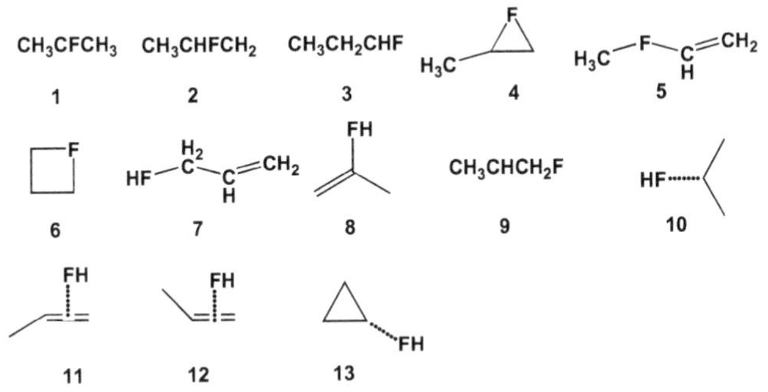

Thermochemistry

Ion (2), being an unsubstituted primary cation does not represent an equilibrium structure. Shaler and Morton studied (3) and (9) at the MP2/6-31G** level of theory and found that they can interconvert.[246] When CF_3^{1+} is reacted with propionaldehyde to generate (3), the resulting ion has enough energy to isomerize to the more stable (1).

Relative Energies of the $[C_3,H_6,F]^+$ Isomers[246,247]

Ions	HF/6-31G//HF/3-21G (kJ/mol)	MP2/6-31G** (kJ/mol)
(1)	0	0
(2)	Not a minimum	
(3)	88	74
(4)	115	
(5)	141	199
(6)	96	121
(7)	140	
(8)	97	150
(9)		112
(10)		101
(11)		207
(12)		215
(13)		217

By experiment, Williamson et al.[248] measured the appearance energy value for the formation of (**1**) from 2-fluoropropane to be 11.23 ± 0.03 eV, which when combined with $\Delta_f H(CH_3CHFCH_3) = -293.5$ kJ/mol[32] and $\Delta_f H(H^\bullet) = 218$ kJ/mol[24] results in $\Delta_f H(\mathbf{1}) = 577$ kJ/mol. The same authors measured the $IE(CH_3CFCH_3{}^{1\bullet}) = 7.14$ eV, leading to $\Delta_f H(CH_3CFCH_3{}^{1\bullet}) = -117$ kJ/mol. From these data, the stabilizing effect of F substitution on the $CH_3CHCH_3{}^{1\bullet}$ radical ($\Delta_f H = 88$ kJ/mol[77]) is large (-205 kJ/mol). More typically, the effect is 180 ± 10 kJ/mol (based on F substitution in other neutral species) and so possibly the IE is not the adiabatic value.

3.7.69 $[C_3,H_3,Cl]^{+\bullet}$ – m/z 74

Isomers

There are three isomers, namely ionized 3-chloropropyne, $CH\equiv CCH_2Cl^{1+\bullet}$, $HCCHCHCl^{1+\bullet}$ and its ring closed version, ionized chlorocyclopropene. The latter two have been identified by calculations at the B3-LYP/6-31G** level of theory.

Identification

The dissociation of ionized 3-chloropropyne, $CH\equiv CCH_2Cl^{1+\bullet}$, has been shown to produce both ionized cyclopropenium and the propargyl ion.[249] At threshold this ionized molecule and its isomer CH_3CCCl generate only the most stable $[C_3,H_3]^+$ ion, cyclopropenium, a dissociation that is accompanied by a significant kinetic energy release of about 0.3 eV. The metastable ion peaks for the reaction are identical, showing that the isomeric ions lose $Cl^\bullet$ via a common transition state.[250]

Thermochemistry

Calculations at the B3-LYP/6-31G** level of theory have determined that $HCCHCHCl^{1+\bullet}$ and its ring closed version, ionized chlorocyclopropene, lie 23 kJ/mol lower and 5 kJ/mol higher, respectively, than ionized 3-chloro-propyne.[249]

3.7.70 $[C_3,H_3,Br]^{+\bullet}$ – m/z 118

The photodissociation of ionized 3-bromopropyne, $CH\equiv CCH_2Br^{1+\bullet}$, exhibits both statistical and nonstatistical behavior, the latter arising from a discrete excited state.[251,252] Calculations at the B3-LYP/6-31G** level of theory have identified the mechanism for $Br^\bullet$ loss to form both the propargyl ion and cy-$C_3H_3{}^{1+}$.[253] Four intermediate isomers in the process leading to cy-$C_3H_3{}^{1+}$ are *cis*- and *trans*-CHCHCHBr$^{1+\bullet}$, $CH_2=C=CHBr^{1+\bullet}$ and ionized bromocy-clopropene. The energies for these ions relative to $CH\equiv CCH_2Br^{1+\bullet}$ at this level of theory are -1.6, -12.5, -91 and $+17$ kJ/mol, respectively. The initial barrier to isomerization of $CH\equiv CCH_2Br^{1+\bullet}$ to *cis*-CHCHCHBr$^{1+\bullet}$ is comparable with the dissociation limit to propargyl ion formation. As with the chloro analog, metastable ionized CH_3CCBr and $CHCCH_2Br$ lose $Br^{\bullet}$ with a large kinetic energy release, about 0.4 eV, to generate cy-$C_3H_3^{]+}$.[250]

3.7.71 $[C_3H_3I]^{+\bullet}$ – m/z 166

In contrast with the chloro and bromo analogs (see above), the ionized CH_3CCI and $CHCCH_2I$ species have the greatest activation energies for halogen atom loss and they generate different, Gaussian metastable ion peaks with small kinetic energy releases. It was argued that the propargyl cation is the $C_3H_3^{]+}$ fragment ion at threshold.[250]

3.7.72 $[C_3,H_5,Cl]^{+\bullet}$ – m/z 76

The photodissociation of four $[C_3,H_5,Cl]^{+\bullet}$ isomers has been investigated by Orth and Dunbar:[254] ionized *trans* 1-chloropropene, *cis* 1-chloropropene, 2-chloropropene and 3-chloropropene. All isomers exhibit fairly distinct photo-dissociation spectra in the 580–640 nm region. These spectra were used as fingerprints for studying the products of the reactions between propene ions, $C_3H_6^{]+\bullet}$ and 1- and 2-chloropropane molecules. The reaction with 1-chloropropane produces a mixture of ionized *trans* and *cis* 1-chloropropene, while the second reaction forms exclusively 2-chloropropene ions. $\Delta_f H$(*trans*-$CH_3CH=CHCl^{]+\bullet}$) was estimated to be 1101 kJ/mol while, based on an IE = 10.05 eV and $\Delta_f H$($ClCH_2CH=CH_2$) = −5.6 kJ/mol,[24] $\Delta_f H$($ClCH_2CH=CH_2^{]+\bullet}$) is 964 kJ/mol.

3.7.73 $[C_3,H_5,Br]^{+\bullet}$ – m/z 120

Isomers

Four stable $[C_3,H_5,Br]^{+\bullet}$ cations have been identified, $BrCH_2CH=CH_2^{]+\bullet}$ (1), $CH_3CH=CHBr^{]+\bullet}$ (2), ionized bromocyclopropane (3) and $CH_3CBr=CH_2^{]+\bullet}$ (4). All four have distinct photodissociation spectra in the 9.2–10.7 μm region.[255]

Identification

Ionization of $BrCH_2CH=CH_2$ by EI generates two isomers, because the resulting reaction of these ions with neutral $BrCH_2CH=CH_2$ shows that 18% of the ions are unreactive. These ions were assigned to (2) because pure (2) does not react, while (3) and (4) do.[256–258] This unreactive fraction decreased on lowering the ionizing electron energy.

Thermochemistry

MP4(SDTQ)/D95** level calculations place the relative energies at 1(0), 2(−39) and 3(17) kJ/mol. The distonic $^{\bullet}CH_2CH_2{}^+CHBr$ ion is a transition state on the surface for the interconversion of (1) and (2) having a relative energy of 73 kJ/mol. A fifth isomer was calculated, an ion–molecule complex between the allyl cation and Br$^{\bullet}$, lying 75 kJ/mol above (1), but is unlikely to be observed as the barrier to (1) is only 4 kJ/mol.

3.7.74 $[C_3,H_7,I]^{+\bullet}$ – m/z 170

Park and Kim were able to demonstrate that the ionized *gauche* and *anti* forms of 1-iodopropane exhibit different dissociation dynamics due to the presence of a repulsive excited state that preferentially forms 2-propyl cations from *gauche* 1-iodopropane ions and protonated cyclopropane from anti-1-iodopropane ions.[259,260]

3.7.75 $[C_3,H,N,O]^{+\bullet}$ – m/z 67

Isomers

The $[C_3,H,N,O]^{+\bullet}$ pair of ions comprises ionized cyanoketene, NCCHCO$^{1+\bullet}$ and HNCCCO$^{1+\bullet}$.

Identification

The two ions can be generated from their respective methylated precursors CH$_3$NCCHCO^{1+} and CH$_3$NHCCCO^{1+}.[261] The CID mass spectrum of ionized cyanoketene is distinguishable from that of HNCCCO$^{1+\bullet}$ by the presence of m/z 41 and more intense m/z 28 and 53.[261,262]

Thermochemistry

$\Delta_f H$(NCCHCO$^{1+\bullet}$) can be estimated by correlation principles to be 1058 kJ/mol,[263] in line with an upper limit of 1050 kJ/mol that results from combining the $\Delta_f H$ of neutral cyanoketene (78 ± 12 kJ/mol, see discussion below on the thermochemistry of $[C_3,H_2,N,O]^+$) with a vertical IE from photoelectron spectroscopy (10.07 eV).[24]

3.7.76 $[C_3,H_2,N,O]^+$ – m/z 68

Isomers

The $[C_3,H_2,N,O]$ family of ions comprises C-, N- and O-protonated HNCCO, HNCCHCO^{1+} (1), H$_2$NCCCO^{1+} (2) and HNCCCOH^{1+} (3), respectively, C-protonated cyanoketene, NCCH$_2$CO^{1+} (4) and the cyclic isomer cy–CHCCNHC(O)–$^{1+}$ (5). Three other isomers have only been investigated at relatively low levels of theory, namely HC=NCHCO^{1+} (6), C=NCH$_2$CO$^+$ (7) and HC(N)CHCO^{1+} (8). Calculations predict that C-protonation of HNCCCO (which can also be thought of as N-protonation of cyanoketene) is most favorable (see relative energies below).[261]

Identification

Ions (1) and (2) have been generated by the dissociative ionization of 4-chlorouracil and 5-amino-4-methoxycarbonylisoxazole, respectively.[261] The CID mass spectrum of (1) obtained in this manner is identical to that obtained from protonating cyanoketene and is distinguishable from that of (2) by the ratios of m/z 51:53 and m/z 25:28 (1:0.8 and 1:2 for (1) and 1:10 and 1:0.3 for (2)).[261] Only ion (4) has an intense m/z 42 in its CID mass spectrum.[261]

Thermochemistry

$\Delta_f H(1) = 824 \pm 12$ kJ/mol can be derived from the PA of cyanoketene (784 kJ/mol)[24] and $\Delta_f H(\text{NCCHCO}) = 78 \pm 12$ kJ/mol. The latter value, $\Delta_f H(\text{NCCHCO})$, comes from the EI/AE data for the dissociation of the meta-stable ionized ester $NCCH_2C(O)OC(CH_3)_3$ to give $CH_3^{|\bullet} + (CH_3)_2COH^{|+} +$ CNCHCO. Using the following ancillary thermochemical data leads to the above $\Delta_f H(\text{NCCHCO})$ value: $AE = 11.4 \pm 0.1$ eV,[263] $\Delta_f H(NCCH_2C(O)OC(CH_3)_3) = -374$ kJ/mol (revised additivity using $C-(H)_2(CO)(CN) = 109$ kJ/mol, derived from the $\Delta_f H(C_6H_5C(O)CH_2CN) = 70$ kJ/mol[32]), $\Delta_f H(CH_3^{|\bullet}) = 146$ kJ/mol,[24] $\Delta_f H((CH_3)_2COH^{|+}) = 501$ kJ/mol (refer to the section on $[C_3,H_7,O]^+$). Note that the effect of CN substitution on ketene is now reasonable, going from -52 to $+78$ kJ/mol, $\Delta = 130$ kJ/mol. (Compare $\Delta_f H(C_2H_4) = +52$ kJ/mol, $\Delta_f H(CH_2CHCN) = 180$ kJ/mol, $\Delta = 128$ kJ/mol.)

The ion, to which structure (4) was originally given,[263] was generated by the loss of $CH_3O^\bullet$ ($\Delta_f H = 17 \pm 4$ kJ/mol)[77] from ionized $NCCH_2C(O)OCH_3$ ($\Delta_f H = -261$ kJ/mol, using the $C-(H)_2(CO)(CN)$ group additivity term, see above). The dissociation produced no metastable ion peak and so it is reasonable to assume that the radical $CH_3O^\bullet$ is generated, i.e., a rearrangement to yield the more stable $^\bullet CH_2OH$ species is not taking place. The appearance energy (11.45 ± 0.09 eV)[263] leads to $\Delta_f H(4) = 827$ kJ/mol. Not only is this very close to $\Delta_f H(1)$, but the result implies that the PA of cyanoketene is *the same* at both carbon and nitrogen. This seems unlikely if the ion generated from 4-chlorouracil has lost $Cl^\bullet$ and an HNCO molecule to produce N-protonated cyanoketene, or rather, an ion with identical MS characteristics to directly prepared, protonated cyanoketene. It is therefore possible that the above disso-ciative ionization of the ester yields the more stable structure (1).

The relative energies obtained for (1), (4) and (6)–(8) at the modest CI/6-31G*//HF/3-21G level of theory were 0, 40, 77, 164 and 363 kJ/mol, mol, respectively.[264] The QCISD/6-311 + G(2d,p)//MP2/6-31G* relative energies for (1), (2), (3) and (5) are 0, 53, 127 and 489 kJ/mol.[261]

3.7.77 $[C_3,H_3,N,O]^{+\bullet}$ – m/z 69

Isomers

The $[C_3,H_3,N,O]^{+\bullet}$ family of ions comprises 14 isomeric species, most of which have only been studied by computations.

| 1 | 2 | 3 | 4 | 5 |

NH_2CHCCO $HNCHCHCO$ $HNCCHCHO$ $NCCH_2CHO$

| 6 | 7 | 8 | 9 | 10 |

$HCNCHCHO$ $HOCH=CHCN$ $HC=NCH_2CO$ $HNCCH_2CO$

| 11 | 12 | 13 | 14 |

Identification

Flammang and coworkers[264,265] have studied the CID mass spectra of isomers (1)–(7), the results of which are summarized below.

Partial CID Mass Spectra for Some of the $[C_3,H_3,N,O]^{+\bullet}$ Isomers

	m/z								
Ions	38	39	40	41	42	43	52	53	54
(1)	17	29	91	100	15	2	2	1	1
(2)	12	19	86	100	24	—	1	—	1
(3)	4	14	14	14	14	96	3	1	100
(4)	—	1	4	12	7	2	4	100	—
(5)	9	16	42	100	—	3	1	3	9
(6)	14	28	100	99	80	—	16	61	—
(7)	—	—	60	53	100	—	—	7	—

Ions (1) and (2), iso-oxazole and oxazole respectively, produce similar mass spectra that are distinct from those of the other linear isomers. As can be seen from the table, isomers (3), (4), (6) and (7) are easily distinguished based on the intensities of m/z 54, 53 and 43–40. Ion (5) is the only isomer having m/z 41 as base peak in its He CID mass spectrum. There is a pronounced doubly charged ion at m/z 33.5 when O_2 is used as the collision gas.

Thermochemistry

The $\Delta_f H$ of (1), (2) and (4) are derived from neutral molecule $\Delta_f H$ values and adiabatic IEs listed in the NIST compendium.[24] $\Delta_f H(1)$ derives from the neutral $\Delta_f H = 78.6$ kJ/mol and an IE $= 9.93$ eV, whence $\Delta_f H(1) = 1037$ kJ/mol. In the case of ion (2), there is a discrepancy between the PE and EI ionization energies. PE spectroscopy yields 9.9 eV while EI gives 9.6 eV. The former value was used with the neutral $\Delta_f H$ of -15.5 kJ/mol to give

$\Delta_f H(2) = 940$ kJ/mol. The relative energies from CI/6-31G*//HF/3-2G calculations must be used with care since the 227 kJ/mol difference between the $\Delta_f H$ of (1) and (2) is clearly too large. For the experimental difference to be as great as 100 kJ/mol too low, signifies an (unlikely) error of at least 1 eV in the IE for one of the compounds.

Heats of Formation of the $[C_3, H_3, N, O]^{+\bullet}$ Isomers

Ions	$\Delta_f H(expt)^a$ (kJ/mol)	Relative Energy[b] (kJ/mol) CI/6-31G*//HF/3-2G
(1)	1037(?)	227
(2)	940	0
(3)		142
(4)	1023[c]	149
(5)		
(6)		−18
(7)		−7
(8)		−48
(9)		134
(10)		235
(11)		−12
(12)		13
(13)		17
(14)		−8

[a]NIST.[24]
[b]Flammang et al.[264]
[c]Lias et al.[95]

3.7.78 $[C_3, H_4, N, O]^+$ – m/z 70

Isomers

The $[C_3, H_4, N, O]^+$ family of ions has not been systematically investigated. Carvalho et al.[266] found that ionized N-bromosuccinimide dissociates to a mixture of $[C_2, N, O_2]$ ions (~64%) and $[C_3, H_4, N, O]$ ions (~36%). The low energy (20 eV) CID mass spectra (employing Ar as the target gas) of m/z 70 ions from ionized N-bromosuccinimide exhibit peaks at m/z 42 (due to $[C_2, N, O_2]$ ions) and 43 and 44 due to $[C_3, H_4, N, O]$ ions. No structure was postulated for the latter ions.

Thermochemistry

The $\Delta_f H$ of $[C_3, H_4, N, O]$ ions can be derived from the PA values for $[C_3, H_3, N, O]$ molecules. Measured PA values of 848.6 and 876.4 kJ/mol for isoxazole ($\Delta_f H = -15.5$ kJ/mol) and oxazole ($\Delta_f H = 78.6$ kJ/mol), respectively, yield $\Delta_f H$ values for the protonated species of 760.4 and 638.1 kJ/mol.[24] In an extended study of neutral [C_3,H_3,N,O] molecule $\Delta_f H$ and PA values, Elrod[267] determined that isoxazole preferred to protonate at the nitrogen atom (G2MS level of theory). Elrod then assumed N protonation for 30 other cyclic and linear [C_3,H_3,N,O] molecules. We have chosen not to reproduce these data here because for each molecule, the site of protonation was assumed and not tested further. The reader is directed to Elrod's study for N-centered PA values for this collection of [C_3,H_3,N,O] molecules. Combined with his calculated $\Delta_f H$ values, these PA give ionic $\Delta_f H$ values.

3.7.79   [C_3,H_5,N,O]$^{+\bullet}$ – m/z 71

Isomers

There are thermochemical data for the molecular ions of acrylamide, $CH_2CHC(O)NH_2^{]+\bullet}$ (**1**), methoxyacetonitrile, $CH_3OCH_2CN^{]+\bullet}$ (**2**), cyclic 2-azetidinone, $cy–C(H)_2C(O)N(H)C(H)_2–^{]+\bullet}$ (**3**) and ethyl isocyanate, $CH_3CH_2NCO^{]+\bullet}$ (**4**).

Thermochemistry

The ionization energies for acrylamide, methoxyacetonitrile and cyclic 2-azetidinone have been measured and are, respectively, 9.5, 10.75 and 9.78 eV.[24] Together with the corresponding neutral $\Delta_f H$ values,[24] -130.2 ± 1.7, -35.65 ± 0.66 and -96.0 ± 0.9 kJ/mol, the ionic $\Delta_f H$ values are (**1**) 786 kJ/mol, (**2**) 1002 kJ/mol and (**3**) 848 kJ/mol respectively. The IE for ethyl isocyanate, CH_3CH_2NCO, is given as 10.32 eV,[29] but there is no value for $\Delta_f H$ for the neutral. However, this can be estimated to be -152 kJ/mol, from the known $\Delta_f H$ for the methyl analog, -130 kJ/mol[55] and using an additivity term of -32 kJ/mol,[123] resulting in $\Delta_f H(\mathbf{4}) = 844$ kJ/mol.

3.7.80   [C_3,H_6,N,O]$^+$ – m/z 72

The [C_3,H_6,N,O]$^+$ group of ions has not been systematically investigated. Protonation of the three [C_3,H_5,N,O] species acrylamide, methoxyacetonitrile and 2-azetidinone yield ionic species with $\Delta_f H$ values of 529, 736 and 582 kJ/mol, respectively. These values are obtained by combining the PA and neutral $\Delta_f H$ data (PA($CH_2CHCONH_2$) = 870.7 kJ/mol, PA(CH_3OCH_2CN) = 758.1 kJ/mol and PA($cy–C(H)_2C(O)N(H)C(H)_2–$) = 852.6 kJ/mol (for $\Delta_f H$ of molecules refer to the section above).[24] Greenberg et al.[268] have shown from HF/6-31G* calculations that 2-azetidinone is protonated at O.

3.7.81   [C_3,H_7,N,O]$^{+\bullet}$ – m/z 73

The $\Delta_f H$ values for three isomers, ionized N,N-dimethyl formamide (**1**, $HC(O)N(CH_3)_2^{]+\bullet}$), N-methyl acetamide (**2**, $CH_3C(O)N(CH_3)^{]+\bullet}$) and propionamide (**3**, $CH_3CH_2C(O)NH_2^{]+\bullet}$) can be derived from their neutral heats of formation and reported IE data to be $\Delta_f H(\mathbf{1}) = 689 \pm 2$, $\Delta_f H(\mathbf{2}) = 610 \pm 6$ kJ/mol and $\Delta_f H(\mathbf{3}) \approx 648$ kJ/mol, respectively. The ancillary data are $\Delta_f H(N,N\text{-dimethyl formamide}) = -192 \pm 2$ kJ/mol,[32] $\Delta_f H(N\text{-methyl acetamide}) = -248$ kJ/mol, $\Delta_f H(\text{propionamide}) = -259$ kJ/mol and the corresponding IE values, 9.13, 8.9 and 9.4 (estimated) eV, respectively.[24]

3.7.82 $[C_3,H_8,N,O]^+$ – m/z 74

The various $[C_3,H_8,N,O]^+$ ions have not been systematically investigated by mass spectrometry. On the basis of reported PA values for the three preceding molecules, N,N-dimethyl formamide (PA = 888 kJ/mol, $\Delta_f H = -197 \pm 10$ kJ/mol), N-methyl acetamide (PA = 888 kJ/mol, $\Delta_f H = -248$ kJ/mol) and propionamide (PA = 876 kJ/mol, $\Delta_f H = -259$ kJ/mol), $\Delta_f H$ values can be deduced for the three corresponding protonated $[C_3,H_7,N,O]^+$ isomers: $\Delta_f H = ([HC(O)N(CH_3)_2]H^+) = 445$ kJ/mol, $\Delta_f H([CH_3C(O)NH(CH_3)]H^+)$ and $\Delta_f H = ([CH_3CH_2C(O)NH_2]H^+) = 395$ kJ/mol. Calculations at the B3-LYP/6-31 + G* level of theory predict that O-protonated N-methylacetamide is 69 kJ/mol lower in energy than its N-protonated isomer.[269]

3.7.83 $[C_3,H_9,N,O]^{+\bullet}$ – m/z 75

The $\Delta_f H$ of only one $[C_3,H_9,N,O]^{+\bullet}$ ion is known. From the thermochemical cycle consisting of the G2(MP2)-calculated $\Delta_f H$ of protonated 3-amino-1-propanol (343.4 kJ/mol)[270] and the experimentally reported PA(NH_2 $CH_2CH_2CH_2OH$) = 962.5 kJ/mol,[24] the heat of formation of the molecule 3-amino-1-propanol can be deduced to be $\Delta_f H(NH_2CH_2CH_2CH_2OH) = -224.1$ kJ/mol. This value combined with the reported IE of the molecule (9.0 eV)[24] yields $\Delta_f H(NH_2CH_2CH_2CH_2OH^{1+\bullet}) = 644$ kJ/mol.

3.7.84 $[C_3,H_{10},N,O]^+$ – m/z 76

Isomers

Five isomeric $[C_3,H_{10},N,O]^+$ ions have been described by computations.[270]

Thermochemistry

The relative MP2/6-31G* energies for protonated 3-amino-1-propanol (**1**) and four isomers (ion (**2**)–(**5**), shown above) are 0, +180, +79, +42 and +202 kJ/mol, respectively. However, the calculated barriers to isomerization to (**1**) from ions (**2**), (**4**) and (**5**) are very small (<6 kJ/mol) and so it is unlikely that these ions can be isolated.[270] Ion (**3**) lies 41 kJ/mol below its lowest dissociation limit (loss of H_2O) and a large barrier for interconversion to (**1**) means that it may be possible to isolate this ion.[270]

3.7.85 $[C_3,H_3,N,S]^{+\bullet}$ – m/z 85[271]

The (methylimino)ethenethione radical cation, $CH_3NCCS^{1+\bullet}$, was generated by the losses of CH_3NCO, CO and HCN from ionized 4,6-dimethylthiazolo-[4,5-d]pyrimidine-5,7-dione. The CID MS of this product was in keeping with the proposed connectivity, showing loss of $CH_3^{1\bullet}$ and the fragment ions $C_2S^{1+\bullet}$ and $CS^{1+\bullet}$. (Note that $CS^{1+\bullet}$ is favored over CN^{1+} by some 260 kJ/mol.) No thermochemical data are available for this ion.

3.7.86 $[C_3,H_3,N,Cl]^+$ – m/z 88

A stable ion having the structure $NCCH_2CHCl^{1+}$ was proposed to be generated by synchrotron photoionization of the molecular cluster $NCCH_3 \cdots CH_2CHCl$.[272]

3.7.87 $[C_3,H_2,O,S]^{+\bullet}$ – m/z 86

The cluster ion $C_2H_2 \cdots OCS^{1+\bullet}$, produced by the addition of $OCS^{1+\bullet}$ ions to acetylene, has been studied by a number of groups.[273–275] This metastable adduct ion dissociates by two competing reactions: the loss of CO and the loss of C_2H_2, the former being an exothermic process with an average kinetic energy release of 0.43 eV. The CO loss product ion, $[C_2,H_2,S]^{1+\bullet}$, has most recently been assigned the thiirene structure,[275] although the thioketene radical cation is more favored energetically by about 40 kJ/mol.[273]

Thermochemistry

The overall thermochemistry of this reaction has been established by calculations at the MP2/6-31 + G(d)//MP2/6-31 + G(d) level of theory that evaluated the exothermicity of the reactions:[273]

$$OCS^{1+\bullet} + C_2H_2 \rightarrow CH_2CS^{1+\bullet} + CO \quad \Delta H_r = -213\,\text{kJ/mol}\,(\text{or} -271\,\text{kJ/mol})^{275}$$

$$OCS^{1+\bullet} + C_2H_2 \rightarrow cy-C(H)C(H)S-^{1+\bullet} + CO \quad \Delta H_r = -172\,\text{kJ/mol}$$

$$OCS^{1+\bullet} + C_2H_2 \rightarrow HCCSH^{1+\bullet} + CO \quad \Delta H_r = -79\,\text{kJ/mol}$$

For values for the heat of formation of the $[C_2,H_2,S]^{+\bullet}$ product ions refer to the section specifically pertaining to these ions.

Note that the binding energy between C_2H_2 and $OCS^{]+\bullet}$ was determined to be 84 kJ/mol[273] from which $\Delta_fH(C_2H_2\cdots OCS^{]+\bullet}) = 1084$ kJ/mol is estimated $(\Delta_fH(C_2H_2) = 227$ kJ/mol, $\Delta_fH(OCS) = -138.41$ kJ/mol and IE(OCS) = 11.18 eV).[24]

3.7.88 $[C_3,H_9,P,S]^{+\bullet} - m/z$ 108[276]

Isomers

Three isomers have been considered: $(CH_3)_3PS^{]+\bullet}$ (ionized trimethylphosphine sulfide), $(CH_3)_2(CH_2)PSH^{]+\bullet}$ and $(CH_3)_2PSCH_3^{]+\bullet}$.

Identification

Isomers of $(CH_3)_3PS^{]+\bullet}$ have been examined by experiment and by ab initio calculations.[1] The MI mass spectrum contains three peaks: intense peaks at m/z 75 (loss of $SH^{]\bullet}$) and 93 (loss of $CH_3^{]\bullet}$) and a weak m/z 61 (loss of $CH_3S^\bullet$). It was argued that $(CH_3)_3PS^{]+\bullet}$ does not undergo rearrangement to any other long-lived isomer, but that two rearrangements can generate short-lived intermediate structures. A 1,3-H shift in the ion produces a distonic intermediate $(CH_3)_2(CH_2)PSH^{]+\bullet}$ that can lose $SH^{]\bullet}$ to make a distonic product $((CH_3)_2PCH_2^{]+\bullet})$ and a 1,2-methyl shift produces the isomeric ion $(CH_3)_2PSCH_3^{]+\bullet}$. The product ion structures for m/z 61, 75 and 93 were probed by CID MS. The ions m/z 61 and 75 were respectively assigned to be $CH_3P(H) = CH_2^{]+}$ and $(CH_3)_2P = CH_2^{]+}$, whereas the m/z 93 ion $([C_2,H_6,P,S]^{]+})$ was assigned to be the minimum energy isomer $(CH_3)_2P = S^{]+}$.

Thermochemistry

The potential energy surface for the above processes was explored by computations at the MP2/6-31G** level of theory. The activation energies for the low energy dissociations were calculated but the absence of ancillary data does not permit the estimation of any absolute energies.

REFERENCES

1. Maluendo, S.A., McLean, A.D., Herbst, E. *Astrophys. J.* **1993**, *417*, 181.
2. Bohme, D.K., Raksit, A.B., Fox, A. *J. Am. Chem. Soc.* **1983**, *105*, 5481.
3. Raksit, A.B., Bohme, D.K. *Int. J. Mass Spectrom. Ion Processes* **1983**, *55*, 69.
4. Smith, D., Adams, N.G., Ferguson, E.E. *Int. J. Mass Spectrom. Ion Processes* **1984**, *61*, 15.
5. Smith, D., Adams, N.G. *Int. J. Mass Spectrom. Ion Processes* **1987**, *76*, 307.
6. Prodnuk, S.D., DePuy, C.H., Bierbaum, V.M. *Int. J. Mass Spectrom. Ion Processes* **1990**, *100*, 693.

7. Prodnuk, S.D., Gronert, S., Bierbaum, V.M., DePuy, C.H. *Org. Mass Spectrom.* **1992**, *27*, 416.

8. Baker, J., Chan, S.-C., Wu, K.-Y., Li, W.-K. *J. Mol. Struct. (THEOCHEM)* **1989**, *184*, 391.

9. Parr, A.C., Jason, A.J., Stockbauer, R. *Int. J. Mass Spectrom. Ion Phys.* **1978**, *26*, 23.

10. Parr, A.C., Jason, A.J., Stockbauer, R., McCulloh, K.E. *Int. J. Mass Spectrom. Ion Phys.* **1979**, *30*, 319.

11. Wong, M.W., Radom, L. *J. Am. Chem. Soc.* **1993**, *115*, 1507.

12. Dannacher, J., Heilbronner, E., Stadelmann, J., Vogt, J. *Helv. Chim. Acta* **1979**, 2186.

13. Parr, A.C., Jason, A.J., Stockbauer, R. *Int. J. Mass Spectrom. Ion Processes* **1980**, *61*, 15.

14. Wong, M.W., Radom, L. *Org. Mass Spectrom.* **1989**, *24*, 539.

15. Burgers, P.C., Holmes, J.L., Mommers, A.A., Szulejko, J.E. *J. Am. Chem. Soc.* **1984**, *106*, 521.

16. Radom, L., Hariharan, P.C., Pople, J.A., Schleyer, P.V.R. *J. Am. Chem. Soc.* **1976**, *98*, 10.

17. Li, W.-K., Riggs, N.V. *J. Mol. Struct. (THEOCHEM)* **1992**, *257*, 189.

18. Ausloos, P.J., Lias, S.G. *J. Am. Chem. Soc.* **1981**, *103*, 6505.

19. Fetterolf, D.D., Yost, R.A., Eyler, J.R. *Org. Mass Spectrom.* **1984**, *19*, 104.

20. Ozturk, F., Baykut, G., Moini, M., Eyler, J.R. *J. Phys. Chem.* **1987**, *91*, 4360.

21. Koppel, C., McLafferty, F.W. *Org. Mass Spectrom.* **1984**, *19*, 643.

22. Lossing, F.P. *Can. J. Chem.* **1972**, *50*, 3973.

23. Minsek, D.W., Chen, P. *J. Phys. Chem.* **1990**, *94*, 8399.

24. *NIST Chemistry WebBook, NIST Standard Reference Data Base Number 69*, Gaithersburg, MD: National Institute of Standards and Technology, **2005**.

25. Janoschek, R., Rossi, M.J. *Int. J. Chem. Kinet.* **2002**, *34*, 550.

26. Holmes, J.L., Lossing, F.P. *Can. J. Chem.* **1979**, *57*, 249.

27. Raghavachari, K., Whiteside, R.A., Pople, J.A., Schleyer, P.V.R. *J. Am. Chem. Soc.* **1981**, *103*, 5649.

28. Mommers, A.A., Burgers, P.C., Holmes, J.L., Terlouw, J.K. *Org. Mass Spectrom.* **1984**, *19*, 7.

29. Velzen, P.N.T.V., Hart, W.J.V.D. *Org. Mass Spectrom.* **1981**, *16*, 237.

30. Wagner, W., Levsen, K., Lifshitz, C. *Org. Mass Spectrom.* **1980**, *15*, 271.

31. Hayakawa, S., Endoh, H., Arakawa, K., Morishita, N., Sugiura, T. *Int. J. Mass Spectrom. Ion Processes* **1995**, *151*, 89.

32. Pedley, J.B., Naylor, R.D., Kirby, S.P. *Thermochemical Data of Organic Compounds*, 2nd ed., London: Chapman and Hall, **1986**.

33. Radom, L., Hariharan, P.C., Pople, J.A., Schleyer, P.V.R. *J. Am. Chem. Soc.* **1973**, *95*, 6531.

34. Fairley, D.A., Milligan, D.B., Wheadon, L.M., Freeman, C.G., Maclagan, R.G.A.R., McEwan, M.J. *Int. J. Mass Spectrom.* **1999**, *185–187*, 253.

35. Bowers, M.T., Shuying, L., Kemper, P., Stradling, R., Webb, H., Aue, D.H., Gilbert, J.R., Jennings, K.R. *J. Am. Chem. Soc.* **1980**, *102*, 4830.

36. Burgers, P.C., Holmes, J.L., Mommers, A.A., Szulejko, J.E. *Org. Mass Spectrom.* **1983**, *18*, 596.

37. Traeger, J.C. *Int. J. Mass Spectrom. Ion Processes* **1984**, *58*, 259.

38. Skancke, A. *J. Phys. Chem.* **1995**, *99*, 13886.
39. Gross, M.L., Lin, P.-H. *Org. Mass Spectrom.* **1973**, *7*, 795.
40. Holmes, J.L., Terlouw, J.K. *Org. Mass Spectrom.* **1975**, *10*, 787.
41. Bowen, R.D., Barbalas, M.P., Pagano, F.P., Todd, P.J., McLafferty, F.W. *Org. Mass Spectrom.* **1980**, *15*, 51.
42. Holmes, J.L., Terlouw, J.K., Burgers, P.C., Rye, R.T.B. *Org. Mass Spectrom.* **1980**, *15*, 149.
43. Bouchoux, G., Choret, N., Flammang, R. *Int. J. Mass Spectrom.* **2000**, *195/196*, 225.
44. Gross, M.L., McLafferty, F.W. *J. Am. Chem. Soc.* **1971**, *93*, 1267.
45. Mourgues, P., Leblanc, D., Audier, H.-E., Hammerum, S. *Int. J. Mass Spectrom. Ion Processes* **1992**, *113*, 105.
46. Aubry, C., Polce, M.J., Holmes, J.L., Mayer, P.M., Radom, L. *J. Am. Chem. Soc.* **1997**, *119*, 9039.
47. Koch, W., Schleyer, P.v.R., Buzek, P., Liu, B. *Croatica Chemica Acta* **1992**, *65*, 655.
48. Frash, M.V., Kazansky, V.V., Rigby, A.M., Santen, R.A.v. *J. Phys. Chem. B* **1997**, *101*, 5346.
49. Chiavarino, B., Crestoni, M.E., Fokin, A.A., Fornarini, S. *Chem. Eur. J.* **2001**, *7*, 2916.
50. Dymerski, P.P., Prinstein, R.M., Bente III, P.F., McLafferty, F.W. *J. Am. Chem. Soc.* **1976**, *98*, 6834.
51. McAdoo, D.J., McLafferty, F.W., Bente III, P.F., *J. Am. Chem. Soc.* **1972**, *94*, 2027.
52. Attina, M., Cacace, F., Giacomello, P. *J. Am. Chem. Soc.* **1980**, *102*, 4768.
53. Baer, T. *J. Am. Chem. Soc.* **1980**, *102*, 2482.
54. Park, S.T., Kim, S.K., Kim, M.S. *J. Chem. Phys.* **2001**, *114*, 5568.
55. Esteves, P.M., Mota, C.J.A., Ramyrez-Solys, A., Hernandez-Lamoneda, R. *J. Am. Chem. Soc.* **1998**, *120*, 3213.
56. Hiraoka, K., Kebarle, P. *J. Am. Chem. Soc.* **1976**, *98*, 6119.
57. Ding, Y.H., Huang, X.R., Lu, Z.Y., Fong, J.N. *Chem. Phys. Lett.* **1998**, *284*, 325.
58. Petrie, S., McGrath, K.M., Freeman, C.G., McEwan, M.J. *J. Am. Chem. Soc.* **1992**, *114*, 9130.
59. Harland, P.W. *Int. J. Mass Spectrom. Ion Processes* **1986**, *70*, 231.
60. Harland, P.W., Maclagan, R.G.A.R. *Faraday Trans. 2* **1987**, *83*, 2133.
61. Raksit, A.B., Bohme, D.K. *Int. J. Mass Spectrom. Ion Processes* **1985**, *63*, 217.
62. Scott, G.B.I., Fairley, D.A., Freeman, C.G., McEwan, M.J., Anicich, V.G. *J. Phys. Chem. A* **1999**, *103*, 1073.
63. Zanathy, L., Bock, B., Lentz, D., Preugschat, D., Botschwina, P. *J. Chem. Soc. Chem. Commun.* **1992**, 403.
64. Redondo, P., Ruiz, M.J., Boronat, R., Barrientos, C., Largo, A. *Theor. Chem. Acc.* **2000**, *104*, 199.
65. Barrientos, C., Redondo, P., Largo, A. *J. Phys. Chem. A* **2000**, *104*, 11541.
66. Petrie, S., Chirnside, T.J., Freeman, C.G., McEwan, M.J. *Int. J. Mass Spectrom. Ion Processes* **1991**, *107*, 319.
67. Milligan, D.B., Wilson, P.F., McEwan, M.J., Anicich, V.G. *Int. J. Mass Spectrom. Ion Processes* **1999**, *185/187*, 663.
68. Heerma, W., Sarneel, M.M., Dijkstra, G. *Org. Mass Spectrom.* **1986**, *21*, 681.
69. Gray, G.A. *J. Am. Chem. Soc.* **1968**, *90*, 2177.

70. Gray, G.A. *J. Am. Chem. Soc.* **1968**, *90*, 6002.

71. Oldham, N.J. *Rapid Commun. Mass Spectrom.* **1999**, *13*, 1694.

72. Moneti, G., Pieraccini, G., Favretto, D., Traldi, P. *J. Mass Spectrom.* **1998**, *33*, 1148.

73. Wincel, H., Fokkens, R.H., Nibbering, N.M.M. *Int. J. Mass Spectrom Ion Processes* **1990**, *96*, 321.

74. Wincel, H. *Int. J. Mass Spectrom Ion Processes* **1998**, *175*, 283.

75. Milligan, D.B., Freeman, C.G., Maclagan, R.G.A.R., McEwan, M.J., Wilson, P.F., Anicich, V.G. *J. Am. Soc. Mass Spectrom.* **2001**, *12*, 557.

76. Pottie, R.F., Lossing, F.P. *J. Am. Chem. Soc.* **1961**, *83*, 4737.

77. Luo, Y.-R. *Handbook of Bond Dissociation Energies in Organic Compounds*, Boca Raton: CRC Press, **2003**.

78. Hopkinson, A.C., Lien, M.H. *J. Am. Chem. Soc.* **1986**, *108*, 2843.

79. Salpin, J.-Y., Nguyen, M.T., Bouchoux, G., Gerbaux, P., Flammang, R. *J. Phys. Chem. A* **1999**, *103*, 938.

80. Chess, E.K., Lapp, R.L., Gross, M.L. *Org. Mass Spectrom.* **1982**, *17*, 475.

81. Gerbaux, P., Flammang, R., Nguyen, M.T., Salpin, J.-Y., Bouchoux, G. *J. Phys. Chem. A* **1998**, *102*, 861.

82. Wincel, H., Fokkens, R.H., Nibbering, N.M.M. *Int. J. Mass Spectrom. Ion Processes* **1989**, *91*, 339.

83. Wincel, H., Wlodek, S., Bohme, D.K. *Int. J. Mass Spectrom. Ion Processes* **1988**, *84*, 69.

84. Wilson, P.F., Freeman, C.G., McEwan, M.J. *Int. J. Mass Spectrom. Ion Processes* **1993**, *128*, 83.

85. Deakyne, C.A., Meot-Ner, M. *J. Phys. Chem.* **1990**, *94*, 232.

86. McEwan, M.J., Denison, A.B., Huntress, W.T., Anicich, V.G., Snodgrass, J., Bowers, M.T. *J. Phys. Chem.* **1989**, *93*, 4064.

87. Smith, S.C., Wilson, P.F., Sudkeaw, P., MacIagan, R.G.A.R., McEwan, M.J., Anicich, V.G., Huntress, W.T. *J. Chem. Phys.* **1993**, *98*, 1944.

88. Bouchoux, G., Flament, J.P., Hoppilliard, Y., Tortajada, J., Flammang, R., Maquestiau, A. *J. Am. Chem. Soc.* **1989**, *111*, 5560.

89. Augusti, R., Gozzo, F.C., Moraes, L.A.B., Sparrapan, R., Eberlin, M.N. *J. Org. Chem.* **1998**, *63*, 4889.

90. Bouchoux, G., Alcaraz, C., Dutuit, O., Nguyen, M.T. *J. Am. Chem. Soc.* **1998**, *120*, 152.

91. Bouchoux, G., Nguyen, M.T., Longevialle, P. *J. Am. Chem. Soc.* **1992**, *114*, 10000.

92. Lopez, R., Rio, E.D., Menendez, M.I., Sordo, T.L. *J. Phys. Chem. A* **2002**, *106*, 4616.

93. Liu, G.-X., Ding, Y.-L., Li, Z.-S., Huang, X.-F., Sun, C.-C. *J. Mol. Struct. (THEOCHEM)* **2001**, *548*, 191.

94. Eberlin, M.N., Morgan, N.H., Yang, S.S., Shay, B.J., Cooks, R.G. *J. Am. Soc. Mass Spectrom.* **1995**, *6*, 1.

95. Lias, S.G., Bartmess, J.E., Liebman, J.F., Holmes, J.L., Levin, R.D., Mallard, W.G. *J. Phys. Chem. Ref. Data* **1988**, *17 (Suppl 1)*.

96. Hammerum, S., Henriksen, J., Henriksen, T., Solling, T.I. *Int. J. Mass Spectrom.* **2000**, *195/196*, 459.

97. Uccella, N.A., Howe, I., Williams, D.H. *J. Chem. Soc. B* **1971**, 1933.

98. Levsen, K., McLafferty, F.W. *J. Am. Chem. Soc.* **1974**, *96*, 139.
99. Bouchoux, G., Penaud-Berruyer, F., Tortajada, J. *J. Mass Spectrom.* **1995**, *30*, 723.
100. Bowen, R.D., Williams, D.H. *J. Chem. Soc. Perkin Trans. 2* **1978**, 1064.
101. Bowen, R.D., Williams, D.H., Hvistendahl, G., Kalman, J.R. *Org. Mass Spectrom.* **1978**, *13*, 721.
102. Hudson, C.E., McAdoo, D.J. *J. Am. Soc. Mass Spectrom.* **1998**, *9*, 138.
103. Chalk, A.J., Mayer, P.M., Radom, L. *Int. J. Mass Spectrom.* **2000**, *194*, 181.
104. Lossing, F.P., Lam, Y.-T., Maccoll, A. *Can. J. Chem.* **1981**, *59*, 2228.
105. Janoschek, R., Rossi, M.J. *Thermochemical Properties from G3MP2B3 Calculations* New York: Wiley InterScience, **2004**.
106. Petersen, A.C., Hammerum, S. *Acta Chim. Scan.* **1998**, *52*, 1045.
107. Hammerum, S., Kuck, D., Derrick, P.J. *Tetrahedron Lett.* **1984**, *25*, 893.
108. Hajgato, B., Flammang, R., Veszprem, T., Nguyen, M.T. *Mol. Phys.* **2002**, *100*, 1693.
109. Tilborg, M.W.E.M.V., Thuijl, J.V. *Org. Mass Spectrom.* **1984**, *19*, 569.
110. Tilborg, M.W.E.M.V., Houte, J.J.V., Thuijl, J.V. *Org. Mass Spectrom.* **1984**, *19*, 16.
111. McGibbon, G.A., Heinemann, C., Lavorato, D.J., Schwarz, H. *Angew. Chem.* **1997**, *109*, 1572.
112. Flammang, R., Barbieux-Flammang, M., Le, H.T., Gerbaux, P., Elguero, J., Nguyen, M.T. *Chem. Phys. Lett.* **2001**, *347*, 465.
113. Flammang, R., Elguero, J., Le, H.T., Gerbaux, P., Nguyen, M.T. *Chem. Phys. Lett.* **2002**, *356*, 259.
114. Gerbaux, P., Barbieux-Flammang, M., Haverbeke, Y.V., Flammang, R. *Rapid Commun. Mass Spectrom.* **1999**, *13*, 1707.
115. Nguyen, V.Q., Turecek, F. *J. Mass Spectrom.* **1996**, *31*, 1173.
116. Wilcox, C.F., Attygalle, A.B. *J. Mol. Struct. (THEOCHEM)* **1998**, *434*, 207.
117. Wang, X., Holmes, J.L. Unpublished data **2005**.
118. Burgers, P.C., Holmes, J.L., Lossing, F.P., Mommers, A.A., Povel, F.R., Terlouw, J.K. *Can. J. Chem.* **1982**, *60*, 2246.
119. Terlouw, J.K., Holmes, J.L., Lossing, F.P. *Can. J. Chem.* **1983**, *61*, 1722.
120. Bouchoux, G., Hoppilliard, Y., Flament, J.-P., Terlouw, J.K., Valk, F.V.D. *J. Phys. Chem. A* **1986**, *90*, 1582.
121. Harshbarger, W.R., Kuebler, N.A., Robin, M.R. *J. Chem. Phys.* **1974**, *60*, 345.
122. Terlouw, J.K., Holmes, J.L., Lossing, F.P. *Can. J. Chem.* **1983**, *61*, 1722.
123. Benson, S.W. *Thermochemical Kinetics*, New York: Wiley Interscience, **1976**.
124. Aubry, C., Holmes, J.L., Terlouw, J.K. *J. Phys. Chem. A* **1997**, *101*, 5958.
125. Scott, A.P., Radom, L. *Int. J. Mass Spectrom. Ion Processes* **1997**, *160*, 73.
126. Traeger, J.C. *Int. J. Mass Spectrom.* **2000**, *194*, 261.
127. Chiang, S.-Y., Fang, Y.-S., Bahou, M., Sankaran, K. *J. Chin. Chem. Soc.* **2004**, *51*, 681.
128. Niessen, W.V., Bieri, G., Asbrink, L. *J. Electron. Spectrosc. Relat. Phenom.* **1980**, *21*, 175.
129. Qu, Z.-W., Ding, Y.-H., Li, Z.-S. *J. Mol. Struct. (THEOCHEM)* **1999**, *489*, 195.
130. Holmes, J.L., Terlouw, J.K., Burgers, P.C. *Org. Mass Spectrom.* **1980**, *15*, 140.
131. Bouchoux, G., Hoppilliard, Y., Flament, J.P. *Org. Mass Spectrom.* **1985**, *20*, 560.
132. Lien, A.C., Hopkinson, M.H. *J. Am. Chem. Soc.* **1986**, *108*, 2843.

133. Maclagan, R.G.A.R., McEwan, M.J., Scott, G.B.I. *Chem. Phys. Lett.* **1995**, *240*, 185.
134. Maquestiau, A., Flammang, R., Pauwels, P. *Org. Mass Spectrom.* **1983**, *18*, 547.
135. Traeger, J.C., Hudson, C.E., McAdoo, D.J. *Org. Mass Spectrom.* **1989**, *24*, 230.
136. Turecek, F., Drinkwater, D.E., McLafferty, F.W. *J. Am. Chem. Soc.* **1991**, *113*, 5950.
137. Turecek, F., Drinkwater, D.E., McLafferty, F.W. *J. Am. Chem. Soc.* **1991**, *113*, 5958.
138. Aubry, C., Holmes, J.L., Terlouw, J.K. *J. Phys. Chem. A* **1997**, *101*, 5958.
139. Traeger, J.C. *Int. J. Mass Spectrom.* **2000**, *194*, 261.
140. McKee, M.L., Radom, L. *Org. Mass Spectrom.* **1993**, *28*, 1238.
141. Li, Y., Baer, T. *Int. J. Mass Spectrom.* **2002**, *218*, 19.
142. Bouchoux, G., Flament, J.-P., Hoppilliard, Y. *Nouv. J. Chem.* **1983**, *7*, 385.
143. Bouchoux, G., Flament, J.-P., Hoppilliard, Y. *Nouv. J. Chem.* **1985**, *9*, 453.
144. Mead, T.J., Williams, D.H. *J. Chem. Soc. B Phys. Org.* **1971**, 1654.
145. Hudson, C.E., McAdoo, D.J. *Org. Mass Spectrom.* **1982**, *17*, 366.
146. Zwinselmann, J.J., Harrison, A.G. *Org. Mass Spectrom.* **1984**, *19*, 573.
147. Bouchoux, G., Hoppilliard, Y., Flammang, R., Maquestiau, A., Meyrant, P. *Org. Mass Spectrom.* **1983**, *18*, 340.
148. Turecek, F., Hanus, V., Gaumann, T. *Int. J. Mass Spectrom. Ion Processes* **1986**, *69*, 217.
149. Traeger, J.C. *Org. Mass Spectrom.* **1985**, *20*, 223.
150. McAdoo, D.J., Witiak, D.N. *J. Chem. Soc. Perkin Trans. 2* **1981**, 770.
151. McAdoo, D.J., McLafferty, F.W., Smith, J.S. *J. Am. Chem. Soc.* **1970**, *92*, 6343.
152. Lifshitz, C., Tzidony, E. *Int. J. Mass Spectrom. Ion Processes* **1981**, *39*, 181.
153. Turecek, F., McLafferty, F.W. *J. Am. Chem. Soc.* **1984**, *106*, 2525.
154. Polce, M.J., Wesdemiotis, C. *J. Am. Soc. Mass Spectrom.* **1996**, *7*, 573.
155. McAdoo, D.J., Hudson, C.E., Traeger, J.C. *Int. J. Mass Spectrom. Ion Processes* **1987**, *79*, 183.
156. Van de Sande, C.C., McLafferty, F.W. *J. Am. Chem. Soc.* **1975**, *97*, 4617.
157. Bouma, W.J., MacLeod, J.K., Radom, L. *J. Am. Chem. Soc.* **1980**, *102*, 2246.
158. Turecek, F., Hanus, V. *Org. Mass Spectrom.* **1984**, *19*, 631.
159. McAdoo, D.J., Hudson, C.E. *J. Mass Spectrom.* **1995**, *30*, 492.
160. McAdoo, D.J., Hudson, C.E., Traeger, J.C. *Org. Mass Spectrom.* **1988**, *23*, 760.
161. Thissen, R., Audier, H.E., Chamot-Rooke, J., Mourgues, P. *Eur. J. Mass Spectrom.* **1999**, *5*, 147.
162. Hudson, C.E., McAdoo, D.J. *Org. Mass Spectrom.* **1984**, *19*, 1.
163. Holmes, J.L., Lossing, F.P. *J. Am. Chem. Soc.* **1980**, *102*, 1591.
164. Flammang, R., Nguyen, M.T., Bouchoux, G., Gerbaux, P. *Int. J. Mass Spectrom.* **2000**, *202*, A8.
165. McLafferty, F.W., Sakai, I. *Org. Mass Spectrom.* **1973**, *7*, 971.
166. Nobes, R.H., Radom, L. *Org. Mass Spectrom.* **1984**, *19*, 385.
167. Bouchoux, G., Penaud-Berruyer, F., Audier, H.E., Mourgues, P., Tortajada, J. *J. Mass Spectrom.* **1997**, *32*, 188.
168. Harrison, A.G., Gaumann, T., Stahl, D. *Org. Mass Spectrom.* **1983**, *18*, 517.
169. Hvistendahl, G., Williams, D.H. *J. Am. Chem. Soc.* **1975**, *97*, 3097.
170. Yeo, A.N.H., Williams, D.H. *J. Am. Chem. Soc.* **1971**, *93*, 395.
171. Tsang, C.W., Harrison, A.G. *Org. Mass Spectrom.* **1973**, *7*, 1377.

172. Bowen, R.D., Harrison, A.G. *Org. Mass Spectrom.* **1981**, *16*, 159.
173. Bowen, R.D., Kalman, J.R., Williams, D.H. *J. Am. Chem. Soc.* **1977**, *99*, 5483.
174. Bowen, R.D., Williams, D.H., Hvistendahl, G., Kalman, J.R. *Org. Mass Spectrom.* **1978**, *13*, 721.
175. Holmes, J.L., Rye, R.T.B., Terlouw, J.K. *Org. Mass Spectrom.* **1979**, *14*, 606.
176. Hudson, C.E., McAdoo, D.J. *J. Am. Soc. Mass Spectrom.* **1998**, *9*, 130.
177. Bowen, R.D., Williams, D.H. *Org. Mass Spectrom.* **1977**, *12*, 475.
178. Andersen, P.E., Hammerum, S. *Eur. Mass Spectrom.* **1995**, *1*, 499.
179. Lossing, F.P. *J. Am. Chem. Soc.* **1977**, *99*, 7526.
180. Refaey, K.M.A., Chupka, W.A. *J. Chem. Phys.* **1968**, *48*, 5205.
181. Crow, F.W., Gross, M.L., Bursey, M.M. *Org. Mass Spectrom.* **1981**, *16*, 309.
182. Holmes, J.L., Mommers, A.A., Szulejko, J.E., Terlouw, J.K. *J. Chem. Soc. Chem. Commun.* **1984**, 165.
183. Holmes, J.L., Terlouw, J.K. In *The Encyclopedia of Mass Spectrometry*, Nibbering, N.M.M., Ed., Amsterdam: Elsevier, **2005**, Vol. 4, p 287.
184. Wesdemiotis, C., Danis, P.O., Feng, R., Tso, J., McLafferty, F.W. *J. Am. Chem. Soc.* **1985**, *107*, 8059.
185. Booze, J.A., Baer, T. *J. Phys. Chem.* **1992**, *96*, 5710.
186. Booze, J.A., Baer, T. *J. Phys. Chem.* **1992**, *96*, 5715.
187. Bowen, R.D., Colburn, A.W., Derrick, P.J. *J. Am. Chem. Soc.* **1991**, *113*, 1132.
188. McAdoo, D.J., Hudson, C.E., Ramanujam, V.M.S., George, M. *Org. Mass Spectrom.* **1993**, *28*, 1210.
189. Cao, J.R., George, M., Holmes, J.L. *Org. Mass Spectrom.* **1991**, *26*, 481.
190. McAdoo, D.J., Ahmed, M.S., Hudson, C.E., Giam, C.S. *Int. J. Mass Spectrom. Ion Processes* **1990**, *100*, 579.
191. Bouchoux, G., Choret, N. *Int. J. Mass Spectrom.* **2000**, *201*, 161.
192. Bouchoux, G., Choret, N., Flammang, R. *Int. J. Mass Spectrom.* **2000**, *195/196*, 225.
193. Harrison, A.G. *Org. Mass Spectrom.* **1987**, *22*, 637.
194. Terlouw, J.K., Weiske, T., Schwarz, H., Holmes, J.L. *Org. Mass Spectrom.* **1986**, *21*, 665.
195. Mautner, M., Ross, M.M., Campanat, J.E. *J. Am. Chem. Soc.* **1985**, *107*, 4839.
196. Chen, H., Holmes, J.L. *Int. J. Mass Spectrom. Ion Processes* **1994**, *133*, 111.
197. Tortajada, J., Provot, G., Morizur, J.-P., Gal, J.-F., Maria, P.-C., Flammang, R., Govaert, Y. *Int. J. Mass Spectrom. Ion Processes* **1995**, *141*, 241.
198. Westmore, J.B., Buchannon, W.D., Plaggenborg, L., Wenclawiak, B.W. *J. Am. Soc. Mass Spectrom.* **1998**, *9*, 29.
199. Turecek, F., Vivekananda, S., Sadilek, M., Polasek, M. *J. Am. Chem. Soc.* **2002**, *124*, 13282.
200. Suh, D., Kingsmill, C.A., Ruttink, P.J.A., Terlouw, J.K., Burgers, P.C. *Org. Mass Spectrom.* **1993**, *28*, 1270.
201. Bouchoux, G., Djazi, F., Houriet, R., Rolli, E. *J. Org. Chem.* **1988**, *53*, 3498.
202. McAdoo, D.J., Witiak, D.N. *Org. Mass Spectrom.* **1978**, *13*, 499.
203. Hudson, C.E., McAdoo, D.J. *Int. J. Mass Spectrom.* **2001**, *210/211*, 417.
204. Hudson, C.E., McAdoo, D.J. *Org. Mass Spectrom.* **1992**, *27*, 1384.
205. Baer, T., Mazyar, O.A., Keister, J.W., Mayer, P.M. *Ber. Bunsen-Gesellschaft* **1997**, *101*, 478.

206. Vajda, J.H., Harrison, A.G., Hirota, A., McLafferty, F.W. *J. Am. Chem. Soc.* **1981**, *103*, 36.

207. Mazyar, O.A., Mayer, P.M., Baer, T. *Int. J. Mass Spectrom. Ion Processes* **1997**, *167/168*, 389.

208. Holmes, J.L., Hop, C.E.C.A., Terlouw, J.K. *Org. Mass Spectrom.* **1986**, *21*, 776.

209. Wesdemiotis, C., Feng, R., Williams, E.R., McLafferty, F.W. *Org. Mass Spectrom.* **1986**, *21*, 689.

210. Burgers, P.C., Holmes, J.L., Mommers, A.A., Szulejko, J.E., Terlouw, J.K. *Org. Mass Spectrom.* **1984**, *19*, 492.

211. Heinrich, N., Schmidt, J., Schwarz, H., Apeloig, Y. *J. Am. Chem. Soc.* **1987**, *109*, 1317.

212. Wesdemiotis, C., Csencsits, R., McLafferty, F.W. *Org. Mass Spectrom.* **1985**, *20*, 98.

213. Burgers, P.C., Holmes, J.L., Hop, C.E.C.A., Terlouw, J.K. *Org. Mass Spectrom.* **1986**, *21*, 549.

214. George, M., Kingsmill, C.A., Suh, D., Terlouw, J.K., Holmes, J.L. *J. Am. Chem. Soc.* **1994**, *116*, 7807.

215. Pakarinen, J.M.H., Vainiotalo, P., Pakkanen, T.A., Kenttamaa, H.I. *J. Am. Chem. Soc.* **1993**, *115*, 12431.

216. Wong, T., Warkentin, J., Terlouw, J.K. *Int. J. Mass Spectrom. Ion Processes* **1992**, *115*, 33.

217. Holmes, J.L., Lossing, F.P. *Org. Mass Spectrom.* **1979**, *14*, 513.

218. McLafferty, F.W., Proctor, C. *Org. Mass Spectrom.* **1983**, *18*, 272.

219. Burgers, P.C., Fell, L.M., Milliet, A., Rempp, M., Ruttink, P.J.A., Terlouw, J.K. *Int. J. Mass Spectrom. Ion Processes* **1997**, *167/168*, 291.

220. Van Baar, B.L.M., Burgers, P.C., Holmes, J.L., Terlouw, J.K. *Org. Mass Spectrom.* **1988**, *23*, 355.

221. Takahashi, Y., Higuchi, T., Sekiguchi, O., Fujizuka, A., Nakajima, S., Tajima, S. *Rapid Commun. Mass Spectrom.* **2000**, *14*, 61.

222. Terlouw, J.K., Heerma, W., Burgers, P.C., Holmes, J.L. *Can. J. Chem.* **1984**, *62*, 289.

223. Postma, R., Ruttink, P.J.A., van Duijneveldt, F.B., Terlouw, J.K., Holmes, J.L. *Can. J. Chem.* **1985**, *63*, 2798.

224. Bouchoux, G., Alcaraz, C., Dutuit, O., Nguyen, M.T. *Int. J. Mass Spectrom. Ion Processes* **1994**, *137*, 93.

225. Szulejko, J.E., McMahon, T.B., Troude, V., Bouchoux, G., Audier, H.E. *J. Phys. Chem. A* **1998**, *102*, 1879.

226. Andrews, C.W., Bowen, J.P., Fraser-Reid, B. *J. Chem. Soc. Chem. Commun.* **1989**, 1913.

227. Bouchoux, G., Choret, N., Flammang, R. *J. Phys. Chem. A* **1997**, *101*, 4271.

228. Bouchoux, G., Penaud-Berruyer, F. *J. Phys. Chem. A* **2003**, *107*, 7931.

229. Fell, L.M., Francis, J.T., Holmes, J.L., Terlouw, J.K. *Int. J. Mass Spectrom. Ion Processes* **1997**, *165/166*, 179.

230. Holmes, J.L., Jean, T.S. *Org. Mass Spectrom.* **1970**, *3*, 1505.

231. Sicilia, M.d.C., Mo, O., Yanez, M., Guillemin, J.-C., Gal, J.-F., Maria, P.-C. *Eur. J. Mass Spectrom.* **2003**, *9*, 245.

232. Hop, C.E.C.A. *J. Mass Spectrom.* **1995**, *30*, 1273.

233. Polce, M.J., Wesdemiotis, C. *Rapid Commun. Mass Spectrom.* **1996**, *10*, 235.

234. Lee, H.-L., Li, W.-K., Chiu, S.-W. *J. Mol. Struct. (THEOCHEM)* **2003**, *620*, 107.

235. Butler, J.J., Baer, T. *Org. Mass Spectrom.* **1983**, *18*, 248.
236. Joshi, R.M. *J. Macromol. Sci. Chem. A* **1979**, *13*, 1015.
237. Broer, W.J., Weringa, W.D. *Org. Mass Spectrom.* **1980**, *15*, 229.
238. Broer, W.J., Weringa, W.D. *Org. Mass Spectrom.* **1977**, *12*, 326.
239. Broer, W.J., Weringa, W.D. *Org. Mass Spectrom.* **1979**, *14*, 36.
240. Van de Graaf, B., McLafferty, F.W. *J. Am. Chem. Soc.* **1977**, *99*, 6810.
241. Levsen, K., Hembach, H., Van de Sande, C.C., Monstrey, J. *Tetrahedron* **1977**, *33*, 1785.
242. Wong, M.W., Wentrup, C., Flammang, R. *J. Phys. Chem.* **1995**, *99*, 16849.
243. Sulzle, D., Beye, N., Fanghanel, E., Schwarz, H. *Chem. Ber.* **1990**, *123*, 2069.
244. Dawson, D.F. In Chemistry, Ph.D. Thesis: Collision Induced Emission Spectroscopy of Small Polyatomic Molecules and Tandem Mass Spectrometry of Perfluorinated Organics; University of Ottawa: Ottawa, **2000**, pp. 413.
245. McAllister, M., Tidwell, T.T., Peterson, M.R., Csizmadia, I.G. *J. Org. Chem.* **1991**, *56*, 575.
246. Shaler, T.A., Morton, T.H. *J. Am. Chem. Soc.* **1991**, *113*, 6771.
247. Stams, D.A., Johri, K.K., Morton, T.H. *J. Am. Chem. Soc.* **1988**, *110*, 699.
248. Williamson, A.D., LeBreton, P.R., Beauchamp, J.L. *J. Am. Chem. Soc.* **1976**, *98*, 2705.
249. Won, D.S., Choe, J.C., Kim, M.S. *Rapid Commun. Mass Spectrom.* **2000**, *14*, 1110.
250. Holmes, J.L., Lossing, F.P. *Can. J. Chem.* **1979**, *57*, 249.
251. Kim, D.Y., Choe, J.C., Kim, M.S. *J. Chem. Phys.* **2000**, *113*, 1714.
252. Krailler, R.E., Russell, D.H. *Int. J. Mass Spectrom. Ion Processes* **1985**, *66*, 339.
253. Kim, D.Y., Choe, J.C., Kim, M.S. *J. Phys. Chem. A* **1999**, *103*, 4602.
254. Orth, R.G., Dunbar, R.C. *J. Am. Chem. Soc.* **1982**, *104*, 5617.
255. Gaumann, T., Riveros, J.M., Zhu, Z. *Helv. Chim. Acta* **1990**, *73*, 1215.
256. Gaumann, T., Zhu, Z., Kida, M.C., Riveros, J.M. *J. Am. Soc. Mass Spectrom.* **1991**, *2*, 372.
257. Riveros, J.M., Galembeck, S.E. *Int. J. Mass Spectrom. Ion Processes* **1983**, *47*, 183.
258. Morgon, N.H., Giroldo, T., Linnert, H.V., Riveros, J.M. *J. Phys. Chem.* **1996**, *110*, 18048.
259. Park, S.T., Kim, S.K., Kim, M.S. *Nature (London)* **2002**, *415*, 306.
260. Park, S.T., Kim, M.S. *J. Chem. Phys.* **2002**, *117*, 124.
261. Flammang, R., Haverbeke, Y.V., Wong, M.W., Ruhmann, A., Wentrup, C. *J. Phys. Chem.* **1994**, *98*, 4814.
262. Moloney, D.W.J., Wong, M.W., Flammang, R., Wentrup, C. *J. Org. Chem.* **1997**, *62*, 4240.
263. Holmes, J.L., Mayer, P.M., Vasseur, M., Burgers, P.C. *J. Phys. Chem.* **1993**, *97*, 4865.
264. Flammang, R., Plisnier, M., Bouchoux, G., Hoppilliard, Y., Humbert, S., Wentrup, C. *Org. Mass Spectrom.* **1992**, *27*, 317.
265. Flammang, R., Haverbeke, Y.V., Laurent, S., Barbieux-Flammang, M., Wong, M.W., Wentrup, C. *J. Phys. Chem.* **1994**, *98*, 5801.
266. Carvalho, M.C., Moraes, L.A.B., Kascheres, C., Eberlin, M.N. *J. Mass Spectrom.* **1997**, *32*, 1137.
267. Elrod, M.J. *Int. J. Mass Spectrom.* **2003**, *228*, 91.

268. Greenberg, A., Hsing, H.-J., Liebman, J.F. *J. Mol. Struct. (THEOCHEM)* **1995**, *338*, 83.
269. Kulhanek, P., Schlag, E.W., Koca, J. *J. Phys. Chem. A* **2003**, *107*, 5789.
270. Bouchoux, G., Choret, N., Penaud-Berruyer, F., Flammang, R. *Int. J. Mass Spectrom.* **2002**, *217*, 195.
271. Flammang, R., Landu, D., Laurent, S., Barbieux-Flammang, M., Kappe, C.O., Wong, M.W., Wentrup, C. *J. Am. Chem. Soc.* **1994**, *116*, **2005**.
272. Sheng, L., Luo, Z., Qi, F., Gao, H., Zhang, Y., Huang, M.-B. *Chem. Phys. Lett.* **1995**, *233*, 347.
273. Graul, S.T., Bowers, M.T. *J. Chem. Phys.* **1991**, *95*, 8328.
274. Chiu, Y.-H., Fu, H., Huang, J.-T., Anderson, S.L. *J. Chem. Phys.* **1996**, *105*, 3089.
275. Felician, M., Armentrout, P.B. *Z. Phys. Chem.* **2000**, *214*, 1035.
276. Keck, H., Tommes, P. *J. Mass Spectrom.* **1999**, *34*, 44.

Index of Empirical Formulae

Milton Keynes UK
Ingram Content Group UK Ltd.
UKHW021857071024
449327UK00021B/1587